ESSENTIALS OF MACHINE OLFACTION AND TASTE

ESSENTIALS OF MACHINE OLFACTION AND TASTE

Edited by

Takamichi Nakamoto
Tokyo Institute of Technology, Japan

WILEY

This edition first published 2016

Published by John Wiley & Sons Singapore Pte Ltd, 1 Fusionopolis Walk, #07-01 Solaris South Tower, Singapore 138628, under exclusive license granted by Higher Education Press Limited Company for all media and languages excluding Chinese and throughout the world excluding Mainland China, and with non-exclusive license for electronic versions in Mainland China.

Registered Office
John Wiley & Sons Singapore Pte Ltd, 1 Fusionopolis Walk, #07-01 Solaris South Tower, Singapore 138628

For details of our global editorial offices, for customer services and for information about how to apply for permission to reuse the copyright material in this book please see our website at www.wiley.com.

Library of Congress Cataloging-in-Publication data applied for

ISBN: 9781118768488

A catalogue record for this book is available from the British Library.

Set in 10/12pt Times by SPi Global, Pondicherry, India

1 2016

Contents

Preface xi

About the Contributors xiii

1 Introduction to Essentials of Machine Olfaction and Tastes 1
Takamichi Nakamoto

2 Physiology of Chemical Sense and its Biosensor Application 3
Ryohei Kanzaki, Kei Nakatani, Takeshi Sakurai, Nobuo Misawa and Hidefumi Mitsuno

2.1 Introduction 3
2.2 Olfaction and Taste of Insects 4
2.2.1 Olfaction 4
2.2.1.1 Anatomy of Olfaction 4
2.2.1.2 Signal Transduction of Odor Signals 6
2.2.1.3 Molecular Biology of Olfaction 7
2.2.2 Taste 8
2.2.2.1 Anatomy of Taste 8
2.2.2.2 Molecular Biology and Signal Transduction of Taste 9
2.3 Olfaction and Taste of Vertebrate 11
2.3.1 Olfaction 11
2.3.1.1 Anatomy of Olfaction 11
2.3.1.2 Transduction of Odor Signals 12
2.3.1.3 Molecular Biology of Olfaction 15
2.3.2 Taste 17
2.3.2.1 Anatomy of Taste 17
2.3.2.2 Transduction of Taste Signals 18
2.3.2.3 Molecular Biology of Taste 20

2.4 Cell-Based Sensors and Receptor-Based Sensors 21
2.4.1 Tissue-Based Sensors 23
2.4.2 Cell-Based Sensors 26
2.4.3 Receptor-Based Sensors 30
2.4.3.1 Production of Odorant Receptors 34
2.4.3.2 Immobilization of Odorant Receptors 35
2.4.3.3 Measurement from Odorant Receptors 36
2.4.4 Summary of the Biosensors 41
2.5 Future Prospects 42
References 43

3 Large-Scale Chemical Sensor Arrays for Machine Olfaction **49**
Mara Bernabei, Simone Pantalei and Krishna C. Persaud

3.1 Introduction 49
3.2 Overview of Artificial Olfactory Systems 50
3.3 Common Sensor Technologies Employed in Artificial Olfactory Systems 53
3.3.1 Metal-Oxide Gas Sensors 53
3.3.2 Piezoelectric Sensors 54
3.3.3 Conducting Polymer Sensors 55
3.4 Typical Application of "Electronic Nose" Technologies 58
3.5 A Comparison between Artificial and the Biological Olfaction Systems 58
3.6 A Large-Scale Sensor Array 59
3.6.1 Conducting Polymers 60
3.6.2 Sensor Interrogation Strategy 62
3.6.3 Sensor Substrate 64
3.7 Characterization of the Large-Scale Sensor Array 68
3.7.1 Pure Analyte Study: Classification and Quantification Capability 69
3.7.2 Binary Mixture Study: Segmentation and Background Suppression Capability 75
3.7.3 Polymer Classes: Testing Broad and Overlapping Sensitivity, High Level of Redundancy 76
3.7.4 System Robustness and Long-Term Stability 77
3.8 Conclusions 79
Acknowledgment 80
References 80

4 Taste Sensor: Electronic Tongue with Global Selectivity **87**
Kiyoshi Toko, Yusuke Tahara, Masaaki Habara, Yoshikazu Kobayashi and Hidekazu Ikezaki

4.1 Introduction 87
4.2 Electronic Tongues 90
4.3 Taste Sensor 92
4.3.1 Introduction 92
4.3.2 Principle 93
4.3.3 Response Mechanism 93
4.3.4 Measurement Procedure 97
4.3.5 Sensor Design Techniques 98

4.3.6 Basic Characteristics 103
4.3.6.1 Threshold 106
4.3.6.2 Global Selectivity 106
4.3.6.3 High Correlation with Human Sensory Scores 108
4.3.6.4 Definition of Taste Information 109
4.3.6.5 Detection of Interactions between Taste Substances 110
4.3.7 Sample Preparation 111
4.3.8 Analysis 112
4.4 Taste Substances Adsorbed on the Membrane 116
4.5 Miniaturized Taste Sensor 117
4.6 Pungent Sensor 122
4.7 Application to Foods and Beverages 124
4.7.1 Introduction 124
4.7.2 Beer 124
4.7.3 Coffee 127
4.7.4 Meat 132
4.7.5 Combinatorial Optimization Technique for Ingredients and Qualities Using a GA 134
4.7.5.1 Introduction 134
4.7.5.2 GA 134
4.7.5.3 Constrained Nonlinear Optimization 137
4.7.6 For More Effective Use of "Taste Information" 137
4.7.6.1 Key Concept 138
4.7.6.2 Taste Attributes or Qualities become Understandable and Translatable When They Are Simplified 138
4.7.6.3 Simplification of Large Numbers of Molecules into a Couple of Taste Qualities Allows Mathematical Optimization 140
4.7.6.4 Summary 141
4.8 Application to Medicines 141
4.8.1 Introduction 141
4.8.2 Bitterness Evaluation of APIs and Suppression Effect of Formulations 141
4.8.3 Development of Bitterness Sensor for Pharmaceutical Formulations 143
4.8.3.1 Sensor Design 143
4.8.3.2 Prediction of Bitterness Intensity and Threshold 144
4.8.3.3 Applications to Orally Disintegrating Tablets 146
4.8.3.4 Response Mechanism to APIs 154
4.8.4 Evaluation of Poorly Water-Soluble Drugs 156
4.9 Perspectives 160
References 163

5 Pattern Recognition **175**
Saverio De Vito, Matteo Falasconi and Matteo Pardo

5.1 Introduction 175
5.2 Application Frameworks and Their Challenges 176
5.2.1 Common Challenges 176
5.2.2 Static In-Lab Applications 177
5.2.3 On-Field Applications 178

5.3 Unsupervised Learning and Data Exploration 180
5.3.1 Feature Extraction: Static and Dynamic Characteristics 180
5.3.2 Exploratory Data Analysis 184
5.3.3 Cluster Analysis 189
5.4 Supervised Learning 190
5.4.1 Classification: Detection and Discrimination of Analytes and Mixtures of Volatiles 192
5.4.2 Regression: Machine Olfaction Quantification Problems and Solutions 196
5.4.3 Feature Selection 200
5.5 Advanced Topics 202
5.5.1 System Instability Compensation 202
5.5.2 Calibration Transfer 208
5.6 Conclusions 210
References 211

6 Using Chemical Sensors as "Noses" for Mobile Robots **219**
Hiroshi Ishida, Achim J. Lilienthal, Haruka Matsukura, Victor Hernandez Bennetts and Erik Schaffernicht

6.1 Introduction 219
6.2 Task Descriptions 220
6.2.1 Definitions of Tasks 220
6.2.2 Characteristics of Turbulent Chemical Plumes 222
6.3 Robots and Sensors 224
6.3.1 Sensors for Gas Detection 224
6.3.2 Airflow Sensing 225
6.3.3 Robot Platforms 226
6.4 Characterization of Environments 226
6.5 Case Studies 230
6.5.1 Chemical Trail Following 230
6.5.2 Chemotactic Search versus Anemotactic Approach 232
6.5.3 Attempts to Improve Gas Source Localization Robots 236
6.5.4 Flying, Swimming, and Burrowing Robots 238
6.5.5 Gas Distribution Mapping 239
6.6 Future Prospective 241
Acknowledgment 242
References 242

7 Olfactory Display and Odor Recorder **247**
Takamichi Nakamoto

7.1 Introduction 247
7.2 Principle of Olfactory Display 247
7.2.1 Olfactory Display Device 248
7.2.2 Olfactory Display Related to Spatial Distribution of Odor 250

7.2.3 Temporal Intensity Change of Odor 251
7.2.3.1 Problem of Smell Persistence 251
7.2.3.2 Olfactory Display Using Inkjet Device 254
7.2.4 Multicomponent Olfactory Display 256
7.2.4.1 Mass Flow Controller 256
7.2.4.2 Automatic Sampler 256
7.2.4.3 Solenoid Valve 258
7.2.4.4 Micropumps and Surface Acoustic Wave Atomizer 260
7.2.5 Cross Modality Interaction 261
7.3 Application of Olfactory Display 263
7.3.1 Entertainment 263
7.3.2 Olfactory Art 265
7.3.3 Advertisement 266
7.3.4 Medical Field 266
7.4 Odor Recorder 267
7.4.1 Background of Odor Recorder 267
7.4.2 Principle of Odor Recorder 268
7.4.3 Mixture Quantification Method 271
7.5 Algorithm of Odor Recipe Exploration 274
7.5.1 Odor Approximation 274
7.5.2 MIMO Feedback Method 276
7.5.3 Method to Increase Number of Odor Components 278
7.5.3.1 SVD Method 278
7.5.3.2 Two-Level Quantization Method 280
7.5.4 Dynamic Method 283
7.5.4.1 Real-Time Reference Method 284
7.5.4.2 Concurrent Method 287
7.5.5 Mixture Quantification Using Huge Number of Odor Candidates 289
7.6 Exploration of Odor Components 292
7.6.1 Introduction of Odor Components 292
7.6.2 Procedure for Odor Approximation 293
7.6.3 Simulation of Odor Approximation 295
7.6.4 Experiment on Essential Oil Approximation 297
7.6.5 Comparison of Distance Measure 301
7.6.6 Improvement of Odor Approximation 303
7.7 Teleolfaction 305
7.7.1 Concept of Teleolfaction 305
7.7.2 Implementation of Teleolfaction System 306
7.7.3 Experiment on Teleolfaction 307
7.8 Summary 308
References 309

8 Summary and Future Perspectives **315**
Takamichi Nakamoto

Index **317**

Preface

This book introduces the current technologies related to machine olfaction and taste, which is expected to become key technologies of human interface. The handbook of this field was published more than 10 years ago, and since then, many new technologies have been developed.

One of the new trends in the sensing field is the olfactory biosensor, utilizing olfactory receptors in a living body. Moreover, an olfactory display to present scents to a human is relatively new. In the field of taste sensors, many applications have accumulated. These important technologies have emerged during the last decade.

Thus, it is necessary to review this field thoroughly and to describe it systematically for both beginners and experts. This allows novices to learn the basics and experienced researchers and engineers to learn about the state of the art of machine olfaction and taste systematically.

This book can cover the entire range of artificial chemical sense including both olfaction and taste in a modernized way. We divide the field of machine olfaction and taste into six categories: biological matter, odor sensing technology, taste sensing technology, pattern recognition, mobile robots with olfaction, and olfactory display together with odor recorders. Thus, readers can separately learn each area from the basic to the latest research.

The primary audience of this book are people in academia and industries such as electronics, computer science, chemistry, and biology. Especially, people related to multimodal interface in virtual reality or human interface might find cues for their research or commercial products. Moreover, people in academia and industries such as communications, agriculture, medicine, advertisement, and entertainment might find this book beneficial. This book is written with the intention to be a "modern bible" and guide of machine olfaction and taste.

Takamichi Nakamoto

About the Contributors

Victor Hernandez Bennetts is a postdoctoral researcher at the Mobile Robot and Olfaction Lab at the Center for Applied Autonomous Sensor Systems (AASS), Örebro University, Sweden. His main interests are mobile robotics, gas sensing, and artificial olfaction. He obtained his Ph.D. in computer science and his master's degree in robotics and intelligent systems from Örebro University, Sweden in 2015 and 2010, respectively. His Ph.D. thesis addresses the task of creating gas distribution models in realistic environments with mobile robots. His master's thesis addresses gas discrimination with a temperature-modulated array of metal oxide gas sensors.

Mara Bernabei received her Laurea degree in electronics engineering from the University of Rome Tor Vergata and her Ph.D. in chemical engineering from the University of Manchester. She is currently a research associate in the School of Chemical Engineering and Analytical Science in the University of Manchester.

Her research activities are concerned with the development and application of artificial olfactory systems based on chemical and biochemical sensor arrays, and with the synthesis and functionalization of conducting polymers and nanocomposites. Her professional expertise include multivariate data analysis, design and fabrication of electronic systems for the acquisition of (bio)chemical sensors, development of embedded systems based on microcontroller technology, and synthesis, characterization, and functionalization of sensing materials.

Saverio De Vito (MS 1998, Ph.D. 2012) graduated from the University of Naples in Computer Engineering in 1998. He was then with a private ICT engineering company as an R&D team leader for telemedicine, earth observation, and distance learning projects of ESA. In June 2004, he joined ENEA as a full researcher. Since 2005, he is a contract professor of applied informatics at the University of Cassino. His research interests include artificial olfaction and electronic noses, statistical pattern recognition, and smart cyber physical systems. He participated and was responsible for several EU projects in the FP6 and FP7 programs as well as JTIs. He (co)authored more than 35 research papers and has served as a referee for several international journals as well as for IJCNN, ISOEN, and NaBIC. He is currently serving as an MC member in EuNetAir Cost Action. He is a member of the IEEE and the International Association for Pattern Recognition (IAPR)-IC and was a member of the XXVIII Italian research expedition in Antarctica.

Matteo Falasconi received his degree in physics from the University of Pavia in 2000 and Ph.D. degree in material engineering from the University of Brescia in 2005. He is currently a research assistant at the Department of Information Engineering of the University of Brescia and a research associate at CNR-INO. His research interests include chemical sensor devices and statistical data analysis for artificial olfactory systems. His work has focused principally on exploratory data analysis, unsupervised classification (clustering and cluster validity), feature extraction and selection, and adaptive drift correction approaches. He is a coauthor of 50 scientific papers.

Masaaki Habara received his B.E. and M.E. degrees both in electrical engineering from Kanazawa Institute of Technology in 1995 and 1998, respectively, and Ph.D. degree in electrical engineering from Kyushu University in 2002. He worked as a visiting associate professor of the Graduate School of Kyushu University until 2011. He joined Intelligent Sensor Technology, Inc., as a research and development engineer in 2011.

Hidekazu Ikezaki received his B.E. and M.E. degrees from Waseda University in 1984 and 1986, respectively, and Ph.D. degree in electrical engineering from Kyushu University in 1999. He was working for Anritsu Corp., Japan, from 1986 to 2002 and is now a president of Intelligent Sensor Technology, Inc., Japan.

Hiroshi Ishida is currently an associate professor in the Graduate School of Bio-Applications and Systems Engineering, Tokyo University of Agriculture and Technology, Japan. He received M.Eng. and Dr.Eng. degrees in electrical and electronic engineering from Tokyo Institute of Technology, Japan, in 1994 and 1997, respectively. From 1997 to 2004, he was a research associate of Tokyo Institute of Technology. From 1998 to 2000, he visited the School of Chemistry and Biochemistry, Georgia Institute of Technology, United States, as a postdoctoral fellow. In 2007, he was a visiting researcher in the Center for Applied Autonomous Sensor Systems (AASS), Örebro University, Sweden. His research interests include biomimetic olfactory sensing, mobile robot olfaction, and olfactory displays.

Ryohei Kanzaki received his B.S., M.S., and D.Sc. degrees in neurobiology from the Institute of Biological Sciences, University of Tsukuba in 1980, 1983, and 1986, respectively. From 1987 to 1990, he was a postdoctoral research fellow at the Arizona Research Laboratories, Division of Neurobiology, University of Arizona. From 1991 to 2003, he was successively an assistant professor, associate professor, and full professor at the Institute of Biological Sciences, University of Tsukuba. From 2004 to 2006, he was a full professor at the Department of Mechano-Informatics, Graduate School of Information Science and Technology, The University of Tokyo. Since 2006, he is a full professor at the Research Center for Advanced Science and Technology (RCAST), The University of Tokyo. He is the vice-director of RCAST (2013–) and the president of the Japanese Society for Comparative Physiology and Biochemistry (2012–2015).

Yoshikazu Kobayashi received his B.E. and M.E. degrees from Gunma University in 1996 and 1998, respectively, and Ph.D. degree in electrical engineering from Kyushu University in 2010. He was working for Anritsu Corp. from 1998 to 2002 and is now working for Intelligent Sensor Technology, Inc., Japan.

Achim J. Lilienthal is a professor of computer science and the head of the Mobile Robotics and Olfaction Lab at the Center for Applied Autonomous Sensor Systems (AASS), Örebro University, Sweden. His main research interests are mobile robot olfaction, rich 3D perception, robot vision, and safe navigation for autonomous transport robots. He obtained his Ph.D. in computer science from Tübingen University, Germany, and his M.Sc. and B.Sc. in physics from the University of Konstanz, Germany. His Ph.D. thesis addresses gas distribution mapping and gas source localization with a mobile robot. His M.Sc. thesis is concerned with an investigation of the structure of charged C60 clusters using gas-phase ion chromatography.

Haruka Matsukura is an assistant professor at the Graduate School of Bio-Applications and Systems Engineering, Tokyo University of Agriculture and Technology, Japan. She received M.Eng. and Dr.Eng. degrees in mechanical engineering from Tokyo University of Agriculture and Technology, Japan, in 2010 and 2013, respectively. From 2010 to 2013, she was a research fellow (DC) of Japan Society for the Promotion of Science. Her research interests include olfactory displays and gas sensing systems.

Nobuo Misawa received his B.Eng. and M.Eng. degrees in 2001 and 2003, respectively, from Tokyo University of Agriculture and Technology, Tokyo, Japan. He received his Ph.D. degree from the Graduate University for Advanced Studies, Aichi, Japan, in 2006. From 2006 to 2010, he was a project assistant professor in The University of Tokyo. Since 2010, he has been a tenure track assistant professor of the Electronics-Inspired Interdisciplinary Research Institute in Toyohashi University of Technology.

Hidefumi Mitsuno received his B.Eng. degree in 2000 in engineering from Kansai University, Osaka, Japan, and M.Agr. degree in 2004 from Kyoto University, Kyoto, Japan. He received his Ph.D. degree in agriculture from Kyoto University, Kyoto, Japan, in 2009. From 2007 to 2013, he was a project researcher at the Research Center for Advanced Science and Technology (RCAST), The University of Tokyo, Tokyo, Japan. Since 2013, he has been a project research associate at the RCAST, The University of Tokyo.

Takamichi Nakamoto received his B.E. and M.E. degrees in 1982 and 1984, respectively, and his Ph.D. degree in electrical and electronic engineering from Tokyo Institute of Technology, Tokyo, Japan. He worked for Hitachi in the area of VLSI design automation from 1984 to 1987. In 1987, he joined Tokyo Institute of Technology as a research associate. In 1993, he became an associate professor with the Department of Electrical and Electronics Engineering, Tokyo Institute of Technology. From 1996 to 1997, he was a visiting scientist at Pacific Northwest Laboratories, Richland, WA, United States. He is currently a professor with the Precision and Intelligence Laboratory, Tokyo Institute of Technology.

Kei Nakatani received his B.Sc. and M.Sc. in 1976 and 1978, respectively, from Okayama University, Okayama, Japan. In 1982, he received his Ph.D. degree in physiology from Tokyo

Women's Medical College, Tokyo, Japan. He worked for the Department of Physiology, Tokyo Women's Medical College from 1978 to 1982. In 1982, he moved to the Department of Physiology and Biophysics, University of Texas at Galveston, TX, United States, as a postdoctoral fellow. In 1986, he joined the Department of Neuroscience, Johns Hopkins University, Baltimore, MD, United States, as a research associate. In 1992, he moved back to Japan and became an associate professor with the Institute of Biological Science, University of Tsukuba, Tsukuba, Japan. He is currently a professor with the Faculty of Life and Biological Sciences, University of Tsukuba.

Matteo Pardo received a degree in physics in 1996 and a Ph.D. in computer engineering in 2000. Since 2002, he is a researcher of the Italian National Research Council. In 2008–2010, he was at the Max Planck Institute for Molecular Genetics in Berlin with a Von Humboldt fellowship for experienced researchers. His research interest is data analysis and pattern recognition for artificial olfaction and genomics. He has been the technical chair of the International Symposium on Olfaction and Electronic Nose 2009. Since September 2011, he is seconded at the Italian Embassy in Berlin as science attaché. He serves in the Council of the European Southern Observatory. He is the counsellor for innovation at the Italian Chamber of Commerce in Germany.

Simone Pantalei received his Laurea degree in electronics engineering and his Ph.D. in engineering of sensorial and learning systems from the University of Rome Tor Vergata, Italy, in 2002 and 2007, respectively.

Until 2012, he worked as a researcher in the Institute for Microelectronics and Microsystems of the Italian National Research Council, where his research interests included finite element method simulations, data analyses and clean room processes applied to chemical sensors, gas sensing, and electronic nose systems.

He was a research associate at the University of Manchester, United Kingdom, where his main research activity focused on the application of electronic nose and tongue systems to the anaerobic digestion of food waste at an industrial scale and joined Multisensor Systems Ltd, UK in 2015.

Krishna C. Persaud is currently professor of chemoreception at the School of Chemical Engineering at the University of Manchester, United Kingdom. He graduated from the University of Newcastle upon Tyne with a B.Sc. honors degree in biochemistry and went on to the University of Warwick to graduate with an M.Sc. in molecular enzymology and a Ph.D. in olfactory biochemistry. He is a fellow of the Royal Society of Chemistry and a fellow of the Institute of Measurement and Control.

His research interests focus on the area of chemical senses from physiology to chemistry. This has included the development of gas sensor arrays for sensing odors based on conducting polymers that became commercialized by Aromascan plc, Osmetech plc, now part of Genmark Diagnostics Inc. He is the founder of Multisensor Systems Ltd and is currently the executive secretary and treasurer of the European Chemoreception Organisation (ECRO).

Takeshi Sakurai received his B.Agr. and M.Agr. degrees in 1999 and 2001, respectively, and his Ph.D. degree in agriculture from Kyoto University, Kyoto, Japan, in 2005. From 2005 to 2007, he was a postdoctoral researcher at Prof. Kanzaki's Lab in The University of Tokyo. In

2007, he became a project research associate at the Research Center for Advanced Science and Technology (RCAST), The University of Tokyo. Since 2013, he has been a project lecturer at the RCAST, The University of Tokyo.

Erik Schaffernicht is a postdoctoral researcher at the Mobile Robot and Olfaction Lab at the Center for Applied Autonomous Sensor Systems (AASS), Örebro University, Sweden. He received his diploma degree in computer science (2006) and his doctoral degree in machine learning (2012) both from the Ilmenau University of Technology, Germany. His research interests include machine learning applications for robotics and environmental mapping with gas-sensitive robots.

Yusuke Tahara received his Ph.D. degree in engineering from Iwate University, Japan, in 2010. From 2010 to 2012, He was a postdoctoral fellow at Kyushu University. Since 2012, he has been an assistant professor at Kyushu University. His main research interests focus on developing taste sensors and biosensors. He is a member of the Institute of Electrical Engineers of Japan and the Japan Society of Applied Physics.

Kiyoshi Toko received his B.E., M.E., and Ph.D. degrees in electrical engineering from Kyushu University in 1975, 1977, and 1982, respectively. He was a dean of the Graduate School of Information Science and Electrical Engineering of Kyushu University from 2008 to 2011 and is now a distinguished professor and a director of the Research and Development Center for Taste and Odor Sensing in Kyushu University. He is a member of the Japan Society of Applied Physics, the Institute of Electrical Engineers of Japan, and the Japanese Association for the Study of Taste and Smell.

1

Introduction to Essentials of Machine Olfaction and Tastes

Takamichi Nakamoto
Precision and Intelligence Laboratory, Tokyo Institute of Technology, Tokyo, Japan

Although there are a variety of sophisticated machines for visual and auditory senses, machines for chemical senses such as olfaction and taste are still prematured. However, they are very important since they are deeply related to our primitive but fundamental capabilities. We can search for foods in daily life as well as can avoid danger using olfactory sense. Although animal's capability is nowadays superior to human ones, we still have them. Those chemical senses cannot be ignored in our daily life. Nowadays we can create cyberspace made up of visual and auditory senses. However, that cyberspace still lacks reality since olfactory and gustatory senses are not included.

The first machine olfaction was proposed about 30 years ago. Then, it was extended and an electronic nose community appeared. Although many papers have been already published, its application toward to industry is still limited. Its sensitivity, selectivity, and robustness against disturbance should be much improved for the actual application. A variety of applications are waiting for its progress. This book describes the current effort of sensing part of machine olfaction.

Machine olfaction has another part such as olfactory display. It works as an actuator in olfaction. An olfactory display is relatively new compared with odor sensing technology. Researchers in virtual reality have focused on the olfactory display to realize cyberspace with chemical sense. Although researcher population of olfactory display is still small, it gradually spreads into the world.

A human olfactory interface has both odor sensing and olfactory display. It is now growing up in the field of human interface. Utilizing those two techniques, odor recorder and teleolfaction system are being studied.

In contrast to olfaction, a taste sensor has been applied to a certain application area. Especially, medical field is its good customer. However, we still wait for simple easy-to-use taste sensor to include taste sense in cyberspace. The attempt to realize it will be later shown.

Essentials of Machine Olfaction and Taste, First Edition. Edited by Takamichi Nakamoto.

Companion website: www.wiley.com/go/nakamoto/olfaction

Recently we often hear the world such as cyber-physical system. Cyber-physical system enables cyberspace with physical senses. However, we have never heard the word "cyber-chemical system." We can have cyber-chemical system if the technologies of machine olfaction and taste are easily available.

This book describes the essential parts of machine olfaction and taste. Chapter 2 describes olfactory mechanism of a living body. Utilizing it, olfactory biosensor is being developed. Chapter 2 also explains the olfactory biosensor.

Chapter 3 shows odor sensing technology. It explains the basics of artificial sensors. Moreover, a large-scale sensor array in the same way as biological one is being studied. This trend in electronic nose is introduced in Chapter 3.

Chapter 4 shows the taste sensor. It describes the principle and its application toward foods and medicines. This chapter explains the latest research review as well as the fundamentals of taste sensor.

Chapter 5 describes the current pattern recognition technologies available in electronic noses. The pattern of many ORN responses is recognized by an olfactory neuron system. Thus, the output pattern of the array of sensors with partially overlapping specificities is recognized in machine olfaction. Chapter 5 describes the basics of pattern recognition technologies together with its advanced technologies.

Chapter 6 explains mobile robot technology with chemical senses. It can search for the target chemical in the field. Its sensor, algorithm to look for the target and the experiment is shown in this chapter.

Chapter 7 shows olfactory display and odor recorder. Various types of olfactory displays are systematically described. Moreover, the review of odor recorder is shown in this chapter.

Chapter 8 is the summary and describes the perspective of machine olfaction and taste.

Each chapter covers an essential part of machine olfaction and taste. It describes basic part at first and then extends their contents to the advanced technology.

2

Physiology of Chemical Sense and its Biosensor Application

Ryohei Kanzaki[1], Kei Nakatani[2], Takeshi Sakurai[1], Nobuo Misawa[3] and Hidefumi Mitsuno[1]

[1] *Research Center for Advanced Science and Technology, The University of Tokyo, Tokyo, Japan*

[2] *Faculty of Life and Environmental Sciences, University of Tsukuba, Tsukuba, Japan*

[3] *Electronics-inspired Interdisciplinary Research Institute, Toyohashi University of Technology, Toyohashi, Japan*

2.1 Introduction

Odorant sensors for detecting various types of odorants are currently required increasingly for several applications, such as disease diagnosis, food administration, and risk management associated with detection of explosives and drugs. Odorant sensors based on metal-ode semiconductor devices, quartz crystal microbalances (QCM), or surface acoustic wave (SAW) detectors have been developed for a variety of odorant-detection applications. However, the performance of these sensors is still inferior to the olfactory systems of living organisms in terms of selectivity, sensitivity, and response time.

In animals olfaction plays a key role in the release of appropriate behavior under complexly changing environment. Animals extract adequate information from numerous odorants in their surroundings and respond in many aspects of the animal's life including foraging, prey detection, finding hosts, and mating. Odor information is detected by olfactory receptor neurons (ORNs) in an olfactory organ and properly processed in neural networks in the brain and finally translated into the appropriate behavioral responses, mechanisms of which are critically important in the development of advanced odor sensors and odor tracking robots based on biological systems as well as neuroscience and neuroethology.

Transduction mechanisms in olfaction have been revealed in vertebrates and insects. Odorants are detected at the surface of the olfactory epithelium which contains ORNs in vertebrates. Insects have ORNs in the sensillum of antennae. Odorant signals detected at the

Essentials of Machine Olfaction and Taste, First Edition. Edited by Takamichi Nakamoto.

Companion website: www.wiley.com/go/nakamoto/olfaction

membrane of the ORNs are converted into electric signals and transmitted to the brain. In vertebrates the transduction is mediated by complex signal transduction pathways through G proteins, adenylyl cyclase, cyclic adenosine monophosphate, and cyclic nucleotide-gated ion channels (i.e., G protein-coupled receptors (GPCRs)). By contrast, insect odorant receptors, coupled with an olfactory receptor coreceptor (Orco), form ligand-gated ion channels (i.e., ionotropic receptors) that control all-in-one odorant reception and ion influx. Moreover, the insect ionotropic receptors can selectively detect various types of odorants covering a wide range of chemical functional groups, including alcohols, aldehydes, ketones, acids, hydrocarbons, and aromatic compounds.

Since the transduction mechanisms in animals have been revealed, odorant receptors would be valuable odorant sensors with high selectivity, high sensitivity, and good response time compared to conventional sensors. So far a number of cell-based odorant sensors have been studied and proposed by using recent advanced gene engineering techniques. Among those cell-based sensors, insects are equipped with sophisticated molecular mechanisms that involve initial activation of odorant receptors. The insect odorant receptors would be valuable odorant sensors with high selectivity, high sensitivity, and good response time and could be assembled into a compact chip to develop portable odorant sensors.

In this chapter, first, transduction mechanisms of insect and vertebrate ORNs are introduced. In addition transduction mechanisms of insect and vertebrate gustatory signals are also introduced. Based on these findings, various kinds of biological components such as tissues, sensory neurons, proteins, and genes regarding olfaction in living organisms have been utilized for application to olfactory sensors. Different types of olfactory sensors, that is, tissue-based sensors, cell-based sensors, and receptor-based sensors, are then introduced.

2.2 Olfaction and Taste of Insects

2.2.1 Olfaction

2.2.1.1 Anatomy of Olfaction

Structure of Olfactory Sensillum

Insects detect odorants with a pair of antennae on their head and, in some dipteran species, a pair of maxillary palp extending from the base of the maxilla (Figure 2.1a). ORNs are housed in cuticular specialization, named olfactory sensillum, on these olfactory organs (Figure 2.1b, c). Olfactory sensillum has numerous minute pores (10–100 nm), named olfactory pores [57, 128], which allow odorant molecules to enter inside the sensillum. ORNs are bipolar neurons that extend their dendrites, the site of odorant reception, into the sensillum and project their axons into the antennal lobe (AL), the first olfactory center of the brain in insects [41]. Cell bodies and inner dendrites of ORNs are surrounded by three accessory cells: the tormogen, trichogen, and thecogen cells (Figure 2.1c). These cells isolate a lymph space surrounding the outer dendrites of the ORNs from the hemolymph. Differences in chemical composition of the sensillum lymph and the hemolymph generate a standing electrical potential difference, the transepithelial potential (see Refs. [58, 98]). Odorant stimulation generates a receptor potential in the outer dendritic membrane, which can induce the generation of action potentials in a more proximally located spike-generating zone. Olfactory sensilla are classified to several types according to their outer shape (s. trichodea, s. basiconica, s. coeloconica, s. placodea,

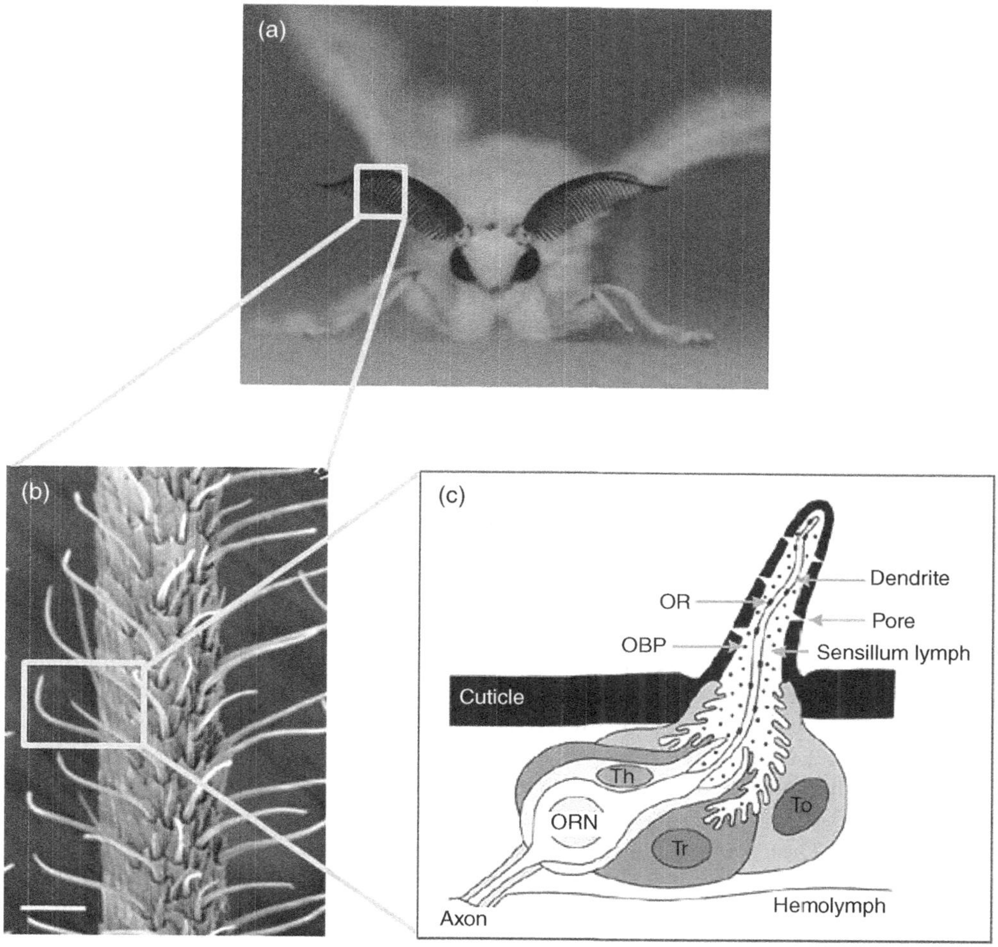

Figure 2.1 Main olfactory sensory organs of the silkmoth *Bombyx mori*. (a) A male silkmoth with its prominent antennae optimized for odorant detection. (b) Scanning electron micrograph of an antenna. Scale bar: 25 μm. (c) Schematic diagram of an olfactory sensillum. ORNs are surrounded by three types of accessory cell: tormogen (To), trichogen (Tr), and thecogen cells (Th). (Figure (c) is reproduced with permission from Ref. [48]. © Springer.)

and so on). In some cases, the type of sensillum is well correlated with its function. For example, s. trichodea in male moths house ORNs specifically tuned to conspecific sex pheromones [52, 53], while others house ORNs for so-called general odorants such as from foods or plants [108].

Detection of Odorants by ORNs

Odorant molecules in the air are first absorbed on the cuticular surface of the sensillum, and then they diffuse inside the sensillum through olfactory pores and pore tubules [52, 54, 55]. Since most volatile odorants are hydrophobic in nature, it is difficult to efficiently pass sensillum lymph to dendritic membrane of ORNs. Therefore, mechanism to facilitate solubilization of odorants into aqueous lymph layer is important to achieve sensitive detection of odorants. For this, insects utilize small (about 15 kDa) soluble globular proteins named odorant binding protein (OBPs) [141] that bind odorants and transport them to dendritic membrane of ORNs. Mechanisms of odorant binding and release by OBPs are well studied using the silkmoth (*Bombyx mori*) pheromone-binding protein 1 (BmPBP1) that binds sex pheromone components of that species. BmPBP1 has two different conformations that reversibly change in a pH-dependent manner [149]. At neutral pH, odorant binding pocket located inside of proteins is open for binding odorant, while at acidic pH C-terminal loop domain of PBP occupies this binding pocket [42, 73]. This conformational transition is believed to occur around dendritic membrane due to lower pH around cellular membrane, resulting in the release of odorant from internal binding pocket around ORs. Then, odorants are detected by OR complex that activate chemoelectrical transduction machinery on dendritic membrane of ORNs.

2.2.1.2 Signal Transduction of Odor Signals

Upon binding to OR, the information of odorants is converted into electrical signals in ORNs. Earlier studies have reported rapid and transient increase of G protein-mediated second messenger, inositol triphosphate (IP_3), in the antennal homogenates after pheromone stimulation [11]. Expression of heterotrimeric G protein in ORNs and activity of its effector enzyme in antennal homogenate were also demonstrated, suggesting that odorant signals are transduced into electrical signals via heterotrimeric G protein-mediated second messenger cascade (Figure 2.2a) [48, 67]. However, recent physiological analysis of ORs revealed that insect ORs form heteromeric complex with their coreceptor Orco (originally named as Or83b in *Drosophila melanogaster*) and function as an odorant-gated ion channel (Figure 2.2b, c) [121, 126, 147]. Orco is originally isolated as a member of insect ORs and has the following unique characteristics [68, 100, 144]: (i) Orco is exceptionally conserved across insect species, while conventional ORs are highly divergent within and across species. (ii) Orco is expressed in most ORNs, while conventional ORs are expressed in specific subsets of ORNs.

Sato *et al.* coexpressed BmOR1 with BmOrco and other combinations of members of the Orco family with ORs in heterologous expression systems. Examination of the electrophysiological properties of an Orco/OR complex revealed that it acts as a pheromone/odorant-gated nonselective cation channel (Figure 2.2b) [121]. Interestingly, there was no evidence for an elevation of second messenger levels upon stimulation with ligands, indicating no involvement of a G protein-mediated cascade in the activation of Orco/OR complexes. Later pharmacological analysis of cultured cells coexpressing Orco and ORs from *D. melanogaster* supported this conclusion [126]. In the meanwhile, Wicher *et al.* found that fast transient and slow prolonged ion currents occur in cultured cells coexpressing DmOrco and *D. melanogaster* ORs upon stimulation with appropriate ligands for the expressed ORs (Figure 2.2c) [147]. They proposed that fast currents result from direct activation of Orco by ORs and slow currents occur via G protein-mediated activation of Orco. Both studies indicated that odorant signals are mediated by odorant-induced channel activity of ORs/Orco complex or Orco but different in terms of involvement of G protein-coupled pathway. Further studies will be required to reach consensus about the roles of G protein-mediated second messenger system on reception of odorants.

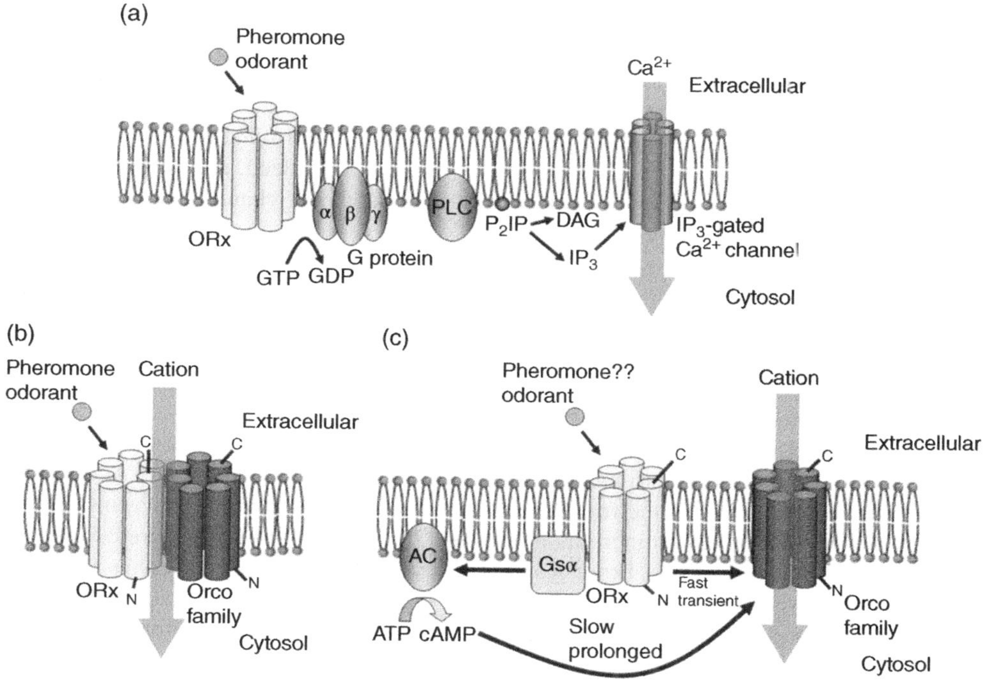

Figure 2.2 Proposed signal transduction mechanisms. (a) A classical model of insect olfactory transduction that involves a G protein-mediated PLC–IP_3 pathway. (b) Alternative model where the odorant receptor (OR) forms a heteromeric odorant-gated nonselective cation channel with an Orco family protein. (c) Alternative model that postulates two pathways. An ionotropic pathway involves the direct activation of Orco by an OR resulting in a rapid but transient cation influx. A metabotropic pathway is coupled to the G protein and induces slow but prolonged cation currents. (Reproduced with permission from Ref. [117]. © 2014, Sakurai, Namiki and Kanzaki.)

In this regard, phosphorylation of ORs by protein kinase C activated by presumably G protein-mediated second messengers reportedly enhances responses to odorants [36, 148], suggesting that second messenger system may not directly activate but modulate activity of Or/Orco channel through phosphorylation of ORs. More recently, it was reported that latency of electrophysiological responses of antennae of several insect species is as fast as several millisecond order [132]. This response speed is in accordance with the range of ionotropic pathway, indicating at least fast response is mediated by ionotropic activity of OR and Orco complex.

2.2.1.3 Molecular Biology of Olfaction

The insect OR gene family was first identified from the fruit fly by bioinformatics-based methods as well as large-scale screenings of olfactory tissue-specific genes [25, 34, 143]. Sixty OR genes are found in whole genome sequence of the fruit fly [144]. After that OR genes have been identified from various insect species. The number of OR genes considerably varied between species ranging from 10 in the body louse to more than 300 in ants. Amino acid sequence comparison revealed that insect ORs form a unique gene family with no obvious homology with any other proteins including ORs from vertebrates. Although insect ORs possess seven-transmembrane domain characteristic to GPCR family, they have a reverse membrane topology compared to GPCRs with their N-terminal on the cytoplasmic side and C-terminal on the extracellular side [7, 49, 72, 89]. Indeed, recent physiological studies demonstrated that insect ORs form odorant-gated ion channel with Orco (see Section 2.2.1.2 in detail).

Response Profiles of ORs

In an OR and Orco channel complex, OR is responsible for ligand binding and determines response profiles of ORNs [121]. By now, response spectrum of more than 100 ORs has been determined by using "empty neuron" expression system in the fly antennae and/or heterologous cell expression systems such as *Xenopus* oocytes. In principle, each OR can bind different odorants and each odorant can be recognized by multiple ORs. Response spectrum of individual OR continuously distributed from narrowly to broadly tuned one [17, 38, 145]. Comprehensive analysis using the fruit fly has uncovered the relationship between ORs and ORNs as well as ORNs and glomerulus in the AL [28, 33]. Similar to vertebrate olfactory systems, most ORNs selectively express one of many ORs, and ORNs expressing the same OR project into a single defined glomerulus in the AL. Since each OR normally responds to various odorants and each odorant is detected by various ORs, odorant information is represented as a combination of activated glomeruli in the AL.

2.2.2 *Taste*

In the gustatory system, the sense of taste is essential for the animals to evaluate which food is good to eat and which food should be avoided. Compared to most mammals that can discriminate five basic tastes (see Section 2.3.2.1), insect basic tastes are divided into four categories: bitter, sweet (sugar), salty, and water. In addition to these tastants, insect gustatory system can detect uncanonical taste substances such as fatty acids, sour tastes, and chemicals unrelated to food such as contact pheromones. In this section, the mechanisms of taste detection in insects are briefly described.

2.2.2.1 Anatomy of Taste

Taste Organ

One of the striking features of the insect gustatory system is that taste organs are not restricted to mouth part but are distributed in multiple body parts. For example, in adult fruit flies, four appendages—the proboscis, legs, anterior wing margins, and ovipositor—possess gustatory function (Figure 2.3a) [79, 129, 142]. The proboscis is a long appendage extending from the head and comprises external taste organ named labella that is located at the apical end of the proboscis and three internal organs—the labral sense organs (LSOs), the dorsal cibarial sense organ (DCSO), and the ventral cibarial sense organ (VCSO) that are located along the pharynx (Figure 2.3b) [129]. These organs play roles in determining whether to ingest or expel food and thus can be regarded as the functional equivalent to the mammalian tongue. Taste organs

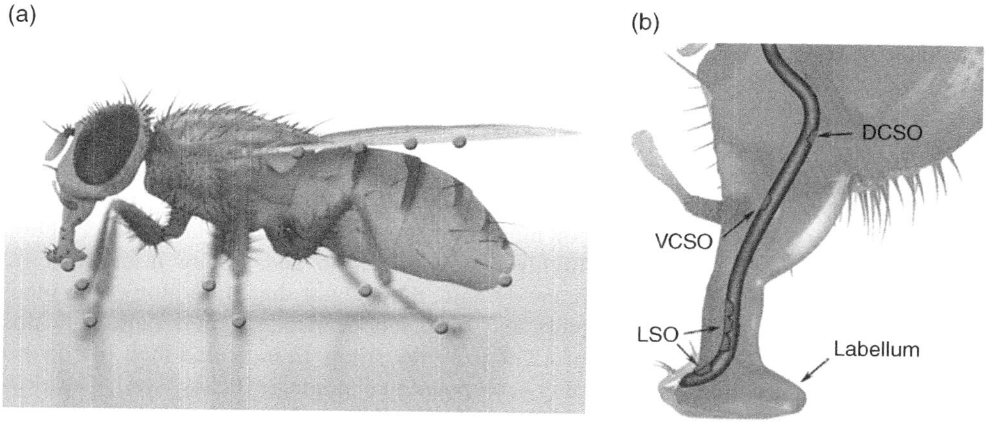

Figure 2.3 Taste organs of adult fruit fly, *D. melanogaster*. (a) Distribution of taste organs on adult fly body. (b) Taste organs in mouth part. (Reproduced with permission from Ref. [79]. © Elsevier.)

are present on distal segment of legs, tarsi. Taste sensors on tarsi carry out initial sampling of potential food and evaluate the quality of it. Tarsal taste organ on male forelegs also plays a role in detecting contact pheromones that promote or inhibit courtship behavior of males [10]. Taste organs on wings are indicated to participate in the detection of microbe-derived lipopolysaccharides that induce glooming behavior to remove microbe from fly's cuticle [157]. The ovipositor taste organ provides information of nutrient conditions to identify location suitable for egg laying [159].

Structure of Taste Sensillum and Gustatory Receptor Neurons
Taste substances (tastants) are detected by sensory neurons referred to as gustatory receptor neurons (GRNs) housed in taste sensillum on those organs. GRNs are bipolar neurons that extend their dendrite into the shaft of taste sensillum and project their axon to the suboesophageal ganglion, taste center in insect brain. In contrast to olfactory sensillum that has many pores on its cuticle, taste sensillum has a single pore at the apical end of the sensillum from which tastants enter into taste sensillum. Typically, there are one to four GRNs and one mechanosensory neuron in individual sensillum. Each GRN is tuned to substances of one of four basic taste categories.

2.2.2.2 Molecular Biology and Signal Transduction of Taste

Recent studies have revealed that tastants are detected by various types of receptors expressed in GRNs. Basically, the types of receptor correspond to taste categories. In this section, the types of receptors and signal transduction activated by interaction of tastants with receptors are summarized.

Bitter and Sweet (Sugar) Taste
Bitter and sweet tastes are detected by the large receptor family named gustatory receptor (GR) that is the major class of insect taste receptor. GR was first discovered from the fruit

fly by bioinformatics approach of nearly completed *Drosophila* genome sequences to seek candidate genes that can encode seven-transmembrane domain receptor [26, 123]. By these analyses, 43 GRs that belong to the novel membrane protein family and expressed selectively in subsets of GRNs were reported. Later analysis revealed that there are 68 GRs in whole genome sequences of *Drosophila*[114]. By now, GR family genes are reported from various insect species of different orders such as mosquito, moth, beetle, wasp, bee, aphid, and louse. Number of GR genes is different between species from 10 in honeybee *Apis melifera* to 114 in disease vector mosquito *Aedes aegypti*. Similar to insect ORs, amino acid sequences of GRs are highly divergent within and among insect species. Although GRs have seven-transmembrane domain, they do not share homology with other known GPCRs and membrane receptors and form independent receptor family. In this sense, membrane topology analysis of a GR from the silkmoth indicated that, like insect ORs, GR has inverted membrane topology compared to canonical GPCRs [164]. The closest relative of GR family is the insect OR family. Phylogenetic analysis revealed that the emergence of the GR family precedes that of the OR family, suggesting that ORs may evolve from GRs [114, 140].

GRs basically function as bitter and sweet taste receptors. Members of the GR family function as receptors for other categories such as an amino acid (L-canavanine) [29], nonvolatile contact pheromones [10], and CO_2[130]. Surprisingly, it is reported that member of GR family is also involved in the detection of nonchemical signals including light [155] and temperature [91], indicating highly divergent roles of GRs.

In many cases multiple GRs are expressed in individual GRNs, suggesting that GRs form heteromeric complex to exert their functions. In the meanwhile, at least 2 GRs tuned to fructose can be functionally reconstructed in heterologous expression system [122]. Thus, mode of action of GRs is still largely unknown.

Signal transduction pathway following activation of GRs is also a major open question in the insect gustatory system. Recent studies raise the possibility that GR signaling is meditated by both G protein-coupled metabotropic pathway and ionotropic activity of GRs. Regarding metabotropic pathway, the expression of heteromeric G protein in GRNs has been shown [134], and mutation or knockdown of effector gene of G proteins reduced physiological and/or behavioral responses to bitter and sweet tastants [29, 50, 60, 136]. The involvement of G protein-coupled pathway is also evident for tastants in other categories but mediated by GRs, including CO_2[160] and an amino acid (L-canavanine) [29]. Regarding ionotropic activity of GRs, at least one GR for fructose appears to function as a tastant-gated nonselective cation channel, independent of a G protein-coupled pathway [122].

Salty Taste

Insects equip two types of salt GRNs: one tuned to high salt and the other to low salt. In *Drosophila* larvae, two epithelial Na channel (ENaC) family members, PPK11 and PPK19, are required for response to low salt [81], whereas in adult flies member of IR family IR76b is required for low-salt detection [165], which encodes continuously open-state Na^+ leak channel. Because Na^+ concentration in sensillum lymph is much lower than that of the hemolymph, influx of Na^+ through ENaC and IR76b occurs when insects take food containing low salt. This influx depolarizes the GRNs. Identification of receptors responsible for detection of high salt is a major question of salt detection in insects.

Sour Taste

Unlike mammals, insects do not possess sour taste cells. Instead, it was shown that subsets of bitter GRNs mediate carboxylic acid signals [21]. In addition, the activity of sweet GRNs is inhibited by acids [21]. However, receptors for these acids have not molecularly identified yet.

Water Taste

GRNs that respond to low osmolality are used for detecting water. A member of the degenerin/epithelial Na channel (DEG/ENaC) family, named PPK28, responds to low osmolality and is necessary and sufficient for water sensitivity [15, 23].

2.3 Olfaction and Taste of Vertebrate

2.3.1 Olfaction

2.3.1.1 Anatomy of Olfaction

Terrestrial vertebrates utilize two kinds of olfactory information, volatile odorants and pheromones.

A main olfactory system, which is responsible for volatile odorant detection, consists of ORNs and the olfactory bulb (Figure 2.4a). ORNs are located in the olfactory epithelium lining the nasal cavity and are surrounded by glia-like supporting cells. There are also basal

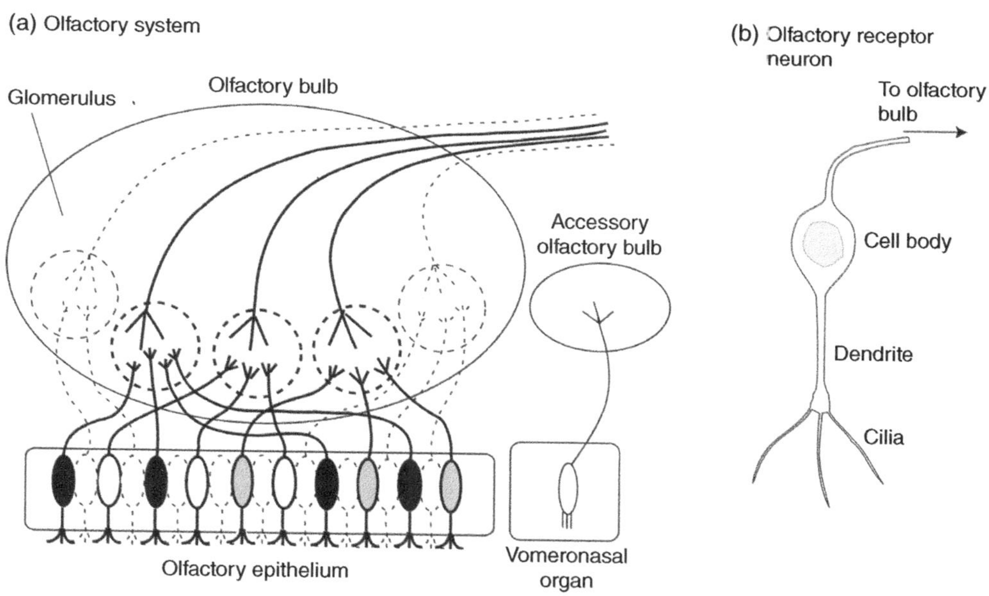

Figure 2.4 (a) Schematic diagram of vertebrate olfactory system. Each ORN in the main olfactory epithelium expresses only one odorant receptor gene and sends its axon terminal into one of the glomeruli in the olfactory bulb. ORNs expressing a given type of odorant receptor converge to a common glomerulus. Vomeronasal sensory neurons in vomeronasal organ project their axon into the accessory olfactory bulb. (b) Schematic drawing of an ORN. Odorant receptors and signal transduction machineries are expressed in the olfactory cilia

cells in the epithelium, which are kinds of progenitor cells that differentiate into ORNs or supporting cells. The olfactory epithelium has been divided into four zones based on odorant receptor expression. A specific type of ORN is located in one of the four zones [111, 137].

Each ORN extends a single dendrite to the surface of olfactory epithelium, and the dendritic knob projects 5–20 fine cilia in the mucus layer where odorant molecules can dissolve and bind to. The ORN extends an axon from the other end of the cell body and projects directly to the olfactory bulb in the forebrain where it synapses on the distal primary dendrites of second-order neurons, mitral cells, and tufted cells (Figure 2.4b).

As detailed in Section 2.3.1.3, ORNs express one of about 1000 odorant receptor proteins in the mouse. Although ORNs that express a particular odorant receptor distributed broadly in a given zone of olfactory epithelium, the axons from particular type of ORNs project to common glomeruli in the olfactory bulb (Figure 2.4a). This means that individual glomerulus represents a single type of receptor [97]. There are 5–10 million ORNs in the olfactory epithelium, and their axons converged to about 2000 glomeruli in the olfactory bulb. It has been estimated that several thousand ORN axons synapse onto the dendrites of only 5–25 mitral cells in each glomerulus [32], indicating that a considerable amount of convergence of olfactory information occurs in the glomeruli.

The vomeronasal organ (VNO), which is responsible for mediating pheromone information, is located at the base of the nasal septum in the mouse. Vomeronasal sensory neurons (VSNs) project axons via vomeronasal nerve to the accessory olfactory bulb, which is located on the dorsal posterior part of the main olfactory bulb (Figure 2.4a).

2.3.1.2 Transduction of Odor Signals

Signal Transduction Cascade in ORNs

The ORN depolarizes in response to an application of volatile odorants to olfactory cilia, which triggers action potentials to transmit the olfactory information along the olfactory receptor axon to the glomeruli in the olfactory bulb. Signal transduction is the process by which chemical information is transformed into an electrical signal within a cell. The signal transduction mechanism of ORN has been extensively studied in the past decades.

Figure 2.5 shows a schematic drawing of the signal transduction cascade of ORNs. A signal transduction takes place within the olfactory cilia. It initiates when volatile odorant molecules bind to an odorant receptor. Odorant-bound receptor activates ORN-specific GTP-binding protein, G_{olf}, which then activates adenylyl cyclase. As a result, second messenger cAMPs are produced from ATP by enzymatic activity of adenylyl cyclase and diffuse within cilia. cAMP binds to and opens cyclic nucleotide-gated (CNG) channels to generate an inward current that causes a depolarization of the cell. CNG channels are non-selective cation channels first found in retinal photoreceptors, which are opened by direct binding of cyclic nucleotide (cAMP in the case of ORNs). In physiological conditions, Na^+ and Ca^{2+} enter the cell through the CNG channels to carry positive charge. The Ca^{2+} influx causes an opening of calcium-gated chloride channels, resulting in an inward current (Cl^- efflux) that causes a depolarization.

It is interesting that both cationic current through CNG channels and anionic current through calcium-gated chloride channels are responsible for odor response, unlike, for example, phototransduction. The chloride component is reported to be as large as the cationic component [71].

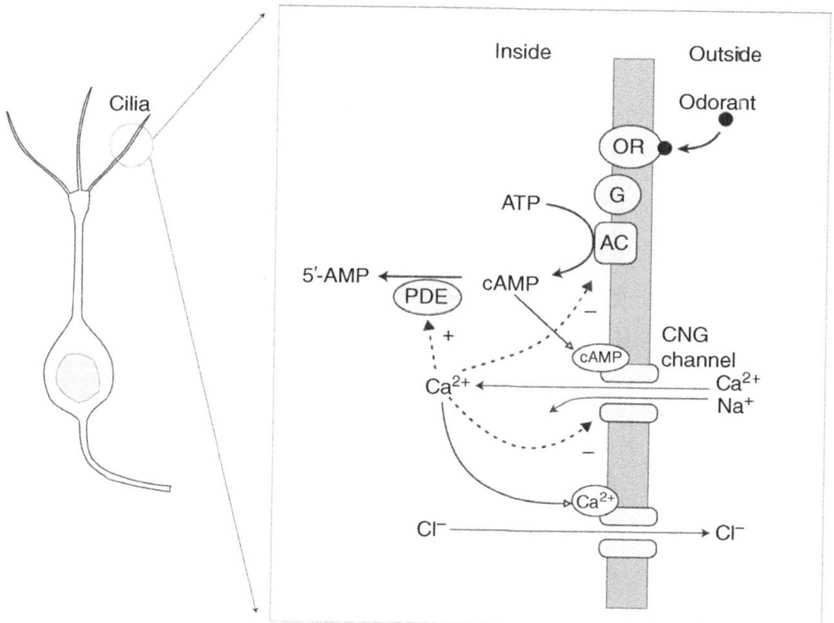

Figure 2.5 Schematic diagram of signal transduction cascade in ORNs. AC, adenylyl cyclase; CNG channel, cyclic nucleotide-gated channel; G, G protein; OR, odorant receptor; PDE, phosphodiesterase. Dashed lines indicate the effects of Ca^{2+} on AC, CNG channel, and PDE. +, facilitation; –, inhibition

Since the olfactory epithelium is exposed to the external environment, that is, freshwater, the cation concentration in the mucus may not be consistent. Therefore, it is important that the inward anionic current would compensate a reduction of cationic current due to a reduction of cation concentration [71].

Odor Adaptation

Many studies have shown that ORNs quickly adapt to odor stimulation. When the prolonged odor stimulus is applied, the depolarization in response to odor stimulation decreases with time although the odor stimulation still exists [70]. Another manifestation of odor adaptation has been demonstrated by double-pulse experiments. When a pair of brief, identical odorant pulses are applied to the cell, the response amplitude induced by the second pulse is remarkably small in comparison with the first response, if the interval between the pulses are sufficiently short. It recovers as the interpulse interval increases (Figure 2.6a) [69, 70].

The dose–response relationship of odor responses shows that the dynamic response range under control conditions is extremely narrow. On the other hand, under the adapted states, the dynamic range of the odor responses shifts and broadens (Figure 2.6b) [69]. The adaptation is important for the ORN to work over a wide range of odorant concentration. When background odors are present, the sensitivity to odor stimulus decreases so that higher concentration of odors can be detected without saturating.

The next question is how the odor adaptation occurs. Kurahashi and Shibuya [70] reported that removal of external Ca^{2+} almost completely abolished the response decay during a

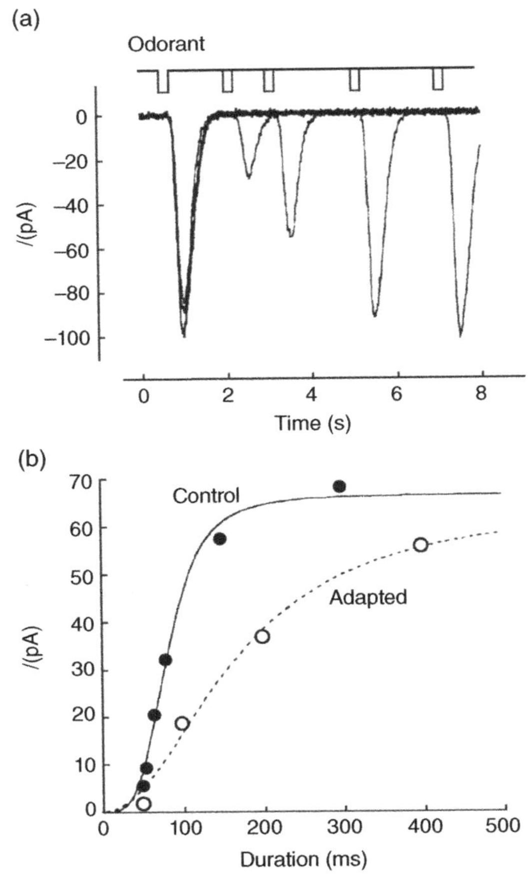

Figure 2.6 Odorant adaptation in an ORN. (a) Pulses of the odorant at the same concentration and duration were applied to the cell at different intervals. Response amplitude to the second pulse reduces depending on the intervals. (b) Relations between stimulus and response of an isolated ORN under control (filled circles) and adapted (open circles) conditions. (Reproduced with permission from Ref. [69]. © Macmillan Publishers Ltd.)

prolonged odorant stimulus, indicating that Ca^{2+} entry is responsible for odor adaptation. The molecular mechanism of odor adaptation mediated by Ca^{2+} has been studied since then.

It has been reported that odorant-induced increase in intracellular Ca^{2+} concentration reduces the cAMP sensitivity of CNG channel [66]. Two categories of adaptation mechanisms mediated by Ca^{2+} have been proposed. One is a direct action of Ca^{2+} on the CNG channel. Ca^{2+} entering through CNG channel binds to calmodulin which is the common calcium-binding protein, and the Ca^{2+}/calmodulin directly inhibits the channel itself [22]. As a result, the open probability of CNG channels decreases, causing a reduction of inward current, and the cell hyperpolarizes.

When ORNs are stimulated for a long period, the decay time course of response stimulated by odor is faster than that stimulated by the photolysis release of caged cAMP, indicating that adaptation of odor response occurs upstream to adenylyl cyclase [133]. One candidate is type III adenylyl cyclase which is expressed in ORNs. Activity of adenylyl cyclase is inhibited by Ca^{2+}/calmodulin-dependent protein kinase and thereby reduction of cAMP production occurred [146]. Another candidate is phosphodiesterase which is a cAMP catabolic enzyme. Ca^{2+}/calmodulin-dependent phosphodiesterase (CAM-PDE) is expressed in ORNs, and CAM-PDE activity is elevated by Ca^{2+} stimulation. It is suggested that Ca^{2+} stimulation of CAM-PDE is necessary for odor adaptation [9, 156].

Signal Transduction Cascade in VSNs

VNO is responsible for detecting pheromones that are chemical substances produced and released by an animals and send information to other animals of mainly the same species. The pheromone molecules are received by receptor proteins expressed in the VSN. The signal transduction mechanism in VSNs is totally different from that in ORNs. Recent studies by molecular genetics have revealed that mouse VSNs express two distinct families of GPCRs, V1R [30] and V2R [40, 92, 116]. Evidences of downstream signal transduction mechanism in VSNs have also been reported. Transient receptor potential channel 2 (TRPC2) is exclusively expressed in VSNs, and the protein is highly localized to VSN sensory microvilli in which the sensory transduction is supposed to take place [78]. The TRPC2 is gated by the lipid messenger diacylglycerol (DAG) that is independent of Ca^{2+} or protein kinase C [88]. From the above evidences, the signal transduction scheme can be proposed as follows. The pheromone molecules are received by V1R or V2R, which then activate phospholipase C (PLC). DAG is synthesized from PIP2 by enzymatic activity of PLC and directly opens the TRPC2 to produce electrical signal. Recent study has shown that increase in intracellular Ca^{2+} concentration is caused by an opening of TRPC2 and such a Ca^{2+} increase regulates the opening of TRPC2 via Ca^{2+} calmodulin, which functions as negative feedback. This negative feedback may be an underlying mechanism of sensory adaptation in VSNs [127].

Recently, it is revealed that Ca^{2+}-activated chloride channels are involved in signaling in VSNs. Stimulus-induced opening of TRPC2 allows Ca^{2+} entering the cell. An increase in Ca^{2+} leads to opening of the Ca^{2+}-activated chloride channels, which amplifies the sensory responses in VSNs [61, 158].

2.3.1.3 Molecular Biology of Olfaction

Cloning of Odorant Receptor Gene

The odorant receptor gene has been cloned for the first time from rat olfactory epithelium using polymerase chain reaction by Buck and Axel [12]. They discovered a large gene family composed of about 1000 different genes, the largest gene family in mammals, which are responsible for the animal to recognize thousands of complex odors. The odorant receptor gene may account for about 2% of the genome. In humans, about 350 odorant receptor genes have been cloned. It has been widely believed that each ORN expresses only one of the odorant receptor genes, which is known as one receptor–one neuron rule [125].

Odorant receptor protein is a member of the GPCR family. Like other GPCRs, odorant receptor contains seven hydrophobic transmembrane domains (Figure 2.7). There is highly conserved pair of cysteines in the unusually long second extracellular loop, which are specific characteristics for odorant receptors. One of the most significant features of odorant receptors

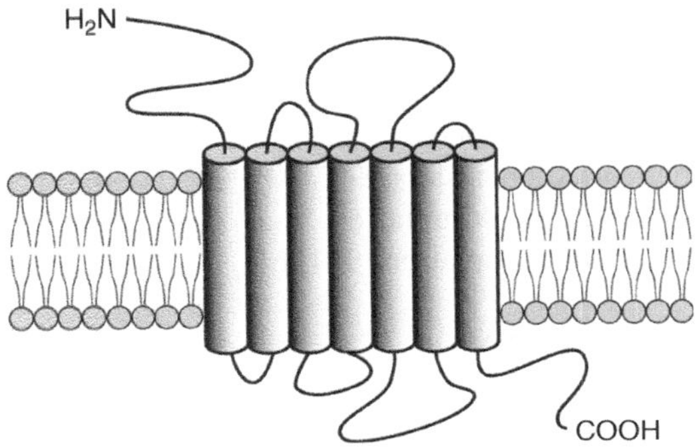

Figure 2.7 Characteristic structures of the odorant receptor. Each odorant receptor contains seven α-helical transmembrane domains, shown as cylinders

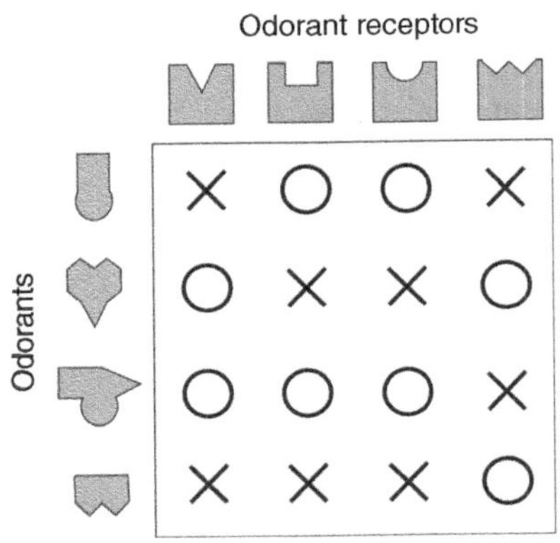

Figure 2.8 Combinatorial coding of odorants. A single odorant receptor recognizes multiple odorants, whereas a single odorant can activate multiple odorant receptors

is their large variability in amino acid sequence in the third, fourth, and fifth transmembrane domains. It is suggested that the binding of an odorant molecule occurs in an odorant binding pocket that is reportedly formed by the third, fifth, and sixth transmembrane domains. These features may account for the odorant receptors to recognize structurally diverse odorants.

Coding of Odor Information

It has long been known that humans can discriminate more than 10 000 odors. Recent study estimated that humans can even discriminate at least one trillion olfactory stimuli [13]. There arises a question whether we can discriminate so many odorants with 350 odorant receptors. Although each ORN expresses single odorant gene, odor molecules are recognized by more than one ORN. The odorant receptors also recognize multiple odor molecules (Figure 2.8). For example, mOR-EG, a mouse odorant receptor that was isolated from a eugenol-responsive ORN, recognizes 22 odorants, whereas some other receptors recognize only a small number of odorants [56]. Physiological experiments showed that a single odorant elicited electrical response in multiple ORNs and the response amplitudes varied. The recognition of an odorant depends on which receptors are activated and to what extent. It has been revealed that each odorant receptor recognizes a specific structural feature in individual odor molecules. Therefore, each odorant or odorant mixture is encoded by multiple odorant receptors (Figure 2.8). The combinatorial nature of the olfactory code underlies the reason how humans can discriminate a huge number of odors.

2.3.2 Taste

2.3.2.1 Anatomy of Taste

In the gustatory system, the sense of taste is essential for the animals to evaluate which food is good to eat and which food should be avoided. Humans and most of other mammals can discriminate five basic tastes—sweet, bitter, sour, salty, and umami—unlike the olfactory system that can discriminate thousands of odorants. Sweet, salty, and umami are appetitive tastes, and bitter and sour are aversive tastes. Foods with sweet, salty, and umami taste are usually required for energy, ionic balance, and synthesizing proteins, while bitter foods are very likely poisonous and sour foods are unripen or spoiled.

Tastants are detected by taste receptor cells in taste buds in which 50–150 taste receptor cells clustered. Taste buds are located on the tongue, and there are three types of structures called papillae (Figure 2.9a). Fungiform papillae that are located in anterior two-third of the tongue contain one or a few taste buds. Foliate papillae that are located in the posterior edge of the tongue and circumvallate papillae that are situated on the very back of the tongue both contain hundreds of taste buds. The taste bud is embedded in the epithelium. The taste pore is a small opening at the surface of the tongue where taste receptor cells are exposed to taste stimuli (Figure 2.9b). Each taste receptor cell is spindle shaped and extends its microvilli to the taste pore, allowing tastants to bind to the taste receptor proteins in microvilli (Figure 2.9c).

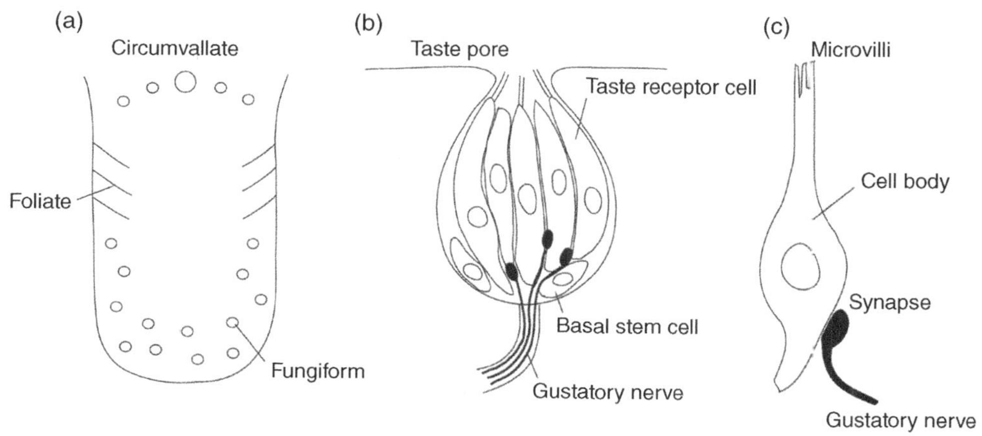

Figure 2.9 Schematic drawing of tongue, taste buds, and taste receptors. (a) There are three types of papillae: circumvallate papillae, foliate papillae, and fungiform papillae. (b) Each taste bud contains 50–150 taste receptor cells that extend their microvilli to the taste pore. (c) Structure of single taste receptor cell. Microvilli are located at the apical end of the cell

On binding tastants to the taste receptor protein, the cell depolarizes through signal transduction pathway. The taste signal is then transmitted to the gustatory nerve via synapse and transmitted to the brain.

2.3.2.2 Transduction of Taste Signals

Tastants are highly diverse in terms of their chemical structure. Salty and sour stimuli are simple ions such as Na^+ and H^+, while sweet, bitter, and umami substances are more complex, such as saccharides, alkaloids, and proteins. Therefore, the structures of receptor proteins and signal transduction mechanisms are varied among the five basic tastes.

Signal Transduction Cascade in Bitter, Sweet, and Umami Taste Receptor Cells

There have been many reports about taste transduction cascade in various vertebrate species, and the proposed hypotheses have wide diversity of signaling pathways (e.g., Kinnamon [62]). However, recent results have demonstrated that sweet, bitter, and umami taste receptors have a common signal transduction cascade (Figure 2.10a) [162].

The receptor proteins of sweet, bitter, and umami are GPCRs. The gustatory signaling starts with the binding of ligands to the receptor proteins followed by a conformational change in the receptor proteins. Activated receptors then activate a taste receptor cell-specific G protein

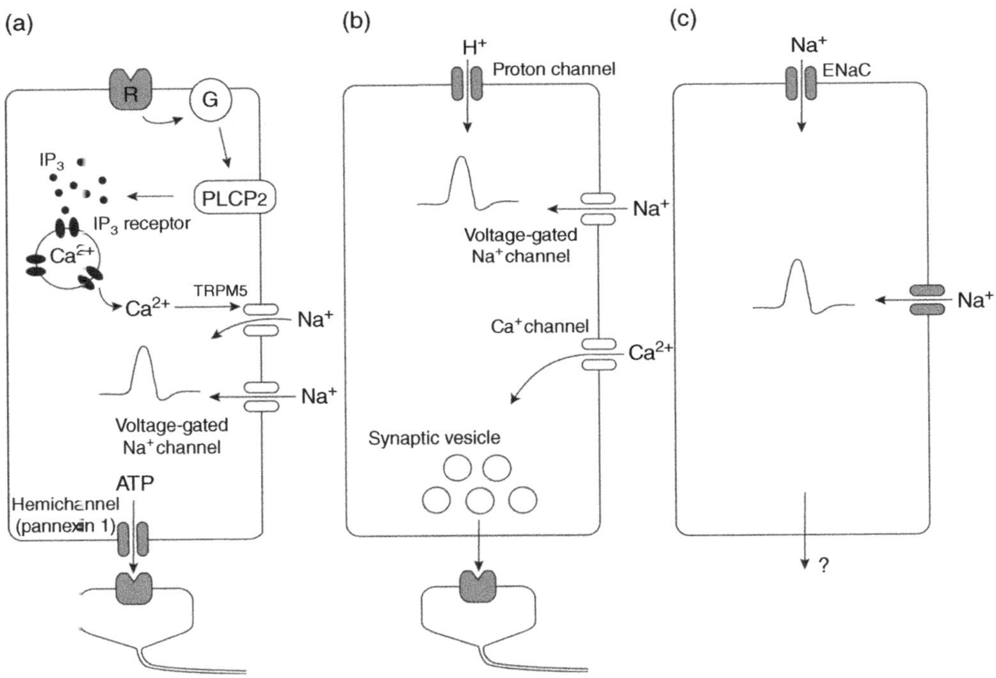

Figure 2.10 Schematic drawing of signal transduction cascades of sweet, bitter, umami (a), sour (b), and salty (c) tastes. ENaC, epithelial sodium channel; G, G protein; IP_3, inositol triphosphate; PLC, phospholipase C; R, receptor protein

gustducin that activates phospholipase C β_2 (PLC-β_2). Gustducin has high sequence homology to photoreceptor specific G protein transducin that is also expressed in taste buds [93]. IP_3 and DAG are generated from the membrane lipid phosphatidylinositol 4,5-bisphosphate (PIP_2) by an enzymatic activity of PLC-β_2. IP_3, a water-soluble second messenger, diffused in the cytosol and binds to IP_3 receptor and causes an opening of the IP_3 receptor/channel leading to the release of Ca^{2+} from intracellular Ca^{2+} stores. The released Ca^{2+} gates TRPM5, a TRP channel family, in the plasma membrane [106, 163]. As a result, Na^+ enters the cell through TMPM5 channel to generate depolarization, which leads to an opening of hemichannel at the basal end of the cell. It has been reported that the taste receptor cells can elicit action potentials that are induced by TMPM5 channel-mediated depolarization. The recent study suggested that the action potentials may be required to open the hemichannels [99].

Signal Transduction Cascade in Sour and Salty Taste Receptor Cells

Tastants of sour and salty are simple ions, H^+ and Na^+, respectively. Therefore, the signal transduction cascades of sour and salty are relatively simple compared to those of sweet, bitter, and umami.

A number of candidates of sour receptors have been proposed over the years. For example, acid-sensing ion channels (ASICs) found in rats are cation channels that are activated by extracellular protons. However, they are unlikely to be a common sour receptor because they are not expressed in mouse taste buds [112]. Other candidates include hyperpolarization-activated cyclic nucleotide-gated channels (HCNs), K^+ channels, and TRP channel PKD2L1 (and/or PKD1L3), but there had been no direct evidences that those candidates are sour receptors [115].

More recently, Chang *et al.* [20] reported that responses of the PKD2L1-expressing taste cell to sour stimulus are mediated by a proton conductance and not mediated by Na^+-permeable channels as previously thought. In this model, protons enter into the sour cell through the proton channel at the apical end of the cell, which causes depolarization leading to a generation of action potentials. As a result, voltage-gated Ca^{2+} channels open, resulting in a rise in intracellular Ca^{2+} concentration at the basal end of the cell followed by a neuro-transmitter release (Figure 2.10b).

Another possible signal transduction pathway is mediated by intracellular acidification caused by membrane-permeable acids. It has been reported that several two-pore domain potassium (K_2P) leak channels are sensitive to acidification [113]. Blocking of K_2P channel by intracellular acidification would generate membrane depolarization, which enhances the depolarization caused by proton channels. This may explain why weak membrane-permeable acids taste sourer than strong acids, such as HCl.

The signal transduction mechanism of salty taste receptor has not been determined yet. A candidate of Na^+-permeable salty receptor is amiloride-sensitive epithelial Na^+ channel (ENaC). Upon application of Na salt, Na^+ passively enters the cell through the ion channel, generating membrane depolarization of the cell (Figure 2.10c).

Synaptic Transmission from Taste Receptor Cells to Second-Order Neurons

The sweet, bitter, and umami taste receptors have a unique mechanism for synaptic transmission from the receptors to second-order neurons. Unlike the conventional chemical synapses, those receptor cells do not express voltage-gated Ca^{2+} channels and synaptic vesicles. The recent studies showed that ATP is released by taste stimulation from type II taste receptor cell that is thought to be sweet, bitter, or umami receptor, suggesting that ATP is a neurotransmitter of these cells. It is suggested that an unconventional depolarization-activated ATP release channel, most likely pannexin 1 hemichannel, was involved in the ATP release [99].

Ionotropic purinergic receptors P2X2 and P2X3 were first discovered in the afferent nerves that innervate taste buds by Bo *et al.* [8]. Double knockout of P2X2 and P2X3 eliminated taste responses in the taste nerves, and stimulation of taste buds *in vitro* evoked ATP, confirming that ATP is the neurotransmitter between the taste receptors and the taste nerves for sweet, bitter, and umami tastes [31].

Synaptic transmission from sour or salty taste cells to taste nerves is not clear yet. Responses to sour and salty stimuli were also abolished in P2X2/P2X3 double knockout mice, although ATP release has not been detected in type III cells that are thought to be sour or salty receptor cells. The mechanisms of synaptic transmission for these cells remain unclear.

2.3.2.3 Molecular Biology of Taste

Bitter, Sweet, and Umami Receptors (G Protein-Coupled Taste Receptors)

The taste receptors for bitter, sweet, and umami have been cloned recently and best understood among the five tastes. These receptors are taste-specific GPCRs which are expressed in each subset of taste receptor cells. There are two classes of GPCRs, T1Rs and T2Rs. Compared to other GPCRs like neurotransmitter receptors, the binding affinity of taste GPCR is generally low (in mM order), which is consistent with the physiological concentration of nutrients in foods.

In 2000, a novel family of GPCRs, T2R family, was first identified from genomic databases, and T2Rs are responsible for detecting bitter taste. T1Rs are A-type GPCRs that are similar to the opsins and the odorant receptors (Figure 2.11). Chandrashekar *et al.* [18] showed that specific T2Rs function as bitter taste receptors using a heterologous expression system. The T2R family comprises about 30 in mammals, and each taste receptor cell expresses a large repertoire of T2Rs, which can explain why taste receptors can detect many structurally diverse chemicals that are bitter to humans [1, 18]. It may be reasonable that the bitter taste receptor cells express most of T2Rs, which means that bitter taste receptor cells cannot distinguish bitter chemicals, because bitter sense evolved to avoid toxic substances. However, physiological study of bitter taste receptor cells showed that most of them were activated by only one out of five bitter chemicals tested, suggesting that bitter-sensitive taste receptor cells could discriminate bitter compounds [14].

Another class of GPCR was T1Rs that are responsible for detecting sweet and umami tastes. T1Rs are C-type GPCRs with large N-terminal domains (Figure 2.11). Three different subunits—T1R1, T1R2, and T1R3—have been identified [101]. T1Rs can only function as a heterodimer. The receptor with a combination of T1R2 and T1R3 can detect sweet substances [101], whereas a combination of T1R1 and T1R3 can detect most of the standard amino acids that cause umami taste (Figure 2.11) [102].

Sour and Salty Receptors

Although candidates for sour receptors including ASICs, HCN1 and HCN4 channels, K^+ channels, and the TRP channels PKD2L1 and PKD1L3 have been proposed, there have been almost no evidences for these candidates. More recently, Huang *et al.* [44] showed that

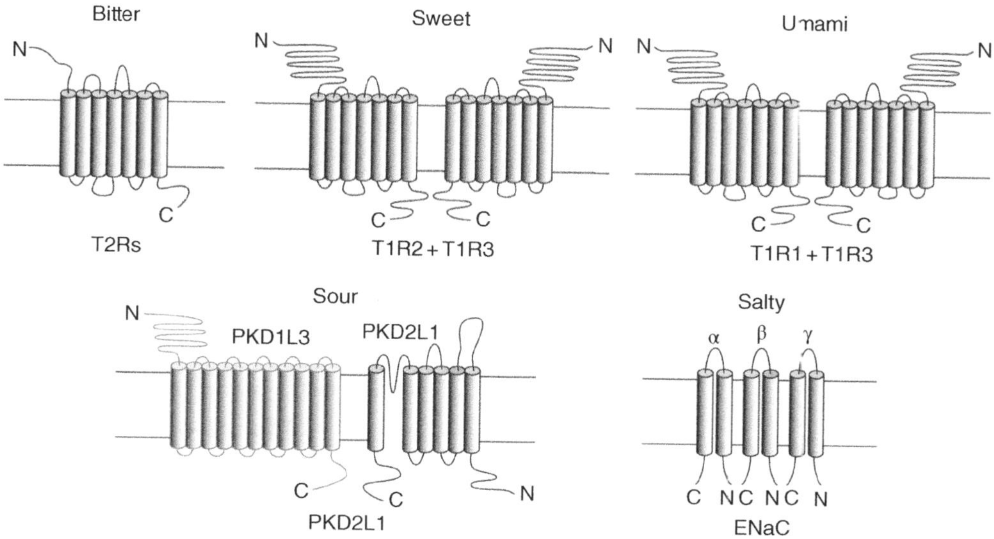

Figure 2.11 Characteristic structures of the taste receptors for five basic tastes. Bitter, sweet, and umami receptors are G protein-coupled receptors. Bitter receptors are T2Rs, while sweet and umami receptors are another class of GPCR, T1Rs. The heteromeric complex T1R2 + T1R3 recognizes sweet substances, while the complex T1R1 + T1R3 recognizes umami substances. Sour receptor PKD2L1 is coexpressed with PKD2L1

the mice lacking the taste receptor cell that expresses PKD2L1 did not respond to sour stimuli, indicating that PKD2L1 is serving as a sour receptor. It is also demonstrated that PKD1L3 is coexpressed with PKD2L1 in circumvallate and foliate taste receptor cells (Figure 2.11) [47].

Two types of salty taste, low and high concentrations of NaCl, have been proposed from behavioral responses. Low salt triggers attractive response, whereas high salt triggers aversive response. Since the attractive response is inhibited by amiloride that is a potent inhibitor of the epithelial sodium channel (ENaC), ENaC has been considered as a candidate for low-salt receptor. This channel is constituted of α-, β-, and γ-subunits (Figure 2.11) [16]. Recently, Chandrashekar *et al.* [19] showed that the genetically engineered mice with taste receptor cells lacking ENaCα completely eliminated salt attraction and sodium taste responses, indicating ENaCs function as low-salt receptors. The mechanism of high-salt reception is not known.

2.4 Cell-Based Sensors and Receptor-Based Sensors

Natural living organisms have equipped the sophisticated olfactory systems with their evolutions. The systems possess the capability to detect environmental odorants with higher performance than we expected. Recently, the mechanisms of the olfactory system have been gradually elucidated from long years of efforts by many researchers as mentioned in earlier sections. Based on these findings, we have become able to utilize various kinds of biological components, such as tissues, sensory neurons, proteins, and genes regarding olfaction in living organisms, for application to biosensors (Figures 2.12 and 2.13). Application of these biological

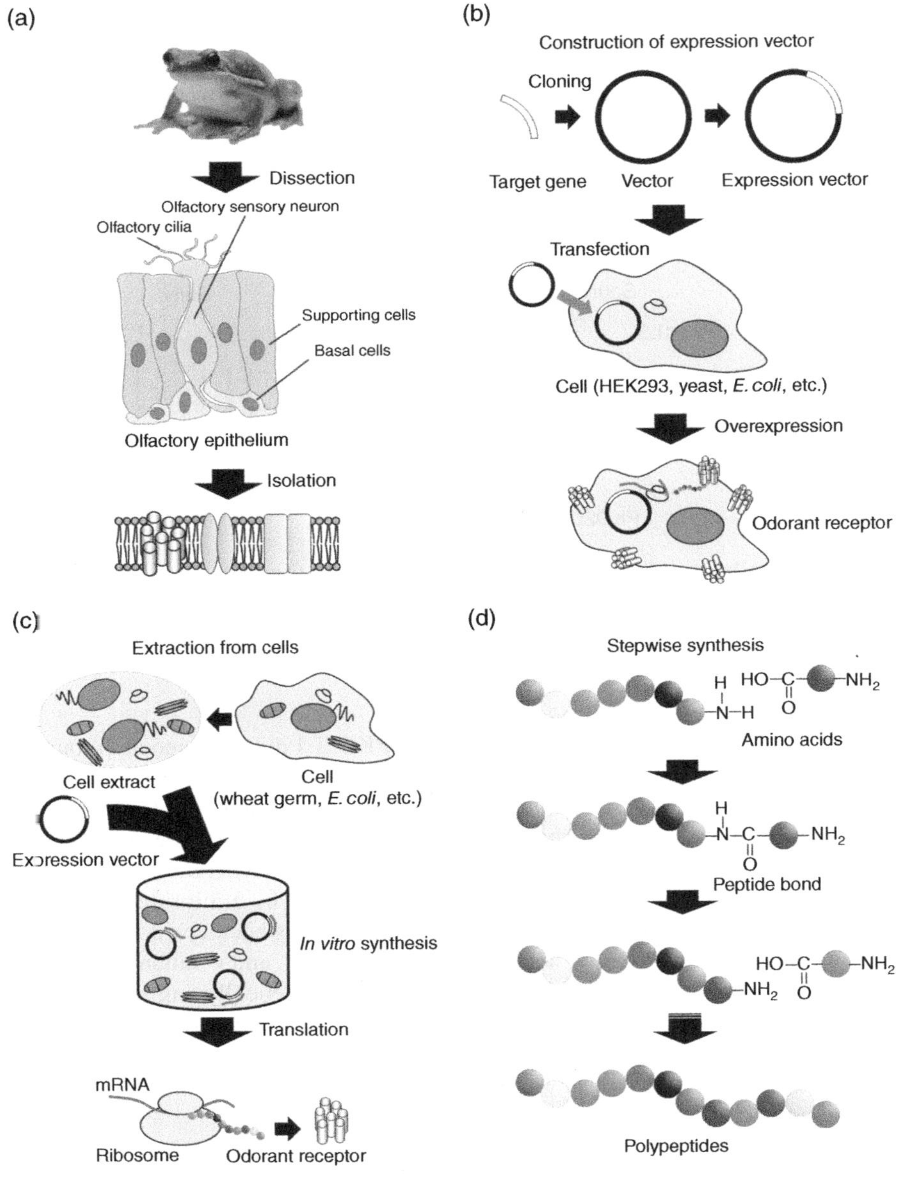

Figure 2.12 Schematic drawings of odorant receptor-production methods for cell-based or receptor-based sensors. (a) Isolation from living organs. (b) Cell-based production. (c) Cell-free production. (d) Chemical production

functions would lead to the development of odorant sensors with higher performances superior to the existing odor sensors in terms of sensitivity and selectivity. In this section, three types of biosensors, that is, tissue-based sensors, cell-based sensors, and receptor-based sensors, are introduced.

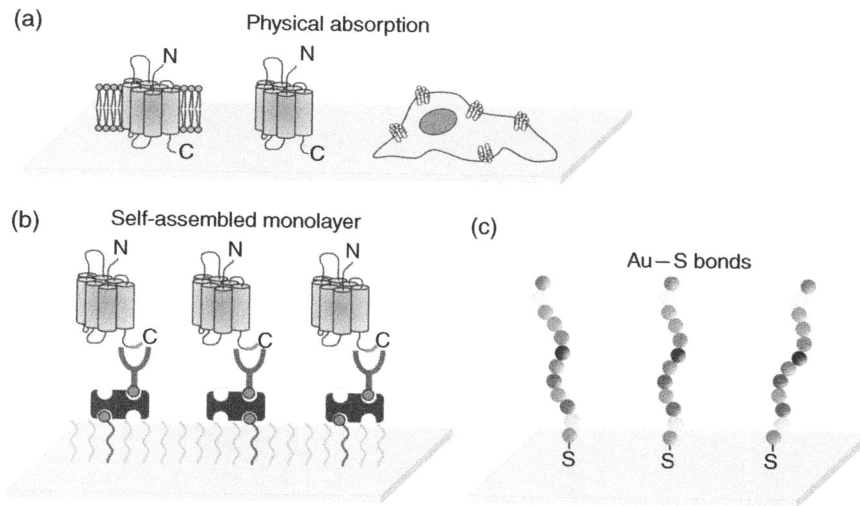

Figure 2.13 Schematic drawing of immobilization methods of odorant receptors, cells expressing odorant receptors, and polypeptides on the surface (gold) of transducers. (a) Physical absorption. (b) Self-assembled monolayer. (c) Au—S bonds

2.4.1 Tissue-Based Sensors

Surgically resected olfactory epithelium and mucosa tissue are directly applicable to electronic integrated devices for making biomimetic electric nose. Liu *et al.* showed that extracellular potential of rat's isolated olfactory epithelium and mucosa tissue could be detected *in vitro* by a microelectrode array (MEA) device and a light-addressable potentiometric sensor (LAPS) [83, 84]. As schematically shown in representative examples (Figures 2.14 and 2.15), in these systems, tissues contact with the conductive sensor device without injury, and thus the electrical signals from living olfactory cells in tissues could be measured. In contrast to needlelike electrodes, the noninvasive nature of MEA and LAPS for olfactory epithelium and mucosa tissue is due to their planar sensing station. They demonstrated that the biohybrid sensing systems could reflect characteristic firing patterns of olfactory epithelium to some volatile organic compounds (VOCs).

Organs of living body are also useful for sensitive detection of several volatile compounds (VCs). For instance, Park *et al.* utilized electroantennograms (EAGs) of some insects, vinegar fly, moth, and wasp (*D. melanogaster*, *Heliothis virescens*, *Helicoverpa zea*, *Ostrinia nubilalis*, and *Microplitis croceipes*) which possessed their own EAG response profiles to 20 different VCs [105]. They successfully recorded EAG responses to several VCs using antenna array consisting of four different insect's antennae mounted on the Quadro-probe EAG recording system (Figure 2.16). Rains *et al.* developed a portable device, Wasp Hound®, employing a trained living wasp (*M. croceipes*) as the sensor element [110]. The device is composed of a ventilated chamber as an insect cage equipped with a camera. It could quantitatively clarify the searching behavior of trained wasps and successfully detected the behavioral responses to the target odor.

Though higher organism's olfaction mechanism has still missing piece of the puzzle, vertebrates are very usable for easy odorant detection such as a typical case of sniffer dog. It's well known that canines have been used for detecting illicit drugs, several explosives, and human

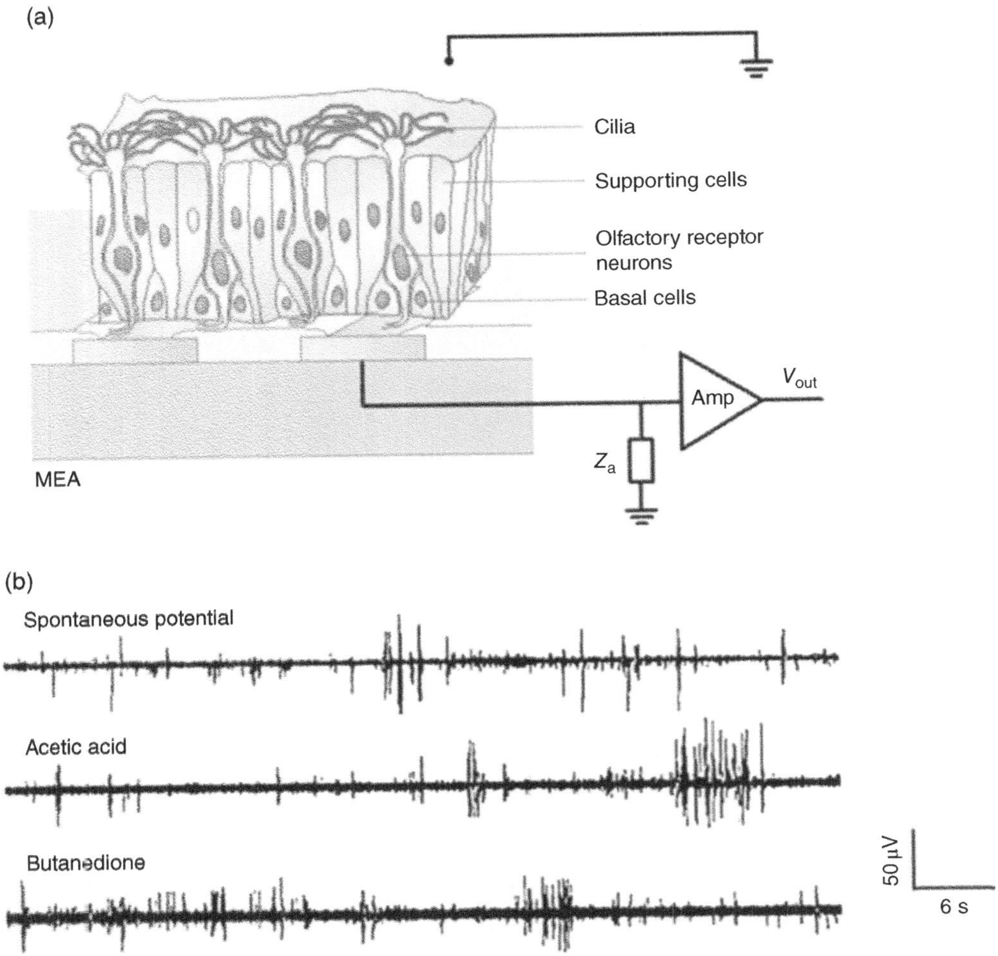

Figure 2.14 Study example of extracellular potentials recording by MEA. (a) Schematic view of olfactory receptor neurons in the epithelium on MEA. (b) Electrophysiological signals induced by the chemical substances. From Ref. [84]. Reproduced with permission from Elsevier

identification. Gazit and Terkel reported that canines could detect explosives with a probability of about 90% in actual outdoor space [35]. Similarly, rats have been applied to VC detection especially for exploring landmine and explosives from the 1970s onward [103, 107]. However, the fact remains that there are disadvantages in that the sensitivities are affected by their physical conditions and their trainings cost money and time.

As summarized in Table 2.1, rat was frequently used for tissue-based odorant sensors. The reasons for this trend may be came from the fact that rat is very easy experimental sample and an abundance of the anatomical insight. At the same time, utilization of some insects is also seen in late years.

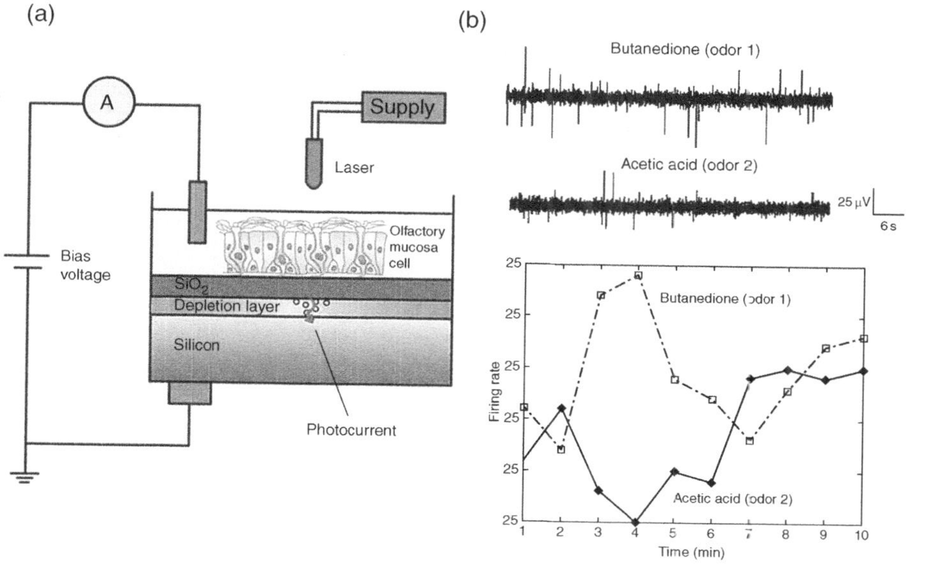

Figure 2.15 LAPS system applied to olfactory mucosa tissue cells. (a) Outline drawing of the LAPS system. (b) Firing patterns and the time courses evoked by chemical stimuli. From Ref. [85]. Reproduced with permission from Elsevier

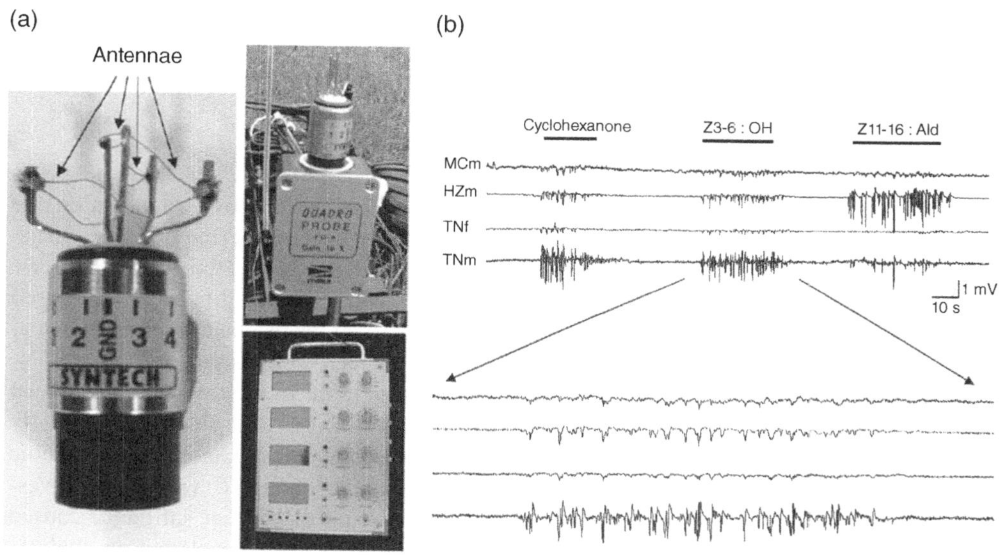

Figure 2.16 (a) Pictures of portable EAG recording system. (b) Simultaneously recorded four antennae EAG responses and the partial expansion patterns triggered by three different stimuli. From Ref. [105]. Reproduced with permission from Oxford University Press

Table 2.1 Summaries of tissue-based sensors

Tissue/organ/bion	Tested compound	Transducer	Tested concentration	Literature
Olfactory epithelium (Sprague-Dawley rat)	Butanedione, acetic acid	MEA	25 μmol/ml	[84]
	Ethyl ether, acetic acid, butanedione, acetone	MEA	10 μM	[86]
	Butanedione, acetic acid	LAPS	25 μmol/l	[83]
	Butanedione, acetic acid	LAPS	25 μmol/l	[85]
Olfactory bulb (Wistar rat)	Glutamic acid	MEA	10 μM to 5 mM	[24]
Antennae (insects: *D. melanogaster*, *Heliothis virescens*, *Helicoverpa zea*, *O. nubilalis*, *Microplitis croceipes*)	Z11-16:Ald, Z3-6:OH, hexanoic acid, benzyl acetate, 2-methyl-5-nitroaniline, cyclohexanone, α-pinene, *cis*-nerolidol, *trans*-nerolidol, β-caryophyllene, β-ocimene, (*R*)-(+)-limonene, methyl jasmonate, 2-diisopropylaminoethanol, indole, 2,2-thiodiethanol, 1-heptanol, 1-octanol, 1-nonanol, 1-decanol	EAG	10 or 100 μg/μl	[105]
Insect (*M. croceipes*, *H. virescens*, *H. zea*)	3-Octanone, myrcene	Behavior observation by camera	25–26 μmol/l	[110]
Canine (Belgian Malinois, Labrador Retriever)	C-4 explosive	—	30 g	[35]
Rat (*C. gambianus*; giant African pouched rat)	TNT and landmines	—	—	[107]

2.4.2 Cell-Based Sensors

As compared to tissue, isolated cell seemed to be more suitable for biosensing application due to easy isolation and transformation processes by developed bioengineering of recent years. Most studies of cell-based sensor were actively performed after entering the 2000s. Here, we mention some typical cultivated neurons and cells which have been utilized as cell-based odorant sensors. In either example, researchers used electrical, or resonant, or optical detection system as the transducer of chemical signal to acquire each output signal.

Insect's ORN and olfactory sensory neuron (OSN) which are in their antenna have been used for chemical biosensing applications. Huotari reported that the antenna of blowfly (*Calliphora vicina*) was specifically sensitive to 1,4-diaminobutane, hexanol-1, and butanoic acid [45]. In that study, *in vivo* action potentials from ORN in antenna of *C. vicina* were investigated by EAG, and the results showed that a higher amount of odorant substance caused saturation of the corresponding ORN response. Huotari and Lantto developed an analysis system for extracellular action potentials [46]. They applied the system for measurement of action potentials of *C. vicina*, mosquitoes (*Aedes communis*), pine weevils (*Hylobius abietis*), and trogossitid beetle (*Trogossita japonica*). They eventually revealed that the relationship between action potential responses of ORNs and odor concentrations obeyed the power law. Liu *et al.* and Wu *et al.* used LAPS for monitoring of OSNs' extracellular potentials [82, 151]. The OSNs were cultivated on the surface of sensor chip, and the cells could be maintained for 1 week in LAPS device. Corcelli *et al.* studied OSNs' responses to two typical explosives, cyclotrimethylenetrinitramine (RDX) and trinitrotoluene (TNT) [27]. Electro-olfactogram recording and calcium imaging of rat olfactory mucosa were utilized for sensing of RDX and TNT. In addition, cilia from pig olfactory epithelia were also used for monitoring of cyclic adenosine monophosphate levels following exposure to odorants and explosives. This practical study implied that explosive substances as well as other odorants present in landmines interacted with olfactory receptors. Xavier *et al.* used a microfluidic device for OSN array [154]. By using calcium imaging, they could detect and analyze odorant responses of about 2900 OSNs for four different fruity/floral smells (vanilla, rose, berry, and banana) in microwells simultaneously. Their approach was based on a large-scale fluorescent investigation of many OSNs which were trapped in the micro chamber. For detection of cultured ORN responses *in vitro*, MEA equipped with gas intake system was also used by Ling *et al.* [80]. Limonene and isoamyl acetate odorants were tested in that system. They analyzed the firing spikes of ORNs and extracted and sorted the different spikes from multiple neuron recordings. Tanada *et al.* showed odorant sensor by means of expressing odorant receptors of insects into dissociated neural cultures of rats [135]. The hybrid system had advantages of easy functional expression of odorant receptors, prolonged lifetime, and amplification of weak ionic currents of odorant receptors.

As presented earlier, native ORN and OSN are directly usable as a biosensing device. Equally, host cells which could express different species' olfactory receptors have been employed for artificial odorant sensing. As a typical cell for such an expression system, a certain cell line of human embryonic kidney cells, namely, HEK293, is frequently used for the expression of several olfactory receptors. Using HEK293 cells, Ko and Park investigated the expression of rat olfactory receptor I7 (ORI7) [64] and showed that the HEK293 cells expressing ORI7 were usable for octanol detection through QCM [63]. Furthermore, Ko and Park reported that intracellular Ca^{2+} sensing molecule, yellow cameleon-2, could conjugate with olfactory receptors' response in HEK293 [65]. In 2009, Lee *et al.* presented that using planar microelectrodes, electrical signals could be obtained from HEK293 cells expressing ORI7 and the olfactory signals could be enhanced by electrical stimulation [74, 75]. In addition, they demonstrated real-time monitoring of cells' responses to odorant (heptanal, octanal, nonanal, decanal, and helional) stimuli using surface plasmon resonance (SPR) as shown in Figure 2.17 [76]. Oh *et al.* also used HEK293 cells expressing four different kinds of human olfactory receptors (hORs)—hOR3A1, hOR1A1, hOR1D2, and hOR1G1—for odorant screening [104]. They cultured the cells in polyethylene glycol diacrylate microwells and visualized the cells' response to odorants through fluorescent observation. Recently, Sato and Takeuchi demonstrated direct chemical vapor detection using HEK293 spheroids in hydrogel micro chambers [120]. As the transduction of cells' response to odorants, they used electrophysiological measurement. It is particularly worth noting that gas-phase odorants could be detected directly by HEK293 spheroids.

Some researchers focused on using host cells such as yeast, *Xenopus laevis* oocyte, and Sf21 cell derived from noctuid moth (*Spodoptera frugiperda*) for odorant sensing. Minic *et al.* successfully expressed ORI7 in budding yeast (*Saccharomyces cerevisiae*) and applied it to odorant screening [94]. They used luciferase reporter for detection of odorant binding events

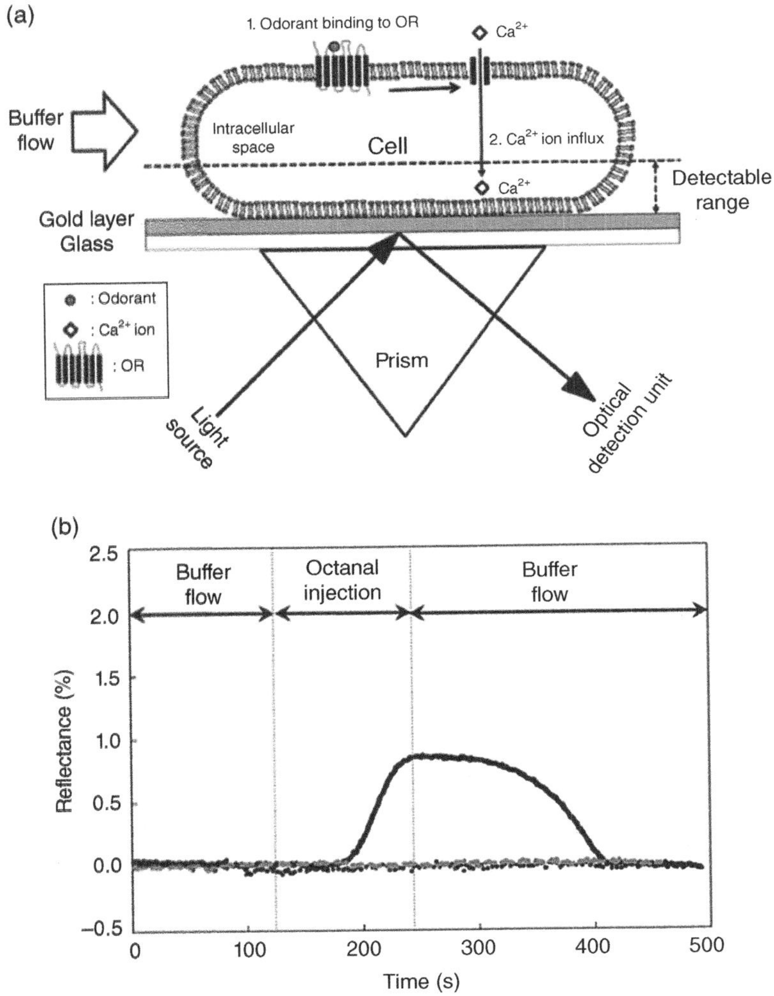

Figure 2.17 (a) Principle of SPR-based odorant detection using cells expressing odorant receptors. (b) SPR response to the odorant (octanol) stimulus. From Ref. [76]. Reproduced with permission from Elsevier

(Figure 2.18). Marrakchi *et al.* also used same species yeast expressing human olfactory receptor OR17-40 [90]. The yeasts were immobilized on integrated planar microelectrodes (Figure 2.19a) and the system measured the yeast conductance which reflected olfactory receptor activation by helional (Figure 2.19b). For explosive detection, genetically modified yeast was applied by Radhika *et al.* [109]. Coexpressed green fluorescent protein with rat olfactory receptor Olfr226 they newly identified was used as the probe for 2,4-dinitrotoluene detection. Misawa *et al.* showed that *Xenopus* oocytes expressing insect olfactory receptors could be sensor elements for odorant sensing [95]. They used small fluidic device integrated with electrodes for two-electrode voltage clamping. In that study, they demonstrated that the

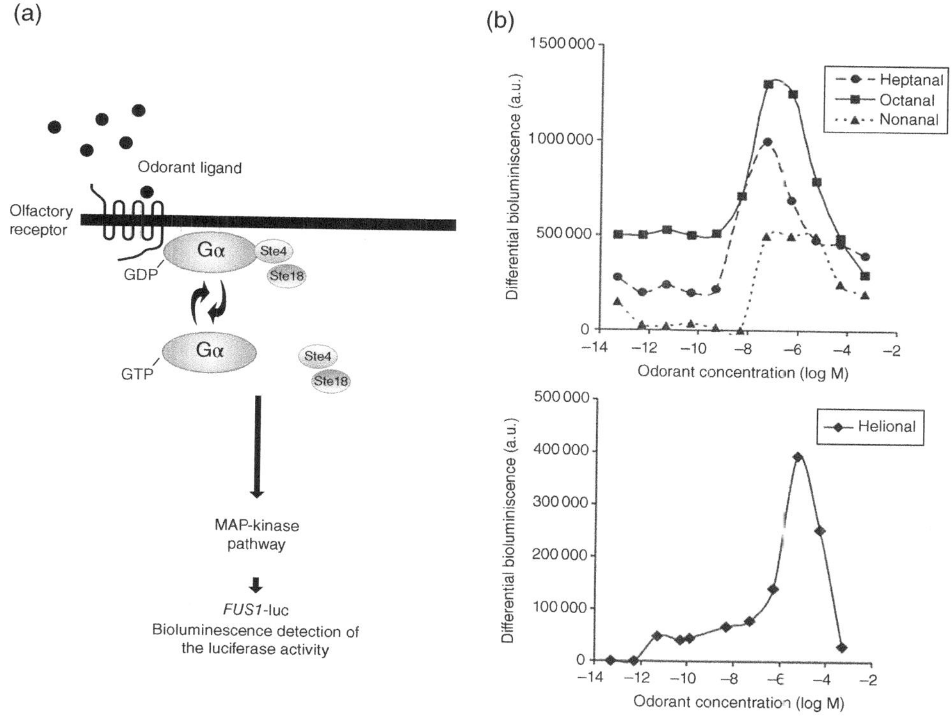

Figure 2.18 Odorant detection through a bioluminescence in genetically modified yeast. (a) Illustration of signal transduction pathway to yield odorant-induced luciferase activity. (b) Differential bioluminescence dose–responses on odorant stimulation. From Ref. [94]. Reproduced with permission from John Wiley & Sons, Ltd

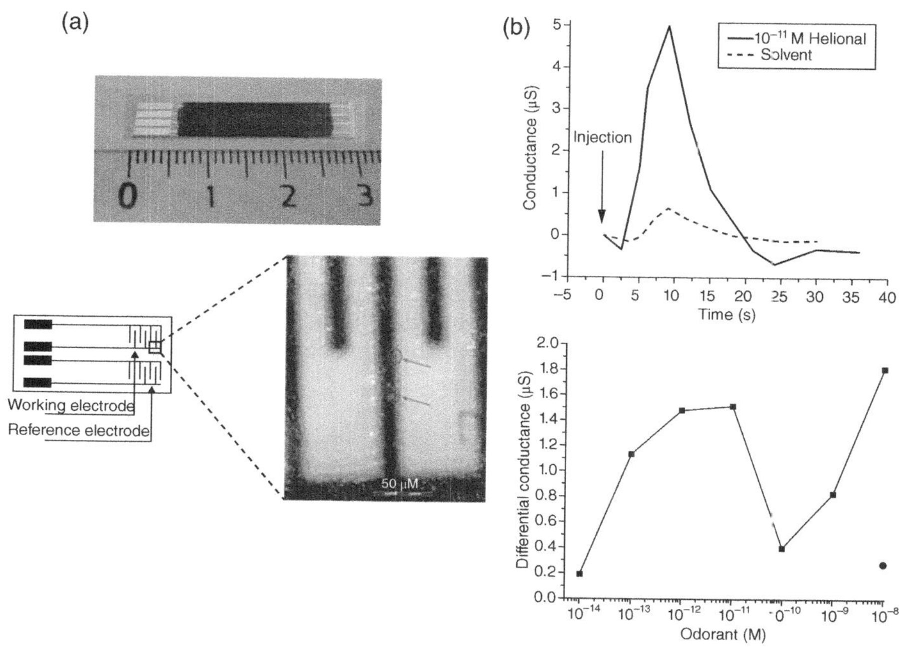

Figure 2.19 (a) Pictures and schematic presentation of interdigitated microelectrodes with deposited yeast cells. (b) The sensor signals and dose–response to helional. From Ref. [90]. Reproduced with permission from Springer Science +Business Media

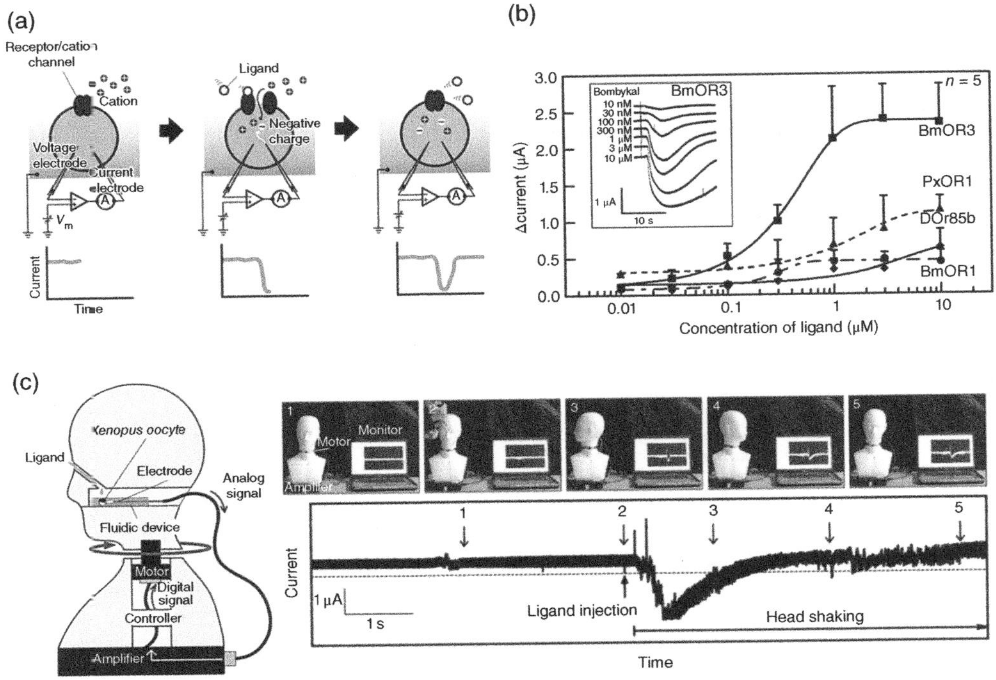

Figure 2.20 (a) Mechanism of TEVC-based odorant sensor using *Xenopus* oocytes expressing odorant receptors. (b) Dose-dependent increases in amplitude of each ligand-induced current. (c) Head-shaking robotic system with the cell-based odorant sensor device. From Ref. [95]. Reproduced with permission from National Academy of Sciences, USA

sensor device could be integrated with simple robotic system (Figure 2.20). Mitsuno *et al.* verified odorant detection using Sf21 cell lines coexpressing insect odorant receptors and Ca^{2+}-sensitive fluorescent protein named GCaMP3 (Figure 2.21) [96]. The Sf21 cell lines could express odorant receptors stably and detect odorants with consistent responsiveness for at least 2 months.

Concerning the variety of tested (or targeted) odorant compounds, it would appear that cell-based sensors are presently superior to tissue-based sensors as you can see in Table 2.2 As mentioned in the opening sentence, cell-based odorant sensors owe the wide range of object substance to accumulation of knowledge about olfactory mechanisms with advancing recent genetic technologies of expression systems.

2.4.3 *Receptor-Based Sensors*

As one type of olfactory biosensors, odorant receptor proteins themselves have been utilized as sensing elements for detecting target odorants. In general, this type of biosensors consists of functional odorant receptor proteins and transducers, which enable us to acquire

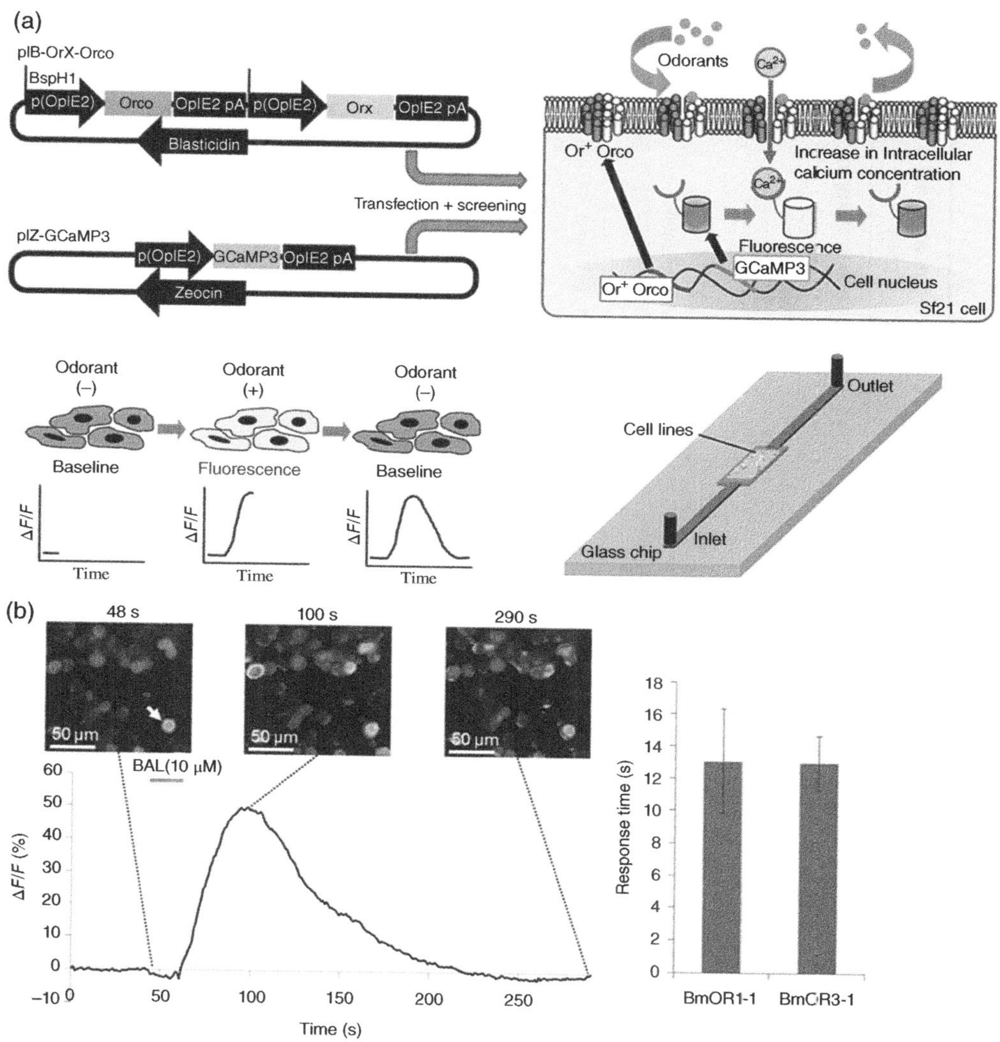

Figure 2.21 (a) Outline of odorant sensor using Sf21 cells expressing odorant receptors and Ca^{2+} sensitive fluorescent protein GCaMP3. (b) The fluorescent response and response time of cell lines. From Ref. [96]. Reproduced with permission from Elsevier

signals associated with interactions between odorant receptor proteins and odorants. Production of functional odorant receptor proteins (Figure 2.12) and their immobilization on the surface of transducers (Figure 2.13) are crucial for the development of receptor-based sensors. In this part, production and immobilization methods of various odorant receptors are summarized, and their application examples for biosensors are introduced.

Table 2.2 Summaries of cell-based sensors

Cell	Odorant receptor	Tested compound or ligand	Transducer	Tested concentration/ detection range	Literature
ORN (*Calliphora vicina*)	—	1,4-Diaminobutane, hexanol-1, butanoic acid	EAG	A few ppb to 500 ppm	[45]
ORN (*C. vicina, Aedes communis, H. abietis, Trogossita japonica*)	—	Hexanol, butyric acid, diaminobutane	EAG	0.1–100 μg	[46]
ORN	—	Acetic acid	LAPS	1–50 μM	[82]
OSN (Sprague-Dawley rat)	—	Mixture of acetic acid, octanol, cineole, hexanol, and 2-heptanone	LAPS	0.1 mM for each	[151]
OSN (pig and Sprague-Dawley rat)	—	TNT, RDX, amyl acetate, benzene, naphthalene, hexachloroethane, styrene, diphenylamine, benzothiazole, toluene, chlorobenzene	EOG and Ca^{2+}-related fluorescent detection (Fura-2/AM)	1 mM or 30 and 300 μM	[27]
OSN (Swiss-Webster mouse)	—	Vanillin, geraniol, benzyl acetate, ethyl butyrate	Ca^{2+}-related fluorescent detection (Fluo4-AM)	200 μM	[154]
ORN (Sprague-Dawley rat)	—	Limonene, isoamyl acetate	MEA	0.4–19 μM	[80]
Neuronal cell (Wistar rat)	BmOR1	Bombykol	Ca^{2+}-related fluorescent detection (EGFP)	1, 10, and 100 μM	[135]
HEK293	I7	Octanal	Ca^{2+}-related fluorescent detection (Fura PE3-AM)	0.1 μM to 10 mM	[64]
	I7	Hexanal, heptanal, octanal, nonanal, decanal	QCM	10^{-8} to 10 mM	[63]
	ODR-10, I7	Diacetyl, octanal	Ca^{2+}-related fluorescent detection (yellow cameleon-2)	0.1 and 1 mM	[65]
	I7	Octanal, helional	MEA	10 mM	[74]

	I7	Octanal	Planar electrode and Ca^{2+}-related fluorescent detection (Fura PE3-AM)	1, 5, and 10 mM	[75]
	I7	Heptanal, octanal, nonanal, decanal, helional	SPR	0.1, 1, 10, and 100 mM	[76]
	hOR3A1, hOR1A1, hOR1D2, hOR1G1	Helional, β-citronellol, bourgeonal, geraniol	cAMP-related fluorescent detection (ZsGreen)	50 nM to 500 μM	[104]
	GPROR2, Or47a	2-Methylphenol, benzaldehyde	Extracellular field potential recording	0.1 μM to 100 mM	[120]
Saccharomyces cerevisiae	I7, OR17-40	Heptanal, octanal, nonanal, helional	G_α subunit-related bioluminescence detection (FUS1-luc)	5×10^{-14} to 5×10^{-4} M	[94]
	OR17-40	Helional	Interdigitated microelectrodes	10^{-14} to 10^{-8} M	[90]
	I7, β_2AR, modified I7 for vanillin and citronellal (named VanR and CiteR), Olfr226	Octanol, octanal, heptanal, hexanal, isoproterenol, vanillin, citronellal, DNT	Ca^{2+}-related fluorescent detection (GFP)	1–50 μM	[109]
Xenopus oocyte	BmOR1, BmOR3, PxOR1, DOr85b	Bombykol, bombykal, A11-16:Ald, 2-heptanone	TEVC	0.01–10 μM	[95]
Sf21	BmOR1, BmOR3	Bombykol, bombykal	Ca^{2+}-related fluorescent detection (GCaMP3)	100–10 000 nM	[96]

2.4.3.1 Production of Odorant Receptors

For application of receptor-based biosensors, appropriate and functional odorant receptors need to be produced to be used as sensing elements for the detection of target odorants. Since the activity of odorant receptors affects various performances of biosensors, such as sensitivity, selectivity, and stability, the production of odorant receptors is an important step in the development of receptor-based biosensors. Several methods for production of odorant receptors have been reported: the extraction from olfactory organs (i.e., olfactory epithelium), overexpression in heterologous cell lines (i.e., bacteria, yeast, and mammalian cell line), cell-free production, and chemical synthesis (Figure 2.12).

Isolation of odorant receptors from the natural olfactory tissues of target living organisms is an effective method for the natural state of odorant receptors (Figure 2.12a). Since target odorant receptors are expressed in the ORNs, the odorant receptors are able to be isolated from olfactory tissues including ORNs. Wu showed that odorant receptors were able to isolate from the dissected olfactory epithelium of bullfrogs to coat onto a sensor array [150]. Since the advantage in this method is to maintain the natural structure of odorant receptors from native ORNs, isolated odorant receptors would exhibit molecular recognition to natural ligands. However, this method is inefficient for collection of large amount of odorant receptors, and it is difficult to isolate desired odorant receptors.

Heterologous expression systems have been commonly utilized for the production of large amount of desired odorant receptor (Figure 2.12b). In general, target odorant receptors genes are genetically subcloned into specific plasmid vectors to construct expression vectors for overexpression in heterologous cells. The expression vectors containing target odorant receptor genes are introduced into the heterologous cell by using transfection methods, such as lipofection, electroporation, microinjection, viral infection, and so on. In cells odorant receptors are yielded from the expression vectors, resulting in the collection of large amount of odorant receptors from the cells. So far, bacteria (*Escherichia coli*) [59, 131, 161], mammalian cell lines (human embryo kidney cell; HEK293 cell) [152, 153], and yeast (*S. cerevisiae*) [5, 6, 43, 124, 138, 139] have been often utilized as heterologous cells. This method has the following advantages: collection of large amount of desired odorant receptors, production of odorant receptors with odorant response profiles similar to in *vivo* profiles and purification of target odorant receptors with affinity tags. Therefore, various kinds of odorant receptors derived from vertebrates and invertebrates have been produced for analyzing their function and utilizing them as sensing elements by using this heterologous expression method [37]. However as disadvantages of this method, the codon usage of target odorant receptor genes might be considered for efficient expression, and proper heterologous cells might be required for getting active odorant receptors.

As an alternative method, cell-free protein production system is recently utilized for the production of membrane proteins containing odorant receptors (Figure 2.12c). Extracts from various living organisms such as wheat germ, *E. coli*, rabbit, and insect are commercially available. Since the extracts contain all cellular components for transcription and translation (RNA polymerase, ribosomes, tRNA, amino acids, energy source, and so on), target proteins are synthesized by mixing the extracts and expression vector containing a target gene *in vitro*. Kaiser *et al*. reported that the human odorant receptors were able to be produced by using cell-free production employing extract from wheat germ, demonstrating that the odorant receptors were available as sensing elements [51]. Similarly, Hamada *et al*. reported that silkmoth's pheromone receptor was synthesized in liposomes including the extract from

E. coli and expression vectors to produce functional odorant receptors [39]. This method as well as heterologous expression system yields and obtains large amount of heterologous proteins. However, there are a few cases for applications to biosensors.

Peptide synthesis is likely to be available as a method, for producing odorant binding sites of odorant receptors as sensing elements (Figure 2.12d). In this method, the binding sites in target odorant receptors were predicted by computational simulation and synthesized by peptide synthesis based on chemical reaction to obtain the sensing element for biosensor. Sankaran *et al.* have computationally simulated secondary structure from amino acid sequence of two mouse odorant receptors to predict odorant binding sites based on comparison to other receptor proteins and docking simulation. Based on the predicted binding sites, the polypeptides were chemically synthesized for sensing elements [118]. However, odorant binding sites of odorant receptors from vertebrates and invertebrates remain unclear. In the future, further knowledge regarding odorant binding sites in odorant receptors from various organisms could facilitate development of sensing elements utilizing polypeptides of odorant binding site.

2.4.3.2 Immobilization of Odorant Receptors

In order to acquire signals associated with interactions between odorant receptors and odorants, odorant receptors and their peptides need to be immobilized onto the surface of transducers. At this time, the odorant receptors have to be maintained on the surface at the state of native structure and with their odorant response profiles. So far, three types of methods are mainly used for immobilization of odorant receptors or their peptides onto the surface of transducers: physical adsorption, self-assembled monolayer (SAM) with biotin/avidin interaction, and Au—S bonding (Figure 2.13).

Physical absorption method has been commonly used for immobilization of odorant receptors onto the surface of transducers (Figure 2.13a). In this method, odorant receptor solution that is produced or isolated by the above methods (see Section 2.4.3.1) is generally spread on the electrode of the transducer (i.e., crystal electrode) without special coating materials, and the solution is completely dried in a desiccator. Due to simplicity of procedure, many researchers have utilized this method [59, 131, 150]. However, since unnecessary membrane proteins and other proteins were also absorbed, the biosensors' selectivity and specificity would be affected.

As a typical linker between solid surface and biomolecule, SAM of alkanethiol molecules has been widely used in the development of several biosensors. Since thiol group (—SH) strongly interacts with gold (Au), biomolecules can be easily immobilized on Au surface through the linker like "Au—S—$(CH_2)_n$—protein" using linear carbon chain whose one end is —SH and opposite end is carboxyl group or amino group, for instance. The distance between Au surface and the immobilized biomolecule is controllable due to SAM property of homogeneous thickness arising from hydrophobic interaction among the linear molecules that form the monolayer. Nowadays, there are many commercially available SAM reagents whose functional groups are preliminarily activated for tethering of biomolecules including some odorant receptors and the partial peptide chains such as odorant binding site. In addition, SAM that is combined with specific antibodies is able specifically and stably to immobilize desired odorant receptors. Therefore, SAM with specific antibodies is currently one of the methods that are often utilized for immobilizing odorant receptors onto the surface of transducers (Figure 2.13b). Odorant receptors with affinity tags are produced by using the heterologous expression or cell-free production (see Section 2.4.3.1). The odorant receptors with tags are immobilized upon the surface of transducers that is coated by SAM with the antibodies for recognizing the tags. Vidic *et al.* reported that "nanosome" isolated from yeast expressing odorant receptors was immobilized on the gold surface of SPR device according to this method [138]. Hou *et al.* also successfully immobilized membrane fraction from yeast expressing an odorant receptor onto the electrochemical impedance spectroscopy (EIS) to measure the interaction between the receptor and odorants [43].

The methods utilizing Au—S bond were also used for immobilization of odorant receptors and polypeptides (Figure 2.13c). Amino acid sequences in proteins or odorant receptors include one type of amino acids, cysteine, whose chemical structure has a thiol group. A thiol group couples to Au to form strong bonds as described earlier. Utilizing this principle, odorant receptors and peptides are able to be fixed onto the surface of transducers. Sankaran *et al.* reported that the chemically synthesized polypeptide was immobilized by using this method [118].

2.4.3.3 Measurement from Odorant Receptors

Signals associated with interactions between odorant receptors and odorants have been measured by transducers, such as field-effect transistors (FET), EIS, QCM, SPR, and SAW. Receptor-based biosensors have been fabricated by various combinations of these transducers with various types of odorant receptors that were produced with the above methods. These examples are summarized in Table 2.3.

FET has been commonly used for acquirement of signals from purified odorant receptors. One of the important merits for using FET is to acquire weak signal of interaction between odorant receptors and odorants due to its innate signal amplification. In human odorant receptor-based bioelectronic nose, Kim *et al.* reported a single wall carbon nanotube (swCNT)-FET that was coated with human odorant receptors, hOR2AG1, which is selectively activated by amyl butyrate (Figure 2.22) [59]. Membrane fraction including the OR was collected from *E. coli* and immobilized onto swCNT-FET. The swCNT-FET sensor exhibited ultrahigh sensitivity at the scale of 100 fM and selectively detected amyl butyrate without detection of other similar chemicals. Similarly, Yoon *et al.* reported biosensors using carboxylated polypyrrole nanotubes (CPNT)-FET and hOR2AG1 [161]. They demonstrated that the chemical immobilization strategy with amino silane (3-aminopropyltrimethoxysilane) enabled them to maintain stable electrical contact in CPNT-FET and quantitatively control immobilization of hOR. The hOR-conjugated CPNT-FET sensor achieved high sensitivity to amyl butyrate at 40 fM and selective detection among similar chemicals. Not only odorant receptors but also peptides as sensing molecules were able to immobilize on the surfaces of swCNT-FET. Lim *et al.* successfully developed odorant receptor-derived peptide (ORP)-conjugated swCNT-FET, which sensitively and selectively detected trimethylamine in real time at concentration as low as 10 fM [77]. The ORP-conjugated swCNT-FET was also demonstrated to be able to determine the quality of three types of seafood and distinguish spoiled seafood.

EIS has been recently reported to be able to be used for odorant detection by immobilizing odorant receptors [2, 3, 43]. EIS consists of three electrodes: working electrodes (odorant receptors), reference electrodes, and counter electrodes. Hou *et al.* immobilized yeast-expressed ORI7 on a gold electrode (working electrode) by SAM and biotin/avidin system and detected odorant-dependent signal change by EIS [43]. They showed that the heptanal and

Table 2.3 Summaries of receptor-based sensors

Odorant receptors	Cell/production methods	Ligands	Transducer	Sensitivity/detection range/ detection limit	Literature
Isolated OR proteins (bullfrog)	Bullfrog (*Rana* spp.)/isolation	*n*-Caproic acid, isoamyl acetate, *n*-decyl alcohol, β-ionone, linalool, ethyl caproate	Piezoelectric crystal electrode	10^{-6} to 10^{-7} g (sensitivity)	[150]
hOR2AG1 (human)	Bacteria (*Escherichia coli*)/cell based	Amyl butyrate	swCNT-FET	100 fM (detection limit)	[59]
	Bacteria (*E. coli*)/cell based	Amyl butyrate	CPNT-FET	40 fM (detection limit)	[161]
OR17-40 (human), ORI7 (rat)	Yeast (*Saccharomyces cerevisiae*)/cell based	Helional, cassione, piperonyl acetate, 3,4-methylenedioxyphenylacetone, 3,4-methylenedioxypropiophenone (OR17-40), octanal (ORI7)	SPR	Approx. 5×10^{-10} to 5×10^{-6} M (detection range (figure))	[138]
OR17-40 (human)	Yeast (*S. cerevisiae*)/cell based	Helional	SPR	—	[139]
	Yeast (*S. cerevisiae*)/cell based	Helional	SPR	10^{-11} to 10^{-5} M (tested range)	[5, 6]
OR17-4 (human)	Extract from wheat germ/ cell-free	Undecanal	SPR	1.2–100 μM	[51]
ORI7 (rat)	Yeast (*S. cerevisiae*)/cell based	Octanal, heptanal, helional	EIS	10^{-13} to 10^{-4} M (detection range)	[43]
	Yeast (*S. cerevisiae*)/cell based	Octanal, helional	EIS	—	[2–4]
ODR-10 (*C. elegans*)	Bacteria (*E. coli*)/cell based	Diacetyl (2,3-butanedione)	QCM	10^{-12} to 10^{-5} M (detection range)	[131]
	Human breast cancer cell (MCF-7)/cell based	Diacetyl (2,3-butanedione)	SAW	1.2×10^{-11} mM (detection limit), 10^{-10} to 10^{-4} mM (detection range)	[152, 153]
Polypeptides (OR binding sites)	Chemical synthesis	Acetic acid, butyric acid, ammonia, dimethylamine, chlorobenzene, benzene	QCM	0.9 mg/ml to 108 mg/ml (acetic acid)	[87]
	Chemical synthesis	1-hexanol, 1-pentanol	QCM	2–3 ppm (1-hexanol), 3–5 ppm (1-pentanol) (detection limit)	[119]
	Chemical synthesis	Trimethylamine	swCNT-FET	10 fM to 1 μM (100 μM)	[77]

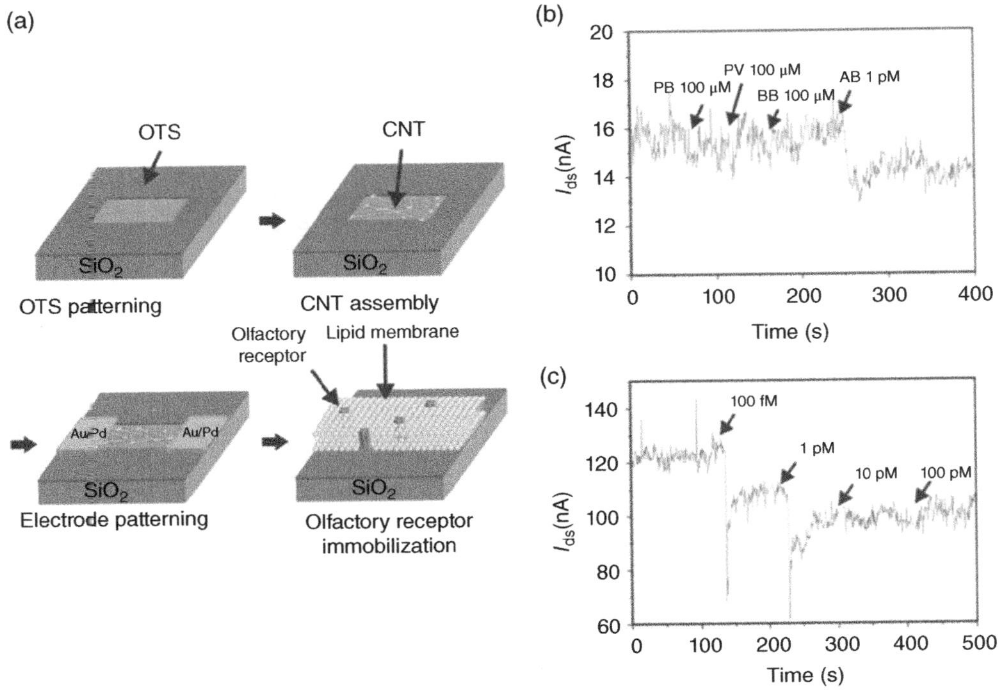

Figure 2.22 Human odorant receptors (hOR2AG1)-functionalized swCNT-FET sensor. (a) Fabrication of hOR2AG1-functionalized swCNT-FET sensor. OCT, octadecyltrichlorosilane. (b) Odorant selectivity of the swCNT-FET sensor. AB, amyl butyrate; BB, butyl butyrate; PB, propyl butyrate; PV, pentyl valerate. (c) Responses of the swCNT-FET sensor to indicate concentrations of AB. From Ref. [59]. Reproduced with permission from John Wiley & Sons, Ltd

octanal were successfully detected using functionalized electrodes. More recently, Alfinito *et al.* showed that the differences in polarization resistance associated with inactive and ligand-bound states of rat receptor ORI7 were measured using EIS and the presence or absence of odorants was analyzed by Nyquist plot (Figure 2.23) [3]. From this theory, they described that electrical properties of a single sensing protein are one possibility to be available as sensing elements.

QCM has been commonly used for odor sensors. Signals of odorant receptors can be obtained from QCM-coated odorant receptors. Sung *et al.* reported that crude membrane extracts of *E. coli* expressing *Caenorhabditis elegans* odorant receptor, ODR-10, were coated onto the surface of QCM by physical absorption (Figure 2.24). Odorant receptors (ODRs) in *C. elegans* belong to GPCR family as same as mammalian odorant receptors. The sensors were demonstrated to selectively detect diacetyl (2,3-butanedione), which is a ligand of ODR-10, with a dynamic range of 10^{-12} to 10^{-5} M [131]. Sankaran *et al.* combined QCM and polypeptides that were chemically synthesized based on computational simulation for prediction of odorant binding sites [118]. This biosensor was developed to detect 1-hexanol and 1-pentanol with detection limits of 2–3 and 3–5 ppm, respectively, which could be applied into the detection of bacterial pathogens in packaged beef. Similarly, a QCM sensor

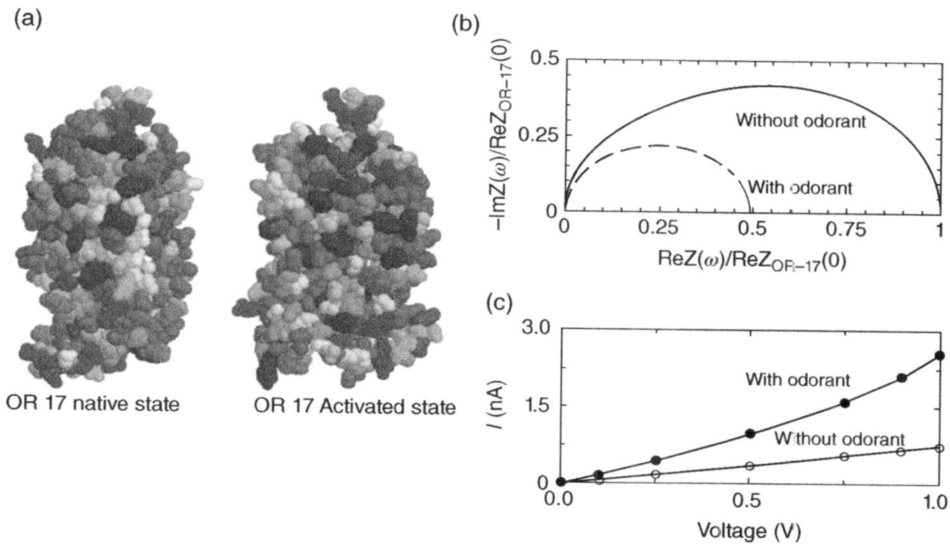

Figure 2.23 (a) Conformational changes of rat ORI7 in the native and activated state. (b) Nyquist plot of the ORI7 with or without the odorant. (c) *I–V* characteristics of the ORI7 with or without the odorant. From Ref. [3]. Reproduced with permission from Elsevier

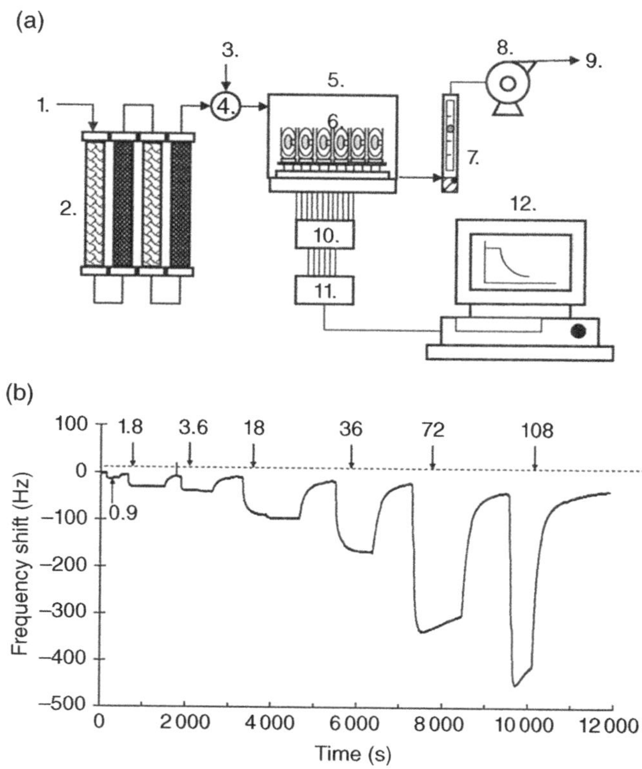

Figure 2.24 (a) Schematic drawing of QCM biosensor. (1) Carrier gas inlet, (2) silica gel and activated carbon, (3) sample inlet, (4) three-way valve, (5) detection chamber, (6) multiarray sensors, (7) flow meter, (8) vacuum pump, (9) waste, (10) oscillator circuit, (11) frequency counter, and (12) computer. (b) Frequency shifts of the sensing system to various concentrations of acetic acid. From Ref. [87]. Reproduced with permission from Elsevier

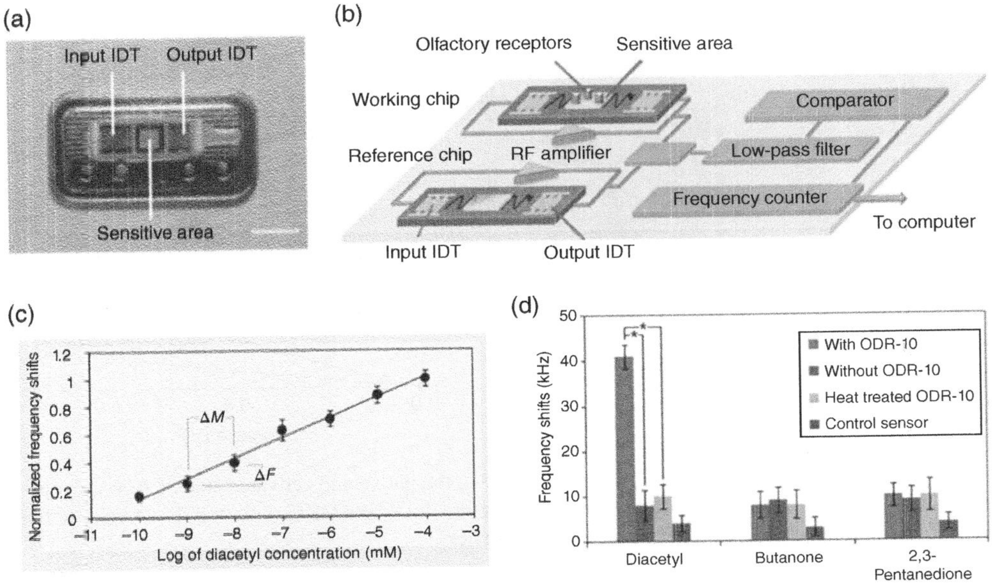

Figure 2.25 SAW device. (a) A photograph of a SAW chip. IDTs, interdigital transducers. Scale bar: 5 mm. (b) Schematic diagram of measurement system of odorant receptor (ODR10a)-functionalized SAW chip. (c) Dose–responses of the SAW chip to various concentrations of diacetyl. (d) Responses of SAW chip immobilized with various membranes to different odorants. From Ref. [153]. Reproduced with permission from Elsevier

combined with chemically synthesized OBP was developed to detect alcohols at room temperature, and the estimated detection limit was about 1–3 ppm [119]. In addition, Lu *et al.* developed a six-chip sensor module with QCMs containing synthetic polypeptides together with conducting polymers to achieve simultaneous detection and identification of various classes of VOCs [87].

SAW sensors were also demonstrated to be possible to be utilized as transducers for measuring signals from odorant receptors (Figure 2.25). Wu *et al.* reported that the SAW sensor-coated membrane fraction of human breast cancer cells, MCF-7 cells, expressing ODR-10 exhibited 10-fold more sensitive (10^{-13} M) to natural ligand, diacetyl, than those of QCM sensors [152]. Furthermore, they demonstrated the SAW sensors with better performances by improving the immobilization efficiency of the odorant receptors onto the surface of SAW chip with SAM absorption to achieve higher sensitivity (10^{-15} M) and longer stability (within 7 days) than their previous work (Figure 2.25) [153].

SPR-based sensing was one of the most popular methods to detect several biomolecules' interactions. Vidic *et al.* showed that nanosomes prepared from OR-expressed yeast were used for detecting odorants through a SPR [138]. In this work, rat ORI7 or human OR1740 with Gα protein was coexpressed in yeast, *S. cerevisiae*, and the yeast-derived nanosomes were immobilized with carboxymethyl-modified dextran polymer hydrogel on the surface of SPR sensor chip. Under the condition of GTPγS existence, which is nonhydrolyzable analogue of GTP, the detection of odorants was achieved by measuring the amount of mobilization of

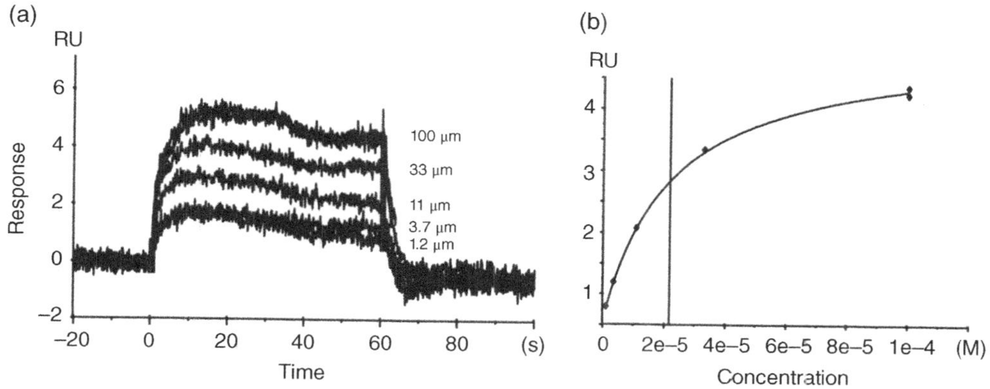

Figure 2.26 Detection of odorant interaction with hOR17-4 on surface plasmon resonance. (a) Responses to undecanal at indicated concentrations. (b) Dose-dependent curve. From Ref. [51]. Reproduced with permission from National Academy of Sciences, USA

Gα protein along with GTPγS. According to this method, helional and cassione were selectively detected in a dose-dependent manner. Vidic *et al.* also demonstrated the immobilization method of nanosome on the gold surface of SPR device by SAMs and biotin/neutravidin to construct nanosome patterning using microcontact printing [139]. In reports of SPR for OR-based biosensors, it was also reported that cell-free produced odorant receptors were applicable (Figure 2.26) [51]. Human odorant receptors, hOR17-4, were synthesized using wheat germ cell-free protein production system and immobilized on a sensor chip of SPR apparatus, Biacore T100 (GE Healthcare). The OR-conjugated SPR sensor dose-dependently detected undecanal, which is known as a ligand of hOR17-4.

2.4.4 *Summary of the Biosensors*

As mentioned in these earlier sections, several odorant sensors based on living systems (bion, organ, tissue, cell, and receptor) have been developed in recent years. There are currently many application studies of human odorant receptors for such a biosensing. On the one hand, some insect odorant receptors are being used as the odorant sensor elements these days. Behind the trend is the fact that insect olfactory systems can be partly reconstructed at cellular level with the recent progress in revealing of the olfactory mechanism. In addition, peptides whose structural motifs have been predicted as odorant binding site are also applied to odorant sensors recently.

Optical-based, resonance-based, and electrical device-based detection systems are mainly used as transducers between the biomaterials and the artificial output interfaces. The combination of these transducers and biomaterials provides many variations of odorant biosensors. Each odorant biosensor has advantages and disadvantages as summarized in Table 2.4. Although each approach has actually some difficulties such as lifetime and laborious handling process of biomaterials, they possess their own advantages in points of the sensitivity and the target odorants as shown in each table of the respective sections. In other words, any odorant biosensors have not yet led to realize the versatile feature of sensitivity,

Table 2.4 Advantages and disadvantages of odorant sensors

Types of biosensor	Advantages	Disadvantages
Tissue-based sensor	Ease to fabricate and immobilize Natural odorant profiles Low cost	Lack of specificity Individual differences Olfactory fatigue Difficulty of natural state and storage Need to kill animals
Cell-based sensor	High sensitivity and selectivity Single type of ORs Nature of membrane for ORs	Difficulty of handling Necessity of culturing Low stability Difficulty of long-term sensing
Receptor-based sensor	High sensitivity and selectivity Longer-term stability Acquirement of receptor activity	Difficulty of purification and isolation

selectivity, portability, robustness, cost-effectiveness, and so on. However, it is expected that further understanding of creature's olfactory system will facilitate a development of odorant biosensors with better performance.

2.5 Future Prospects

According to the elucidation of molecular mechanisms in olfaction of natural living organisms, the mechanism can be used for the olfactory biosensors based on biological molecules. In this chapter, three different types of olfactory biosensors have been introduced: tissue-based biosensor, cell-based biosensors, and receptor-based biosensors. Some researches demonstrated that the performances of these biosensors were superior to those of existing odor sensors in terms of sensitivity and selectivity. However, the olfactory mechanisms of living organisms have been not completely elucidated. Living organisms detect and discriminate various types of environmental odorants with higher performances than expected. For example, recent research demonstrates that biological molecules, such as OBPs, coreceptor, membrane proteins, and membrane transport proteins, function in their olfactory organs and ORNs. In the future, along with the advances in studies on olfaction and taste, the performances of artificial odorant biosensors reach to those of living organisms.

Response characteristics in olfactory biosensors are in accordance with those of the expressed odorant receptors. Therefore, in order to develop odorant biosensors for detecting target odorants, we need to characterize the odorant selectivities of various types of odorant receptors from various living organisms to select the odorant receptors that enable us to detect a target odorant. The database of odorant selectivity in odorant receptors has been constructed regarding to *D. melanogaster* odorant receptors and mammalian odorant receptors. We can use them to select the odorant receptors for the target odorants. The methodology we have described in this chapter, in conjunction with the large repertoire of odorant receptors, enable to develop practical odorant biosensors for various applications, such as food administration, environmental monitoring, and health management.

References

[1] Adler, E., Hoon, M. A., Mueller, K. L., Chandrashekar, J., Ryba, N. J., and Zuker, C. S. (2000) *Cell* 100, 693–702.
[2] Alfinito, E., Pennetta, C., and Reggiani, L. (2010) *Sensors and Actuators B: Chemical* 145(2), 554–558.
[3] Alfinito, E., Millithaler, J. F., Pennetta, C., and Reggiani, L. (2010) *Microelectronics Journal* 41(11), 718–722.
[4] Alfinito, E., Millithaler, J. F., Reggiani, L., Zine, N., and Jaffrezic-Renault, N. (2011) *RSC Advances* 1(1), 123–127.
[5] Benilova, I., Chegel, V. I., Ushenin, Y. V., Vidic, J., Soldatkin, A. P., Martelet, C., Pajot, E., and Jaffrezic-Renault, N. (2008) *European Biophysics Journal* 37, 807–814.
[6] Benilova, I. V., Vidic, J. M., Pajot-Augy, E., Soldatkin, A. P., Martelet, C., and Jafftezic-Renault, N. (2008) *Materials Science and Engineering C* 28(5–6), 633–639.
[7] Benton, R., Sachse, S., Michnick, S. W., and Vosshall, L. B. (2006) *PLoS Biology* 4, e20.
[8] Bo, X., Alavi, A., Xiang, Z., Oglesby, I., Ford, A., and Burnstock, G. (1999) *Neuroreport* 10(5), 1107–1111.
[9] Borisy, F., Ronnett, G., Cunningham, A., Juilfs, D., Beavo, J., and Snyder, S. (1992) *The Journal of Neuroscience* 12(3), 915–923.
[10] Bray, S. and Amrein, H. (2003) *Neuron* 39, 1019–1029.
[11] Breer, H., Boekhoff, I., and Tareilus, E. (1990) *Nature* 345, 65–68.
[12] Buck, L. and Axel, R. (1991) *Cell* 65(1), 175–187.
[13] Bushdid, C., Magnasco, M. O., Vosshall, L. B., and Keller, A. (2014) *Science* 343(6177), 1370–1372.
[14] Caicedo, A. and Roper, S. D. (2001) *Science* 291(5508), 1557–1560.
[15] Cameron, P., Hiroi, M., Ngai, J., and Scott, K. (2010) *Nature* 465, 91–95.
[16] Canessa, C. M., Schild, L., Buell, G., Thorens, B., Gautschi, I., Horisberger, J. D., and Rossier, B. C. (1994) *Nature* 367(6462), 463–467.
[17] Carey, A. F., Wang, G., Su, C.-Y., Zwiebel, L. J., and Carlson, J. R. (2010) *Nature* 464, 66–71.
[18] Chandrashekar, J., Mueller, K. L., Hoon, M. A., Adler, E., Feng, L., Guo, W., Zuker, C. S., and Ryba, N. J. (2000) *Cell* 100(6), 703–711.
[19] Chandrashekar, J., Kuhn, C., Oka, Y., Yarmolinsky, D. A., Hummler, E., Ryba, N. J., and Zuker, C. S. (2010) *Nature* 464(7286), 297–301.
[20] Chang, R. B., Waters, H., and Liman, E. R. (2010) *Proceedings of the National Academy of Sciences of the United States of America* 107(51), 22320–22325.
[21] Charlu, S., Wisotsky, Z., Medina, A., and Dahanukar, A. (2013) *Nature Communications* 4, 2042.
[22] Chen, T. Y. and Yau, K. W. (1994) *Nature* 368(6471), 545–548.
[23] Chen, Z., Wang, Q., and Wang, Z. (2010) *Journal of Neuroscience* 30, 6247–6252.
[24] Chen, Q., Xiao, L., Liu, Q., Ling, S., Yin, Y., Dong, Q., and Wang, P. (2011) *Biosensors and Bioelectronics* 26, 3313–3319.
[25] Clyne, P. J., Warr, C. G., Freeman, M. C., Lessing, D., Kim, J., and Carlson, J. R. (1999) *Neuron* 22, 327–338.
[26] Clyne, P. J., Warr, C. G., and Carlson, J. R. (2000) *Science* 287, 1830–1834.
[27] Corcelli, A., Lobasso, S., Lopalco, P., Dibattista, M., Araneda, R., Peterlin, Z., and Firestein, S. (2010) *Journal of Hazardous Materials* 175, 1096–1100.
[28] Couto, A., Alenius, M., and Dickson, B. J. (2005) *Current Biology* 15, 1535–1547.
[29] Devambez, I., Ali Agha, M., Mitri, C., Bockaert, J., Parmentier, M.-L., Marion-Poll, F., Grau, Y., and Soustelle, L. (2013) *PloS One* 8, e63484.
[30] Dulac, C. and Axel, R. (1995) *Cell* 83(2), 195–206.
[31] Finger, T. E., Danilova, V., Barrows, J., Bartel, D. L., Vigers, A. J., Stone, L., Hellekant, G., and Kinnamon, S. C. (2005) *Science* 310(5753), 1495–1499.
[32] Firestein, S. (2001) *Nature* 413(6852), 211–218.
[33] Fishilevich, E. and Vosshall, L. B. (2005) *Current Biology* 15, 1548–1553.
[34] Gao, Q. and Chess, A. (1999) *Genomics* 60, 31–39.
[35] Gazit, I. and Terkel, J. (2003) *Applied Animal Behaviour Science* 82, 65–73.
[36] Getahun, M. N., Olsson, S. B., Lavista-Llanos, S., Hansson, B. S., and Wicher, D. (2013) *PloS One* 8, e58889.
[37] Glatz, R. and Bailey-Hill, K. (2011) *Progress in Neurobiology* 93, 270–296.
[38] Hallem, E. A. and Carlson, J. R. (2006) *Cell* 125, 143–160.
[39] Hamada, S., Tabuchi, M., Toyota, T., Sakurai, T., Hosoi, T., Nomoto, T., Nakatani, K., Fujinami, M., and Kanzaki, R. (2014) *Chemical Communications* 50, 2958–2961.

[40] Herrada, G. and Dulac, C. (1997) *Cell* 90(4), 763–773.
[41] Hildebrand, J. G. and Shepherd, G. M. (1997) *Annual Review of Neuroscience* 20, 595–631.
[42] Horst, R., Damberger, F., Luginbühl, P., Güntert, P., Peng, G., Nikonova, L., Leal, W. S., and Wüthrich, K. (2001). *Proceedings of the National Academy of Sciences of the United States of America* 98, 14374–14379.
[43] Hou, Y. X., Jaffrezic-Renault, N., Martelet, C., Zhang, A. D., Minic-Vidic, J., Gorojankina, T., Persuy, M. A., Pajot-Augy, E., Salesse, R., Akimov, V., Reggiani, L., Pennetta, C., Alfinito, E., Ruiz, O., Gomila, G., Samitier, J., and Errachid, A. (2007) *Biosensors and Bioelectronics* 22(7), 1550–1555.
[44] Huang, A. L., Chen, X., Hoon, M. A., Chandrashekar, J., Guo, W., Trankner, D., Ryba, N. J., and Zuker, C. S. (2006) *Nature* 442(7105), 934–938.
[45] Huotari, M. J. (2000) *Sensors and Actuators B: Chemical* 71, 212–222.
[46] Huotari, M. and Lantto, V. (2007) *Sensors and Actuators B: Chemical* 127, 284–287.
[47] Ishimaru, Y., Inada, H., Kubota, M., Zhuang, H., Tominaga, M., and Matsunami, H. (2006) *Proceedings of the National Academy of Sciences of the United States of America* 103(33), 12569–12574.
[48] Jacquin-Joly, E. and Merlin, C. (2004) *Journal of Chemical Ecology* 30, 2359–2397.
[49] Jordan, M. D., Anderson, A., Begum, D., Carraher, C., Authier, A., Marshall, S. D. G., Kiely, A., Gatehouse, L. N., Greenwood, D. R., Christie, D. L., Kralicek, A. V., Trowell, S. C., and Newcomb, R. D. (2009) *Chemical Senses* 34, 383–394.
[50] Kain, P., Badsha, F., Hussain, S. M., Nair, A., Hasan, G., and Rodrigues, V. (2010) *Chemical Senses* 35, 663–673.
[51] Kaiser, L. G.-B., Steuerwald, J., Vanberghem, D., Herlihy, M., and Zhang, S. G. (2008) *Proceedings of the National Academy of Sciences of the United States of America* 105(41), 15726–15731.
[52] Kaissling, K.-E. (1987). *R. H. Wright Lectures on Insect Olfaction*. Burnaby: Simon Fraser University.
[53] Kaissling, K.-E., Kasang, G., Bestmann, H., Stransky, W., and Vostrowsky, O. (1978) *Naturwissenschaften* 65, 382–384.
[54] Kanaujia, S. and Kaissling, K. E. (1985) *Journal of Insect Physiology* 31, 71–81.
[55] Kasang, G. and Kaissling, K.-E. (1972). Specificity of primary and secondary olfactory processes in *Bombyx* antennae, in *Fourth International Symposium on Olfaction and Taste*, ed. D.Schneider, Stuttgart: Wissenschaftl. Verlagsgesellsch, pp. 200–206, August 2–4, 1971, Starnberg, Germany.
[56] Katada, S., Hirokawa, T., Oka, Y., Suwa, M., and Touhara, K. (2005) *The Journal of Neuroscience* 25(7), 1806–1815.
[57] Keil, T. A. (1999) Morphology and development of the peripheral olfactory organs. In *Insect Olfaction*, ed. B. S. Hansson, pp. 5–48. Berlin: Springer.
[58] Kijima, H., Okada, Y., Oiki, S., Goshima, S., Nagata, K., Kazawa, T. (1995) *Journal of Comparative Physiology A* 177: 123–133.
[59] Kim, T. H., Lee, S. H., Lee, J., Song, H. S., Oh, E. H., Park, T. H., and Hong, S. (2009) *Advanced Materials* 21(1), 91–94.
[60] Kim, S. H., Lee, Y., Akitake, B., Woodward, O. M., Guggino, W. B., and Montell, C. (2010) *Proceedings of the National Academy of Sciences* 107, 8440–8445.
[61] Kim, S., Ma, L., and Yu, C. R. (2011) *Nature Communications* 2, 365.
[62] Kinnamon, S. C. (2000) *Neuron* 25(3), 507–510.
[63] Ko, H. J. and Park, T. H. (2005) *Biosensors and Bioelectronics* 20, 1327–1332.
[64] Ko, H. J. and Park, T. H. (2006) *Biological Chemistry* 387, 59–68.
[65] Ko, H. J. and Park, T. H. (2007) *Journal of Microbiology and Biotechnology* 17, 928–933.
[66] Kramer, R. H. and Siegelbaum, S. A. (1992) *Neuron* 9(5), 897–906.
[67] Krieger, J. and Breer, H. (1999) *Science* 286, 720–723.
[68] Krieger, J., Klink, O., Mohl, C., Raming, K., and Breer, H. (2003) *Journal of Comparative Physiology A: Neuroethology, Sensory, Neural, and Behavioral Physiology* 189, 519–526.
[69] Kurahashi, T. and Menini, A. (1997) *Nature* 385(6618), 725–729.
[70] Kurahashi, T. and Shibuya, T. (1990) *Brain Research* 515(1–2), 261–268.
[71] Kurahashi, T. and Yau, K. W. (1993) *Nature* 363(6424), 71–74.
[72] Leal, W. S., Tsitoura, P., Andronopoulou, E., Tsikou, D., Agalou, A., Papakonstantinou, M. P., Kotzia, G. A., Labropoulou, V., Swevers, L., Georgoussi, Z., and Iatrou, K. (2010) *PloS One* 5, e15428.
[73] Lee, D., Damberger, F. F., Peng, G., Horst, R., Güntert, P., Nikonova, L., Leal, W. S., and Wüthrich, K. (2002). *FEBS Letters* 531, 314–318.
[74] Lee, S. H., Jeong, S. H., Jun, S. B., Kim, S. J., and Park, T. H. (2009) *Electrophoresis* 30, 328–3288.

[75] Lee, S. H., Jun, S. B., Ko, H. J., Kim, S. J., and Park, T. H. (2009) *Biosensors and Bioelectronics* 24, 2659–2664.
[76] Lee, S. H., Ko, H. J., and Park, T. H. (2009) *Biosensors and Bioelectronics* 25, 55–60.
[77] Lim, J. H., Park, J., Ahn, J. H., Jin, H. J., Hong, S., and Park, T. H. (2013) *Biosensors and Bioelectronics* 39, 244–249.
[78] Liman, E. R., Corey, D. P., and Dulac, C. (1999) *Proceedings of the National Academy of Sciences of the United States of America* 96(10), 5791–5796.
[79] Liman, E. R., Zhang, Y. V., and Montell, C. (2014) *Neuron* 81, 984–1000.
[80] Ling, S., Gao, T., Liu, J., Li, Y., Zhou, J., Li, J., Zhou, C., Tu, C., Han, F., and Ye, X. (2010) *Biosensors and Bioelectronics* 26, 1124–1128.
[81] Liu, L., Leonard, A. S., Motto, D. G., Feller, M. A., Price, M. P., Johnson, W. A., and Welsh, M. J. (2003) *Neuron* 39, 133–146.
[82] Liu, Q., Cai, H., Xu, Y., Li, Y., Li, R., and Wang, P. (2006) *Sensors and Actuators B: Chemical* 22, 318–322.
[83] Liu, Q., Ye, W., Hu, N., Cai, H., Yu, H., and Wang, P. (2010) *Biosensors and Bioelectronics* 26, 1672–1678.
[84] Liu, Q., Ye, W., Xiao, L., Du, L., Hu, N., and Wang, P. (2010) *Biosensors and Bioelectronics* 25, 2212–2217.
[85] Liu, Q., Ye, W., Yu, H., Hu, N., Du, L., Wang, P., and Yang, M. (2010) *Sensors and Actuators B: Chemical* 146, 527–533.
[86] Liu, Q., Hu, N., Zhang, F., Zhang, D., Hsia, K. J., and Wang, P. (2012) *Biomedical Microdevices* 14, 1055–1061.
[87] Lu, H. H., Rao, Y. K., Wu, T. Z., and Tzeng, Y. M. (2009) *Sensors and Actuators B: Chemical* 137(2), 741–746.
[88] Lucas, P., Ukhanov, K., Leinders-Zufall, T., and Zufall, F. (2003) *Neuron* 40(3), 551–561.
[89] Lundin, C., Käll, L., Kreher, S. A., Kapp, K., Sonnhammer, E. L., Carlson, J. R., Von Heijne, G., and Nilsson, I. (2007) *FEBS Letters* 581, 5601–5604.
[90] Marrakchi, M., Vidic, J., Jaffrezic-Renault, N., Martelet, C., and Pajot-Augy, E. (2007) *European Biophysics Journal* 36, 1015–1018.
[91] Masek, P. and Keene, A. C. (2013) *PLoS Genetics* 9, e1003710.
[92] Matsunami, H. and Buck, L. B. (1997) *Cell* 90(4), 775–784.
[93] McLaughlin, S. K., McKinnon, P. J., Spickofsky, N., Danho, W., and Margolskee, R. F. (1994) *Physiology and Behavior* 56(6), 1157–1164.
[94] Minic, J., Persuy, M., Godel, E., Aioun, J., Connerton, I., Salesse, R., and Pajot-Augy, E. (2005) *Federation of European Biochemical Societies Journal* 272, 524–537.
[95] Misawa, N., Mitsuno, H., Kanzaki, R., and Takeuchi, S. (2010) *Proceedings of the National Academy of Sciences of the United States of America* 107, 15340–15344.
[96] Mitsuno, H., Sakurai, T., Namiki, S., Mitsuhashi, H., and Kanzaki, R. (2015) *Biosensors and Bioelectronics* 65, 287–294.
[97] Mori, K., Nagao, H., and Yoshihara, Y. (1999) *Science* 286(5440), 711–715.
[98] Morita, H. and Shiraishi, A. (1985). Chemoreception physiology. In *Comprehensive Insect Physiology, Biochemistry and Pharmacology: Nervous System*, Vol. 6, ed. G. A.Kerkut and L. I.Gilbert, pp. 133–170. Oxford/NewYork: Pergamon Press.
[99] Murata, Y., Yasuo, T., Yoshida, R., Obata, K., Yanagawa, Y., Margolskee, R. F., and Ninomiya, Y. (2010) *Journal of Neurophysiology* 104(2), 896–901.
[100] Nakagawa, T., Sakurai, T., Nishioka, T., and Touhara, K. (2005) *Science* 307, 1638–1642.
[101] Nelson, G., Hoon, M. A., Chandrashekar, J., Zhang, Y., Ryba, N. J., and Zuker, C. S. (2001) *Cell* 106(3), 381–390.
[102] Nelson, G., Chandrashekar, J., Hoon, M. A., Feng, L., Zhao, G., Ryba, N. J., and Zuker, C. S. (2002) *Nature* 416(6877), 199–202.
[103] Nolan, R., Weinstein, S., and Weinstein, C., (1978) Proceedings of the New Concepts of Symposium and Workshop on Detection and Identification of Explosives, Reston, VA, October 30–November 1, 1978, pp. 201–205.
[104] Oh, E. H., Lee, S. H., Lee, S. H., Ko, H. J., and Park, T. H. (2014) *Biosensors and Bioelectronics* 53, 18–25.
[105] Park, K. C., Ochieng, S. A., Zhu, J., and Baker, T. C. (2002) *Chemical Senses* 27, 343–352.
[106] Perez, C. A., Huang, L., Rong, M., Kozak, J. A., Preuss, A. K., Zhang, H., Max, M., and Margolskee, R. F. (2002) *Nature Neuroscience* 5(11), 1169–1176.
[107] Poling, A. (2010) *The Psychological Record* 60, 715–728.
[108] Pophof, B. (1997). *Physiological Entomology* 22, 239–248.

[109] Radhika, V., Proikas-Cezanne, T., Jayaraman, M., Onesime, D., Ha, J. H., and Dhanasekaran, D. N. (2007) *Nature Chemical Biology* 3, 325–330.

[110] Rains, G. C., Utley, S. L., and Lewis, W. J. (2006) *Biotechnology Progress* 22, 2–8.

[111] Ressler, K., Sullivan, S., and Buck, L. (1993) *Cell* 73(3), 597–609.

[112] Richter, T. A., Dvoryanchikov, G. A., Chaudhari, N., and Roper, S. D. (2004) *Journal of Neurophysiology* 92(3), 1928–1936.

[113] Richter, T. A., Dvoryanchikov, G. A., Roper, S. D., and Chaudhari, N. (2004) *The Journal of Neuroscience* 24(16), 4088–4091.

[114] Robertson, H. M. (2003) *Proceedings of the National Academy of Sciences* 100, 14537–14542.

[115] Roper, S. D. (2007) *Pflügers Archives—European Journal of Physiology* 454(5), 759–776.

[116] Ryba, N. J. P. and Tirindelli, R. (1997) *Neuron* 19(2), 371–379.

[117] Sakurai, T., Namiki, S., and Kanzaki, R. (2014) *Frontiers in Physiology* 5: 125.

[118] Sankaran, S., Panigrahi, S., and Mallik, S. (2011) *Biosensors and Bioelectronics* 26, 3103–3109.

[119] Sankaran, S., Panigrahi, S., and Mallik, S. (2011) *Sensors and Actuators B: Chemical* 155(1), 8–18.

[120] Sato, K. and Takeuchi, S. (2014) *Angewandte Chemie International Edition* 53, 11798–11802.

[121] Sato, K., Pellegrino, M., Nakagawa, T., Nakagawa, T., Vosshall, L. B., and Touhara, K. (2008) *Nature* 452, 1002–1006.

[122] Sato, K., Tanaka, K., and Touhara, K. (2011) *Proceedings of the National Academy of Sciences* 108, 11680–11685.

[123] Scott, K., Brady, R., Jr., Cravchik, A., Morozov, P., Rzhetsky, A., Zuker, C., and Axel, R. (2001) *Cell* 104, 661–673.

[124] Segui, S. R., Pla, M., Minic, J., Pajot-Augy, E., Salesse, R., Hou, Y., Jaffrezic-Renault, N., Mills, C. A., Samitier, J., and Errachid, A. (2006) *Analytical Letters* 39, 1735–1745.

[125] Serizawa, S., Miyamichi, K., and Sakano, H. (2004) *Trends in Genetics* 20(12), 648–653.

[126] Smart, R., Kiely, A., Beale, M., Vargas, E., Carraher, C., Kralicek, A. V., Christie, D. L., Chen, C., Newcomb, R. D., and Warr, C. G. (2008) *Insect Biochemistry and Molecular Biology* 38, 770–780.

[127] Spehr, J., Hagendorf, S., Weiss, J., Spehr, M., Leinders-Zufall, T., and Zufall, F. (2009) *The Journal of Neuroscience* 29(7), 2125–2135.

[128] Steinbrecht, R. A. (1999). Olfactory receptors. In *Atlas of Arthropod Sensory Receptors-Dynamic Morphology in Relation to Function*, eds. E. Eguchi and Y. Tominaga, pp. 155–176. Tokyo: Springer.

[129] Stocker, R. (1994) *Cell and Tissue Research* 275, 3–26.

[130] Suh, G. S. B., Wong, A. M., Hergarden, A. C., Wang, J. W., Simon, A. F., Benzer, S., Axel, R., and Anderson, D. J. (2004) *Nature* 431, 854–859.

[131] Sung, J. H., Ko, H. J., and Park, T. H. (2006) *Biosensors and Bioelectronics* 21(10), 1981–1986.

[132] Szyszka, P., Gerkin, R. C., Galizia, C. G., and Smith, B. H. (2014) *Proceedings of the National Academy of Sciences* 111: 16925–16930.

[133] Takeuchi, H. and Kurahashi, T. (2002) *Journal of Physiology* 541(Pt 3), 825–833.

[134] Talluri, S., Bhatt, A., and Smith, D. P. (1995) *Proceedings of the National Academy of Sciences* 92, 11475–11479.

[135] Tanada, N., Sakurai, T., Mitsuno, H., Bakkum, D. J., Kanzaki, R., and Takahashi, H. (2012) *Analyst* 137, 3452–3458.

[136] Ueno, K. and Kidokoro, Y. (2008) *European Journal of Neuroscience* 28, 1956–1966.

[137] Vassar, R., Ngai, J., and Axel, R. (1993) *Cell* 74(2), 309–319.

[138] Vidic, J. M., Grosclaude, J., Persuy, M. A., Aioun, J., Salesse, R., and Pajot-Augy, E. (2006) *Lab on a Chip* 6(8), 1026–1032.

[139] Vidic, J., Pla-Roca, M., Grosclaude, J., Persuy, M. A., Monnerie, R., Caballero, D., Errachid, A., Hou, Y. X., Jaffrezic-Renault, N., Salesse, R., Pajot-Augy, E., and Samitier, J. (2007) *Analytical Chemistry* 79(9), 3280–3290.

[140] Vieira, F. G. and Rozas, J. (2011) *Genome Biology and Evolution* 3, 476–490.

[141] Vogt, R. G. and Riddiford, L. M. (1981) *Nature* 293, 161–163.

[142] Vosshall, L. B. and Stocker, R. F. (2007) *Annual Review of Neuroscience* 30, 505–533.

[143] Vosshall, L. B., Amrein, H., Morozov, P. S., Rzhetsky, A., and Axel, R. (1999) *Cell* 96, 725–736.

[144] Vosshall, L. B., Wong, A. M., and Axel, R. (2000) *Cell* 102, 147–159.

[145] Wang, G., Carey, A. F., Carlson, J. R., and Zwiebel, L. J. (2010) *Proceedings of the National Academy of Sciences* 107, 4418–4423.

[146] Wei, J., Zhao, A. Z., Chan, G. C., Baker, L. P., Impey, S., Beavo, J. A., and Storm, D. R. (1998) *Neuron* 21(3), 495–504.

[147] Wicher, D., Schäfer, R., Bauernfeind, R., Stensmyr, M. C., Heller, R., Heinemann, S. H., and Hansson, B. S. (2008) *Nature* 452, 1007–1011.

[148] Wicher, D., Hansson, B. S., Olsson, S. B., Llanos, S. L., Getahun, M. N., and Sargsyan, V. (2011) *Frontiers in Cellular Neuroscience* 5, 5.

[149] Wojtasek, H. and Leal, W. S. (1999) *Journal of Biological Chemistry* 274, 30950–30956.

[150] Wu, T. Z. (1999) *Biosensors and Bioelectronics* 14(1), 9–18.

[151] Wu, C., Chen, P., Yu, H., Liu, Q., Zong, X., Cai, H., and Wang, P. (2009) *Biosensors and Bioelectronics* 24, 1498–1502.

[152] Wu, C. S., Du, L. P., Wang, D., Wang, L., Zhao, L. H., and Wang, P. (2011) *Biochemical Biophysics Research Communications* 407(1), 18–22.

[153] Wu, C. S., Du, L. P., Wang, D., Wang, L., Zhao, L. H., and Wang, P. (2012) *Biosensors and Bioelectronics* 31(1), 44–48.

[154] Xavier, A. F., Gregory, A. C., Scott, V. V., Lisa, F. H., and Albert, F. (2010) *Lab on a Chip* 10, 1120–1127.

[155] Xiang, Y., Yuan, Q., Vogt, N., Looger, L. L., Jan, L. Y., and Jan, Y. N. (2010) *Nature* 468, 921–926.

[156] Yan, C., Zhao, A., Bentley, J., Loughney, K., Ferguson, K., and Beavo, J. (1995) *Proceedings of the National Academy of Sciences of the United States of America* 92(21), 9677–9681.

[157] Yanagawa, A., Guigue, A. M. A., and Marion-Poll, F. (2014) *Frontiers in Behavioral Neuroscience* 8, 254.

[158] Yang, C. and Delay, R. J. (2010) *The Journal of General Physiology* 135(1), 3–13.

[159] Yang, C.-H., Belawat, P., Hafen, E., Jan, L. Y., and Jan, Y.-N. (2008) *Science* 319, 1679–1683.

[160] Yao, C. A. and Carlson, J. R. (2010) *The Journal of Neuroscience* 30, 4562–4572.

[161] Yoon, H., Lee, S. H., Kwon, O. S., Song, H. S., Oh, E. H., Park, T. H., and Jang, J. (2009) *Angewandte Chemie, International Edition* 48(15), 2755–2758.

[162] Zhang, Y., Hoon, M. A., Chandrashekar, J., Mueller, K. L., Cook, B., Wu, D., Zuker, C. S., and Ryba, N. J. (2003) *Cell* 112(3), 293–301.

[163] Zhang, Z., Zhao, Z., Margolskee, R., and Liman, E. (2007) *The Journal of Neuroscience* 27(21), 5777–5786.

[164] Zhang, H.-J., Anderson, A. R., Trowell, S. C., Luo, A. R., Xiang, Z.-H., and Xia, Q.-Y. (2011) *PloS One* 6, e24111.

[165] Zhang, Y. V., Ni, J., and Montell, C. (2013) *Science* 340, 1334–1338.

3

Large-Scale Chemical Sensor Arrays for Machine Olfaction

Mara Bernabei, Simone Pantalei and Krishna C. Persaud
School of Chemical Engineering and Analytical Science, The University of Manchester, Manchester, UK

3.1 Introduction

Among the five senses, smell is undoubtedly of less importance in the ordinary lives of humans, while it is of vital importance for animals. Animals live in complex environments that contain a great diversity of odorant molecules, at different concentrations, and despite the complexity of the environment, they are able to detect odors, determine the intensity and the source of the odors, and discriminate one particular odor present in a complex background of other odors. Exploiting these capabilities of olfaction, animals are able to find food, mates, predators, and partners. Because of the potentials of the olfactory system, there has been a great interest in developing instruments that mimic it.

Since the appearance of the first artificial nose in the 1980s [80, 84], these devices have been developed to commercial instruments, and they found dedicated applications in different fields, such as in the food and pharmaceutical industries, environmental control, clinical diagnostics, and more [4].

The working principle of these instruments is based on the employment of an array of a few different nonselective chemical sensors. Each sensor in the array is able to detect a range of different odors, not just one, with a sensitivity that is different from the sensitivity shown by the other elements of the array. Such a feature makes the response of the electronic nose dependent on the whole range of chemical information contained in an odor. This is something similar to what occurs in biological olfaction.

Despite the progress that has been achieved during more than 30 years of research in this field, no artificial olfactory system exhibits same performance of the biological olfaction. The latter is superior to any artificial sniffer in specificity, robustness, detection limits, coding capacity, stability over time, response time, size, power consumption, and portability.

Essentials of Machine Olfaction and Taste, First Edition. Edited by Takamichi Nakamoto.

Companion website: www.wiley.com/go/nakamoto/olfaction

Previously, lack of knowledge of the biological system's architecture and signal processing mechanisms performed within the olfactory system has been the limiting factor in the development of e-noses with similar performance.

In the last two decades, the situation has changed. New discoveries have enhanced the understanding of the olfactory structure and coding mechanisms [15, 48, 49, 89, 96–98, 104, 118].

Exploiting these possibilities, it has been possible, recently, to build computational models of the main building blocks of the olfactory system: the olfactory receptor layer, olfactory bulb, and olfactory cortex [11, 49, 92, 93, 97]. These models should be able to carry out the signal processing mechanisms executed by the biological olfaction. However, without realistic sensor data inputs, such models can neither be benchmarked nor completely exploited. Currently, acquisition of such data is really necessary.

Our ideas were to develop a large sensor array arranged in a matrix configuration, mimicking the olfactory epithelium and interfaced with the computational models. The main goal of this research was the implementation in the sensor array of similar features exhibited by the epithelium: a large number of olfactory receptor neuron (ORNs) of multiple types, with broad and overlapping specificity and different dynamic ranges of responses to the same analytes. As in biological olfaction, it is expected that these characteristics would enhance the discriminatory power and the robustness of the artificial olfactory devices, increasing the signal-to-noise (S/N) ratio and allowing detection of previously unknown odors.

To reach this goal many problems had to be solved. There is no large chemical sensor array in existence and no electronic circuits available able to read high-dimensional gas sensing arrays having many electrical connections.

This chapter focuses on the fabrication of a large-scale chemical sensor array.

3.2 Overview of Artificial Olfactory Systems

One of the first studies reporting the possibility of detecting odors dates back to around 1920. At that time Zwaardemaker and Hogewind showed that uncharged drops of pure water are sprayed from a fine jet, while positive drops are obtained if an odorant water solution is sprayed [121]. Although these were preliminary, the introduction of chemical sensors, dated around 1950, is considered as the event that led to the development of artificial olfactory systems. The lack of specificity of these sensors, which was initially considered as a drawback, was then regarded as an advantage for developing devices for odor recognition [72]. Progress in the understanding of the architecture and features of the biological olfactory system made clear that its sensory cells are also nonspecific, and the introduction of chemical sensors provided the scientists the possibility to implement this feature in an artificial olfactory system. In 1982, the first "electronic nose," composed of three gas chemical nonselective sensors, was built by Persaud and Dodd [80]. Since then, the term "electronic nose" started to be used in conferences and papers.

The architecture of the instrument envisaged by Persaud and Dodd was practically the same implemented later on in actual systems. There are a sensor array, composed of nonselective chemical sensors; an odor delivery system; an electronic circuit, able to read the sensors; and a pattern recognition (PARC) engine for odor classification and discrimination. With this architecture, when a chemical environment is analyzed by an artificial sniffer, the odorant molecules present in the environment are delivered to the sensor array by means of the delivery system. Each sensor interacts with the odorant molecules, changing one or more chemical–physical properties. These changes are transduced by the sensors in electronic signals, which are acquired by the readout circuit. As the sensors composing the array have different sensitivity and selectivity, each sensor reacts to the odor differently and generates its own response which is related to odor type and concentration. Hence, the response of an electronic nose is a pattern of electronic signals, the sensor pattern, having lower dimensionality with respect to the chemical pattern characterizing the environment under test. The sensor pattern, which is a sort of "fingerprint" of the sample analyzed, is digitized, transmitted to a computer and processed with a data analysis technique. The data analysis process correlates the sensor response to the chemical pattern and permits odor recognition (see Figure 3.1). The odor recognition process is based on an odor pattern dataset created during an initial training phase of the e-nose by associating sensor patterns to well-known odor samples. The great advantage of these instruments is that they allow the perception of the "aroma" of a sample, instead of the separation of individual chemical components of such aroma, as done by conventional analytical instruments, such as gas chromatography–mass spectrometry (GC/MS).

The sensors that can be successfully used for "electronic noses" must meet a number of criteria. They must display broad selectivity and high sensitivity to different analytes, rapid response time, and short recovery times. These features provide the artificial sniffers with the ability to detect and discriminate a large number of odors in a brief time, as animals do. If the sensor array was to be composed of highly chemically selective sensors, just a few types of

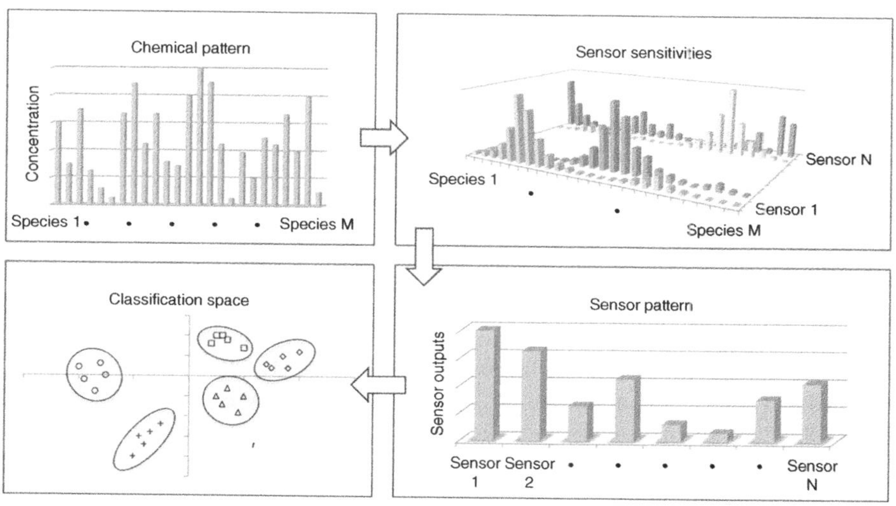

Figure 3.1 Electronic nose working principles. This picture shows the sequences composing an electronic nose measurement. When an e-nose analyzes a chemical environment, the chemical pattern characterizing that environment (top-left graph) is transduced by the sensor array in a pattern of electronic signals (bottom-left graph), called sensor pattern, having a lower dimensionality with respect to the original one. The sensor pattern is recorded in a computer and treated with data analysis techniques that correlate the sensor responses to the original odor classes composing the environment and odor classification is performed

odors could be detected, being those chemical species that the sensors are sensitive to. Using broad specificity sensors, artificial olfactory systems are capable of also detecting new odors.

Besides the features detailed above, the sensors in the array should be stable and robust and able to be fabricated reproducibly, so that the "electronic nose" will be also a robust, stable, and reproducible system. Other sensor requirements regard their sensitivity to environmental variables, such as temperature and humidity. These systems are typically employed in poorly controlled environments. If the array exploited sensors with high sensitivity to physical variables, the response would depend not only on the sample under test but would also be affected by other interfering factors. However, all known chemical sensors exhibit some cross-sensitivity to these environmental variables, and it cannot be easily eliminated. Hence, the best approach is the use of sensors with low sensitivity to temperature and humidity, together with the use of a delivery system capable of controlling these physical variables that may interfere with the sensor behavior. Such solution guarantees that the sensor response is only related to the odor.

Thinking about portable devices to be used in different fields and environments, other desirable requirements for the sensing elements are small size and low power consumption.

Several different artificial olfactory systems have been developed using different types of chemical sensors [2, 6, 9, 14, 19, 22].

A chemical gas sensor is composed of two parts, the chemical interactive material (CIM) and the transducer. The CIM is the sensing layer that captures the odorant molecules and, in response to this interaction, changes one or more of its physical properties. Different sensitive materials can be selected to be employed in the electronic noses. The selection of the CIM is primarily made considering its sensitivity and selectivity toward the analytes that are the target for the specific application [1, 2, 27, 28]. Some of the most commonly used sensing layers are:

- Conductive and nonconductive polymers
- Molecular films
- Catalytic metals
- Metal oxides
- Solid-state ionic conductors

This list is not exhaustive. At the present time, great efforts are being made by scientists to develop new sensing materials more adequate to mimic ORNs. Several papers have appeared recently, reporting the use of olfactory receptors, odorant binding proteins, and the major urinary proteins as interactive sensing layers for e-noses [18, 31, 82].

The transducer is the element of the sensor that transduces the physical variations that happen in the CIM into an electronic signal. For the same sensing layer, different transduction mechanisms can be used, depending on which property of the CIM is translated. Hence, combining different transduction mechanisms with different CIMs, diverse chemical sensors can be built (see Figure 3.2).

In Table 3.1, a classification of chemical sensors based on types of common transducers is reported. Each kind of sensor has its advantages and drawbacks. The choice of the sensor technology to adopt in the artificial olfactory system depends on the particular application. Since different transduction mechanisms detect different aspects of the same odor molecules, in order to increase the number of detectable odors and enhance the discrimination power of the device, hybrid instruments employing several sensor technologies have been also introduced in recent years [52, 56, 61, 63, 105].

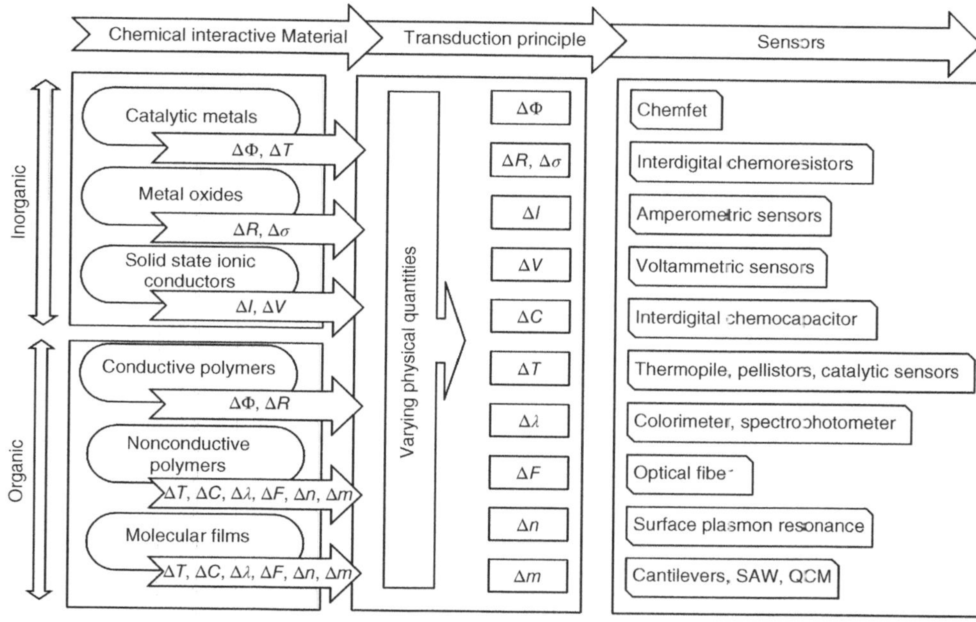

Figure 3.2 Technology diagram of existing electronic nose systems. A sensor is composed of two main parts, the chemical interactive material (CIM), which is the element sensitive to the odorant molecules, and the transducer. The CIM changes one or a few physical features as reaction to the interaction with the odor. The transducer reads the physical variation of the CIM and converts it into an electronic signal. Different CIM and different transducers can be used, giving rise to different sensors. The CIM commonly employed in the e-noses and their physical characteristics that can be taken as sensor responses are reported in the first column of blocks. The transduction principles normally implemented in the sensors are in the second column. The different sensor technologies obtained by combining CIMs and transducers are listed in the third column. So, for example, conductive polymers can be utilized as sensitive material in a ChemFET, where the change of their conductivity, as a consequence of the interaction with an analyte, is detected as a current passing through a transistor, or in chemiresistor sensors, where changes in electrical resistance is detected

3.3 Common Sensor Technologies Employed in Artificial Olfactory Systems

Among the sensor technologies listed previously, there are three main classes that have been largely used in the electronic noses, and most of the commercial artificial systems are based on them. Hence, a brief description of these three families is given in Sections 3.3.1–3.3.3.

3.3.1 Metal-Oxide Gas Sensors

Metal-oxide gas sensors (MOS) are the most extensively used sensors for odor detection, as they have been commercially available for a number of years. They were discovered in 1971, and, since then, they have been used in many applications [3, 7, 41, 42, 74, 78, 91]. Metal-oxide sensors consist of a metal-oxide semiconducting film, such as SnO_2, TiO_2, and ZnO, coated onto a ceramic substrate, usually alumina, and heated by a heating wire. At a certain

Table 3.1 Commonly used sensors. These are grouped based on the mechanism of transduction

Sensor type	Transducing principle	References
ChemFET, light-addressable potentiometric sensors	Work function	[67, 73, 102, 103, 120]
Thermopile, pellistor catalytic sensors	Temperature	[100, 119]
Amperometric gas sensors	Ionic current	[16, 17, 21, 115]
Chemiresistors	Conductivity	[8, 35, 57, 65, 66, 90, 106]
Chemocapacitors	Permittivity	[13, 34, 55, 62, 76, 77]
Colorimeter, spectrometer	Optical spectrum	[5, 43, 53, 54, 70, 95]
Optical fibers	Fluorescence	[64, 101, 107]
Optical fibers, surface plasmon resonance	Refractive index	[20, 71, 75, 99, 116]
Cantilevers, surface acoustic wave, quartz crystal microbalance	Mass	[29, 50, 94]

high temperature, the metal-oxide semiconductor interacts with oxygen in the air. The oxygen is adsorbed on the crystal vacancies, and a sensor resistance value is established at the equilibrium. During a measurement, the odorant molecules react with the oxygen species, modifying such equilibrium and modulating the conductivity. The sensitivity of these sensors is quite good, especially to combustible substances (0.1–100 ppm), and it can be tuned by modulating the working temperature and doping the semiconductor with noble metal catalysts (e.g., Pt, Pd, Al, Au). Their major drawbacks are the high operational temperatures, between 100 and 600°C; their poor specificity; high power consumption; poisoning, caused by sulfur compounds or weak acids; and their high sensitivity to certain compounds such as ethanol, which can hide the effects on the sensor response of other analytes of interest.

3.3.2 *Piezoelectric Sensors*

Piezoelectric sensors, in particular those based on quartz crystal microbalance (QCM) and surface acoustic waves (SAW), have been largely employed in electronic noses [117].

QCM sensors are composed of a disk of piezoelectric quartz placed between two electrodes and covered with a sensing material. When excited with an AC voltage, the crystal vibrates at a characteristic resonance frequency. During a measurement, the sensor is exposed to an analyte which is adsorbed on the sensing material. Hence, the mass on the crystal is increased and causes a shift of the resonance frequency. The following equation, known as Sauerbrey's equation, relates the frequency shift to the adsorbed mass:

$$\Delta f = \frac{2 f_0^2}{A\sqrt{\rho_q \mu_q}} \Delta m \tag{3.1}$$

where Δf is the frequency shift; μ_q is the shear modules of quartz for AT-cut crystal; ρ_q is the density of the quartz; A is the area between electrodes, called piezoelectrically active crystal area; Δm is the adsorbed mass; and f_0 is the resonant frequency. By measuring Δf, it is possible to evaluate the concentration of the analyte. Different sensing materials can be used as coatings, giving the capability to tune the sensors [33].

Like QCM, SAW sensors are based on a frequency shift. They are built by evaporating onto the surface of a piezoelectric substrate two interdigitated metal electrodes. One is used as input and the other one as output. Applying an electric signal to the input electrode, an acoustic wave is generated at operating frequencies between 100 and 400 MHz [79]. The wave propagates along the surface of the piezoelectric and is collected to the output electrode. If a sensitive membrane is placed between the transducers, the adsorption of odorant molecules onto the membrane changes the mass on the wave path, altering the frequency of the wave. The frequency shift is correlated to the amount of the material adsorbed on the membrane by the following equation:

$$\Delta f = \frac{\Delta f_{\mathrm{p}} c_{\mathrm{v}} K_{\mathrm{p}}}{\rho_{\mathrm{p}}} \tag{3.2}$$

where Δf_{p} is the change in frequency caused by the deposition of the sensing membrane, c_{v} is the vapor concentration, K_{p} is the partition coefficient, and ρ_{p} is the density of the membrane used [39, 40]. The substrates normally used for the SAW are ZnO, lithium niobate, and quartz. The sensitivity is tuned by changing the coating.

Both SAW and QCM sensors exhibit poor signal-to-noise performance, high sensitivity to humidity and temperature, viscoelastic effects that interfere with the mass loading at high concentrations of the samples, and very poor reproducibility in sensor manufacture that makes the replacement of damaged sensors rather difficult. Nevertheless, technology has improved dramatically, and they are widely used as chemical sensors.

3.3.3 Conducting Polymer Sensors

Intrinsically conducting polymer (ICP) sensors are based on measuring the resistance of a thin-film polymer coated over two electrodes. When the ICP sensor is exposed to analyte vapors, a chemical interaction occurs on its surface [81]. The vapors are adsorbed or absorbed into the polymer, inducing physical or chemical interactions that cause changes in the electrical conductivity of the film. The sensor response is correlated to the odorant concentration and is reversible when the vapor is removed. ICP sensors are particularly attractive as sensing films because their structure can be easily modified to provide a large variety of materials with different sensitivity to a wide range of vapors and they can be deposited in a variety of ways onto different substrates. Conductive polymer sensors also feature low power consumption, rapid and reversible responses, ease to produce, low poisoning effects, and long lifetime. Their main drawback is a high sensitivity to humidity and temperature. Conducting polymers sensors are discussed in more detail in subsequent sections of this chapter. In fact, because of their features, they were considered the best option for building a unique artificial olfactory system able to mimic the biological system.

In Table 3.2 are listed some commercially available devices based broadly on "e-nose" concepts. The sensor technologies that they implement are also specified. It is evident from the table that most of the commercial artificial systems are based on the sensors just described.

Table 3.2 Some commercially available systems based broadly on "e-nose" concepts

Manufacturer	Technologies	Models
Alpha MOS (http://www.alpha-mos.com)	MOS and mass spectrometry	Fox 2000, 3000, and 4000 and Gelmini are e-noses based on MOS sensor arrays Prometheus combines fingerprint mass spectrometry and MOS sensors
AIRSENSE Analytics (http://www.airsense.com)	MOS, electrochemical (EC) sensors, ion mobility spectrometer (IMS), photoionization detector (PID)	i-PEN, PEN2, PEN3, GDA, and GDA2 use MOS and EC sensors
Applied Sensor (http://www.appliedsensor.com)	MOS, MOSFET QCM, IR	iAQ–engine and iAQ-2000 are based on MOS technology
Microsensor Systems (http://www.microsensorsystems.com)	SAW, PID, EC, and catalytic sensors	HAZMATCAD uses SAW sensor arrays and EC cells to identify nerve and blister agents, as well as agents that attack the blood or cause choking Safesite® Multithreat detector utilizes SAW, cadmium zinc telluride (CZT), PID, EC, and catalytic bead technologies for monitoring toxic gases, oxygen, and VOCs Sirius is a portable PID multigas detector Other models are described in the website
Smiths Group or Cyrano Sciences, Inc. (http://www.smithsdetection.com)	Carbon black polymers	Cyranose 320, commercialized in 2000, is an efficient and affordable way to perform on-site analysis. It is sold by Smiths Detection group, which acquired Cyrano Sciences, Inc.
Electronic Sensor Technology (http://www.estcal.com)	GC, SAW	4200 portable zNose, 4300 battery-operated zNose, 4500 portable zNose, and 7100 benchtop zNose are commercial devices that employ gas chromatographers coupled with SAW sensors
RST Rostock System-Technik GmbH (http://www.rst-rostock.de)	MOS	SamDetect FF2, SamDetect FF2D, and SamDetect GFD1 are arrays of 6 MOS sensors

Multisensor Systems (http://www.multisensor.co.uk)	MOS, CP	MS100, MS1100, and MS1200 are used for the detection of VOCs and hydrocarbon in clean and wastewater applications and in environmental monitoring, industrial effluent containment, chemical processing, and landfill site applications MS2000 Total Trihalomethanes (THM) provides an efficient measurement of the levels of trihalomethanes in drinking water with an accuracy of +/−10 ppb The MS3000 is an ammonia monitor designed for both wastewater and raw/clean water applications
MS Tech (http://www.ms-tech.co.il)	QCM	QCM sensors are the sensor employed in DUOSCAN™, a portable narcotics and contraband detector, and EXPLOSCAN™, a handheld explosives trace and vapor detector
Sysca AG	MOS	ARTINOS® series are electronic noses based on MOS sensors. They are able to detect and identify single gases as well as complex odors down to the ppb range

Major manufacturers, together with the specification of the technology that each company employs, and examples of the commercial devices produced are given. The list of the models from each producer is not exhaustive, but comprises only a few examples.

3.4 Typical Application of "Electronic Nose" Technologies

The possibility of exploiting different sensor technologies with different specificity makes electronic nose concepts useful in a very wide variety of applications, especially those where odor analysis is important.

Its use in food applications has been extensively examined. It has been tested for inspecting quality and freshness of the food, monitoring manufacturing and cooking processes, investigating shelf life, and evaluating packaging contaminants. The Fox 4000, a commercial nose comprising 18 MOS and produced by Alpha MOS, France, has been demonstrated to be able to detect bacterial growth in milk and shelf life. PEN2, a ten MOS sensor device, was used by Gomez *et al.* [36–38] for detecting volatile compounds emitted from unripe, half-ripe and full-ripe tomatoes. The device was able to discriminate the differences in maturity of the samples with 100% of correct classification.

Many applications of electronic noses in medicine, environment, and other fields have been reviewed [110].

Environmental applications of e-noses are increasing in recent years because of the consciousness of the effect of the pollution on human health [109, 113]. Such technology, with its simplicity and rapidity, can be successfully employed for indoor measurements in workplaces, automobiles, and housing buildings; for outdoor environmental control, in order to evaluate the pollution level; for measurements of exhaust gas streams at the source; and for evaluating the water pollution.

A real outcome in this field is represented by Multisensor Systems Ltd., United Kingdom (http://www.multisensor.co.uk), a company that was born at the University of Manchester, United Kingdom. It produces and sells instruments based on MOS and polymeric sensors that are used for detecting VOCs in clean water and wastewater, air, and land.

The introduction of the electronic noses in medicine, as early and noninvasive diagnostic tool, could be a real challenge. It is very well known that the detection of VOCs in human biological samples could allow a nonpainful diagnosis. So far, this route has not been practicable with standard analytical instruments, as GC or GC/MS, because of their high cost and the long time needed for such analysis. The electronic nose can be potentially used for detecting VOCs from the headspace of human fluids, resulting in a fast and noninvasive diagnosis [108, 111, 112, 114]. Moreover, it could be very useful in those illnesses for which markers are not yet known. A large variety of clinical studies have been done in the last 10 years, and the capability of the e-noses to diagnose human diseases has been demonstrated [24–26, 30, 32].

3.5 A Comparison between Artificial and the Biological Olfaction Systems

Despite the success of this technology in odor analysis, the performance of today's artificial systems is far from that found in biological olfaction. The better performance of the latter is due to the great number of receptor neurons and to the unique architecture of the olfactory pathway, where three main elements—the olfactory epithelium, the olfactory bulb, and the olfactory cortex—are completely integrated. So far, the electronic noses that have been developed are not able to mimic two important features: redundancy and synergistic behavior.

In the artificial olfactory devices, similarly to the brain, the signals coming from the sensors are processed in order to classify odors. However, this process is carried out using mathematical tools. An initial sensor signal preprocessing is often performed. Different preprocessing metrics can be used with different aims. For example, the time dependence of the signal can be removed in order to reduce the amount of data that needs to be handled during the phase of PARC. Linearization of the sensor response can be performed when the intensity of the odors is important, considering that the sensor response is in general nonlinear with increasing concentration of analyte. Data reduction techniques, such as principal component analysis (PCA), imitating the massive convergence of the ORNs to the glomeruli, can also be applied to the whole sensor array response, in order to reduce the data dimensionality.

The data resulting from this first processing step or coming directly from the sensors without any pretreatment are subjected to PARC analysis to obtain discrimination and/or classification. During this phase, the sensor patterns are compared to known odor patterns, which have been stored in a knowledge base during a previous training stage of the device, and, using any of the many classification techniques available, identification of the stimulus is performed.

It is evident that the whole procedure is based on mathematical processing of the signal and, contrary to the biological olfaction, no synergistic behavior is exhibited by current "electronic noses."

Moreover, the existing sensor arrays implement the broad and overlapping sensitivity feature exhibited by the receptors, but it has not yet been possible to achieve the same dimensionality and redundancy level available in the olfactory epithelium. Current chemical sensor arrays, either homogeneous or heterogeneous, have few sensing elements when compared to natural noses. Many attempts to increase the sensor number and their diversity have been made during recent years, but there are many limiting factors that have not yet permitted the achievement of this objective. In fact, a large array of electronic sensors leads to an unmanageable number of electrical connections; to high power consumption; to long response times, which is something inadmissible in applications of artificial olfaction; to large instruments; and, as a consequence of all the previous factors, to the reduction of the portability of the system. A recent publication [51] reported the fabrication of a device containing three sensor arrays of 300 elements using 24 different sensitive materials, still far from the dimensionality and the variety of ORNs. At this time, a large-scale chemical sensor array operating directly in the electrical domain does not exist.

We can say that there is a parallelism between human and artificial olfaction, but the performance of the biological system is superior to that of any existing artificial device as it has not been so far possible to make an apparatus with the same receptors' features and able to perform the same information processing carried out in the brain.

Currently, different computational models have been built in recent years to mimic biological olfaction signal processing. However, without realistic sensor data inputs, such models can neither be benchmarked nor completely exploited to build devices with performances comparable with those of the biological systems. Hence, the development of an "artificial epithelium" capable of generating real sensory inputs is really necessary.

3.6 A Large-Scale Sensor Array

In the development of biomimetic chemical sensors, it is beneficial to understand how the biological model works and how an animal's brain analyzes the information provided by these sensory inputs. In the last few decades, great progress has been achieved in the understanding of both the anatomical structure and the signal processing of the olfactory system. This, together with new and more precise technologies available for the fabrication of chemical sensors, leads to new opportunities for mimicking such a system, based on biologically inspired computational solutions and very large arrays of sensors. The basic idea was to find inspiration in natural olfactory systems and convert this into biologically meaningful questions that are helpful for technological development of artificial sensing systems. We are involved in the development of signal and data processing systems inspired by the neuronal processing of the biological olfactory pathway. Our ideas were to develop a large array of polymeric sensors arranged in a matrix configuration mimicking the olfactory epithelium and interfaced with computational models of the main building blocks of the olfactory system: the olfactory receptor layer, olfactory bulb, and olfactory cortex.

The objective was achieved by fabricating 16384 sensing elements with 24 different sensing materials deposited onto interdigitated electrodes having different geometries. A custom-made scanning circuit was designed and built for the interrogation of the sensors. This circuit is able to acquire data from up to 65536 sensors, arranged in 16 modules, each composed of 4096 elements. The arrays are read in parallel for reducing the data acquisition time. Conducting polymers were selected as sensing materials, as they have several attractive features. Conducting organic polymer sensors are chemiresistors that have rapid adsorption and desorption kinetics at room temperature and have high sensitivities, typically able to detect volatile chemicals at low concentrations, especially if these analytes are polar [81]. Conducting polymer sensors are of the reversible type and no heater elements are required; therefore, the sensors operate with low power consumption. They can be intrinsic or extrinsic. Here we deal with intrinsic conducting polymers that are doped with an electron-donating (p-type) or electron-withdrawing (n-type) species. Moreover, there are many monomers that may be used to make conducting polymers, mainly pyrrole, thiophene, aniline, and their derivatives, but they all have in common a conjugated system of π-electrons along their polymer backbone. They are easy to develop and various methods of manufacture can be used. Specificity of the sensors toward different classes of compounds can be closely tailored by altering the functional group chemistry of the sensors.

3.6.1 Conducting Polymers

ICPs can show low ionization potentials, the energy required to remove an electron (easily oxidized), and high electron affinities, the energy gained from the addition of an electron (easily reduced). Hence, conducting polymers can be more easily oxidized or reduced than conventional polymers. Most conventional polymers have full valence bands and empty conduction bands, which are separated from each other by a wide energy gap. In contrast, conjugated polymers have narrower band gaps, and doping can change their band structures by either taking electrons from the valence band (p-doping) or adding electrons to the conduction band (n-doping). In the case of polypyrrole, for instance, the backbone is neutral in the reduced state and positively charged in the oxidized state. The doping of conducting polymers is mainly performed chemically or electrochemically. A dopant is required to have a sufficient oxidizing or reducing ability. Oxidative doping is called p-type and reductive doping, n-type. The redox reactions can be generalized as follows:

$$\begin{aligned} &(\mathrm{CH})_n + \text{oxidant}(\mathrm{CH})_n^{+} + \text{oxidant derivative} \\ &(\mathrm{CH})_n + \text{reductant}(\mathrm{CH})_n^{-} + \text{reducing derivative} \end{aligned} \tag{3.3}$$

Oxidation of polypyrrole involves the removal of an electron resulting in a conjugated radical ion which is delocalized over a section of the polymer. The defect is called a polaron and has a positive charge and a spin of ½. The polaron achieves stability by polarizing the surrounding environment. The polaron may be further oxidized to form a bipolaron, or oxidation may occur on another neutral polymer unit to form another polaron, but it is more energetically favorable for bipolaron formation. Polarons and bipolarons are mobile across the polymer chain via rearrangement [85, 86]. The positive charges are considered the mobile charge carriers, and their concentration and mobility determine the polymer's conductivity, σ, as defined in the following equation:

$$\sigma = en\mu \quad (3.4)$$

where e = charge of an electron = 1.6022×10^{-19} C
n = charge carrier concentration (cm^{-3})
μ = carrier mobility ($cm^2V^{-1}s^{-1}$) = limiting factor of conductivity

Apart from the effect of dopants, the conduction process can be dictated also by hopping mechanisms. Hopping is intrachain, interchain, and interparticle motion of charges in a polymer matrix. The intrachain movement or motion of the charge carrier through a single chain depends on the efficient conjugation. Interchain movement, jumping from one chain to another, is determined by the geometry of the stacking or ordering of the polymer chains. The mobility also depends on the movement of electrical charges from particle to particle. Variable-range hopping process in which electrons are delocalized on monomer rings is a well-described conductivity mechanism observed as nearest-neighbor hopping and variable-range hopping [85].

Besides the doping process described previously, a secondary doping can happen between the primary-doped polymer and a second dopant. Such process has been proposed to explain why the conductivity of conducting polymers is changed by organic vapors. MacDiarmid *et al.* [68, 69] reported that electrically neutral analyte gases act as secondary dopant and change the polymer's properties. These changes are believed to be due to the alteration of the molecular conformation of the polymer from compact to expanded coils. The amount of charge involved in the process and its distribution within the polymer are governed by the analyte's electron affinity and the polymer's electron affinity, which is given by its work function. Janata and Josowicz [58] reported that several possible interactions in a polymer chemiresistor can happen between a gaseous analyte and the sensing material that can account for the signal modulation read from a sensor. The various sites of interaction and the processes are summarized here.

Partial electron transfer can occur between a conducting polymer and a secondary dopant (gas), which can act as a Lewis acid or base, altering the number of the charge carriers in the polymer backbone and, therefore, its conductivity.

It has been also proved that electron acceptors, such as NO_2 and I_2, are able to remove electrons from the aromatic rings of the polymers, increasing the conductivity of *p*-type materials and decreasing the conductivity of *n*-type materials. On the contrary, electron donors generate an increase in the conductivity of the *n*-types and a decrease in the conductivity of *p*-type polymers. Many different kinds of polymeric sensors are based on this process [6, 8, 59].

While in the previous interaction mechanisms the conductivity modulation of the polymer is explained by a change in the number of the charges, a different interaction between the analyte and the polymer can happen in the bulk of the polymer, causing a modification in the conductivity due to the alteration of the bulk mobility of carriers. The bulk mobility change results from conformational changes of the polymer backbone. Due to the slow penetration of gases into the conducting polymer, the responses occurring from change in bulk mobility create long response times which are often accompanied by hysteresis.

The affinity of a conducting polymer sensor to a volatile analyte can also be increased by introducing specific binding sites [58]. In fact, electron transfer can happen also on a component mixed with the polymer, on the counterion introduced during the doping process, and on the side chains added to the main ring of the polymer. The introduction of catalysts in the polymer can be used to make sensors able to detect inert gases. Hence, it is possible to create a variety of materials that have different chemical sensing properties dependent on the type of dopant used, degree of doping, type of side chains added, and materials combined to the polymer.

For our project, we were able to tune almost 30 different sensing materials. Of these, 24 were deposited on the four substrates designed for the large array and are listed in Table 3.3.

3.6.2 Sensor Interrogation Strategy

Obtaining signals from a large-scale array where small changes in electrical resistance are to be measured is not easy. The large number of chemiresistors would lead to an unmanageable number of electrical connections. To overcome the problem, it was considered more convenient to arrange the elements in a matrix configuration where each sensor shares one terminal

Table 3.3 List of the polymers deposited onto the sensor elements

Class number	Deposited polymer	Class number	Deposited polymer
1	PANI/CB	14	PANI/CSA/PVP
2	PEDOT	15	PANI/pTSA/PS
3	PANI long chain	16	PANI/pTSA
4	PPY doped	17	PANI/CSA/PEO
5	PDDT/$FeCl_3$	18	PANI/pTSA/PEO
6	PANIPOL	19	PANI/CSA/PS
7	P3HT	20	PANI/PTSA/PVP
8	PPY/CB	21	PANI/CSA/PMMA
9	PDDT/$FeCl_3$ filtered	22	PANI/DBSA/PS
10	PANI/CSA	23	PANI/DBSA/PMMA
11	P3DT/$FeCl_3$	24	PANI/DBSA/PEO
12	P3OT/$FeCl_3$	25	ADS406PT/$FeCl_3$
13	P3BT/$FeCl_3$		

Twenty-four polymers were synthesized and tested. In the list there are 25 materials, as the material numbered as 9 is the same as that numbered 5, but filtered with a 0.2 μm filter. All the polymers synthesized belong to the three main families of ICPs commonly used in the sensor field: polyaniline, polypyrrole, and polythiophene. Their sensitivities and selectivities were tuned by adding side chains to the main polymer ring using different dopant species and preparing copolymers.

with all belonging to the same row and the second terminal with elements belonging to the same column. Hence, N sensors, assembled in an $n \times m$ matrix, require $n+m$ wiring connections, instead of $2N$. Each component in the array can be individually addressed by using two multiplexers, one for selecting the appropriate column of the matrix and the other one for selecting the row. The adoption of this matrix configuration is not without consequence. It causes electrical crosstalk between the elements in the array, making the response of any single sensor dependent on the behavior of all other sensors in the net. Several techniques to read resistive arrays, while compensating for the crosstalk arising between adjacent elements, have been described in the literature [23, 87]. All these approaches differ in terms of performance, reading error, crosstalk, circuit complexity, and sensitivity to parameters of multiplexers and to other components of the electronic circuits [100]. The simplified scheme of the readout circuit adopted within the project to read the large array is shown in Figure 3.4. It uses buffers to feed back the output voltage to the unselected rows, while the unselected columns are floating and the scanning circuit behaves as an ideal voltage divider, where one sensor at a time is sequentially connected to the output node and to the ground. The equivalent voltage divider circuit is shown in Figure 3.3 for the sensor resistance R_{ij}. In the scheme the parasitic capacitance, C_p, which acts in parallel to R_{ij} is also shown.

Another issue we dealt with was the time needed to acquire the sensor data. In the case of large numbers of sensors, this can be very long and, therefore, make the instrument unusable for real-life applications. In our case, the sensor sampling rate resulted mainly depending on the circuit transient time constant, τ, which is directly proportional to the equivalent parasitic capacitance, C_p, and to $R_L//R_{ij}$, $\tau = C_p * (R_L//R_{ij})$ (see Figure 3.3). To reduce the transient τ, the parasitic capacitance or the value of the sensor resistance can be reduced. Constraints on the

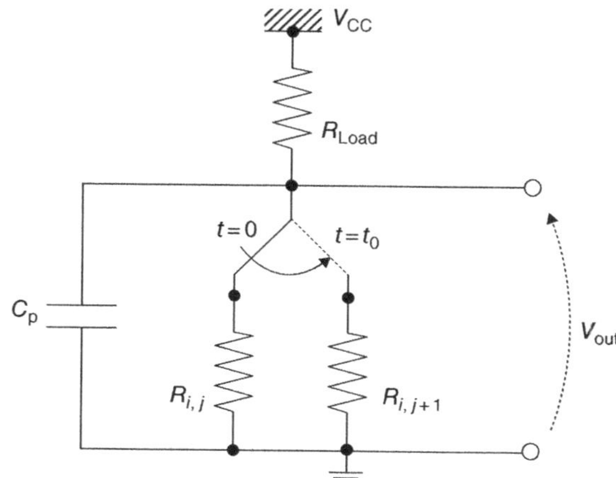

Figure 3.3 Ideal behavior of the scanning circuit when no unwanted resistive paths exist in parallel to the element being read, R_{ij}. At $t=0$, the circuit in Figure 3.4 selects the ij sensor element. Hence, R_{ij} has a terminal connected to the load resistance, R_{LOAD}, and to the voltage V_{cc}, while the other terminal is connected to the ground. C_p is the parasitic capacitance that acts in parallel to R_{ij}, increasing the time needed to the sensor to reach a steady state. All of the other elements in the array are disconnected. The V_{out} is only due to the current flowing through R_{ij}. When the R_{ij} has been acquired, at $t=t_0$, the next element read is the $R_{i,j+1}$, as the array is scanned row by row

weight and the dimensions of the final device did not allow reduction of this time by increasing the level of interdigitation of the metal electrodes, which would have reduced the resistance read. Hence, the reduction of the parasitic capacitance was the easiest and efficient way to increase the sensor sampling rate. As C_p is mainly due to the multiplexer and is directly proportional to the number of the multiplexed lines, a modular arrangement of the sensors was adopted. The set of $N = 16384$ sensors was subdivided into four smaller arrays of 64×64 elements each, and an identical acquisition system was built for each array. Hence, four acquisition modules capable of reading the smaller arrays at the same time were fabricated and implemented into a device. Each acquisition module has a logic unit, a field-programmable gate array (FPGA), which is programmed to carry out the algorithm for scanning the array sequentially. This modular solution reduces the time needed for acquiring data from the entire sensor array because of the reduction of the parasitic capacitance of the single acquisition unit, due to the smaller dimension of the multiplexers needed for selecting the sensor and because of the parallelization of the acquisition. In this way, each array of sensors can be scanned and read in less than a second, and, because of the parallelization, any N number of arrays implemented into the system required the same time to be read.

In order to allow the array to operate as close as possible to the optimum electrical parameters, each sensor should be also adapted to the load resistor used to read it. In this condition, the maximum transfer of the active power is achieved, and the S/N ratio is increased. When the number of the sensors is large, it is impossible to adapt all the sensors. In our case, each FPGA in the acquisition module stores 4096 R_{LOAD} values. They are the best load resistances calculated for each sensor, by choosing among seven metallic film resistors having fixed values ranging from 1 kΩ to 1 MΩ, placed on each sensor module and decoded with a four-bit binary code. The evaluation and the memorization of the best R_{LOAD} are performed just one time, at the start-up of the device, during an initial phase called "calibration." At the end of the calibration process, the FPGA stores 4096 4-bit codes in its internal RAM. These data are then used during the scanning of the sensor array, when the R_{LOAD} value calculated for the sensor under measurement is connected, by the R_{LOAD} analogue multiplexer and the column multiplexer, to the sensing element (Figure 3.4). As the calibration is performed just once before the acquisition of the array, it does not result in an increased acquisition time.

3.6.3 Sensor Substrate

One of the interesting aspects of the biological olfactory system is that receptors of the same type often show significant differences in the dynamic range of the responses produced to different concentrations of the same odorant. This may be due to differences in age of different receptor neurons—which are renewed constantly in the biological system. It has been conjectured that this variability may significantly add information content to the patterns of responses generated across the entire olfactory mucosa [44–47, 60, 88]. In our case we imposed significant variability on to the arrays (i) by varying the electrode spacing (which affects the linear-to-nonlinear ratio of the voltage divider and the diffusion of the analyte) and (ii) by varying the form factor (which affects the number of "squares" in parallel) so to accommodate a larger number of sensing polymers with widely varying range of polymer resistivity.

The sensor dies were designed to contain 4096 sensor elements in an area of 2.1 cm^2—with a range of sensor geometries and zones as shown in Figure 3.5.

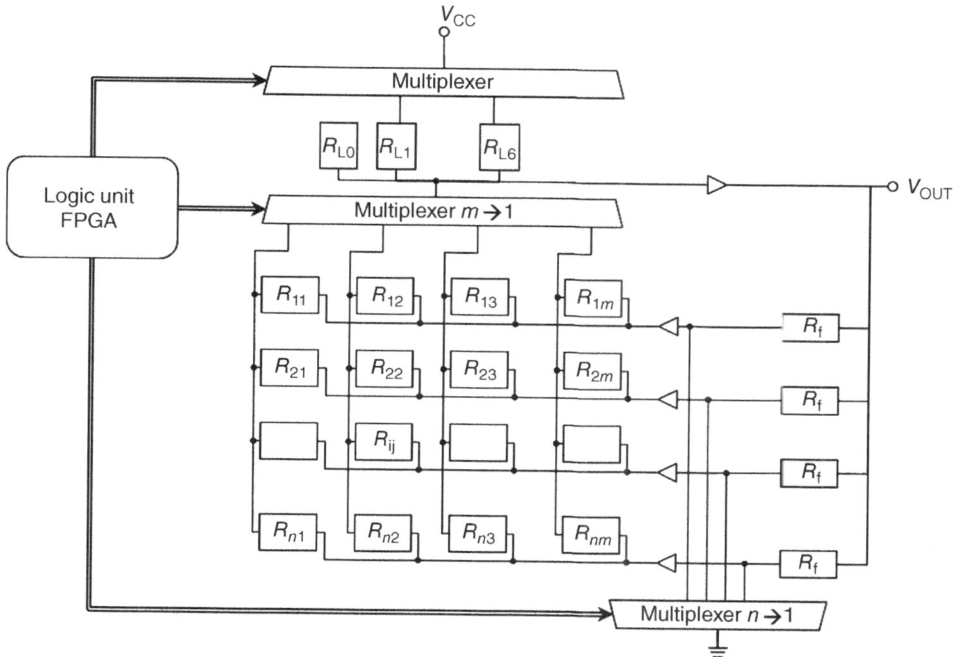

Figure 3.4 The picture shows the matrix arrangement of the chemiresistors in the array, together with the simplified scheme of the electronic readout circuit developed for scanning and reading the array. The designed circuit uses buffers to feed back the output voltage to the unselected rows, while the unselected columns are floating and the scanning circuit behaves as an ideal voltage divider, where one sensor at a time is sequentially connected to the output node and to the ground. The multiplexer on the top select the best load resistance for each sensor, choosing among seven metallic film resistors. The *m-to-*1 multiplexer selects the column of the sensor which is read, while the *n-to-*1 multiplexer selects its row. When the element R_{ij} is acquired, the column multiplexer selects the *j*th column and connects it to the load resistance addressed by the load multiplexer and then to power. The row multiplexer connects the *i*th row to the ground, while all the other nonselected rows are connected to the output voltage by means of feedback buffers. Hence, the selected sensor is correctly powered, while no current can pass through the nonselected resistances, because of the zero voltage difference across them. This is illustrated in the figure by considering the semiselected element R_{22}. It has one terminal virtually connected at V_{out} by means of the feedback line and the first terminal connected at V_{out} through the column multiplexer. Hence, no current can pass through R_{22}, and the entire parasitic path is disconnected from the sensor read, R_{ij}

Gap sizes between the interdigitated electrode transducers ranged from 10 to 40 μm and sensor element areas ranged from 7 200 to 54 400 μm^2. In Table 3.4, the electrical and geometrical characteristics of the sensors fabricated on each sensor module are reported.

The sensor contacts were made using anisotropic bonding with an electrically conducting tape to a printed circuit board. This board routes the connections from the sensor die to a second board that implements the scanning algorithm for reading the electrical resistance of the sensors, as shown in Figure 3.4. A PTFE chamber for delivery of vapors to the array was clamped firmly to the board as shown in Figure 3.6. Details of the sensor dies and mounting

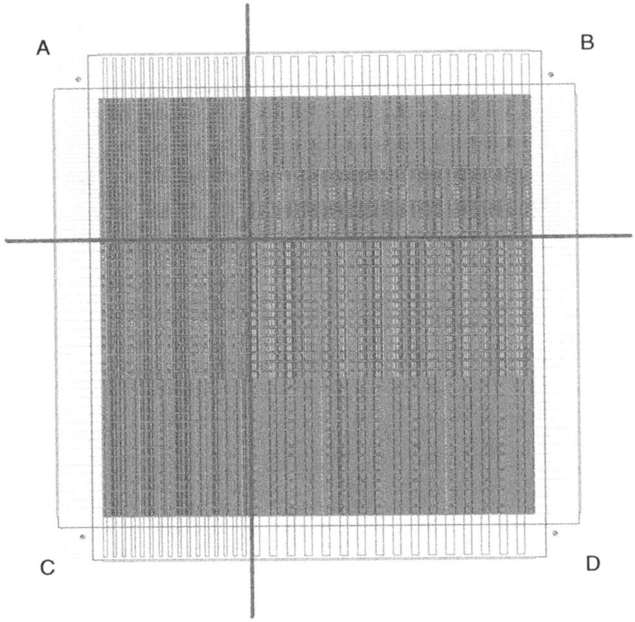

Figure 3.5 Design of the 64×64 sensor die. The array is divided in four regions (indicated from A to D). Another two regions can be identified within B, C, and D, bringing to seven the total regions the die is subdivided into. Each region encompasses different interdigitated geometries. The size of the die is 2.1 cm×2.1 cm

are given in previous publications [10–12]. On a silicon wafer, the sensor electrodes of Cr/Au were deposited to form a matrix by vapor deposition methods using photolithographic masks to the individual electrode geometries. SU-8 photoresist was used to ensure electrical isolation between metal layers deposited. In Figure 3.6 the sensor submodule completely assembled is shown.

At the end of our study, using the approaches detailed previously, we were able to build an innovative modular instrument capable of hosting and acquiring data from 65 536 resistive chemical sensors by reading in parallel 16 sensor arrays, each composed of 4096 elements. The scanning circuit developed for reading each array is able to compensate for the crosstalk introduced by the matrix arrangement of the sensors and permits the sensing elements to always operate under optimum conditions. Independent power supplies are provided to each scanning circuit, so that the 16 arrays are completely independent from each other. This provides the system of a good stability and high level of versatility, as any number of sensors between zero and 65 536 in blocks of 4096 can be read. The modular approach also makes possible to change or replace a single sensor array without interfering with the others, making the system more robust.

At the moment four sensor arrays are connected into the artificial olfactory system, giving a total of 16 384 sensors consisting of 24 different types. The device has been provided with similar features shown by ORNs: broad and overlapping specificity to chemical species, different

Table 3.4 List of the different geometries of the interdigitated transducers implemented into the sensor array

Type	Element size (W×H)	# of elements	Parallel squares between electrodes	Active surface between electrodes (μm^2)	Electrode gap (μm)
1	220 μm × 220 μm	1024	18	7200	20
2	420 μm × 220 μm	256	15	24000	40
3	420 μm × 220 μm	512	48	19200	20
4	420 μm × 220 μm	256	204	20400	10
5	220 μm × 420 μm	256	15	24000	40
6	220 μm × 420 μm	512	48	19200	20
7	220 μm × 420 μm	256	224	22400	10
8	420 μm × 420 μm	256	40	64000	40
9	420 μm × 420 μm	512	128	51200	20
10	420 μm × 420 μm	256	544	54400	10

Ten different interdigitated geometries were realized. On the second column of the table, the dimensions of each electrode type are reported. The third column shows the number of the parallel squares resulting in-between column and row electrodes. The fourth column reports the active surface for each transducer geometry. In the last column of the table, the gap size for each type of the interdigitated electrodes is pointed out. This variability of the interdigitated transducer was introduced in order to increase the sensor dynamic range variability. The latter contributes to improve the performance of the final sensor system as it enriches the information which is useful in data analysis phase.

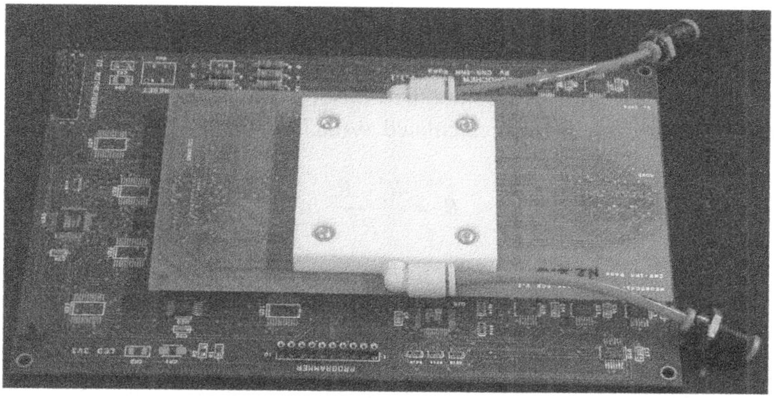

Figure 3.6 Picture of the sensor submodule completed assembly. The PTFE measurement chamber encloses the sensor array, which is connected by means of an anisotropic conducting tape to a first board. This is a simple board that routes the connections from the sensor die to a second bigger board, clearly visible at the bottom of the module. This last board implements the scanning algorithm for reading the electrical resistance of the 4096 sensors

dynamic responses, and a high level of redundancy. This should allow it to exhibit unique performance, never seen before in any other artificial system. To test the system, laboratory experiments have been carried out and some results are presented in the following sections.

3.7 Characterization of the Large-Scale Sensor Array

The large-scale sensor array was built with the aim of mimicking the ORNs and generating real sensory inputs to be analyzed by algorithms directly abstracted from neuroscience models. Hence, immediately after its fabrication, laboratory experiments were started with the objective to test the functionality of the electronic circuit developed for reading it; verify that it had the desired features; assess the capability of the redundant sensor system in terms of odor classification and quantification, background suppression, and segmentation; evaluate the advantages obtained by using a large number of sensors of multiple types; and produce data to validate the neuromorphic models of olfaction.

A chemical analyte delivery system based on permeation tubes and bubblers coupled to mass flow controllers (MFC) for vapor dilution was constructed for rigorous testing of the sensor array. It allowed the delivery of analytes at concentrations between 1 ppm and several thousands of parts per million as well as binary mixtures over a range of different concentrations. Humidity and temperature sensors enabled the monitoring of these physical parameters that may affect the sensor response. Permeation tubes were used for generating low concentrations of gases and bubblers for high concentrations.

The delivery module was utilized for performing measurements with four sensor arrays, which were fed simultaneously, by generating four identical parallel analyte streams. To realize this, the flow leaving the delivery system was separated into two streams passing through two identical paths. Each of them was subsequently divided in another two indistinguishable paths so that four identical paths were achieved and the main analyte flow resulted split into four identical flows delivered to the four sensor arrays.

The sensor data acquired by interrogating the large array were preprocessed before analysis. First of all, the sensor signals were filtered by means of a moving average filter with a 30 s window. This process produced a smoother sensor signal, reducing its fluctuations. Then, the sensor response to an analyte was calculated as the fractional variation of its resistance, R_r:

$$R_r = \frac{R_s - R_0}{R_0} \tag{3.5}$$

where R_0 is the value of the sensor resistance at the baseline, while R_S is the resistance at steady state, which is at end of the measurement, when the sensor response reaches the steady state after exposure to the analyte. Finally, autoscaling was performed in order to make all the sensors equally weighted. On the preprocessed data, standard techniques of chemometric analysis, in particular PCA and partial least squares regression analysis (PLSRA), were performed with the aim of testing the improved capability of the large array of sensors. In order to understand the better ability of this unique array in respect to the traditional systems, the sensor data need to be analyzed with standard data processing techniques. The introduction of new neuromorphic data processing models itself should improve the sensor performance, while we want to take into account only the advantages introduced by the features that we have given to the sensor array. Some of the data collected have also been processed by neuromorphic models of the olfactory system developed within a European project, NEUROCHEM. The results of this analysis have been already published [83].

Several experiments with pure analytes are ongoing. The data collected by these measurements are used to evaluate the classification and quantification capabilities of the array and to determine the range of concentrations detectable, exposure time, and recovery time for the different analytes. Studies on binary mixtures are also in progress to assess the ability of the sensors in terms of segmentation and background suppression.

An investigation is also going on to prove that the variability introduced by fabricating transducers with different geometries and by using different polymers, dopants, and copolymers provides the array with high redundancy level and permits to the sensors having same sensing material but different transducer to originate different sensor dynamic ranges or transient responses.

In the following section, the outcomes of some of these studies are presented and discussed.

3.7.1 Pure Analyte Study: Classification and Quantification Capability

The data collected by exposing the sensors to pure analytes were analyzed to assess the discrimination and quantification capabilities of the sensing device.

The results here presented are related to two kinds of fatty acids, butyric acid and acetic acid. These are only selected results from a larger research program aimed at characterizing the sensor system with respect to carboxylic acids. Carboxylic acids find great use in different fields. They are employed for producing chemical compounds, pharmaceuticals, and food additives. Hence, thinking of possible applications of our device, it has been decided to characterize the responses to these acids.

Tables 3.5 and 3.6 detail, respectively, the concentrations of butyric acid and acetic acid generated and delivered to the sensors for such experiments. In the same tables, the numbers of repetitions carried out for each concentration are reported. One of the datasets for 196.1 ppm of butyric acid was not used in the analysis, as problems during the measurements generated an outlier.

The data collected by these experiments were preprocessed as described previously. Then, an initial PCA was carried out on the preprocessed features. The score plots produced are shown in Figure 3.7 for butyric acid and Figure 3.8 for acetic acid.

In the score plot regarding the butyric acid, it is clear that the discrimination of the different concentrations happens through the first principal component. In particular, along the PC1 direction it is possible to individuate three main clusters: the first one is composed of the classes corresponding to 23.0, 49.0, and 74.8 ppm; the second one is determined by the concentrations ranging between 107 and 165.7 ppm; and the last one comprises the highest concentrations, which are 196 and 226 ppm. Hence, from this initial exploratory analysis, it seems that the system is not able to discriminate concentrations of butyric acid lower than 100 ppm.

Looking at the score plot of the acetic acid, we can observe that also in this case, the first principal component is the most significant for distinguishing the different concentrations.

To verify the PCA results, partial least squares (PLS) regression models were built for both the analytes. Considering the large number of sensors used to build the model, overfitting

Table 3.5 Measurement protocol for butyric acid

Butyric acid (equilibrium vapor pressure)		
Concentrations (ppm)	Number of repetitions	Symbol
23.0	5	Down triangle
49.0	5	Asterisk
74.8	5	Square
107.7	5	Cross
134.6	5	Diamond
165.7	5	Up triangle
196.1	5	Star
226.4	5	Circle

The concentrations of butyric acid delivered to the sensor array are specified on the first column. The percentage error associated with the concentrations was calculated to be 8%. The second column reports the number of replicates for each concentration. The third column depicts the corresponding symbol used in Figure 3.7 to indicate each concentration.

Table 3.6 Measurement protocol for acetic acid

Acetic acid (equilibrium vapor pressure)		
Concentrations (ppm)	Number of repetitions	Symbol
119.6	9	Down triangle
696.5	18	Square
2167.8	9	Cross
2876.0	8	Diamond
3605.3	9	Asterisk

The concentrations of acetic acid delivered to the sensor array are specified on the first column. The percentage error associated to the concentrations was calculated to be 8%. The second column reports the number of replicates for each concentration. The third column depicts the corresponding symbol used in Figure 3.8 to indicate each concentration.

could arise if the number of latent variables (LVs) is not chosen accurately. In fact, by increasing the number of the LVs, a better fit of the model to the data is achieved. However, the prediction capability of the model itself is reduced, as it is too specific for the data analyzed and it is not able to predict new data. In order to avoid overfitting, the number of LVs to be used for generating the model was optimized with a cross-validation technique. These techniques perform a partition of the dataset available into two subsets, a training set or calibration set and a validation set. The first group is used for a training step during which the model is generated, and the second group is used for testing the model. The partitioning of the dataset is iteratively performed, changing, at each iterative step, the data composing the two subsets. In this way, all the data are used for both training and validation phases. The validity of the

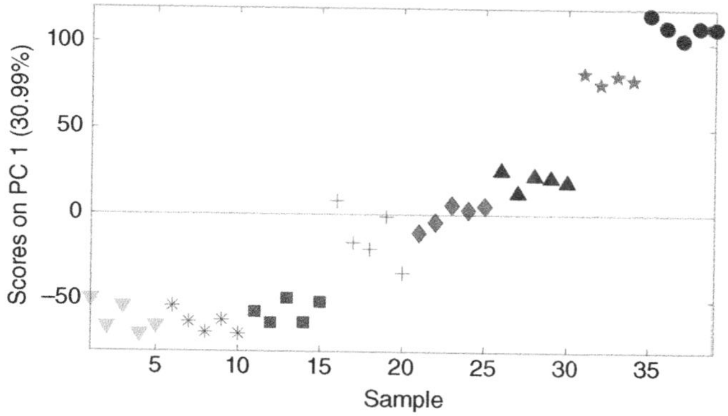

Figure 3.7 Score plot for butyric acid data. An initial exploratory principal component analysis was performed on the data collected by delivering to the sensors butyric acid vapors at the different concentrations listed in the Table 3.5. The score plot resulting from this analysis is reported in this picture. The first principal component is the only one represented in the plot, as it is evident that the information regarding the different concentrations of the butyric acid is given by only this component. It seems that concentrations lower than 100 ppm are not discriminated. In fact, the first three concentrations analyzed are not discriminated at all

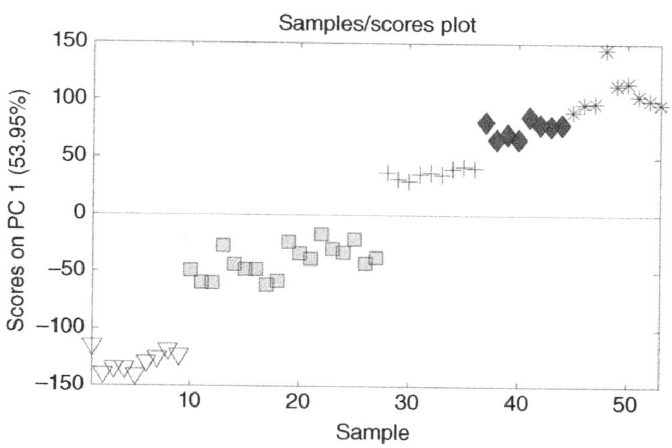

Figure 3.8 Score plot for acetic acid data. This picture shows the score plot achieved by applying the principal component analysis on the data collected by exposing the sensors to acetic acid vapors. The concentrations tested are listed on the table above. The first principal component contains the information regarding the different concentrations of the acetic acid

created model is evaluated using two kinds of error—the Root Mean Square Error of Calibration (RMSEC), which is the prediction error performed on the calibration set, and the Root Mean Square Error of Cross-Validation (RMSECV), calculated on the validation set. These two errors take into account the error performed by the model in predicting a data belonging firstly to the calibration set and secondly to the training set. They are calculated with the following formulae:

$$\text{RMSEC} = \sqrt{\sum_i \frac{\left(y_i^{\text{LS}} - y_i\right)^2}{N}} \tag{3.6}$$

$$\text{RMSECV} = \sqrt{\sum_i \frac{\left(y_i^{\text{LS}} - y_i\right)^2}{K}} \tag{3.7}$$

where y_i^{LS} is the predicted value of y, K is the dimension of the validation set, and N is the dimension of the calibration set. The RMSEC decreases with the increasing of the number of LVs used for building the model, indicating a better fit of the model to the experimental data. The RMSECV decreases for the first LVs; then it reaches a minimum and it starts to increase. The minimum of the RMSECV curve represents the trade-off between the ability of the model to adapt to the experimental data and to predict new data.

Hence, the number of the LVs that minimizes the RMSECV is the optimum for use in model building, avoiding the overfitting. The cross-validation technique used in the present work is the leave-one-out approach. In this case the validation set is composed of a single element, while the calibration set is composed of $N-1$ elements, where N is the dimension of the whole dataset. The cross-validation is repeated until all data have been used once as the validation set.

The characteristics of the PLS regression models built for the butyric acid and acetic acid are shown in the score plots in Figures 3.9 and 3.10, respectively.

The butyric acid model was optimized with seven LVs and exhibits a very good linear correlation between the predicted analyte concentration and the measured one. The errors associated to the model are quite low, 1.2 ppm in calibration and 4.8 ppm in validation.

The PLS analysis of the butyric acid data partially confirms the PCA results in terms of limit of detection (LOD). In fact, the prediction values of the concentrations of 23.0 and 49.0 ppm are practically in the same range, meaning that there is no discrimination. Regarding the third class, 74.8 ppm, it partially overlays in the same range of the first two classes, while in the previous PCA there was entire overlap. The fourth class, 107.7 ppm, is completely separated from the first three classes. Hence, considering also the errors, we can estimate an LOD for butyric acid below 100 ppm.

The acetic acid model was built using six LVs and the RMS errors associated to the model are 45.0 ppm in calibration and 132.1 ppm in validation. The model shows a good linear correlation between the predicted and measured analyte concentrations. In the case of the acetic acid, the concentrations analyzed do not allow to have a good estimation of the LOD. Considering the RMSECV, the spread of the first class of acetic acid, and the linearity behavior of the model, we can expect an LOD around the first concentration tested, which is 120 ppm. To have a better estimation of the LOD, new datasets at concentrations lower and around 120 ppm need to be generated. If we consider the RMSEC and RMSECV values, it seems that the performances of the sensor system are worst when dealing with acetic acid, with respect to the butyric acid. However, it needs to be considered that the range of the concentrations generated for the acetic acid is wider than that generated for the butyric acid. At high concentrations the effect of the error in the analyte delivered is greater than that at low concentrations.

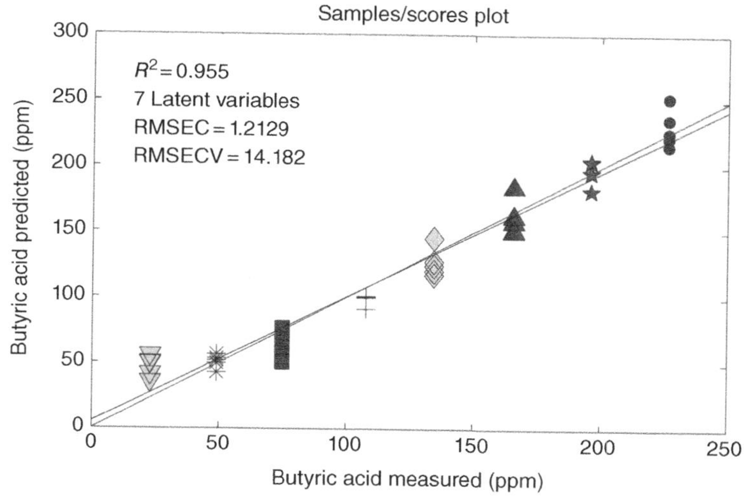

Figure 3.9 A partial least squares (PLS) regression model for predicting different concentrations of butyric acid was built by using seven latent variables (LVs). This is the number of LVs that produces the minimum value of the RMSECV. The errors associated to the model are 1.2 and 4.82 ppm, respectively. The PLS model validates the PCA results and permits to estimate a limit of detection lower than 100 ppm. The model also exhibits a very good linear correlation between the predicted analyte concentration and the measured one

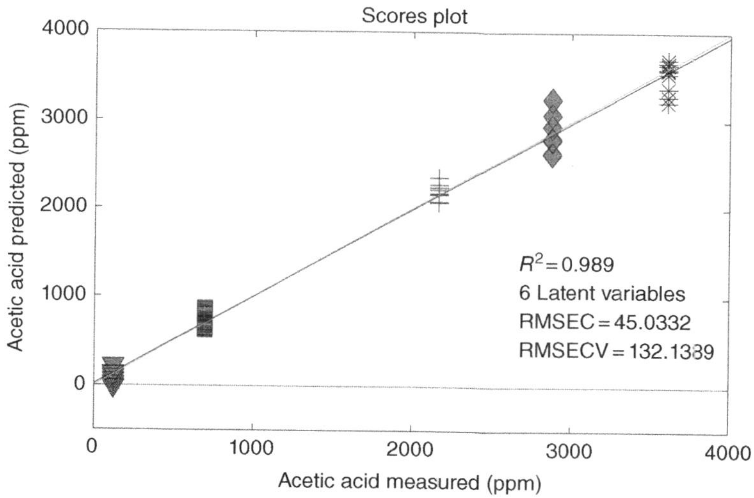

Figure 3.10 The figure shows the score plot achieved with a partial least squares regression model built for predicting the acetic acid concentrations. Six latent variables are used for building the model while minimizing the RMSECV. The RMSEC and RMSECV achieved by the model are 45.03 and 132.13, respectively. The model shows a good linear correlation between the predicted and the measured concentrations

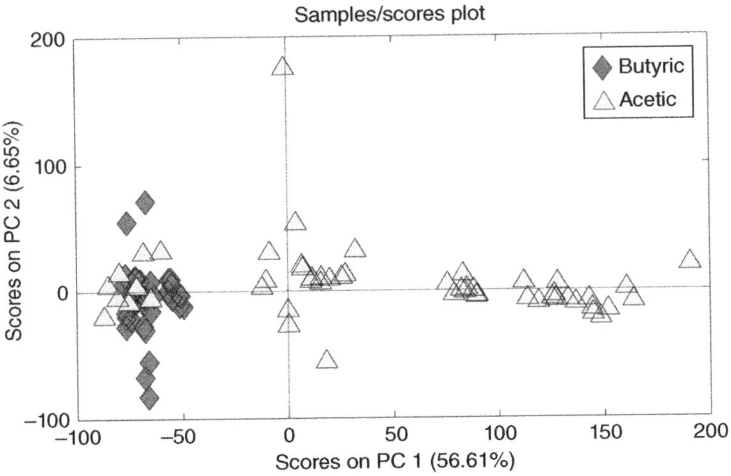

Figure 3.11 This picture shows the score plot achieved by carrying out a principal component analysis on the two datasets created by generating the different concentrations of both butyric and acetic acid vapors. Unfortunately, the analysis does not provide significant results and does not allow us to make meaningful assertions, because of the different concentration ranges of the two analytes. If we look at the scores of the acetic acid, it is clear the direction along which the data are distributed. However, the butyric acid's scores produce a single cluster together with the low concentrations of the acetic acid, and it is not possible to identify any direction for the distribution of the butyric acid

Consequently, the interclass variance is wider at high concentrations and gives a greater contribution to the RMSEC and RMSECV.

Unfortunately, the different concentration ranges of the two analytes tested so far made it rather complicated to carry out an analysis similar to the one performed with the pure analytes, but considering the two datasets simultaneously. The score plots achieved do not have significant results and do not allow meaningful assertions to be made. If we look at the score plot resulting from the PCA (see Figure 3.11), it is clear the direction along which the acetic acid data are distributed. However, the butyric acid's scores produce a single cluster together with the low concentrations of the acetic acid, and it is not possible to identify any direction for the distribution of the butyric acid.

A better result is achieved if we carry out a PLS analysis. The PLS model is shown in Figure 3.12. In this case, even if we cannot discriminate the different concentrations of the butyric acid, it seems that the scores of the butyric acid are going to be distributed along a direction that is almost orthogonal to the one across which the acetic acid data are distributed.

As it has been mentioned previously, these are the first results of a wider research on the fatty acids. Ongoing measurements of butyric and acetic acid are being performed in order to have a complete understanding of the sensor array performance when dealing with these two acids.

We have previously tested another two analytes, ethanol and 2-butanone. In this case, we achieved good results when we processed the data of the pure analytes singly and together. In particular, the PCA score plot indicates that data of the two different analytes lie on directions that are rather orthogonal between them, confirming the capability of the system to clearly discriminate between the two odors. Also, the distance of a single observable point from the origin of the two odor vectors were correlated to the concentration of the analyte, indicating

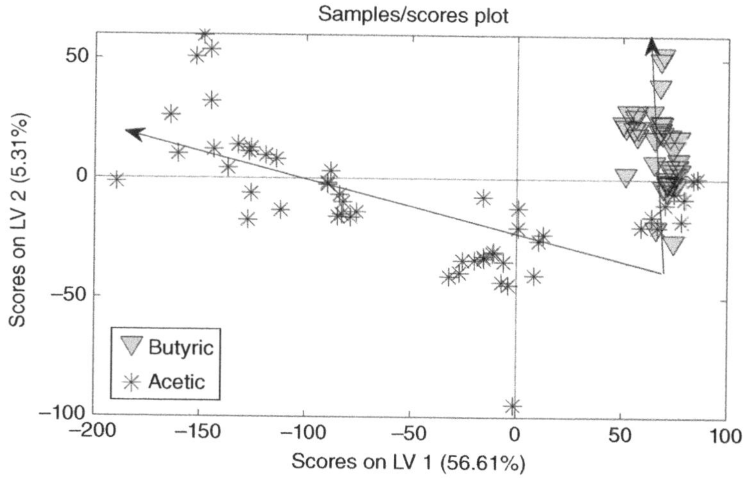

Figure 3.12 The partial least squares analysis performed on the dataset of the butyric together with the acetic acid dataset produced the model shown in this picture. In this case, even if we cannot discriminate the different concentrations of the butyric acid, it seems that the scores of the butyric are going to be distributed along a direction that is almost orthogonal to the one across which the acetic acid data are distributed

the possibility of correctly quantifying the odor intensity in terms of concentration. The details of this study have been already published and can be found in Ref. [12]. The results shown above and those shown previously on 2-butanone and ethanol demonstrate the capability of the system to classify and quantify different analytes, depending on an LOD which is characteristic for each substance.

3.7.2 Binary Mixture Study: Segmentation and Background Suppression Capability

The identification of the components of a mixture and the discrimination of an odor in a complex background of other odors are tasks that the biological olfactory systems perform constantly. Hence, they are considered important for the artificial olfactory systems, but they are not easy to achieve.

The performance of the large array in terms of segmentation of binary mixtures and background suppression has been tested and proved [12].

For assaying the segmentation capabilities, mixtures of ethanol and 2-butanone were generated in a way that the relative concentrations of the two analytes are varied at fixed percentages of a certain maximum component concentration. The results demonstrate the ability of the array to discriminate the mixtures from the pure analytes and to recognize different concentration ratios of the two components in the mixture. We can say that the large array should be able to recognize not only the presence of a single analyte in a mixture but also variations of its concentration. Hence, the odor segmentation process seems to be possible with the developed device.

The background suppression capability was verified by generating binary mixtures where the concentration of ethanol is kept constant and high, while 2-butanone concentration is varied, starting from zero. The dataset built was processed by PCA and PLS and proved that the system is able not only to detect 2-butanone in the presence of high background but also the different concentrations of 2-butanone.

More details on this study on segmentation and background suppression can be found in a series of publications [10–12, 83].

3.7.3 Polymer Classes: Testing Broad and Overlapping Sensitivity, High Level of Redundancy

The different polymers and the diversity of the characteristics of the transducer geometries contribute to generate a spread pattern of sensor responses. This is of fundamental importance in the realization of a device mimicking the natural olfaction. In order to verify the effectiveness of this assumption in the realized electronic nose, one can analyze the response of all the sensors and group their responses depending on the polymer class they belong to.

Figure 3.13 depicts a boxplot representing the steady-state ($\Delta R/R_0$) response of all the 16384 sensors to an input of 13734 ppm of ethanol. Each separate box represents the statistical information about the sensor responses for each class of polymers. The numbers on the x-axis represent the polymer type, as coded in Table 3.3. The line included in each box depicts the median of the sensor response for that particular type of polymer. Each box extends from the 25th to the 75th percentile of the sensor responses, while the whiskers indicate the most extreme sensor responses not considered as outliers.

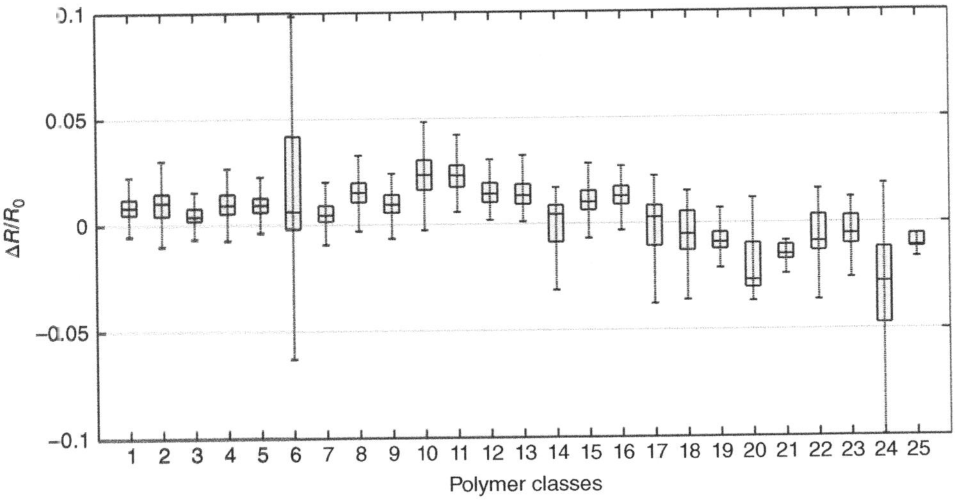

Figure 3.13 Boxplot of the steady-state response of the sensors to a concentration of 40 ppm of ethanol. The responses of all the sensors made with the same conducting polymer have been grouped together to originate a boxplot, which is related to that polymer. The line included in each box depicts the mean of the sensor response for that particular type of polymer. Each box extends from the 25th to the 75th percentile of the sensor responses, while the whiskers indicate the most extreme sensor responses not considered as outliers. The number on the x-axis indicates the polymer type as coded in Table 3.3

As one can expect, there is no one single polymer type that gives a dramatically higher response with respect to all the others. This is to be expected, meaning that all the polymers possess a broad selectivity to the chemical species. Additionally, looking at the statistical information expressed on the boxplot, it is to be assumed that the distribution of the sensor responses of a single polymer class is statistically different from the distribution of the sensor response of any other polymer class (as verified by performing an ANOVA test on different response classes). This is of extreme importance for the system, as it contributes to create a given and peculiar pattern of response for different chemical species.

These two characteristics of the chosen sensor set will give the device the capability of discriminating between different chemical species, even if none of the sensors is highly selective for those particular analytes, and also of quantifying their concentrations.

3.7.4 *System Robustness and Long-Term Stability*

In the biological olfactory system, the first representation of an odor, which is a pattern of signals generated by the large number of redundant ORNs, is transduced into a second spatial pattern, generated at the glomerular level. The convergence ratio from the first level to the second one is very high; in mammals it is around 25 000 ORNs per glomerulus, and it allows the amplification of weak signals, as well as an averaging of the receptor input. This leads to an increase in the S/N ratio and, consequently, in the robustness and stability of the system.

We are expecting that the large number of sensors and the high degree of redundancy implemented in our large sensor array provide it with similar advantages.

The four sensor arrays built are each composed of 4096 elements. However, not all of these are functional. A study aimed at determining the distribution of the sensor resistances within the arrays indicated that there are a certain number of unresponsive or noisy sensors. This is to be attributed in particular to faulty sensors, where the electrical bonding failed to correctly connect rows and/or column connections. However, the analyses showed in the previous sections were achieved considering the data generated by all the sensors, even those that were not working properly. The good results achieved indicate that the performance of the system is not significantly degraded by the presence of faulty sensors. As expected, the large number of redundant responses provides averaged sensor responses that compensate for the defects.

To investigate the long-term stability of the large-scale sensor array, we have recently carried out measurements of ethanol that are replications of some experiments carried out after an interval of three years. Figures 3.14 and 3.15 show the PLSRA plots related to the old data and the new data, respectively. The concentrations generated for the analysis extended over a wide range starting from 340 ppm up to 13 740 ppm. There were at least eight replicate measurements for each concentration.

The old model was optimized with 6 LVs and generates errors of 41.8 ppm in calibration and 371.4 ppm in validation. The new model was optimized with eight LVs, and it exhibits an error in calibration of 60.6 ppm and an error in validation of 649.1 ppm. The new model shows higher errors with respect to the old model. Hence, we can say that the large sensor array is affected by an aging effect that has degraded its performance but this does not compromise its capabilities. In fact, it shows good regression capability, very good linear behavior, and very good repeatability of the measurements, even after 3 years (see Figure 3.15). This result is a real success, in particular considering that one of the major drawbacks perceived with conducting polymer sensors is their poor long-term stability.

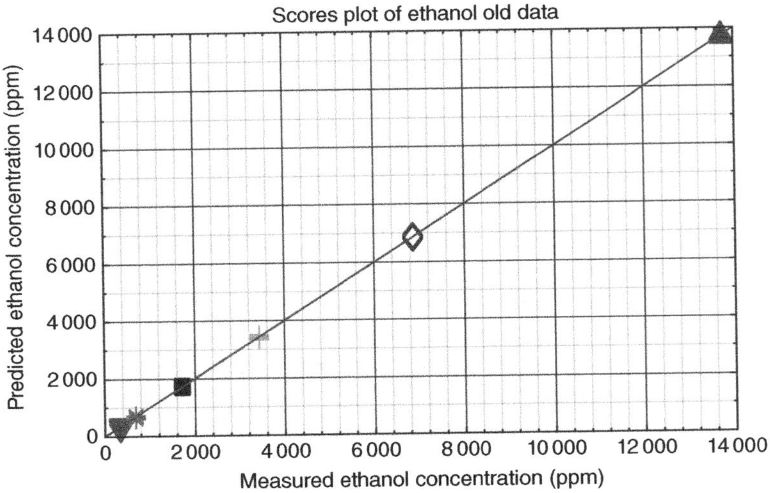

Figure 3.14 A partial least squares regression model built for predicting different concentrations of ethanol. This is an "old" dataset generated 3 years ago, immediately after the fabrication of the device. The concentrations generated for the analysis extended over a wide range from 340 ppm up to 13 740 ppm. At least eight replicate measurements for each concentration were carried out. The model was optimized with six latent variables and generates errors of 41.8 ppm in calibration and 371.4 ppm in validation. It shows an excellent linear behavior and reproducibility

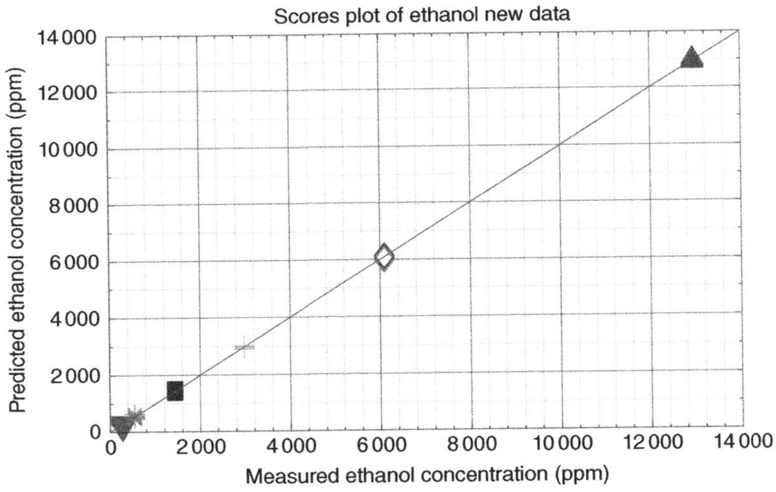

Figure 3.15 A partial least squares regression model built by analyzing a new ethanol dataset. It was built with data collected 3 years after that shown in Figure 3.14. The "new" dataset considers the same range of concentrations as the old dataset. The values of the concentrations tested are slightly different from those of Figure 3.14 due to different working temperatures. The new model was optimized with eight latent variables, and it exhibits an error in calibration of 60.6 ppm and an error in validation of 649.1 pm. Comparing Figures 3.14 and 3.15, we can say that there is an aging effect acting on the large sensor array, which increases the validation and calibration errors. However, even after three years the system exhibits good regression capability, very good linear behavior, and very good repeatability of the measurements. Hence, the aging effect is not compromising the functioning of the system. This is a characteristic due to the large number of sensors providing redundancy to the system

3.8 Conclusions

The main aim of our study was the development of a large-scale sensor array that, mimicking biological olfactory sensor characteristics, could be interfaced with computational models of biological olfaction. So far, sensor arrays featuring such ability do not exist, while the acquisition of realistic sensory data is really necessary to create realistic models of chemosensory processing in biology. Attempts to increase the number of the sensors in the core array of the artificial electronic noses and to provide such systems with features similar to those of the biological epithelium have been done in the electrical and optical domain. However, none of them led to the fabrication of what can be defined as an "artificial epithelium." We believe that we have succeeded in our aim to create such a system.

One of the major problems that impeded the fabrication of large arrays of sensors in the electrical domain has been the unmanageable number of the electrical connections needed for interfacing numerous elements to a readout circuit. In this project this drawback was overcome by arranging the sensors in a matrix configuration and developing a custom-made scanning circuit that, exploiting the concept of a virtual open circuit, allows interrogation of the array while compensating for the crosstalk generated by the arrangement.

A large number of sensors also imply long acquisition times, which is not compatible with most sensor applications. The sensor acquisition time should be rapid if a device aimed at real-life applications needs to be built. In order to reduce such times, we modified the original plan of building a unique big array and developed smaller subarrays. The subarrays were implemented in a modular system able to read them in parallel. This modular system can interrogate in parallel up to 16 smaller arrays, each consisting of 4096 elements. Hence, any number of sensors between zero and 65 536 and multiples of 4096 can be acquired. Because of the parallel reading and the reduction of the parasitic capacitance of the smaller scanning circuits, 65 536 sensors can be read 16-fold faster than in the case of a single array.

The modular approach also supplies to the system good stability and high level of versatility. In fact, as each module was equipped with its own power and communication line, it can be easily replaced without interfering with the functionality of the other arrays. Hence, a damaged array can be easily substituted with a similar one, or new sensor types can be tested by simply connecting a new subarray to the device.

To provide the array with high degree of redundancy, as exhibited by the ORNs in the epithelium layer, conducting polymers were chosen as sensing materials. Such choice allowed the development of up to 30 different sensing coatings. Their specificities and sensitivities were tuned by adding functional groups to the main ring using different dopant species and preparing copolymers. Of those available, 24 polymer types were deposited on the four sensor arrays that have been tested in this study.

In the biological olfactory system, receptors of the same type give different dynamic ranges of responses to the same analytes. Hence, to increase further the variability in the sensor die, various transducer geometries were patterned on the array.

Experiments with different analytes have been carried out to test the capability of the sensor system and to characterize it. The analysis of the data collected by means of this experiment confirmed that the sensor array is able to:

- Inherently emulate the broad specificities and also selectivities that are found in biological systems
- Classify and quantify odors
- Segment binary mixtures and suppress a noisy background

The complete characterization of the device is a long process that is not yet finished. Different aspects of the sensor system need to be examined to have a comprehensive understanding of its potential for real applications. However, these studies have shown that the high degree of redundancy implemented into the system makes it robust. In fact, the data analysis presented was carried out considering the responses of all the sensors, including faulty sensors. Despite this, the array exhibited very good performance, showing that it is capable of compensating for defective sensors.

Moreover, a comparison of data from the large sensor array before and after 3 years of its fabrication highlighted a weak aging effect that has not compromised the capability of the device. Hence, the high level of redundancy compensates the lack of stability exhibited by the traditional devices.

It should not be forgotten the great advantage presented by this device: it is the only system able to generate realistic data input for computational hypotheses of biological chemosensory systems. It is the only system available currently that can be interfaced with neuromorphic models to test their performance. This will ultimately lead to the production of artificial sensing systems with improved sensitivity and specificity.

Acknowledgment

The work described was supported by NEUROCHEM (grant 216916), a project funded by the European Commission under FP7, Bio-ICT Convergence—FET program.

References

[1] Abe, H., Kanaya, S., Takahashi, Y., & Sasaki, S. I. (1988), Extended studies of the automated odor-sensing system based on plural semiconductor gas sensors with computerized pattern-recognition techniques, *Analytica Chimica Acta*, vol. 215, no. 1–2, pp. 155–168.

[2] Abe, H., Yoshimura, T., Kanaya, S., Takahashi, Y., Miyashita, Y., & Sasaki, S. I. (1987), Automated odor-sensing system based on plural semiconductor gas sensors and computerized pattern-recognition techniques, *Analytica Chimica Acta*, vol. 194, pp. 1–9.

[3] Abe, S., Choi, U. S., Shimanoe, K., & Yamazoe, N. (2005), Influences of ball-milling time on gas-sensing properties of Co_3O_4–SnO_2 composites, *Sensors and Actuators B: Chemical*, vol. 107, no. 2, pp. 516–522.

[4] Alam, H. & Saeed, S. H. (2013), Modern applications of electronic noses: A review, *International Journal of Electrical and Computer Engineering (IJECE)*, vol. 3, no. 1, pp. 52–63.

[5] Ballantine, J., Callahan, D., Maclay, G. J., & Stetter, J. R. (1992), An optical waveguide acid vapor sensor, *Talanta*, vol. 39, no. 12, pp. 1657–1667.

[6] Barisci, J. N., Wallace, G. G., Andrews, M. K., Partridge, A. C., & Harris, P. D. (2002), Conducting polymer sensors for monitoring aromatic hydrocarbons using an electronic nose, *Sensors and Actuators B: Chemical*, vol. 84, no. 2–3, pp. 252–257.

[7] Barsan, N., Koziej, D., & Weimar, U. (2007), Metal oxide-based gas sensor research: How to? *Sensors and Actuators B: Chemical*, vol. 121, no. 1, pp. 18–35.

[8] Bartlett, P. N. & Ling-Chung, S. K.(1989), Conducting polymer gas sensors. Part III: Results for four different polymers and five different vapours, *Sensors and Actuators*, vol. 20, no. 3, pp. 287–292.

[9] Bastos, A. C. & Magan, N. (2006), Potential of a an electronic nose for the early detection and differentiation between Streptomyces in potable water, *Sensors and Actuators B: Chemical*, vol. 116, no. 1–2, pp. 151–155.

[10] Beccherelli, R., Zampetti, E., Pantalei, S., Bernabei, M., & Persaud, K. (2009), Very large chemical sensor array for mimicking biological olfaction. *AIP Conference Proceedings*, vol. 1137, pp. 155–158.

[11] Beccherelli, R., Zampetti, E., Pantalei, S., Bernabei, M., & Persaud, K. (2010), Design of a very large chemical sensor system for mimicking biological olfaction, *Sensors and Actuators B: Chemical*, vol. 146, no. 2, pp. 446–452.

[12] Bernabei, M., Persaud, K. C., Pantalei, S., Zampetti, E., & Beccherelli, R. (2012), Large-scale chemical sensor array testing biological olfaction concepts, *IEEE Sensors Journal*, vol. 12, no. 11, pp. 3174–3183.

[13] Botsialas, A., Oikonomou, P., Goustouridis, D., Ganetsos, T., Raptis, I., & Sanopoulou, M. (2013), A miniaturized chemocapacitor system for the detection of volatile organic compounds, *Sensors and Actuators B: Chemical*, vol. 177, pp. 776–784.

[14] Brahim-Belhouari, S., Bermak, A., Shi, M. H., & Chan, P. C. H. (2005), Fast and robust gas identification system using an integrated gas sensor technology and Gaussian mixture models, *IEEE Sensors Journal*, vol. 5, no. 6, pp. 1433–1444.

[15] Buck, L. & Axel, R. (1991), A novel multigene family may encode odorant receptors: A molecular basis for odor recognition, *Cell*, vol. 65, no. 1, pp. 175–187.

[16] Cao, Y., Smith, P., & Heeger, A. J. (1992), Counter-ion induced processibility of conducting polyaniline and of conducting polyblends of polyaniline in bulk polymers, *Synthetic Metals*, vol. 48, no. 1, pp. 91–97.

[17] Cao, Y., Smith, P., & Heeger, A. J. (1993), Counter-ion induced processibility of conducting polyaniline, *Synthetic Metals*, vol. 57, no. 1, pp. 3514–3519.

[18] Capone, S., De Pascali, C., Francioso, L., Siciliano, P., Persaud, K., & Pisanelli, A. (2009), Electrical characterization of a Pig Odorant Binding Protein by Impedance Spectroscopy, Sensors, 25–28 October 2009, IEEE, pp. 1758–1762.

[19] Casalinuovo, I. A., Di Pierro, D., Coletta, M., & Di Francesco, P. (2006), Application of electronic noses for disease diagnosis and food spoilage detection, *Sensors*, vol. 6, no. 11, pp. 1428–1439.

[20] Chadwick, B. & Gal, M.1993, Enhanced optical detection of hydrogen using the excitation of surface plasmons in palladium, *Applied Surface Science*, vol. 68, no. 1, pp. 135–138.

[21] Chang, S. C., Stetter, J. R., & Cha, C. S. (1993), Amperometric gas sensors, *Talanta*, vol. 40, no. 4, pp. 461–477.

[22] Covington, J. A., Gardner, J. W., Briand, D., & de Rooij, N. F. (2001), A polymer gate FET sensor array for detecting organic vapours, *Sensors and Actuators B: Chemical*, vol. 77, no. 1–2, pp. 155–162.

[23] D'Alessio, T. (1999), Measurement errors in the scanning of piezoresistive sensors arrays, *Sensors and Actuators A: Physical*, vol. 72, no. 1, pp. 71–76.

[24] D'Amico, A., Bono, R., Pennazza, G., Santonico, M., Mantini, G., Bernabei, M., Zarlenga, M., Roscioni, C., Martinelli, E., Paolesse, R., & Di Natale, C.2008, Identification of melanoma with a gas sensor array, *Skin Research and Technology*, vol. 14, no. 2, pp. 226–236.

[25] D'Amico, A., Di Natale, C., Paolesse, R., Macagnano, A., Martinelli, E., Pennazza, G., Santonico, M., Bernabei, M., Roscioni, C., Galluccio, G., Bono, R., Agrò, E. F., & Rullo, S. (2008), Olfactory systems for medical applications, *Sensors and Actuators B: Chemical*, vol. 130, no. 1, pp. 458–465.

[26] Di Francesco, F., Fuoco, R., Trivella, M. G., & Ceccarini, A. (2005), Breath analysis: Trends in techniques and clinical applications, *Microchemical Journal*, vol. 79, no. 1–2, pp. 405–410.

[27] Di Natale, C., Macagnano, A., Paolesse, R., Mantini, G., Tarizzo, E., D'Amico, A., Sinesio, F., Bucarelli, F. M., Moneta, E., & Quaglia, G. B. (1998), Electronic nose and sensorial analysis: Comparison of performances in selected cases, *Sensors and Actuators B: Chemical*, vol. 50, no. 3, p. 246.

[28] Di Natale, C., Macagnano, A., Paolesse, R., Tarizzo, E., D'Amico, A., Davide, F., Boschi, T., Faccio, M., Ferri, G., Sinesio, F., Bucarelli, F. M., Moneta, E., & Quaglia, G. B. (1997), A comparison between an electronic nose and human olfaction in a selected case study, Proceedings of the International Conference on Solid State Sensors and Actuators, Chicago, IL, USA, June 16–19, 1997, vol. 2, pp. 1335–1338.

[29] Di Natale, C., Macagnano, A., Repole, G., Saggio, G., D'Amico, A., Paolesse, R., & Boschi, T. (1998), The exploitation of metalloporphyrins as chemically interactive material in chemical sensors, *Materials Science and Engineering C*, vol. 5, no. 3–4, pp. 209–215.

[30] Di Natale, C., Macagnano, A., Martinelli, E., Paolesse, R., D'Arcangelo, G., Roscioni, C., Finazzi-Agrò, A., & D'Amico, A. (2003), Lung cancer identification by the analysis of breath by means of an array of non-selective gas sensors, *Biosensors and Bioelectronics*, vol. 18, no. 10, pp. 1209–1218.

[31] Di Pietrantonio, F., Cannata, D., Benetti, M., Verona, E., Varriale, A., Staiano, M., & D'Auria, S. (2013), Detection of odorant molecules via surface acoustic wave biosensor array based on odorant-binding proteins, *Biosensors and Bioelectronics*, vol. 41, pp. 328–334.

[32] Dutta, R., Hines, E., Gardner, J., & Boilot, P. (2002), Bacteria classification using Cyranose 320 electronic nose, *BioMedical Engineering OnLine*, vol. 1, no. 1, p. 4.

[33] Ferreira, G. N., da-Silva, A. C., & Tome, B. (2009), Acoustic wave biosensors: Physical models and biological applications of quartz crystal microbalance, *Trends in Biotechnology*, vol. 27, no. 12, pp. 689–697.

[34] Fischerauer, A., Schwarzmueller, C., & Fischerauer, G. (2009), Substrate influence on the characteristics of interdigital-electrode gas sensors, Proceedings of the Sixth International Multi-Conference on Systems, Signals and Devices, Systems, Signals and Devices, 2009. SSD '09, Vols. 1 and 2, pp. 959–963.

[35] Gardner, J. W. & Bartlett, P. N. (1993), Design of conducting polymer gas sensors: Modelling and experiment, *Synthetic Metals*, vol. 57, no. 1, pp. 3665–3670.

[36] Gomez, A. H., Wang, J., Hu, G. X., & Pereira, A. G. (2006), Electronic nose technique potential monitoring mandarin maturity, *Sensors and Actuators B: Chemical*, vol. 113, no. 1, pp. 347–353.

[37] Gomez, A. H., Wang, J., Hu, G., & Pereira, A. G. (2007), Discrimination of storage shelf-life for mandarin by electronic nose technique, *LWT–Food Science and Technology*, vol. 40, no. 4, pp. 681–689.

[38] Gomez, A. H., Wang, J., Hu, G., & Pereira, A. G. (2008), Monitoring storage shelf life of tomato using electronic nose technique, *Journal of Food Engineering*, vol. 85, no. 4, pp. 625–631.

[39] Grate, J. W. & Klusty, M. (1991), Surface acoustic-wave vapor sensors based on resonator devices, *Analytical Chemistry*, vol. 63, no. 17, pp. 1719–1727.

[40] Grate, J. W. & Wise, B. M. (1999), Chemical information from polymer-coated acoustic wave sensor arrays, *Chemical Microsensors and Applications II*, vol. 3857, pp. 170–173.

[41] Grossmann, K., Weimar, U., & Barsan, N. (2013), Semiconducting metal oxides based gas sensors, *Oxide Semiconductors*, vol. 88, pp. 261–282.

[42] Grunwaldt, J. D., Huebner, M., Koziej, D., Barsan, N., & Weimar, U. (2013), The potential of operando XAFS for determining the role and structure of noble metal additives in metal oxide based gas sensors, *Journal of Physics: Conference Series*, vol. 430, p. 012078.

[43] Guan, B., Zhao, J., Lin, H., & Zou, X. (2014), Characterization of volatile organic compounds of vinegars with novel electronic nose system combined with multivariate analysis, *Food Analytical Methods*, vol. 7, no. 5, pp. 1073–1082.

[44] Gutierrez, A. & Marco, M. (2009), *Biologically Inspired Signal Processing for Chemical Sensing* (188). New York, Springer-Verlag. Studies in Computational Intelligence.

[45] Gutierrez, A., Pardo, A., & Marco, S. (2006), Information-theoretic analysis of odor intensity at the olfactory epithelium, *Chemical Senses*, vol. 31, no. 8, p. E3–E4.

[46] Gutierrez-Galvez, A. & Gutierrez-Osuna, R. (2006), Increasing the separability of chemosensor array patterns with Hebbian/anti-Hebbian learning, *Sensors and Actuators B: Chemical*, vol. 116, no. 1–2, pp. 29–35.

[47] Gutierrez-Osuna, R. & Nagle, H. T. (1999), A method for evaluating data-preprocessing techniques for odour classification with an array of gas sensors, *IEEE Transactions on Systems Man and Cybernetics Part B (Cybernetics)*, vol. 29, no. 5, p. 626.

[48] Haberly, L. B. (2001), Parallel-distributed processing in olfactory cortex: New insights from morphological and physiological analysis of neuronal circuitry, *Chemical Senses*, vol. 26, no. 5, pp. 551–576.

[49] Haberly, L. B. & Bower, J. M. (1989), Olfactory cortex: Model circuit for study of associative memory?*Trends in Neurosciences*, vol. 12, no. 7, pp. 258–264.

[50] Hamacher, T., Niess, J., Schulze Lammers, P., Diekmann, B., & Boeker, P. (2003), Online measurement of odorous gases close to the odour threshold with a QMB sensor system with an integrated preconcentration unit, *Sensors and Actuators B: Chemical*, vol. 95, no. 1–3, pp. 39–45.

[51] Harun, F., Taylor, J., Covington, J., & Gardner, J. (2009), An electronic nose employing dual-channel odour separation columns with large chemosensor arrays for advanced odour discrimination, *Sensors and Actuators B: Chemical*, vol. 141, no. 1, pp. 134–140.

[52] Hu, W. L., Li, R., & Wang, P. (1998), A novel method for wine analysis based on sensor fusion technique, Technical Digest of the Seventh International Meeting on Chemical Sensors, Beijing, China, July 27–30, 1998, pp. 142–144.

[53] Huang, X., Zou, X., Shi, J., Guo, Y., Zhao, J., Zhang, J., & Hao, L. (2014), Determination of pork spoilage by colorimetric gas sensor array based on natural pigments, *Food Chemistry*, vol. 145, pp. 549–554.

[54] Huo, D., Wu, Y., Yang, M., Fa, H., Luo, X., & Hou, C. (2014), Discrimination of Chinese green tea according to varieties and grade levels using artificial nose and tongue based on colorimetric sensor arrays, *Food Chemistry*, vol. 145, pp. 639–645.

[55] Igreja, R. & Dias, C. J. (2006), Dielectric response of interdigital chemocapacitors: The role of the sensitive layer thickness, *Sensors and Actuators B: Chemical*, vol. 115, no. 1, pp. 69–78.

[56] Im, J., Sengupta, S. K., Baruch, M. F., Granz, C. D., Ammu, S., Manohar, S. K., & Whitten, J. E. (2011), A hybrid chemiresistive sensor system for the detection of organic vapors, *Sensors and Actuators B: Chemical*, vol. 156, no. 2, pp. 715–722.

[57] Ivanov, P., Llobet, E., Vilanova, X., Brezmes, J., Hubalek, J., & Correig, X. (2004), Development of high sensitivity ethanol gas sensors based on Pt-doped SnO_2 surfaces, *Sensors and Actuators B: Chemical*, vol. 99, no. 2–3, pp. 201–206.

[58] Janata, J. & Josowicz, M. (2003), Conducting polymers in electronic chemical sensors, *Nature Materials*, vol. 2, no. 1, pp. 19–24.

[59] Potje-Kamloth, K. (2011), Chemical gas sensors based on organic semiconductor polypyrrole, *Critical Reviews in Analytical Chemistry*, vol. 32, no. 2, pp. 121–140.

[60] Kauer, J. S. & White, J. (2001), Imaging and coding in the olfactory system, *Annual Review of Neuroscience*, vol. 24, pp. 963–979.

[61] Kladsomboon, S., Lutz, M., Pogfay, T., Puntheeranurak, T., & Kerdcharoen, T. (2012), Hybrid optical-electrochemical electronic nose system based on Zn-porphyrin and multi-walled carbon nanotube composite, *Journal of Nanoscience and Nanotechnology*, vol. 12, no. 7, pp. 5240–5244.

[62] Koll, A., Mayer, F., Hagleitner, C., Scheiwiller, D., Brand, O., & Baltes, H. (1998), A flip-flop packaged CMOS chemical microsystem for detection of volatile organic compounds, *SPIE Proceeding Series*, vol. 3328, pp. 223–232.

[63] Kramer, K. E., Rose-Pehrsson, S. L., Johnson, K. J., & Minor, C. P. (2009), Hybrid arrays for chemical sensing, in: Ryan, M. A., Shevade, A. V., Taylor, C. J., Homer, M. L., Blanco, M., & Stetter, J. R. (eds.), *Computational Methods for Sensor Material Selection*. New York, Springer, pp. 265–298.

[64] Kudo, H., Yamashita, T., Miyajima, K., Arakawa, T., & Mitsubayashi, K. (2013), NADH-fluorometric biochemical gas sensor (bio-sniffer) for evaluation of indoor air quality, *IEEE Sensors Journal*, vol. 13, no. 8, pp. 2828–2833.

[65] Kurosawa, S., Kamo, N., Matsui, D., & Kobatake, Y. (1990), Gas sorption to plasma-polymerized copper phthalocyanine film formed on a piezoelectric crystal, *Analytical Chemistry*, vol. 62, no. 4, pp. 353–359.

[66] Lange, U. & Mirsky, V. M. (2011), Chemiresistors based on conducting polymers: A review on measurement techniques, *Analytica Chimica Acta*, vol. 687, no. 2, pp. 105–113.

[67] Lundstrom, I., Shivaraman, S., Svensson, C., & Lundkvist, L. (1975), A hydrogen-sensitive MOS field-effect transistor, *Applied Physics Letters*, vol. 26, no. 2, pp. 55–57.

[68] MacDiarmid, A. G. & Epstein, A. J. (1994), The concept of secondary doping as applied to polyaniline, *Synthetic Metals*, vol. 65, no. 2–3, pp. 103–116.

[69] MacDiarmid, A. G. & Epstein, A. J. (1995), Secondary doping in polyaniline, *Synthetic Metals*, vol. 69, no. 1–3, pp. 85–92.

[70] Markovics, A. & Kovacs, B. (2014), Optical ammonia sensors for environmental applications, *Analytical Letters*, vol. 47, no. 3, pp. 465–477.

[71] Mishra, S. K., Rani, S., & Gupta, B. D. (2014), Surface plasmon resonance based fiber optic hydrogen sulphide gas sensor utilizing nickel oxide doped ITO thin film, *Sensors and Actuators B: Chemical*, vol. 195, pp. 215–222.

[72] Moncrieff, R. W. (1961), An instrument for measuring and classifying odors, *Journal of Applied Physiology*, vol. 16, no. 4, pp. 742–749.

[73] Morita, Y., Nakamura, K. I., & Kim, C. (1996), Langmuir analysis on hydrogen gas response of palladium-gate FET, *Sensors and Actuators B: Chemical*, vol. 33, no. 1-3, pp. 96–99.

[74] Moseley, P. T. (1997), Solid state gas sensors, *Measurement Science and Technology*, vol. 8, no. 3, pp. 223–237.

[75] Nelson, S. G., Johnston, K. S., & Yee, S. S. (1996), High sensitivity surface plasmon resonance sensor based on phase detection, *Sensors and Actuators B: Chemical*, vol. 35, no. 1–3, pp. 187–191.

[76] Oikonomou, P., Botsialas, A., Manoli, K., Goustouridis, D., Valamontes, E., Sanopoulou, M., Raptis, I., & Patsis, G. (2012), Chemocapacitor performance modeling by means of polymer swelling optical measurements, *Sensors and Actuators B: Chemical*, vol. 171, pp. 409–415.

[77] Oikonomou, P., Botsialas, A., Patsis, G., Manoli, K., Goustouridis, D., Valamontes, E., Pantazis, N., Sanopoulou, M., & Raptis, I. (2011), Hybrid integration of microfabricated chemocapacitor arrays with miniaturized read-out electronics towards low-power gas sensing module. *Procedia Engineering*, vol. 25, pp. 1117–1120.

[78] Oprea, A., Barsan, N., & Weimar, U. (2009), Work function changes in gas sensitive materials: Fundamentals and applications, *Sensors and Actuators B: Chemical*, vol. 142, no. 2, pp. 470–493.
[79] Pearce, T. C., Schiffman, S. S., Nagle, H. T., & Gardner, J. W. (2003), *Handbook of Machine Olfaction: Electronic Nose Technology*. Weinheim, John Wiley & Sons, Ltd.
[80] Persaud, K. & Dodd, G. (1982), Analysis of discrimination mechanisms in the mammalian olfactory system using a model nose, *Nature*, vol. 299, no. 5881, pp. 352–355.
[81] Persaud, K. C. (2005), Polymers for chemical sensing, *Materials Today*, vol. 8, no. 4, pp. 38–44.
[82] Persaud, K. C. (2012), Biomimetic olfactory sensors, *IEEE Sensors Journal*, vol. 12, no. 11, pp. 3108–3112.
[83] Persaud, K. C., Marco, S., & Gutierrez-Galvez, A. (2013), *Neuromorphic Olfaction*. Boca Raton, CRC Press. Frontiers in Neuroengineering Series.
[84] Persaud, K. C. & Pelosi, P. (1985), An approach to an artificial nose, *Transactions: American Society for Artificial Internal Organs*, vol. 31, pp. 297–300.
[85] Prigodin, V. N., Samukhin, A. N., & Epstein, A. J. (2004), Variable range hopping in low-dimensional polymer structures, *Synthetic Metals*, vol. 141, no. 1–2, pp. 155–164.
[86] Pron, A. & Rannou, P. (2002), Processible conjugated polymers: From organic semiconductors to organic metals and superconductors, *Progress in Polymer Science*, vol. 27, no. 1, pp. 135–190.
[87] Purbrick, J. A. (1981), A force transducer employing conductive silicone rubber, Proceedings of the First International Conference on Robot Vision and Sensing, Stratford upon Avon, April 1–3, 1981, pp. 73–80.
[88] Raman, B., Sun, P. A., Gutierrez-Galvez, A., & Gutierrez-Osuna, R. (2006), Processing of chemical sensor arrays with a biologically inspired model of olfactory coding, *IEEE Transactions on Neural Networks*, vol. 17, no 4, pp. 1015–1024.
[89] Ressler, K. J., Sullivan, S. L., & Buck, L. B. (1994), Information coding in the olfactory system: Evidence for a stereotyped and highly organized epitope map in the olfactory bulb, *Cell*, vol. 79, no. 7, pp. 1245–1255.
[90] Rivera, D., Alam, M. K., Davis, C. E., & Ho, C. K. (2003), Characterization of the ability of polymeric chemiresistor arrays to quantitate trichloroethylene using partial least squares (PLS): Effects of experimental design, humidity, and temperature, *Sensors and Actuators B: Chemical*, vol. 92, no. 1–2, pp. 110–120.
[91] Roeck, F., Barsan, N., & Weimar, U. (2008), Electronic nose: Current status and future trends, *Chemical Reviews*, vol. 108, no. 2, pp. 705–725.
[92] Rospars, J. P., Sanda, P., Lansky, P., & Duchamp-Viret, P. (2013), Responses of single neurons and neuronal ensembles in frog first- and second-order olfactory neurons, *Brain Research*, vol. 1536, pp. 144–158.
[93] Sandstrom, M., Lansner, A., Hellgren-Kotaleski, J., & Rospars, J. P. (2009), Modeling the response of a population of olfactory receptor neurons to an odorant, *Journal of Computational Neuroscience*, vol. 27, no. 3, pp. 337–355.
[94] Santonico, M., Pittia, P., Pennazza, G., Martinelli, E., Bernabei, M., Paolesse, R., D'Amico, A., Compagnone, D. & Di Natale, C. (2008), Study of the aroma of artificially flavoured custards by chemical sensor array fingerprinting, *Sensors and Actuators B: Chemical*, vol. 133, no. 1, pp. 345–351.
[95] Sarkisov, S. S., Czarick, M., III, Fairchild, B. D., Liang, Y., Kukhtareva, T., & Curley, M. J. (2014), Colorimetric polymer-metal nanocomposite sensor of ammonia for the agricultural industry of confined animal feeding operations, *Optical Engineering*, vol. 53, no. 2, p. 021107.
[96] Shepherd, G. M. (1987), A molecular vocabulary for olfaction, *Annals of the New York Academy of Sciences*, vol. 510, no. 1, pp. 98–103.
[97] Shepherd, G. M. (1991), Computational structure of the olfactory system, in: Davis, J. & Eichenbaum, H. (eds.), *Olfaction: A Model System for Computational Neuroscience*, Cambridge, MA, MIT Press, pp. 3–41.
[98] Shepherd, G. M. (1993), Principles of specificity and redundancy underlying the organization of the olfactory system, *Microscopy Research and Technique*, vol. 24, no. 2, pp. 106–112.
[99] Snyder, D. M. (2014), Optical phase-shift dynamics in surface-modified transparent polymers: Application of wavefront distortion analysis to refractive index (RI)-based sensor development, *Sensors and Actuators B: Chemical*, vol. 195, pp. 473–477.
[100] Someya, T., Kato, Y., Sekitani, T., Iba, S., Noguchi, Y., Murase, Y., Kawaguchi, H., & Sakurai, T. (2005), Conformable, flexible, large-area networks of pressure and thermal sensors with organic transistor active matrixes, *Proceedings of the National Academy of Sciences of the United States of America*, vol. 102, no. 35, pp. 12321–12325.
[101] Sung, T. W. & Lo, Y. L. (2013), Ammonia vapor sensor based on $CdSe/SiO_2$ core–shell nanoparticles embedded in sol–gel matrix, *Sensors and Actuators B: Chemical*, vol. 188, pp. 702–708.

[102] Tanese, M., Marinelli, F., Angione, D., & Torsi, L. (2008), Overview of recent developments in organic thin-film transistor sensor technology, *Nuovo Cimento Della Societa Italiana di Fisica C—Geophysics and Space Physics*, vol. 31, no. 4, pp. 457–473.

[103] Torsi, L., Tafuri, A., Cioffi, N., Gallazzi, M. C., Sassella, A., Sabbatini, L., & Zambonin, P. G. (2003), Regioregular polythiophene field-effect transistors employed as chemical sensors, *Sensors and Actuators B: Chemical*, vol. 93, no. 1–3, pp. 257–262.

[104] Vassar, R., Chao, S. K., Sitcheran, R., Nuñez, J. M., Vosshall, L. B., & Axel, R. (1994), Topographic organization of sensory projections to the olfactory bulb, *Cell*, vol. 79, no. 6, pp. 981–991.

[105] Wang, D., Yu, K., Wang, Y., Hu, Y., Zhao, C., Wang, L., Ying, K., & Wang, P. (2012), A hybrid electronic noses' system based on MOS-SAW detection units intended for lung cancer diagnosis, *Journal of Innovative Optical Health Sciences*, vol. 5, no. 1, pp. 1150006-1–1150006-7.

[106] Wang, L. C., Tang, K. T., Kuo, C. T., Ho, C. L., Lin, S. R., Sung, Y., & Chang, C. P. (2010), Portable electronic nose system with chemiresistor sensors to detect and distinguish chemical warfare agents, *Journal of Micro-Nanolithography Mems and Moems*, vol. 9, no. 3, p. 031010.

[107] White, J., Kauer, J. S., Dickinson, T. A., & Walt, D. R. (1996), Rapid analyte recognition in a device based on optical sensors and the olfactory system, *Analytical Chemistry*, vol. 68, no. 13, pp. 2191–2202.

[108] Wilson, A. & Lester, D. (1997), Use of an electronic-nose device for profiling headspace volatile metabolites to rapidly identify phytopathogenic microbes, *Phytopathology*, vol. 87, no. 6 Suppl, p. S116.

[109] Wilson, A. D. (2013), Diverse applications of electronic-nose technologies in agriculture and forestry, *Sensors*, vol. 13, no. 2, pp. 2295–2348.

[110] Wilson, A. D. & Baietto, M. (2009), Applications and advances in electronic-nose technologies, *Sensors*, vol. 9, no. 7, pp. 5099–5148.

[111] Wilson, A. D. & Baietto, M. (2011), Advances in electronic-nose technologies developed for biomedical applications, *Sensors*, vol. 11, no. 1, pp. 1105–1176.

[112] Wilson, A. D. (2012), Advanced methods for teaching electronic-nose technologies to diagnosticians and clinical laboratory technicians, *Procedia: Social and Behavioral Sciences*, vol. 46, pp. 4544–4554.

[113] Wilson, A. D. (2012), Review of electronic-nose technologies and algorithms to detect hazardous chemicals in the environment, *Procedia Technology*, vol. 1, pp. 453–463.

[114] Wilson, A. D. (2012), Theoretical and practical considerations for teaching diagnostic electronic-nose technologies to clinical laboratory technicians, *Procedia: Social and Behavioral Sciences*, vol. 31, pp. 262–274.

[115] Winquist, F., Holmin, S., Rülcker, C. K., Wide, P., & Lundström, I. (2000), A hybrid electronic tongue, *Analytica Chimica Acta*, vol. 406, no. 2, pp. 147–157.

[116] Wu, Y., Yao, B., Zhang, A., Rao, Y., Wang, Z., Cheng, Y., Gong, Y., Zhang, W., Chen, Y., & Chiang, K. (2014), Graphene-coated microfiber Bragg grating for high-sensitivity gas sensing, *Optics Letters*, vol. 39, no. 5, pp. 1235–1237.

[117] Wyszynski, B. & Nakamoto, T. (2009), Linking biological and artificial olfaction: Biomimetic quartz crystal microbalance odor sensors, *IEEJ Transactions on Electrical and Electronic Engineering*, vol. 4, no. 3, pp. 334–338.

[118] Yokoi, M., Mori, K., & Nakanishi, S. (1995), Refinement of odor molecule tuning by dendrodendritic synaptic inhibition in the olfactory bulb, *Proceedings of the National Academy of Sciences*, vol. 92, no. 8, pp. 3371–3375.

[119] Zemel, J. N. (1996), Future directions for thermal information sensors, *Sensors and Actuators A: Physical*, vol. 56, no. 1–2, pp. 57–62.

[120] Zhang, Q. T., Wang, P., Li, J. P., & Gao, X. G. (2000), Diagnosis of diabetes by image detection of breath using gas-sensitive laps, *Biosensors and Bioelectronics*, vol. 15, no. 5–6, pp. 249–256.

[121] Zwaardemaker, H. & Hogewind, F. (1919), On spray-electricity and waterfall electricity, *KNAW Proceedings*, vol. 22, no. I, pp. 429–437.

4

Taste Sensor: Electronic Tongue with Global Selectivity

Kiyoshi Toko[1,2], Yusuke Tahara[2], Masaaki Habara[3], Yoshikazu Kobayashi[3] and Hidekazu Ikezaki[3]
[1] *Research and Development Center for Taste and Odor Sensing, Kyushu University, Fukuoka, Japan*
[2] *Graduate School of Information Science and Electrical Engineering, Kyushu University, Fukuoka, Japan*
[3] *Intelligent Sensor Technology, Inc., Kanagawa, Japan*

4.1 Introduction

Electronic tongues are defined as sensors for measuring liquid using several kinds of nonspecific and low-selectivity sensors; as a result, multivariate analyses such as principal component analysis (PCA) and partial least squares (PLS) are usually performed on the outputs of sensor arrays. The taste sensor focused on in this chapter is a kind of electronic tongue, but the concrete purpose and method of sensing are very different from those of other electronic tongues. Many review papers for electronic tongues (abbreviated sometimes as "e-tongues") have been published so far [1–25], and hence results obtained from the taste sensor are focused on here.

The main method of evaluating the taste of items in food and pharmaceutical industries is a sensory test, in which experienced evaluators, called sensory panelists, actually taste samples to evaluate them; however, this method has some problems such as low objectivity and reproducibility as well as the great stress imposed on the panelists. To resolve these problems, an objective method for discriminating and quantifying the taste of foods and medicines has been developed. The taste receptors of humans do not necessarily recognize individual chemical substances, as mentioned in Section 4.3. Each of the receptors for the five basic tastes simultaneously receives multiple chemical substances, showing a semiselective property. In chemical analysis methods such as liquid and gas chromatographies, therefore, it is practically impossible to measure the taste of foods containing several hundred types of taste

Essentials of Machine Olfaction and Taste, First Edition. Edited by Takamichi Nakamoto.

Companion website: www.wiley.com/go/nakamoto/olfaction

substances. It should be noted that there are interactions between different tastes and between taste substances. For example, the bitterness of coffee is suppressed by adding sugar and a synergetic enhancing effect for umami can be obtained by mixing two kinds of umami substances represented by amino acids and nucleotide-derived substances.

The first concepts of a taste sensor appeared in 1990 [26] after a patent for the taste sensor was applied in 1989. The developed taste sensor was equipped with multichannel electrodes using a lipid/polymer membrane for the transducer to measure the taste in liquid. The term "electronic tongue" appeared in 1995 [27]. The taste sensor developed by Toko's group can be considered to be an electronic tongue with global selectivity. Here, global selectivity is defined as the decomposition of the characteristics of a chemical substance into taste qualities and their quantification, rather than the discrimination of individual chemical substances, by mimicking the human tongue, on which the taste of foods is decomposed into individual types of taste by each taste receptor. In contrast, ion sensors or enzyme sensors aim at the detection of each chemical substance with high selectivity, which means a one-to-one correspondence with a particular chemical substance. The taste sensor is developed in order to discriminate and identify foods or beverages and, furthermore, quantify the taste using the scale of taste. On the basis of this concept, the taste sensor has been commercialized as taste sensing systems SA402B and TS-5000Z, which are the world's first commercialized electronic tongue systems. Currently, over 350 systems have been sold and are utilized in food and pharmaceutical companies throughout the world.

Let us briefly explain the physiological knowledge of taste. The taste we perceive is composed of five kinds of taste qualities: sourness, saltiness, sweetness, bitterness, and umami [28–30]. Additional qualities are pungency and astringency, which are related to the sense of pain (Table 4.1). The fifth taste, "umami," is sometimes called savoriness (strictly speaking, it is not the same) and is acknowledged as an independent taste found by a Japanese scientist [28]. Sweetness is produced by sugar, glucose, and artificial sweetener and becomes our energy source. Saltiness is produced by cations such as sodium ions; however, anions also contribute to saltiness because we perceive different saltiness if anion species are changed in aqueous solution. Sourness is due to hydrogen ions generated from acetic acid, hydrochloric acid, and citric acid, for example. It activates metabolism, but generally is a signal of rot.

Table 4.1 Chemical substances showing five basic taste qualities, pungency, and astringency related to the sense of pain and the meaning of each taste

Taste	Main materials	Meaning, characteristics
Sweetness	Sucrose, glucose, artificial sweetener	Source of energy
Saltiness	Cations represented by sodium ions	Supply of minerals necessary for fluid balance
Sourness	Dissociated hydrogen ions from acetic acid, hydrochloric acids, citric acids, and so on	Activation of metabolism, signal of decay
Bitterness	Caffeine, theobromine, quinine, humulone	Warning of toxicity
Umami	Monosodium glutamate (MSG), disodium inosinate (IMP), disodium guanylate (GMP)	Supply of essential amino acids and nucleotides (ingredient of nucleic acids)
Astringency	Tannin-based compounds	Through mucosal protein surfaces and bitter taste receptors
Pungency	Capsaicin, allyl isothiocyanate, piperine	Through the thermal, pain receptors

There are many kinds of chemical substances tasting bitter, as represented by caffeine, theobromine, quinine, and isohumulon. Bitterness gives a warning of toxicity. Umami is the taste produced typically by monosodium glutamate (MSG) contained in seaweeds. Other typical umami substances are disodium inosinate (IMP) mainly contained in fish and meat and disodium guanylate (GMP) contained in mushrooms. Umami plays the role of supplying indispensable amino acids and nucleotides to our bodies.

Chemical compounds with pungency are received at sensory receptors that are sensitive to temperature and pain. In other words, they are not received at taste receptors in gustatory cells. The major pungent compounds are allyl isothiocyanate, capsaicin, and piperine. Astringency is caused by polyphenol, mainly tannin. These compounds are said to mainly stimulate pain receptors.

We have many sandy grains called papillae on the tongue. One papilla contains several to a few hundred sensory organs called taste buds, which are composed of approximately 50–100 gustatory cells. Bitterness receptors, a group of taste-2 receptors (T2Rs) present in taste cells, were discovered in 2000 [31–33] followed by the discovery of sweetness receptors (T1R2+T1R3) [34] and umami receptors (T1R1+T1R3) [35]. Each taste receptor receives multiple chemical substances constituting a single taste. Namely, taste receptors exhibit semiselectivity rather than rigid and high selectivity. As for sourness and saltiness, the reception mechanisms have not yet been completely clarified; polycystic kidney disease 2-like 1 protein (PKD2L) [36, 37] and epithelial sodium channel (ENaC) [38] have been identified as the candidate receptors, respectively. The reception of taste substances leads to the release of neurotransmitters, and hence, taste information perceived in this way is transmitted to taste nerves and finally reaches the gustatory area in the brain.

The measurement of taste is very different from that of other quantities such as light and sound waves. Among the five senses, light is received by of the sense of sight, sound waves by the sense of hearing, and pressure or temperature by the sense of touch. These quantities are single physical quantities. In other words, these are properties of an object or attributes such as length and weight. These quantities are independent of humans. Even if we do not exist, these quantities can exist. The situation is very different in the senses of taste and smell. In these senses, an enormous variety of chemical substances are received by receptors and recognized by the brain, as aforementioned. Chemical substances have no taste or smell. Taste and smell have meaning when we interpret them. Even if we measure the attributes of chemical substances, we cannot express taste or smell. It is necessary to consider subjectivity and objectivity to gain clues for creating a taste sensor (Figure 4.1). Five senses are recognized comprehensively in the brain. The term "taste" that we normally use refers to a subjective recognition occurring in the brain. It is difficult to quantify this type of taste using current scientific tools. A very important fact in order to understand the nature of gustatory sense is that the state of potential response of neurons to the taste of various materials shows little change throughout primary to tertiary neurons. This implies that information of a basic taste has already been almost completely compiled in the primary neurons. The information is merely transmitted to the brain through the secondary and tertiary neurons. On the basis of this knowledge, let us hypothesize that we measure the electric signal in the gustatory nerve; then we can judge the taste from the recorded electrical signals. At this stage, the taste is nothing but objective data. We can quantify and know this "taste." The taste sensor should be capable of measuring and quantifying this objective "taste."

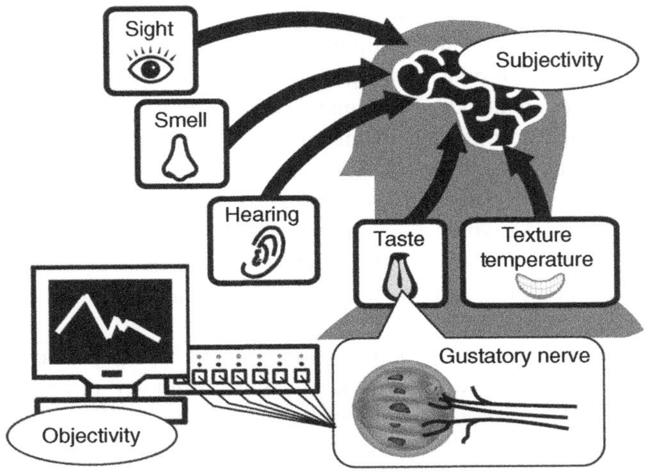

Figure 4.1 Subjectivity and objectivity of taste

4.2 Electronic Tongues

As mentioned previously, the features of electronic tongues based on sensor arrays to measure liquid are (i) low selectivity and high cross-selectivity instead of high selectivity and (ii) the capability of statistically analyzing the outputs from multiple sensors. These features, that is, low selectivity, high cross-selectivity, and statistical analysis, are similar to those of electronic noses for measuring gases. The development of electronic noses started in 1982 [39–43], and new measurement technologies have been generated, together with electronic tongues. Table 4.2 shows a summary of e-tongues and taste sensing studies. Many review papers [1–25] have been published, and hence we explain briefly typical e-tongues here.

A St. Petersburg–Rome collaboration team reported e-tongues with arrays of ion-selective electrodes (ISE) and the analysis of pollution in water in 1997 [44]. Porphyrins and related macrocycles are known as ionophores in ion-sensitive electrodes, and D'Amico and Di Natale reported porphyrin-based e-tongues for making analytical chemistry applications for liquids [45, 48] and reported an example of an application to quality management of wines [49].

Winquist and Lundström reported a voltammetric e-tongue in 1997 [52] and then developed a hybrid e-tongue by combining the technologies for potentiometry, voltammetry, and conductivity measurement [15, 21, 53, 62]. Six different types of metallic electrode were used as the working electrodes in voltammetric measurements to obtain different potential responses, and PCA was used to analyze the obtained data and discriminate foods [15, 47]. The discrimination of beer classes in the aging process was successful using a cyclic-voltammetric e-tongue including three enzymatic biosensors, and the reduction of the variables of electrochemical signals was also performed for wine characterization using a genetic algorithm (GA) combined with PLS [51, 63].

The group of Vlasov and Legin in St. Petersburg reported an e-tongue with solid-state crystalline ISE based on chalcogenide glass [46, 64] and discriminated and analyzed foods and beverages such as Italian wine [50] and mineral water by PCA and analysis using neural network techniques [20]. They also reported the quantification of the bitter taste of structurally diverse active pharmaceutical ingredients (APIs) using an e-tongue under parameter-limited conditions [54].

Table 4.2 Summary of e-tongues and taste sensing studies

Type of study	Type of sensing principle	Data processing	References
Environmental application (polluted water)	Sensor array of ion-selective electrodes (ionophores), potentiometric sensor	Multiple linear regression (MLR) Partial least squares (PLS) Nonlinear regression (NLLS) Backpropagation neural network (BPNN) Principal component analysis (PCA)	[44–46]
	Voltammetric sensor (some metals electrodes)	PCA	[47]
Array of miniaturized potentiometric sensor	Sensor array of ion-selective electrodes (ionophores), potentiometric sensor	PLS	[48]
Application for quality management of wines	Potentiometric sensor, ion-selective electrodes (ionophores)	PCA, PLS	[49, 50]
Monitoring the aging beers and wines	Enzymatic biosensors, cyclic voltammetry	PCA, neural network	[51]
Classification of various fruit drinks and milk	Voltammetric sensor	PCA	[52, 53]
Hybrid e-tongue	Potentiometry, voltammetry, and conductivity measurements	PCA, artificial neural net	[21]
Bitter taste assessment of pharmaceutical	Potentiometric sensor array (27 cross-sensitive sensors)	PLS	[54]
SH-SAW-based e-tongue (discrimination between liquids of different basis tastes)	SH-SAW	PCA	[55]
Biological tissue such as taste cells and receptors	Biosensor recording of extracellular potentials	—	[56–58]
Colorimetric cross-sensitive sensor array		PCA, hierarchical clustering analysis	[59–61]

Gardner's group reported not only e-noses but also e-tongues based on a dual shear horizontal surface acoustic wave (SH-SAW), which was successful in discriminating between liquids of different basic tastes [55]. Although the sensor category does not belong to e-tongues, a microfluidic device was constructed for on-site determination of fish freshness using the adenosine-5-triphosphate (ATP) concentration as an indicator of freshness by Suzuki's group [65]. The combination of this kind of device with e-tongues will contribute largely to quality monitoring of foods.

Recently, researchers have been attempting to develop biosensors using biological tissue such as taste cells and receptors by applying genetic engineering technologies [56–58]. Whereas the reproducibility and stability of usable sensors should be improved, this kind of e-tongue might be promising as a future sensing device. Suslick's group analyzed liquids such as beer and soft drinks using a colorimetric cross-sensitive sensor array that comprises multiple chemically responsive dyes [59–61], and furthermore, tried to identify and characterize lung cancer by measuring exhaled breath using the same principle and device [66].

Electronic tongues are aimed at discriminating and analyzing foods and beverages and are well known as sensing technologies that greatly contribute to quality management. Aissy Inc., Japan, a venture from Keio University, provides accurate analysis using its original taste sensors and services useful for the development of new products and marketing in the food industry [67].

There are two types of commercialized electronic tongue in the world [7, 8, 68]. One is the taste sensing systems SA402B and TS-5000Z (Intelligent Sensor Technology Inc., Japan), which is usually called the taste sensor, and another is the Astree II e-tongue (Alpha MOS, France).

4.3 Taste Sensor

4.3.1 Introduction

Objective taste evaluation has been attracting attention in various fields, such as the food, beverages, and pharmaceutical industries. As mentioned in Section 4.2, a promising candidate for the taste evaluation is an e-tongue. It is equipped with different types of sensors, including ISE, ionophore-based electrodes, and enzymatic biosensors. These electrodes have extremely high selectivity to a given ion, but low selectivity to similar taste substances. Resultant data are analyzed by multivariate analyses, such as PCA, discriminant analysis (DA), and artificial neural network (ANN). However, these techniques cannot adequately predict the taste unless "supervised data," such as the kind and/or amount of chemical components contained or experimental conditions, are obtained before the analysis. This means that the use of electronic tongues may be limited only to quality control in which a sample is compared with a control sample to evaluate whether they are the same or not in terms of a certain point.

We have been developing a taste sensing system whose concept is quite different from that of the aforementioned electronic tongues. In the 1980s, Toko and coresearchers at Kyushu University, Japan, studied electrochemical phenomena of lipid membranes and found that a membrane doped with a dioleyl phosphate (DOPH) responded to some tastants [69–83], and in 1990, they reported that multichannel lipid/polymer membrane sensors exhibited different output patterns to different taste substances [26, 84]. Following those results Toko's group not only conducted further research on the multichannel sensors [85–105], but also applied them to foods and beverages, including beer [85, 91, 106, 107], coffee [108], sake [109, 110], miso (soybean paste) [111], milk [112], green tea [113], amino acids [114–116], wine [117], soy sauce [118], salts [119–121], and beef [122]. The results clearly demonstrate that these membranes act as "taste sensors." This section describes all aspects from the response mechanism of taste sensors to the details of the analysis methods.

4.3.2 *Principle*

Membranes of taste sensors consist of lipids, plasticizers, and polyvinyl chloride (PVC). The following is the procedure for fabricating a taste sensor electrode:

1. The necessary types and amount of lipid(s) and plasticizer(s) are added to tetrahydrofuran (THF) and mixed for 1 h.
2. PVC is added to the solution and then mixed for another 1 h.
3. The mixture is poured in to a petri dish to dry it at room temperature for 3 days.
4. A lipid/polymer membrane approximately 200 μm thick is created in the dish and then attached to the surface of a sensor probe using a solution of 800 mg PVC and 10 ml THF as an adhesive.
5. After 2 days of drying, the taste sensor electrode is complete.

Figure 4.2 shows the chemical structures of the lipids and plasticizers used [123]. Taste sensor electrodes have various characteristics depending greatly on the types and amounts of lipids and plasticizer incorporated. This will be described further in Section 4.3.6.

Measurement by the taste sensor is based on the potentiometric principle, and therefore, the electrical potentials between working electrodes for taste sensing and a reference electrode are measured. These electrodes use a silver wire coated with silver chloride (AgCl). Before the measurement, these electrodes, which are filled with a solution of 3.33 M potassium chloride and saturated AgCl solution as the inner solution, should be preconditioned for at least 2 days to stabilize the electrical potential. After the preconditioning, the electrodes are connected to an amplifier to measure the electrical potentials difference and are converted to digital code by a digital voltmeter and subsequently fed to a computer (Figure 4.3).

The taste sensing system models SA401, SA402, SA402B, and TS-5000Z were produced by Intelligent Sensor Technology, Inc. in 1993, 1996, 2000, and 2007, respectively (Figure 4.4). They have been widely used in a variety of fields, such as the food, beverage, and pharmaceutical industries, universities, and national research institutes. These instruments can be equipped with up to eight types of taste sensor electrodes, providing information on taste qualities, including saltiness, sourness, and sweetness. This will be presented in detail in Sections 4.3.3 and 4.3.6.

4.3.3 *Response Mechanism*

According to the Gouy–Chapman theory [125, 126], an electrical layer is formed on the surface of the lipid/polymer membrane of a taste sensor when it is immersed in water. Next, the membrane is immersed in a solution containing a taste substance, inducing a change in electrical potential. This means that the interaction between the membrane and the tastant causes the change in electrical potential.

To investigate the interaction further, the charge density at the membrane surface was calculated on the basis of the Gouy–Chapman theory and the Poisson–Boltzmann equation [127, 128] and was compared with the experimental membrane's response results for sodium chloride (salty) [94, 95], hydrochloric acid (sour) [95], quinine hydrochloride

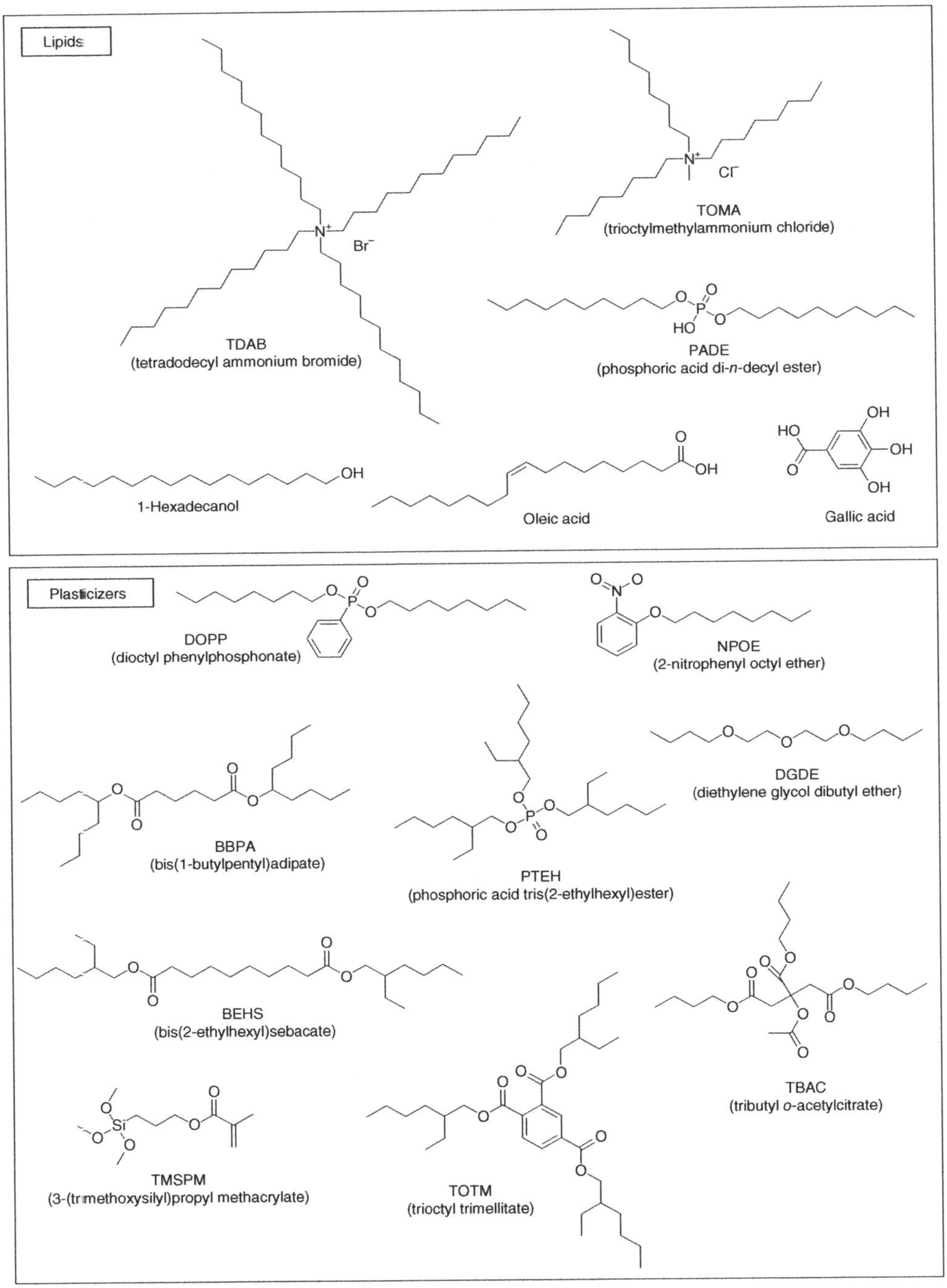

Figure 4.2 Chemical structures of artificial lipids and plasticizers. Republished with permission from Ref. [123]. Copyright 2009, Institute of Electrical Engineers of Japan

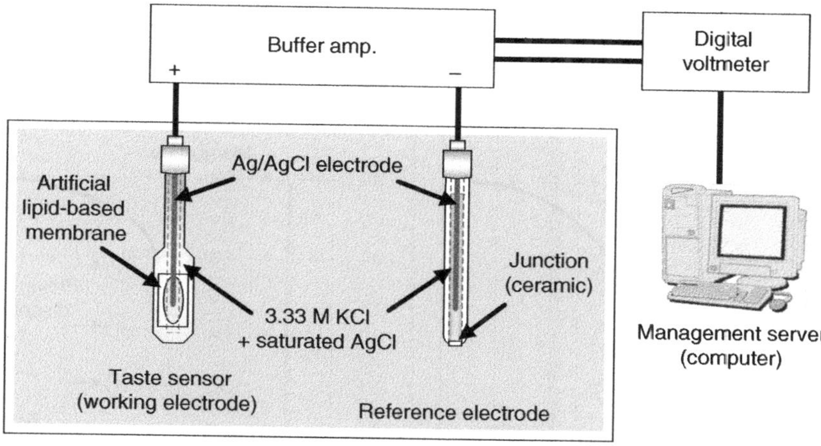

Figure 4.3 Diagram of the taste sensing system. Republished with permission from Ref. [123]. Copyright 2009, Institute of Electrical Engineers of Japan

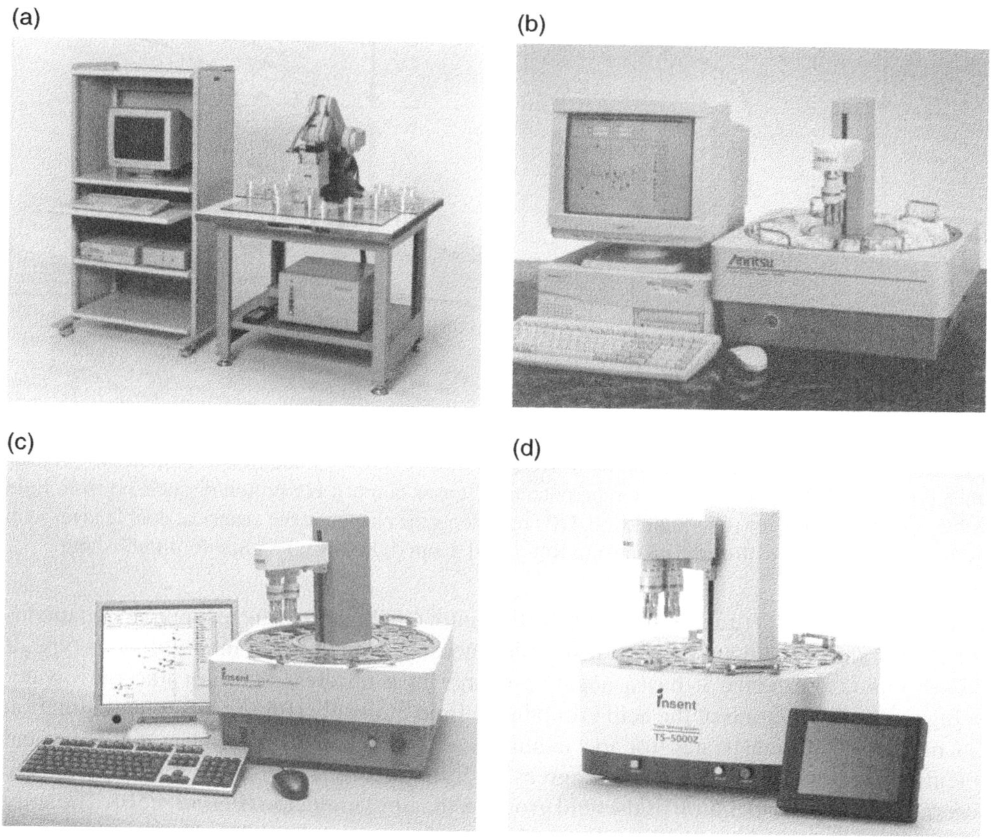

Figure 4.4 Models of the taste sensing system. (a) First model: *SA401*; (b) second model: *SA402*; (c) third model: *SA402B*; and (d) fourth model: *TS-5000Z*. Republished with permission from Ref. [124]. Copyright 2013, Pan Stanford Publishing

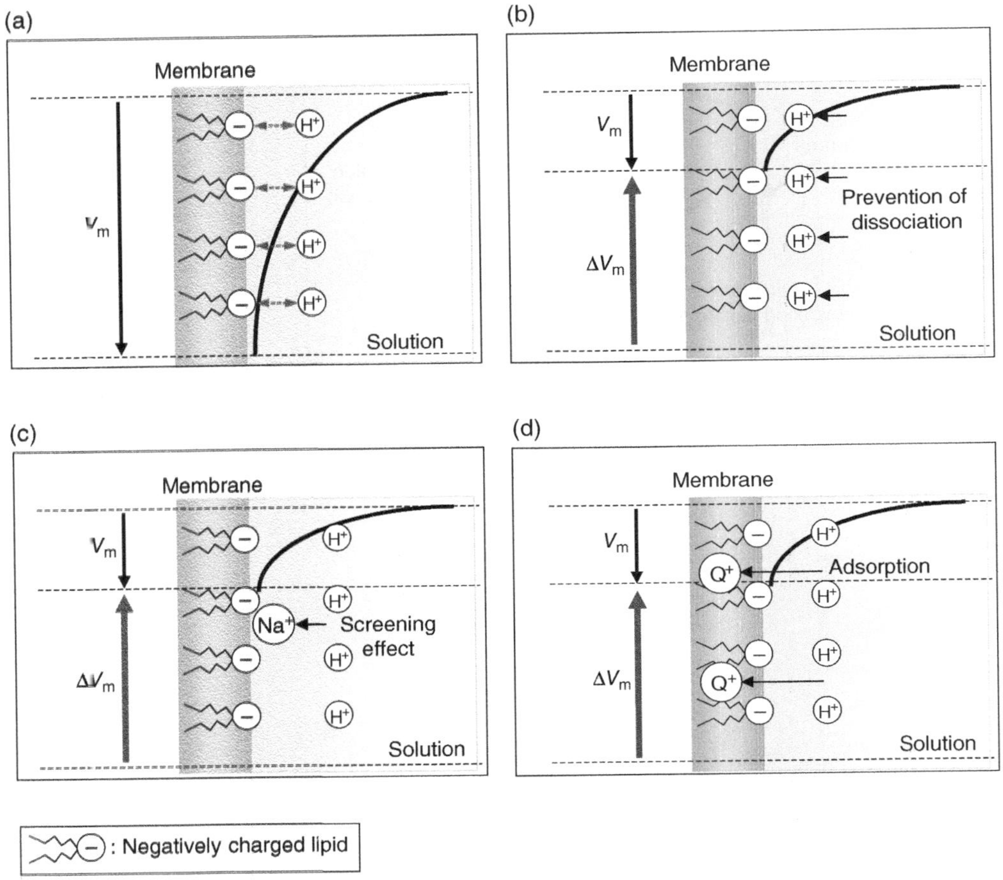

Figure 4.5 Diagrams of response mechanisms of negatively charged membrane to sour, salt and bitter taste substances, normal state (a), addition of HCl, adittion of NaCl (c), addition of quinine. V_{mm}, membrane potential; ΔV_m, change in membrane potential (sensor output); H^+, proton dissociated from lipid molecule; Na^+, sodium ion; Q^+, quinine ion. OBJ represents the change in the electrical double layer with distance. Republished with permission from Ref. [124]. Copyright 2013, Pan Stanford Publishing

(bitter) [95], and MSG (umami, savory) [100]. Figure 4.5 illustrates the response mechanisms of a membrane to the three types of taste substances. This membrane incorporates a type of phospholipid, and hence it has a negative charge in a tasteless solution like pure water owing to the dissociation of the acid group in the lipid molecule (Figure 4.5a). This situation can be called "normal state" for the membrane. The sour material causes the electrical potential to be more positive, which agrees with the theoretical results. This suggests the prevention of the dissociation of the acid group in the lipid molecule (Figure 4.5b). The salty substance also causes a positive change in the electrical potential, and the result is in agreement with the theoretical ones. This demonstrates that potassium chloride directly affects the electrical double layer formed at the membrane surface, which is called the "screening effect" (Figure 4.5c).

The amount of quinine hydrochloride in the membrane immersed in a solution of 1 mM quinine hydrochloride (bitterness) for 1 h was determined by electron spectroscopy for chemical analysis (ESCA) [100]. The result showed that quinine hydrochloride was adequately contained in the membrane, suggesting that it strongly adsorbs on the hydrophobic part of the membrane and then causes a change in the electrical potential. The umami substance is considered to be involved in a unique mechanism because it causes the negatively charged membrane to be more negative. In addition, the ESCA experimental result showed that the umami sample was not contained in the membrane although it was rinsed with pure water before the analysis. These results indicate that MSG interacts with the membrane via the positively charged amino group, while the negatively charged carboxyl group causes the electrical potential to be more negative [100, 104].

4.3.4 Measurement Procedure

As mentioned in Section 4.3.3, interactions of taste substances with lipid/polymer membranes depend on the type of taste quality and can be divided into two groups: (i) electrostatic interaction (salty, sour, and umami substances) and (ii) adsorption (bitter and umami substances).

To detect both types of interactions, the specific measurement procedures have been developed [113, 129]. Figure 4.6 illustrates the measurement procedure based on the change in the electrical potential of membranes over time. First, a taste sensor electrode is immersed in a reference solution of 30 mM potassium chloride and 0.3 mM tartaric acid. This solution is used to measure the potential V_r as a standard output and is referred to as the tasteless sample in the taste measurement. Second, the potential V_s is measured in the sample solution. Third,

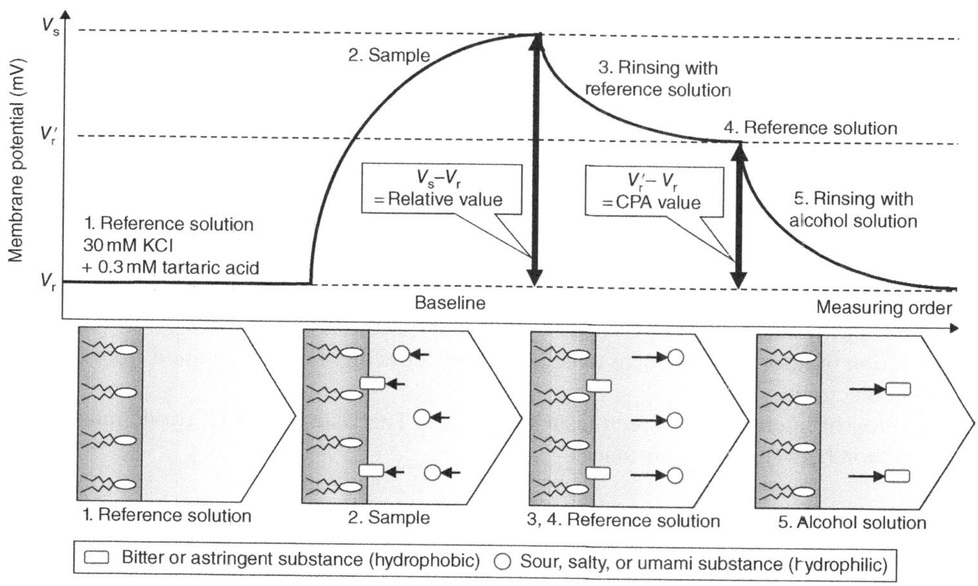

Figure 4.6 Measurement procedure. Republished with permission from Ref. [124]. Copyright 2013, Pan Stanford Publishing

after rinsing the membrane with the reference solution, the potential V_r' is measured again in the reference solution. The difference in the potentials ($V_s - V_r$) is the relative value, which reflects "initial taste" for humans, including saltiness, sourness, and umami. The difference in the potentials ($V_r' - V_r$) indicates the change in the electrical potential owing to the adsorption of samples to the membrane and is called the CPA value (change of membrane potential caused by adsorption of chemical substances). This properly reflects "aftertaste" for humans, including bitterness and astringency. Finally, the taste sensor electrode is rinsed with washing solution containing alcohol to remove the adsorbed chemicals before measuring the next sample.

4.3.5 *Sensor Design Techniques*

Following the research results obtained by Toko's group on the taste sensor, the first model of taste sensing system SA401 was put on the market in 1993. It was capable of the classification of various foods and beverages. However, the taste sensor membranes on the system at that time had some problems to be overcome. Each of the membranes had low selectivity to similar taste substances, and therefore all sensor outputs must be analyzed together by multivariate analysis, such as PCA. PCA is one of the mathematical techniques of reducing a large number of variables to as few alternative variables as possible without loss of information. However, principal components from PCA results are dimensionless data, and hence the resultant data cannot be interpreted as taste information unless information on taste qualities for all samples is given before the analysis. This characteristic allows the classification based on the PCA results to be suited to quality control but unsuited to taste evaluation for the development of food and beverage products.

To overcome this problem, the concept behind the taste sensor has drastically changed to a new one that each sensor membrane should respond more selectively to similar taste substances and recognize different taste qualities by itself. To achieve these points, the requirements below must be fulfilled in view of taste sensing technology [13]:

1. Threshold: The threshold of taste sensor electrode must be the same as the human taste threshold.
2. Global selectivity: The taste sensor electrode must respond consistently to the same taste, similarly to the human tongue.
3. High correlation with human sensory scores: The taste sensor electrode must have high correlation with human sensory scores.
4. Definition of information: There must be a clear defined unit of information from the taste sensor electrode.
5. Detection of interaction between taste substances: The taste sensor electrode must detect interaction between taste substances.

These requirements will be discussed in Section 4.3.6 in more detail. To design the sensor characteristics, we focused on the fact that the physicochemical property of taste substances relies on the type of taste quality [124, 129]. Table 4.3 shows the adsorption ability and taste threshold for four types of taste substances. Salty substances such as potassium chloride have such a low hydrophobicity as to be easily hydrated in water and a relatively high taste threshold

Table 4.3 Physicochemical properties of taste qualities

Taste quality	Adsorption ability	Taste threshold
Saltiness	None	High
Sourness	None	Low
Bitterness	High	Low
Umami	Low	Medium

Source: Republished with permission from Ref. [124]. Copyright 2013, Pan Stanford Publishing.

because they are vital to human life as a mineral source. Sour substances such as citric acid also have low hydrophobicity and a low taste threshold because sourness is a signal of the decomposition of food and beverage to humans. Bitter substances such as quinine hydrochloride have such a high hydrophobicity that they are negligibly hydrated in water and an extremely low taste threshold because bitterness is traditionally recognized as indicating toxicity. Umami substances such as MSG or peptides have a slight aftertaste, which is often called "richness" or "mouthfulness," possibly because of their ability to slightly adsorb on the human tongue. Also, they have a medium level of the taste threshold compared with other taste substances. In this way, taste substances can be classified in accordance with the two physicochemical properties.

As mentioned in Section 4.3.3, the interaction between taste substances and lipid/polymer membranes depends on the type of taste quality. On the basis of the interaction and the classification, we proposed sensor design techniques to optimize the electric charge density and hydrophobicity of lipid/polymer membranes.

Let us consider one method of optimizing the electric charge density [130]. Not only the type but also the amount of lipid incorporated in a membrane significantly affects the sensor characteristics. Figure 4.7 shows the effect of lipid concentration in a membrane on the relative values of a positively charged membrane sensor that contains tetradodecyl ammonium bromide (TDAB) as the lipid and 2-nitrophenyl octyl ether (NPOE) as the plasticizer. In general, the electric charge density increases with increasing the charged lipids concentration. In Figure 4.7, the response to NaCl (saltiness) increases negatively as the lipid concentration increases. This suggests that the screening effect by Cl^- anions causes the electrical potential to be more negative. In contrast, the increment in the electric charge density causes the membrane to be unresponsive to tartaric acid (sourness) and MSG (umami). This is because TDAB's function as an anion exchanger prevents interactions with H^+ cations from such tastants.

A specific nonlinear response to iso-alpha acid (bitterness) is observed; there is a negative peak of the relative value at the lipid concentration of 0.1%. To ascertain the cause of the non-linear response, we also examined the effect of the electric charge density on the membrane potential in the reference solution, which is V_r, as mentioned in Section 4.3.4. The top graph in Figure 4.8 shows the relationship between the lipid concentration in the membrane and the membrane potential for iso-alpha acid, and the bottom graph is part of Figure 4.7. As shown in the upper graph of Figure 4.8, the membrane potential drastically increases at low lipid concentrations with increasing lipid concentration upon the addition of a positively charged lipid into the membrane. However, it becomes almost constant at higher lipid concentrations owing to the limit of the electric charge density. That is why a curve is created in the figure and why only the bitter substance exhibits the nonlinear response. Most bitter substances

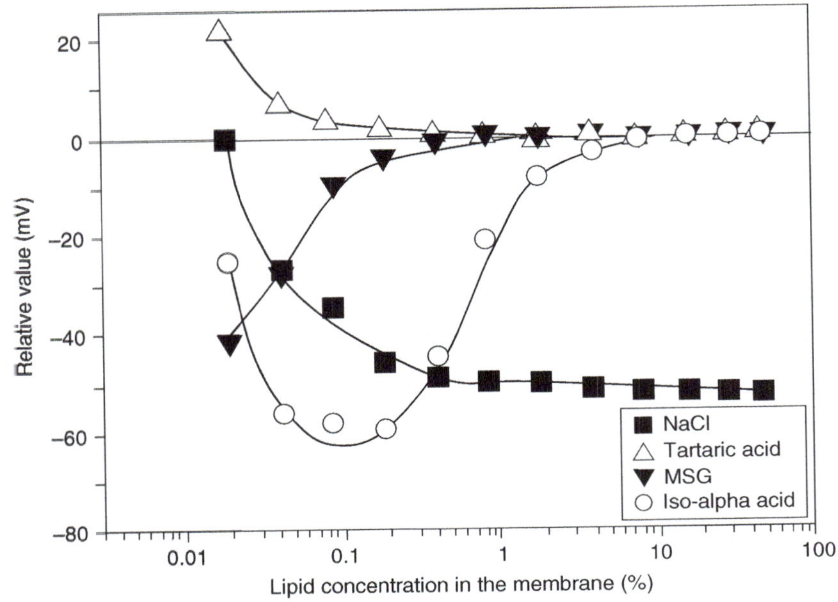

Figure 4.7 Relationship between lipid concentration in membrane and relative value of bitterness sensor. The concentrations of each of samples are 0.01 vol.% iso-alpha acid, 300 mM NaCl, 2.7 mM tartaric acid, 10 mM MSG. All samples include 30 mM KCl and 0.3 mM tartaric acid as the supporting electrolyte. Republished with permission from Ref. [130]. Copyright 1999, Institute of Electrical Engineers of Japan

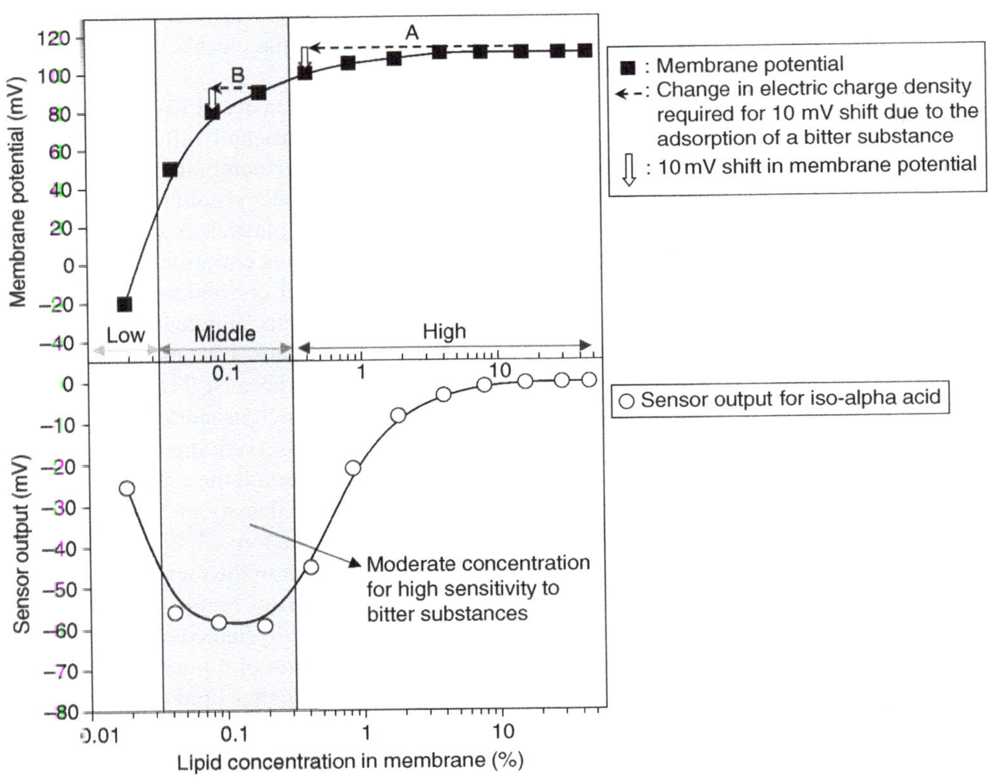

Figure 4.8 Relationships between lipid concentration in the membrane and membrane potential and between lipid concentration in the membrane and relative value for iso-alpha acid. Republished with permission from Ref. [130]. Copyright 1999, Institute of Electrical Engineers of Japan

interact with lipid/polymer membranes by adsorption and cause a change in the electrical potential. That means that the response to bitter substances is determined by the change in the electric charge density.

To understand this, let us consider the changes at three lipid concentrations. In the upper graph of Figure 4.8, downward arrows mean a change in the electrical potential that yields a sensor output of 10mV, while leftward dashed arrows mean a change in the electric charge density necessary for the 10mV sensor output. As arrow A indicates, a considerable change in the electrical potential at a high lipid concentration is necessary to obtain a sensor output of 10mV. However, such a change is practically impossible because an extremely small amount of bitter substances adsorb on the membrane, thus causing a slight change. Therefore, membranes containing a high concentration of lipids exhibit no or less response to bitter substances. In contrast, as arrow B indicates, even a slight change in the electrical potential at moderate lipid concentrations causes a sensor output of 10mV. This means that membranes containing a moderate concentration of the lipid have the highest sensitivity to bitter substances compared with other lipid concentrations. Membranes with a small amount of lipid have low sensitivity because there is no lipid on which bitter substances can adsorb. Therefore, the peak of the relative values represents the most appropriate amount of lipid for high sensitivity and selectivity to bitter substances.

In contrast, Figure 4.7 also shows that membranes with a significant amount of lipid have high sensitivity and selectivity to salty materials. As described in Section 4.3.3, the response mechanism to salty substances involves the screening effect, and hence increasing the amount of lipid to enhance the effect is desirable for good sensitivity and selectivity [129]. The response to umami substances is associated with both the screening effect and adsorption [100], which means that membranes containing moderate amounts of the lipid exhibit high sensitivity and selectivity to such substances. Sour materials cause a change in the electrical potential through the dissociation of the acid group of the lipid molecule, meaning that membranes with high sensitivity and selectivity to such materials must contain an adequate amount of lipid. That appears to be contradictory to Figure 4.7. The lipid TDAB is one of the quaternary ammonium salts that fully dissociate at any pH. Therefore, TDAB is not suited for use as the lipid in the membrane of a sourness sensor, and negatively charged lipids that partially dissociate in water, such as phosphoric acid di-*n*-decyl ester (PADE), are desirable [129].

Next, let us consider another method of optimizing the hydrophobicity of lipid/polymer membranes. Our experimental results show that lipids with short hydrophobic chains are suitable for the sensing material of salty and sour substances [129], while lipids with long hydrophobic chains can be a key element in the bitterness sensor. This implies that lipids with characteristics similar to those of the target tastant are promising sensing materials.

Plasticizers were originally considered to be responsible for physical properties such as strength and stiffness [131]. However, we have revealed that plasticizers can also be promising materials for controlling the hydrophobicity of membranes [123]. Figure 4.9 shows the effects of the type of plasticizer and the amount of lipid on the CPA values for basic taste substances. A negatively charged lipid, PADE, is incorporated in the membranes at different concentrations with eight types of plasticizers, including dioctyl phenylphosphonate (DOPP) and bis (1-butylpentyl) adipate (BBPA). For all the plasticizers, lipid-free membranes have no CPA value for all substances, even quinine hydrochloride. This clearly demonstrates that lipids are necessary as they act as ionic sites where changes in the electrical potential of the membrane occur upon the adsorption of taste substances. This will be discussed further in Section 4.8.3. The method of optimizing the membrane hydrophobicity will aid the development of bitterness sensors.

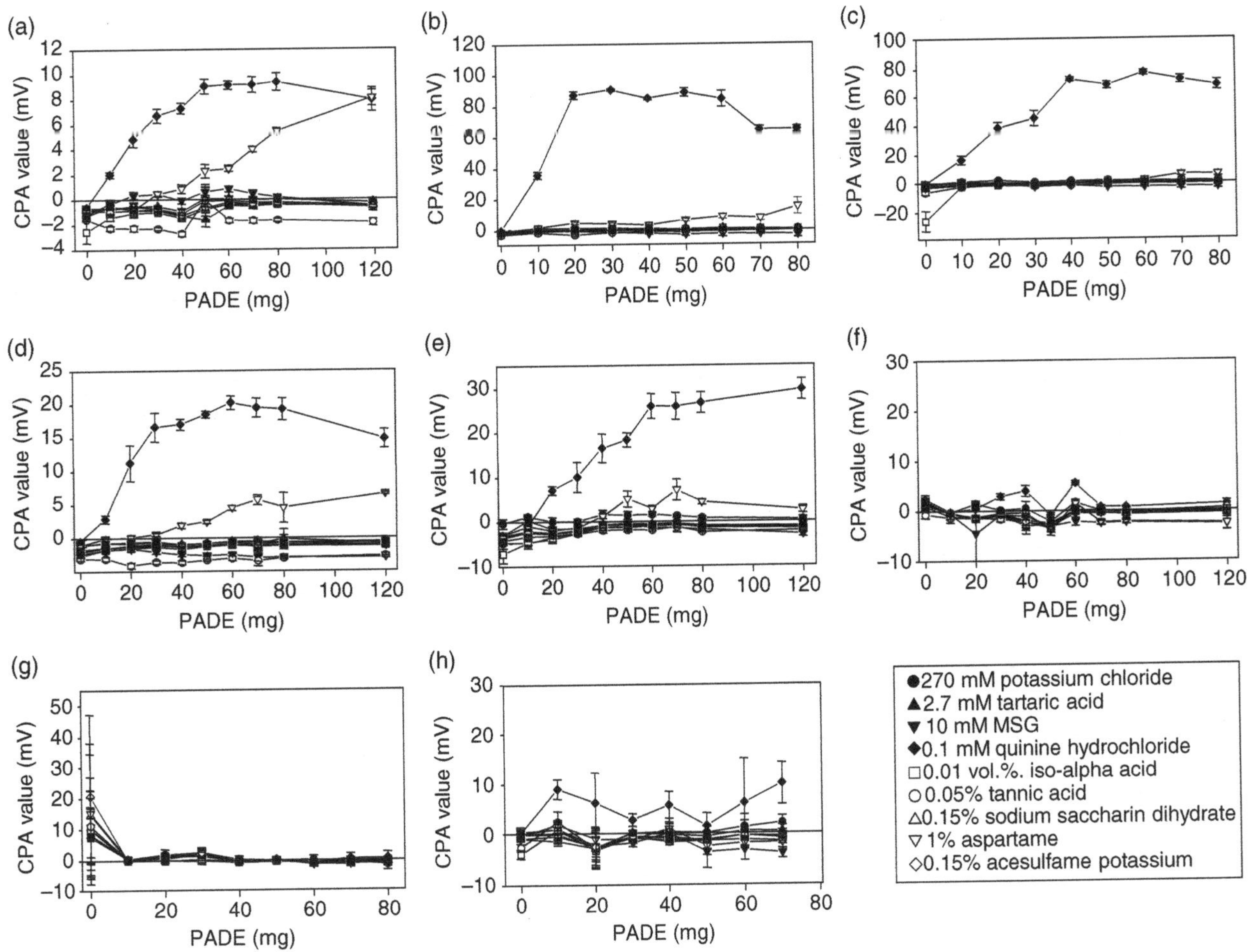

Figure 4.9 Effects of the type of plasticizer and the amount of lipid on the sensor response to substances with the basic tastes. The *x*-axis represents the PADE contents in the membrane while the *y*-axis shows the CPA value. Data are expressed as mean ± SD (n=4). All samples include 30 mM KCl and 0.3 mM tartaric acid as the supporting electrolyte. Lipid: PADE (phosphoric acid di-*n*-decyl ester); Plasticizers: (a) DOPP; (b) BBPA; (c) BEHS; (d) PTEH; (e) TBAC; (f) TMSPM; (g) DGDE; (h) TOTM. Republished with permission from Ref. [123]. Copyright 2009, Institute of Electrical Engineers of Japan

4.3.6 *Basic Characteristics*

With the sensor design techniques, nine types of "next-generation" taste sensor membranes have been developed. The components of the taste sensor membranes are listed in Table 4.4. The membranes can be divided into three groups depending on the type of lipids. Positively charged membranes contain a positively charged lipid, such as a quaternary ammonium salt, in the reference solution and respond negatively to negatively charged samples. Negatively charged membranes contain a negatively charged lipid, such as a phosphoric acid ester, in the reference solution and exhibit a positive response to positively charged analytes. Blend membranes contain both types of lipids.

The taste sensor electrodes have different responses depending on their characteristics and accurately provide information on taste quality, such as saltiness, sourness, bitterness, sweetness, umami, and astringency (Figure 4.10) [124]. The umami sensor AAE has high sensitivity and selectivity to the umami sample of 10 mM MSG solution with the relative value, which is the difference in the potentials ($V_s - V_r$), as defined in Section 4.3.4, of about −80 mV. This sensor membrane is one of the blend membranes that contain two types of lipids at the moderate lipid concentration designed to interact with umami substances through electrostatic and hydrophobic interactions. It also can be used for the evaluation of kokumi taste [132], which is also referred to as "continuity," "mouthfulness," or "thickness" [133]. The saltiness sensor CT0 selectively responds to the salty sample of 270 mM KCl solution with the relative value of about −50 mV. This sensor incorporates a high concentration of a positively charged lipid to act as an anion exchanger.

The sourness sensor CA0 exhibits a selective response to the sour sample of 2.7 mM tartaric acid solution with the relative value of over 30 mV. This sensor contains two types of negatively

Table 4.4 Chemical components of taste sensor membranes

Taste sensor electrode	Artificial lipid	Plasticizer
Umami sensor AAE	Phosphoric acid di(2-ethylhexyl) ester, trioctylmethylammonium chloride	Dioctyl phenylphosphonate
Saltiness sensor CT0	Tetradodecylammonium bromide 1-hexadecanol	Dioctyl phenylphosphonate
Sourness sensor CA0	Phosphoric acid di(2-ethylhexyl) ester, oleic acid, trioctylmethylammonium chloride	Dioctyl phenylphosphonate
Bitterness sensor C00 (for acidic bitter materials)	Tetradodecylammonium bromide	2-Nitrophenyl octyl ether
Bitterness sensor AC0 (for basic bitter materials)	Hexadecanoic acid	Dioctyl phenylphosphonate
Bitterness sensor AN0 (for basic bitter materials)	Phosphoric acid di-*n*-decyl ester	Dioctyl phenylphosphonate
Bitterness sensor BT0 (for bitter hydrochloride salts)	Phosphoric acid di-*n*-decyl ester	Bis(1-butylpentyl) adipate Tributyl *O*-acetylcitrate
Astringency sensor AE1	Tetradodecylammonium bromide	Dioctyl phenylphosphonate
Sweetness sensor GL1	Trimellitic acid, tetradodecylammonium bromide	Dioctyl phenylphosphonate

Source: Republished with permission from Ref. [124]. Copyright 2013, Pan Stanford Publishing.

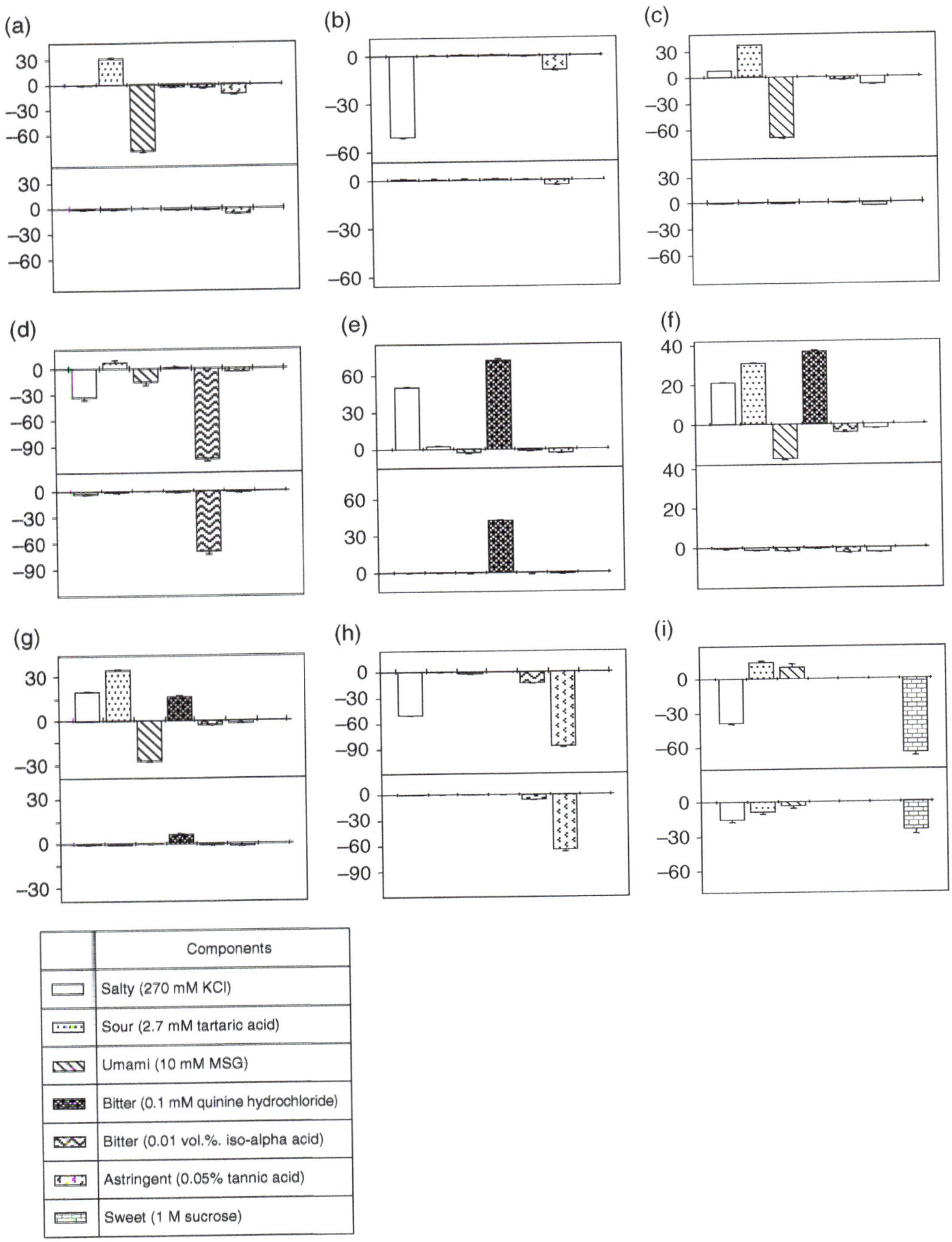

Figure 4.10 Responses of taste sensor electrodes to solutions of basic tastes. Data are expressed as mean ±SD ($n=4$). All samples include 30 mM KCl and 0.3 mM tartaric acid as the supporting electrolyte. (a) Umami sensor AAE; (b) saltiness sensor CT0; (c) sourness sensor CA0; (d) bitterness sensor C00; (e) bitterness sensor BT0; (f) bitterness sensor AC0; (g) bitterness sensor AN0; (h) astringency sensor AE1; (i) sweetness sensor GL1. Top and bottom graphs in each figure represent relative values and CPA values, respectively. Republished with permission from Ref. [124]. Copyright 2013, Pan Stanford Publishing

charged lipids as well as a positively charged lipid to sufficiently reflect the change in the electrical potential owing to the dissociation of the acid group of the negatively charged lipids at acidic pH.

The bitterness sensor C00 has high sensitivity and selectivity to the bitter sample of 0.01 vol.% iso-alpha acid solution with the relative value of over −100 mV and the CPA value of approximately −60 mV. This sensor membrane includes a positively charged lipid that selectively interacts with negatively charged bitter substances, which is also called acidic bitter substances. It can be used for the evaluation of the bitterness of beer [134, 135]. In contrast, the bitterness sensors BT0, AC0, and AN0 contain a negatively charged lipid to interact with basic bitter substances that are positively charged. The bitterness sensor BT0 has been developed to highly and selectively respond to such bitter substances [123] and shows the relative value of approximately 80 mV and the CPA value of approximately 40 mV to the bitter substance of 0.1 mM quinine hydrochloride solution. The details of the sensor BT0 will be discussed in Section 4.8.3.

Another bitterness sensor, AN0, for acidic bitter substances exhibits relatively high sensitivity and selectivity to the bitter quinine hydrochloride with the CPA value of approximately 10 mV. Compared with the sensor BT0, the sensors AC0 and AN0 have lower sensitivity but have been shown to be beneficial for the evaluation of clarithromycin [136, 137] and famotidine [138], respectively.

The astringency sensor AE1 selectively responds to the astringent sample of 0.05% tannic acid solution with the relative value of approximately −90 mV and the CPA value of approximately −60 mV. It is well suited for the evaluation of the astringency of green tea and black tea [139–145].

The sweetness sensor GL1 is a sensor with a different response mechanism from those of other sensors, as shown in Figure 4.5. Most sweetness is provided by sugars and sugar alcohols, which are nonelectrolytes. The nonelectrolytes, in principle, cannot cause a change in the electrical potential, and therefore it was previously thought that taste sensor membranes based on potentiometric measurement cannot detect sweetness.

However, our research group found that DOPP-based membranes preconditioned with a solution of gallic acid or some of its derivatives as a sweetness-sensing material responded to solutions of 1 M sugar or sugar alcohol [146–154]. This may be due to the interaction between the membrane and sweet substances via the binding of a carboxyl group of the sweetness-sensing material to two adjacent hydroxyl groups of sweet substances. However, the research also revealed that the membranes used without rinsing with a washing solution of 30% EtOH and 10 mM KOH before measuring the samples exhibit no response. This possibly indicates the existence of substances that help mediate the interaction. The sweetness sensor GL1 containing trimellitic acid, selected from among candidate sweetness-sensing materials, exhibits the highest response to the sweet sample of 1 M sucrose solution with the relative value of approximately −60 mV.

These advanced taste sensor membranes have been applied not only to various types of foods and beverages, including coffee [155], kamairi-cha [156], matcha [157], wine [158], distilled spirits [159, 160], soup stock [161], milk [162, 163], meat [164–166], sausage [167], rice [168, 169], salts [170, 171], dried squid [172], and food additives [173] but also to water quality inspection [174]. Practical examples will be discussed in Section 4.7 in greater detail.

4.3.6.1 Threshold

Commonly used chemical analyses or ISE have micro-, nano-, and picomolar detection limits for a target, providing valuable quantitative information. However, taste sensor electrode must have the same taste threshold as humans because a too high detection limit leads to low correlation between taste sensor electrodes and human sensory scores.

As discussed in Section 4.3.5, the physicochemical property of taste substances relies on the types of taste qualities because each taste quality has a different meaning to humans. For this reason, the human taste thresholds decrease in the order of saltiness, umami, sourness, and bitterness [175, 176]. Figure 4.11 shows the concentration dependence of taste sensor electrodes for the five basic tastes and astringency [2]. The figure shows that the order of taste thresholds of taste sensor electrodes agrees well with that of humans.

4.3.6.2 Global Selectivity

Global selectivity is a whole new concept for taste sensor electrodes proposed by Professor Toko. In general, traditional ISE must have high selectivity so that they exhibit a high response to a target ion in the presence of interfering ions. However, such characteristics are not suitable for taste sensing because there are many materials that produce similar taste qualities in

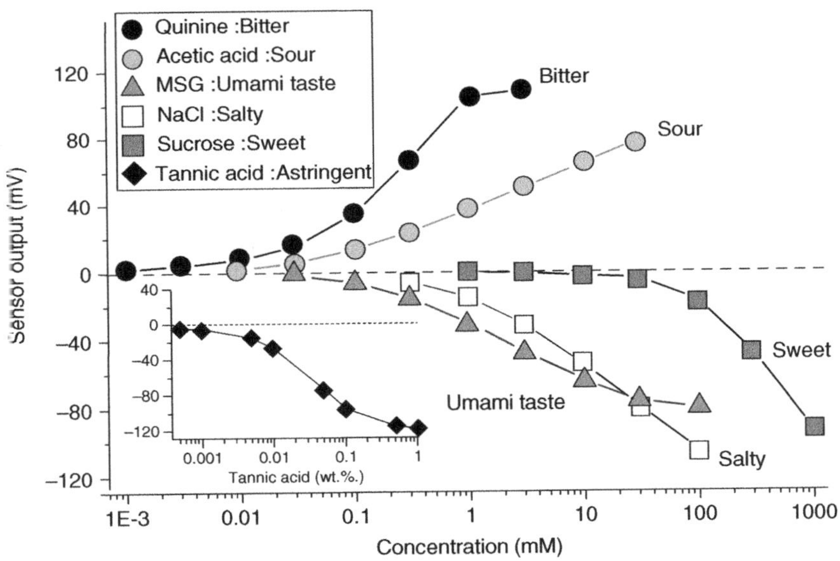

Figure 4.11 Concentration dependence on taste substances for six taste sensor electrodes. The relative values for sensors CA0, AAE, CT0, and GL1 were used for acetic acid, MSG, NaCl, and sucrose, respectively, while the CPA values for sensors BT0 and AE1 were used for quinine hydrochloride and tannic acid, respectively. Only the result for tannic acid uses weight percentage (wt.%.) because its molar weight is unknown. Data are expressed as mean ± SD (n=4). All samples include 1 mM KCl as the supporting electrolyte. Republished with permission from Ref. [2]. Copyright 2013, Pan Stanford Publishing

foods and beverages. Global selectivity is a specific characteristic needed only for taste sensor electrodes that respond consistently to the same taste similarly to the human tongue.

Figure 4.12 shows the sensor responses to similar taste substances [13, 124]. The umami substances MSG, IMP, GMP, and disodium succinate are found in seaweed, meat, and mushroom, respectively. Bitter substances are widely used in drugs in the pharmaceutical field. Astringent substances are often called "polyphenols" and are abundant in wine, black tea, green tea, and coffee. Sweet substances are all mono- or disaccharides. All of the corresponding sensor electrodes exhibit a similar response to similar taste substances, clearly indicating that the taste sensor electrodes have global selectivity.

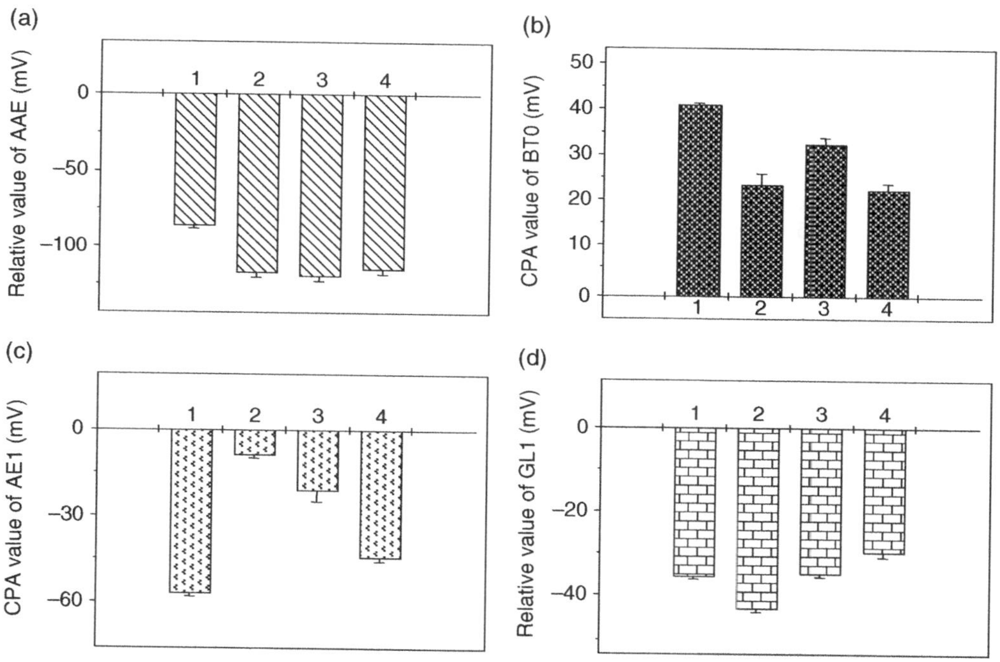

	Taste	Components
	Umami	(1) MSG, (2) IMP, (3) GMP, (4) disodium succinate (all materials are 10 mM)
	Bitter	(1) Quinine, (2) cetirizine, (3) hydroxyzine, (4) bromhexine (all materials are 0.1 mM hydrochloride salts)
	Astringent	(1) 0.05% tannic acid, (2) 10 mM gallic acid, (3) 5 mM caffeic acid, (4) 1 mM epigallocatechin gallate
	Sweet	(1) Sucrose, (2) fructose, (3) maltose, (4) lactose (all materials are 10%)

Figure 4.12 Responses of taste sensor electrodes to similar taste substances. (a) Relative value of umami sensor AAE, (b) CPA value of bitterness sensor BT0, (c) CPA value of astringency sensor AE1, (d) relative value of sweetness sensor GL1. Data are expressed as mean ± SD ($n=4$). All samples include 30 mM KCl and 0.3 mM tartaric acid as the supporting electrolyte. GMP, disodium 5′-guanosine monophosphate; IMP, disodium 5′-inosine monophosphate; MSG, monosodium glutamate. Republished with permission from Ref. [124]. Copyright 2013, Pan Stanford Publishing

4.3.6.3 High Correlation with Human Sensory Scores

By fulfilling the two requirements discussed in Section 4.3.5, that is, the same threshold as humans and global selectivity, the taste sensor data inevitably become correlated with human sensory scores. Figure 4.13 shows, for example, the correlation between the results of taste

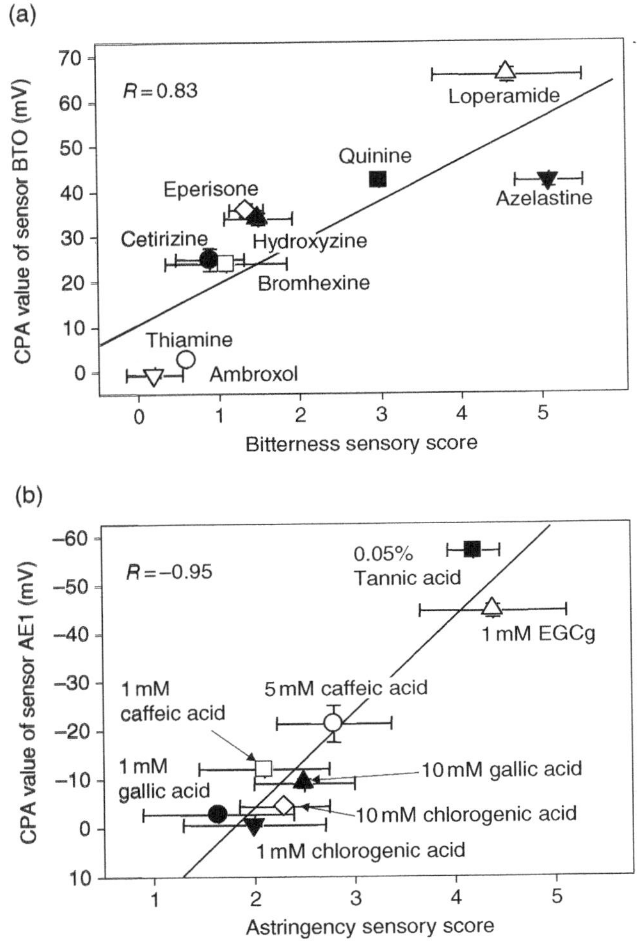

Figure 4.13 Relationship between results of taste sensor electrodes and human sensory scores for similar tastes. Tastes in (a) and (b) were scored by three and eight panelists, respectively. (a) Quinine hydrochloride concentrations of 0.01, 0.03, 0.1, 0.3, and 1.0 mM were used as standards and were assigned scores of 1, 2, 3, 4, and 5, respectively. (b) Tannic acid concentrations of 0.005, 0.011, 0.024, 0.05, and 0.11% were used as standards with assigned scores of 1, 2, 3, 4, and 5, respectively. Error bars in the *x*- and *y*-directions show the standard deviation of panelists' scores and the measurement error ($n=4$), respectively. All samples include 30 mM KCl and 0.3 mM tartaric acid as the supporting electrolyte. EGCg, epigallocatechin gallate. Republished with permission from Ref. [124]. Copyright 2013, Pan Stanford Publishing

sensor electrodes and sensory tests for similar taste substances [13, 124]. In Figure 4.13a, all bitter substances were at the same concentration of 0.1 mM. In a sensory test, azelastine hydrochloride shows the highest bitterness intensity, which is almost equal to the same bitterness as a solution of 1 mM quinine hydrochloride. In contrast, thiamine hydrochloride produces the weakest bitterness intensity and is actually almost tasteless. This difference is due to the threshold of bitterness substances. As shown in Figure 4.13a, the bitterness sensor BT0 can distinguish the difference in the thresholds, and then evaluate the bitterness with high correlation with human sensory scores. By the same token, the astringency sensor AE1 has a good correlation with human sensory scores on astringency, as shown in Figure 4.13b.

4.3.6.4 Definition of Taste Information

The Weber–Fechner law states that the ratio between the initial intensity for human stimuli, such as of the olfactory or gustatory sense, and the discrimination threshold is a constant (the Weber fraction), and the relationship between a stimulus and the corresponding perceived intensity is logarithmic [29, 177]. In this context, the smallest detectable increase for the gustatory sense is about 20% [178]. According to these findings, we have defined "taste information" on the basis of sensor outputs and their characteristics. Let us consider, for example, the NaCl concentration dependence of the saltiness sensor showing a slope of 50 mV/decade (Figure 4.14). When the concentration of 1% increases by 20%, it becomes 1.2%. Furthermore, the concentration of 1.2% increases by 20% to 1.44%. This 20% increase in the concentration equals the smallest detectable increase for humans and is constant on a

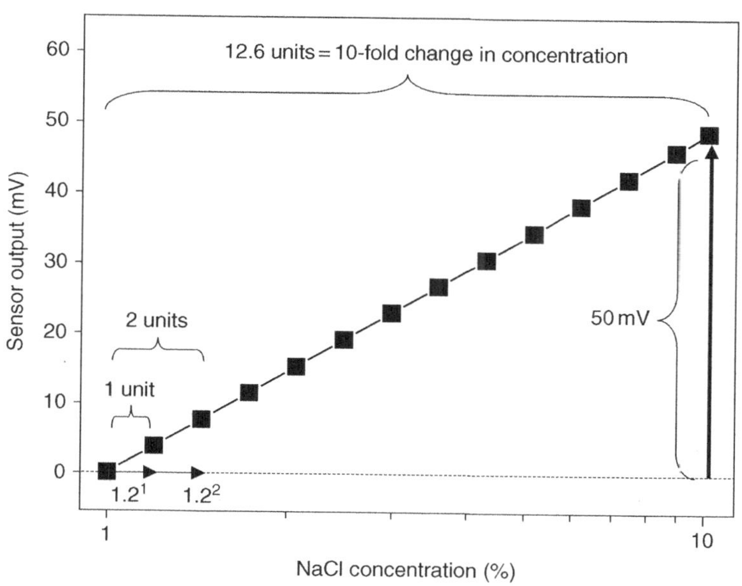

Figure 4.14 Example of conversion factor calculation for saltiness sensor with slope of 50 mV/decade for potassium chloride. Republished with permission from Ref. [124]. Copyright 2013, Pan Stanford Publishing

Table 4.5 Taste information converted from taste sensor electrodes' outputs. The type of standard sample used to calculate the conversion factor depends on the type of taste sensor electrodes

Taste sensor electrode	Taste information from relative value	Taste information from CPA value	Standard sample for calculating conversion factor
Umami sensor AAE	Umami	Richness	10 mM MSG
Saltiness sensor CT0	Saltiness	(None)	270 mM potassium chloride
Sourness sensor CA0	Sourness	(None)	2.7 mM tartaric acid
Bitterness sensor C00	Acidic bitterness	Aftertaste from acidic bitterness	0.01 vol.%. iso-alpha acid
Bitterness sensor AC0	None	Aftertaste from basic bitterness 1	0.1 mM quinine hydrochloride
Bitterness sensor AN0	(None)	Aftertaste from basic bitterness 2	0.1 mM quinine hydrochloride
Bitterness sensor BT0	None	Aftertaste from hydrochloride salts	0.1 mM quinine hydrochloride
Astringency sensor AE1	Astringency	Aftertaste from astringency	0.05% tannic acid
Sweetness sensor GL1	Sweetness	(None)	1 M sucrose

Source: Republished with permission from Ref. [124]. Copyright 2013, Pan Stanford Publishing.

logarithmic scale. Therefore, if the 20% increase is defined as 1 unit, there are 12.6 units in a 10-fold concentration difference on the logarithmic scale [13, 124]. This relationship gives a slope of 0.25 unit/mV, which is called the "conversion factor." In this case, the conversion factor provides taste information "saltiness" by its multiplication with the saltiness sensor output.

Twelve types of taste information from the nine taste sensor electrodes are listed in Table 4.5. Each conversion factor can be calculated as the slope of a 10-fold concentration difference between the reference solution and a corresponding solution, which is one of standard samples.

4.3.6.5 Detection of Interactions between Taste Substances

When you drink coffee, you might add a spoonful of sugar and milk to make the taste milder. We feel as if the bitterness will be reduced although the amount of bitter substances is not changed in the coffee at all. This phenomenon is the result of interactions between taste substances, the suppression effect. In contrast, there is a traditional Japanese cooking technique for extracting as much umami quality as possible by adding dried bonito and seaweed together in boiled water. This is the synergistic effect and is called awase dashi in Japanese.

These effects make it difficult for commonly used analyses to evaluate the taste totally, but are becoming important factors in several fields. For example, in food and beverage industries, appropriate amounts and types of ingredients and chemical materials must be added to manufacture more palatable products. In the pharmaceutical field, the unpleasant taste of drugs is the main issue because most patients, especially children, often refuse oral administration of such drugs, resulting in poor adherence and compliance. Therefore, APIs should be formulated with bitterness-masking agents, such as sucrose.

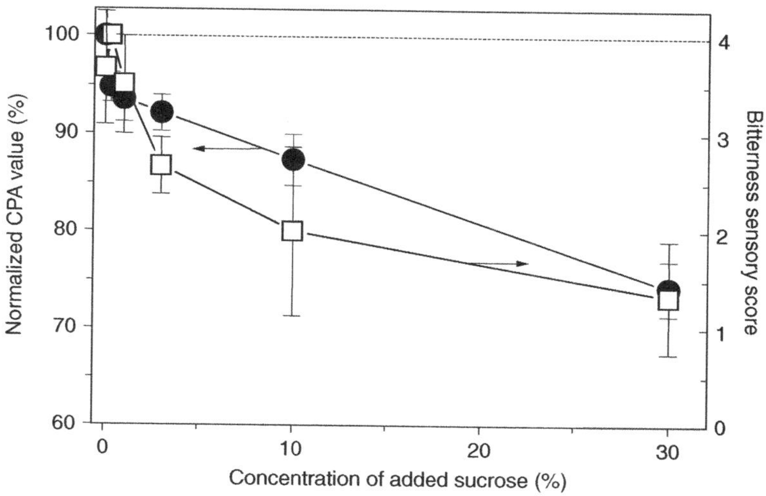

Figure 4.15 Bitterness suppression effect of sucrose on quinine hydrochloride using bitterness sensor BT0. CPA values are normalized to 100 and expressed as mean ± SD (n=4). Error bars for the sensor output and the sensory score show the measurement error (n=4) and the standard deviation of the panelists' scores (n=3), respectively. Quinine hydrochloride concentrations of 0.01, 0.021, 0.047, and 0.1 mM were used as standards for bitterness with assigned scores of 1, 2, 3, and 4, respectively. All samples include 10 mM KCl as the supporting electrolyte. Republished with permission from Ref. [123]. Copyright 2009, Institute of Electrical Engineers of Japan

Taste sensor electrodes can measure such synergistic/suppression effects. For example, Figure 4.15 shows the suppression effect of sucrose on the bitterness of quinine hydrochloride. The CPA value of the bitterness sensor BT0 decreases as the sucrose concentration increases [123]. This agrees well with the result of the sensory test, and hence the taste sensor can be applied in the development of food and beverage products and drug formulations.

4.3.7 Sample Preparation

The taste sensor can evaluate the taste of almost all foods and beverages, but appropriate sample preparation must be carried out according to the type and property of the sample. Basically, liquid samples, such as beer, tea, and coffee, can be measured as they are without any preparation. However, solid foods, such as meat and fish, must be processed by sample preparation prior to the measurement.

Figure 4.16 is an example of the preparation of cheese. To fully extract taste substances from the cheese, hot water of approximately 40°C is desirable; higher temperatures in extraction could cause the original taste of the sample to change or deteriorate. Regarding the dilution rate, approximately five times dilution is appropriate; an overly diluted sample may taste different from the original. Fats included in meat may damage taste sensor membranes, and therefore they should be removed by centrifugation. In preparing powder chemicals, such as drugs, potassium chloride should be added as a supporting electrolyte to pure water to make the membrane potentials of the taste sensor electrodes more stable. Samples based on pure water cannot be used for the taste measurement as pure water has an extremely high specific electrical resistance.

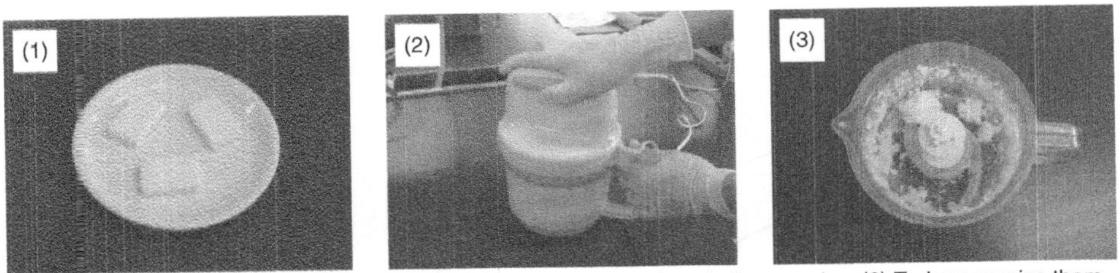

(1) There are differences among individual cheeses, so prepare a few similar samples. (2) To homogenize them, mix them in a food processor for 1 min. (3) Confirm that the cheeses are fully mixed.

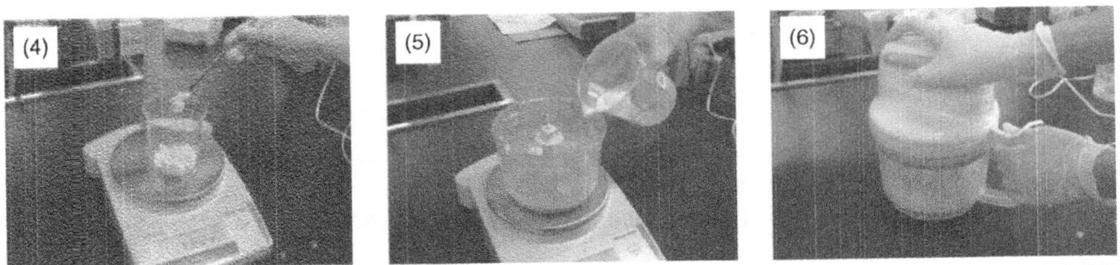

(4) Weigh around 50 g of the cheese, and put it in a food processor. (5) Add exactly four times the amount of 40° pure water (five times dilution). (6) Mix for 1 min to extract taste substances.

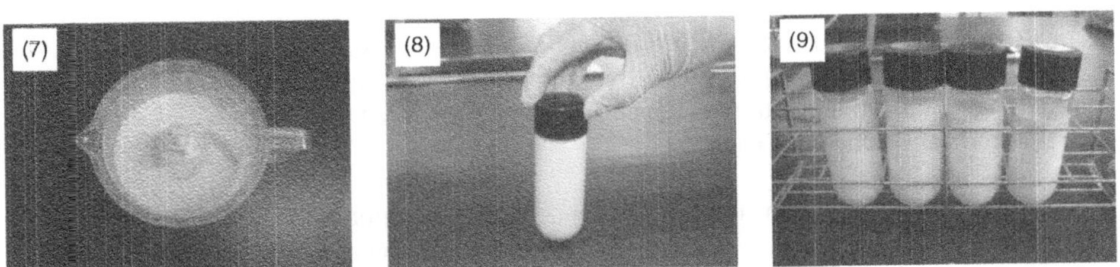

(7) Confirm that the cheese and water are fully mixed. (8) Centrifuge the mixed solution for 10 min at 3000 rpm. (9) After centrifugation, confirm that it is cleanly divided into two phases.

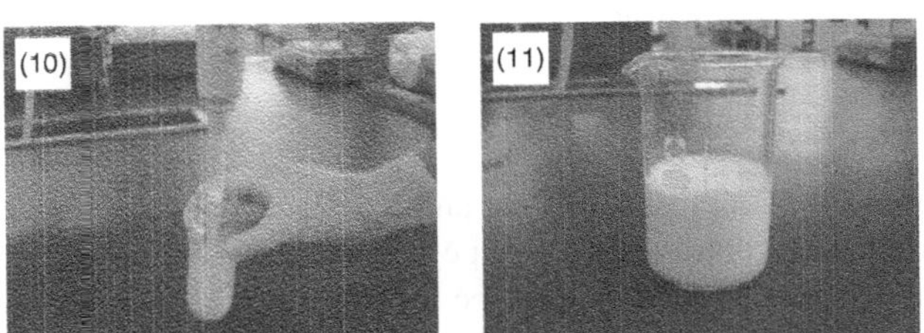

(10) Remove the water phase out of the centrifuged material. (11) Use the solution obtained as a measurement sample.

Figure 4.16 Preparation of cheese (in the case of 5 times dilution)

4.3.8 Analysis

In evaluating taste, measurement data must be treated with five types of data processing using the analysis application. Figure 4.17 shows a flowchart of the analysis method. In taste measurement, a sample generally should be measured four times to confirm the accuracy of sensor outputs. A sensor output in the first cycle tends to be higher than those in any other cycle, which results in low accuracy. Therefore, it must be deleted from the measurement data before the analysis.

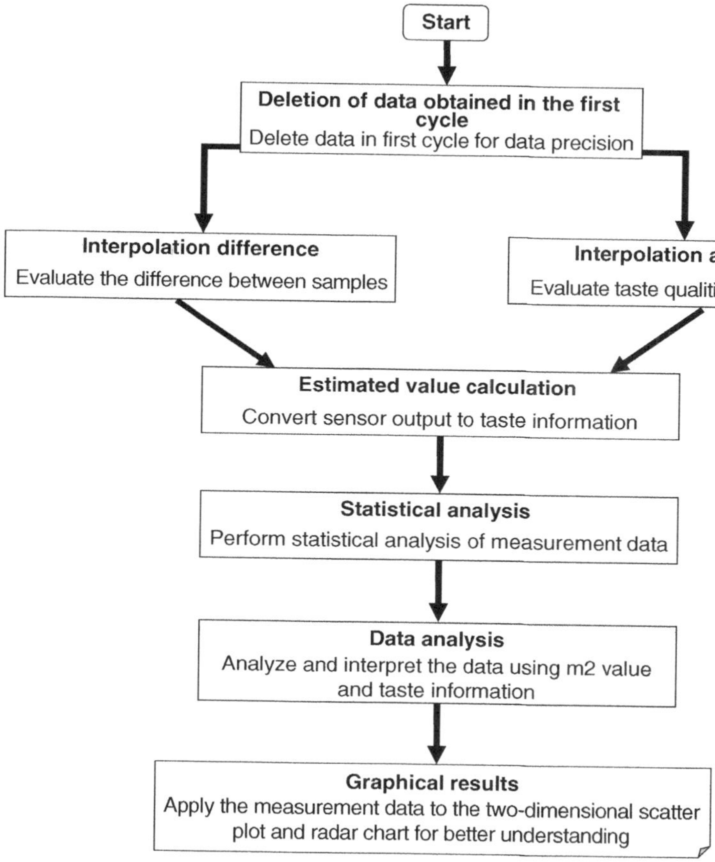

Figure 4.17 Flowchart of the analysis method

In the taste sensing system, all samples will be measured in the order from first sample to last sample. The sensor outputs of the three cycles for all samples change with similar tendencies. This change is called measurement error and must be corrected. Based on the change in sensor outputs of the first sample, which is also regarded a control sample, measurement errors of any other samples can be corrected. Figure 4.18 shows diagrams of how to correct measurement errors. Let us consider an example of a measurement result for three samples with two cycles. From the sample sequence of raw data, the error term E is calculated to correct the measurement error more efficiently. If Y_{11} and Y_{21} are the sensor outputs of the first sample in the first and second cycles, respectively, the cycle error is equal to $(Y_{21} - Y_{11})$. By dividing the cycle error by the number of the samples, the error term in the first cycle is expressed as

$$E = \frac{Y_{21} - Y_{11}}{N} \tag{4.1}$$

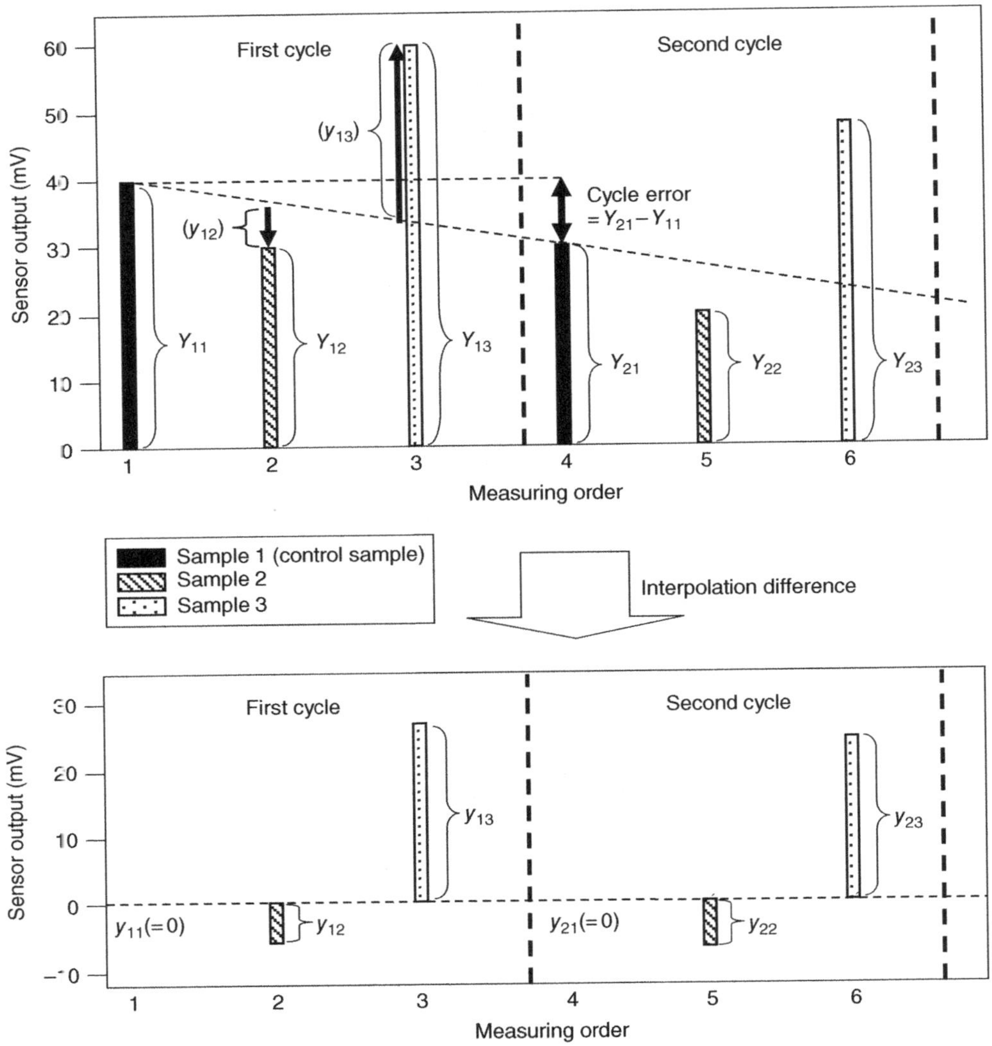

Figure 4.18 Diagrams of measurement error correction by interpolation difference

where N is the number of samples. Thus, if Y_{ki} is the sensor output for the ith sample in the kth cycle before the correction and y_{ki} is the sensor output after the correction, then the error term in the kth cycle, E_k, can be expressed as

$$E_k = \frac{Y_{(k+1)1} - Y_{k1}}{N} \tag{4.2}$$

Therefore, y_{ki} can be given by

$$y_{ki} = Y_{ki} - \left(Y_{k1} - E_{k(i-1)}\right) \tag{4.3}$$

As a result, the sensor outputs of the first sample y_{11} and y_{21} become zero, as shown in the bottom graph of Figure 4.18. This correction method is the interpolation difference, which has been defined by Intelligent Sensor Technology, Inc. Also, as the sensor output of y_{k1} is inevitably changed to zero, a measurement dataset can be connected with different measurement datasets by setting the control samples in the two measurement data to the origin.

After the interpolation difference, since all the sensor outputs of the first sample become zero, it cannot be determined whether the outputs did not exist initially or became zero as a result of the interpolation difference. To solve this problem, interpolation addition has been proposed. In interpolation addition, the output of the first sample in the first cycle, Y_{11}, is added to all values of all samples after the interpolation difference. Therefore, in this case, y_{k1} can be given by

$$y_{ki} = Y_{ki} - \left(Y_{k1} - E_{k(i-1)}\right) + Y_{11} \tag{4.4}$$

After the measurement error correction, sensor outputs can be converted to taste information according to their sensor characteristics, as discussed in Section 4.3.6.4.

Among several types of taste information, it is important to effectively find statistically significant ones to make taste assessment more beneficial. In the analysis of taste assessment, a specific index, error rate $m2$, is given for each item of taste information:

$$m2 = \sqrt{\frac{e_1^2 + e_2^2 + \cdots + e_n^2}{\left(V_1 - \bar{V}\right)^2 + \left(V_2 - \bar{V}\right)^2 + \cdots + \left(V_n - \bar{V}\right)^2}} \times 100 \tag{4.5}$$

where n is the number of samples, e_n is the measurement error of a taste information for the nth sample, V_n is the average taste information for the nth sample, and $\bar{V}$ is the average taste information for all samples. As this equation indicates, $m2$ is the ratio of the sum of measurement errors (numerator) and the sum of variations (denominator) for a given item of taste information and helps users to understand which taste information is statistically significant in the analysis.

Figure 4.19 shows the images of statistical significance between all samples using $m2$. When there is a statistical significance between all samples, the difference between the samples becomes larger in comparison with the sum of measurement errors, as shown in the

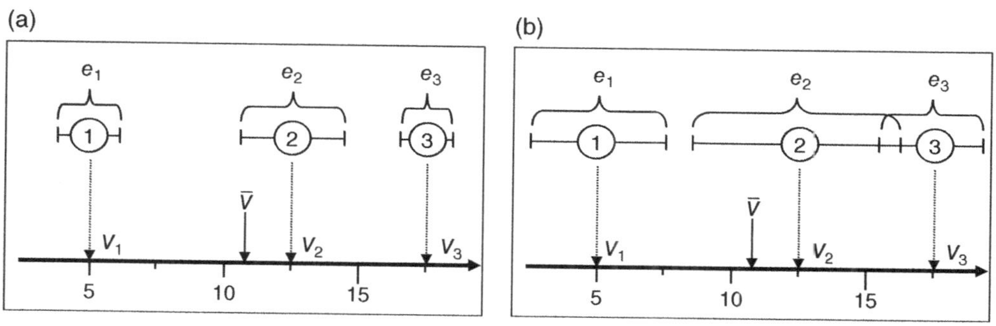

Figure 4.19 Image of statistical significance between all samples. (a) Statistical significance between samples. (b) No statistical significance between samples

left graph of Figure 4.19, subsequently making $m2$ smaller. Meanwhile, when there is no statistical significance, the sum of measurement errors becomes larger in comparison with the difference between samples, thus making $m2$ larger, as shown in the right graph of Figure 4.19. Basically, taste information with $m2$ above 50 should not be used in the analysis.

4.4 Taste Substances Adsorbed on the Membrane

As identified previously, the taste sensor uses lipid/polymer membranes as the taste sensing element and outputs a change in the membrane potential caused by the interaction between the lipid/polymer membrane and the taste substance and has global selectivity. When a bitter or astringent substance is adsorbed onto the lipid/polymer membrane, the membrane potential changes (CPA). The CPA values of the bitterness and astringency sensors are used as the indices of the aftertaste of samples and are in good agreement with the results of sensory tests by panelists [2, 13, 23]. In this section, we show the correlation between the CPA and the amount of adsorbed quinine (basic bitterness), iso-alpha acid (acidic bitterness), or tannic acid (astringency). Quinine is positively charged, and iso-alpha acid and tannic acid are negatively charged in solution.

The CPA value was measured using the taste sensing system (SA 402B, Intelligent Sensor Technologies, Inc.). The amount of taste substances adsorbed onto the lipid/polymer membrane was measured using an ultraviolet–visible spectrophotometer. The measurement procedure is as follows. First, a 5 ml of taste substance solution was added dropwise onto a petri dish on which a lipid/polymer membrane had been formed, and left to stand for 30 s to allow taste substance molecules, that is, quinine, iso-alpha acid, or tannic acid, in the solution to adsorb onto the membrane. After 30 s, a 3 ml of the solution was taken from the petri dish to measure the absorbance of the solution. The concentration of taste substances in the measured solution was calculated from the measured absorbance and calibration curve. The difference between the concentration of the taste substance solution added dropwise and that of the solution calculated from the absorbance was defined as the amount of adsorbed taste substances. This value was divided by the area of the petri dish to obtain the amount of taste substances adsorbed per square centimeter.

The amounts of taste substances adsorbed onto each lipid/polymer membrane are as below. *Quinine*: the amount of adsorbed quinine was approximately 10^{-6} to 10^{-7} g/cm^2 when the quinine concentration was 0.01–1 mM. The CPA value increased as the amount of adsorbed quinine increased (Figure 4.20a). The CPA value initially increased markedly when the amount of adsorbed quinine slightly increased but gradually saturated when the amount of adsorbed quinine further increased. *iso-alpha acid*: the amount of adsorbed iso-alpha acid was approximately 10^{-6} to 10^{-7} g/cm^2 when the quinine concentration was 0.001–0.03 vol.% (Figure 4.20b). The CPA absolute value initially increased markedly when the amount of adsorbed quinine slightly increased but gradually saturated when the amount of adsorbed quinine further increased. *Tannic acid*: the amount of adsorbed tannic acid was approximately 1–5 μg/cm^2 when the tannic acid was 0.001–0.05 wt.% (Figure 4.20c). The CPA absolute value initially increased markedly when the amount of adsorbed tannic acid slightly increased but gradually saturated when the amount of adsorbed quinine further increased.

Our previous studies revealed that the CPA value is generated when the taste substance adsorbs onto the surface of the lipid/polymer membrane, causing the charge density of the membrane surface to change [113, 130]. The adsorption causes the suppression of dissociation and the screening effect by paired ions, resulting in the reduced change in the surface charge density of the membrane [2, 113, 179, 180]. This fundamental finding concerning

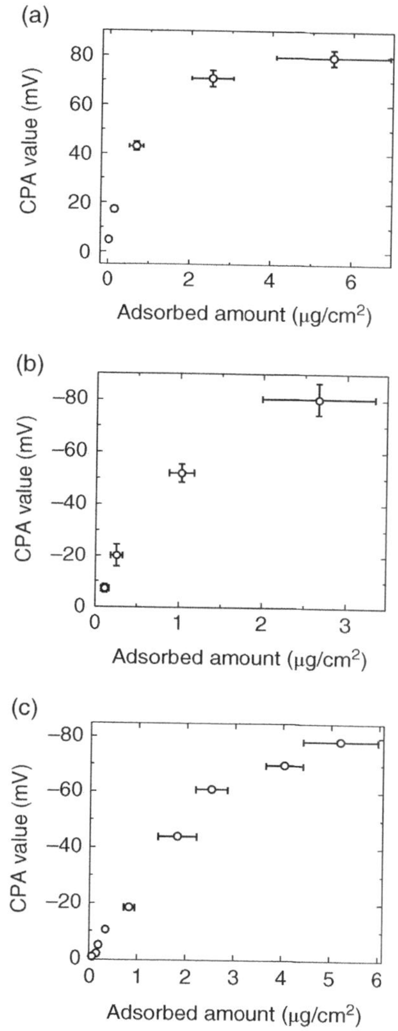

Figure 4.20 Relationships between the CPA value and the amounts of adsorbed quinine (a) [179], iso-alpha acid (b) (From Ref. [2]), and tannic acid (c) (Reprinted with permission from Ref. [180]. Copyright 2014, American Scientific Publishers)

the CPA value will contribute to the development of new lipid/polymer membranes and the further improvement of global selectivity.

4.5 Miniaturized Taste Sensor

Some research groups have proposed portable-type, miniaturized, or disposable taste sensor systems [181–183]. However, practical implementations of these types of taste sensor, which might address the limitations of conventional laboratory analyses, have not yet been reported.

Commercialized taste sensors are the taste sensing systems TS-5000Z and SA402B, which are utilized in the production of foods and medicines, as well as in quality-control applications in food and pharmaceutical companies throughout the world. However, this system is very heavy and expensive. If the sensing system can be reduced in size and cost to an inexpensive device capable of in-field evaluation, it would contribute significantly to the industries and to research fields.

We have developed a miniaturized flat-type sensor chip (40 mm × 26 mm × 2.2 mm) and portable taste sensor (80 mm × 25 mm × 20 mm) [184–186]. Figure 4.21a shows the fabricated sensor chip. The fabricated sensor chip consists of Ti/Ag electrodes patterned onto polycarbonate substrates using Ag/AgCl ink (Ag/AgCl Ink, BAS Inc., Japan), a strip of double-sided adhesive tape (polyimide), and a partition (polycarbonate). Polycarbonate and polyimide are well-known low-cost engineering plastics. The commercialized sensor probe consists of saturated KCl solution as inner solution and lipid/polymer membrane (Figure 4.3). On the other hands, the working electrode on the sensor chip consists of a poly-(hydroxyethyl methacrylate) hydrogel (pHEMA) layer with KCl as an electrolyte layer membrane and the lipid/polymer membrane layer (Figure 4.21b). The chemical components of the lipid/polymer membrane are the same as the components of a commercialized taste sensor (Table 4.3). The reference electrode consists of a PVC membrane layer and the pHEMA layer with KCl as the reference electrode of Figure 4.3. A small hole (φ 0.5 mm) is made through the PVC layer and the pHEMA layer of the reference electrode as a liquid junction (Figure 4.21c).

The working electrode and the reference electrode on the chip need to be both physically and chemically highly durable against washing processes using washing solutions including ethanol. Moreover, the electrode potential of the reference electrode needs to be stable during the measurement period. To evaluate the performance of the sensor chips, experiments were performed using the TS-5000Z taste sensing system, and comparisons were made with a conventional taste sensor probe (Intelligent Sensor Technology Inc., Japan). The fabricated taste sensor chip was connected to this taste sensing system. Figure 4.22 shows responses of the

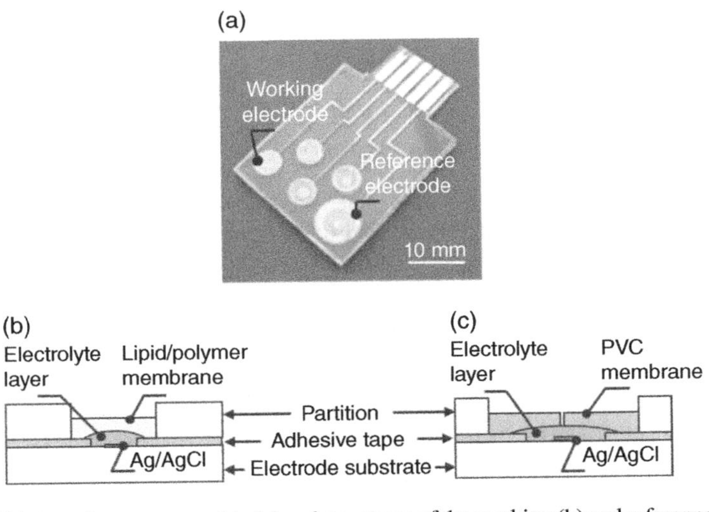

Figure 4.21 Fabricated taste sensor chip (a) and structures of the working (b) and reference (c) electrodes. (From Ref. [185])

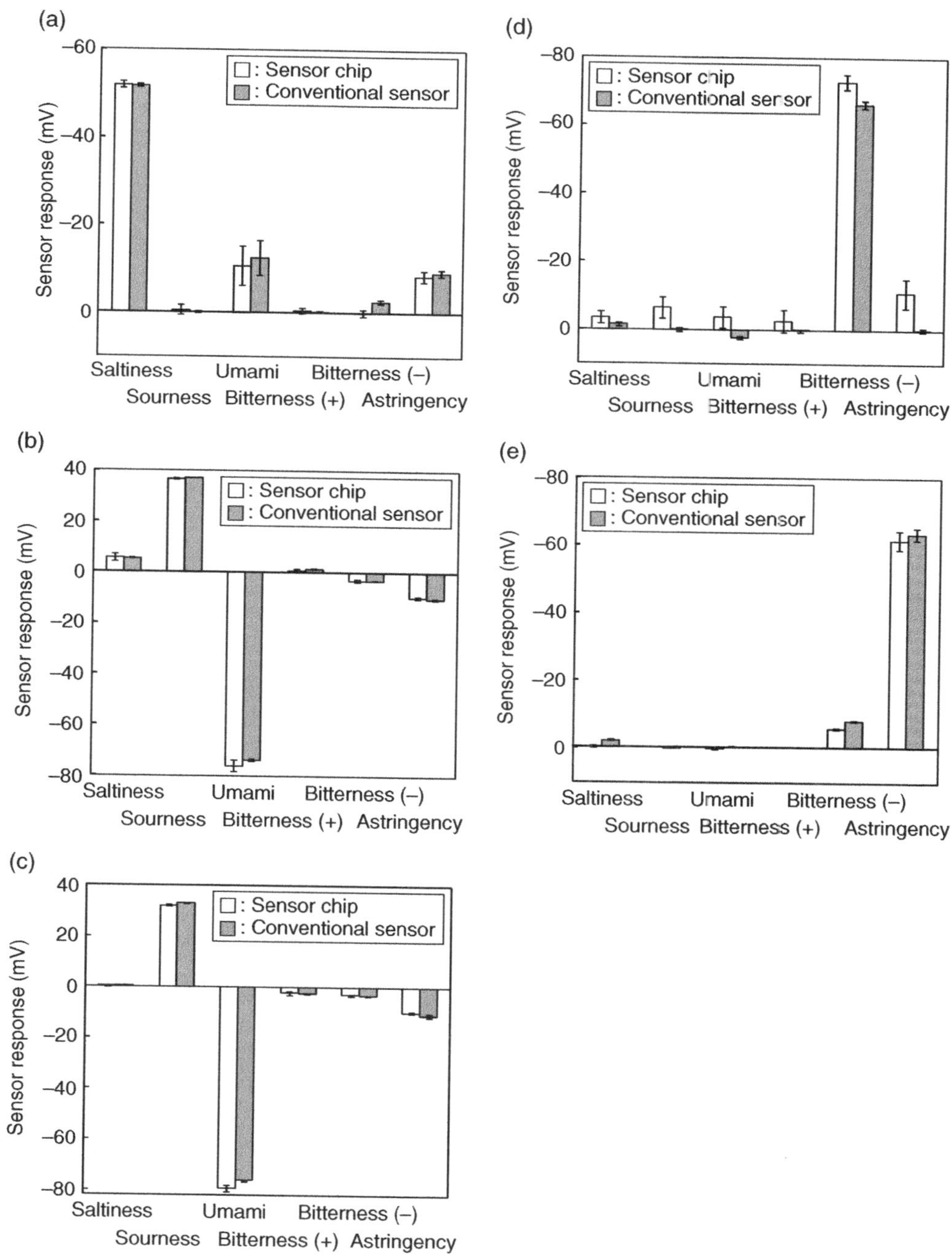

Figure 4.22 Responses of the sensor chip to five basic taste samples. (a) Saltiness sensor, (b) sourness sensor, (c) umami sensor, (d) bitterness sensor, and (e) astringency sensor. Reprinted with permission from Ref. [186]. Copyright 2014, IEEE

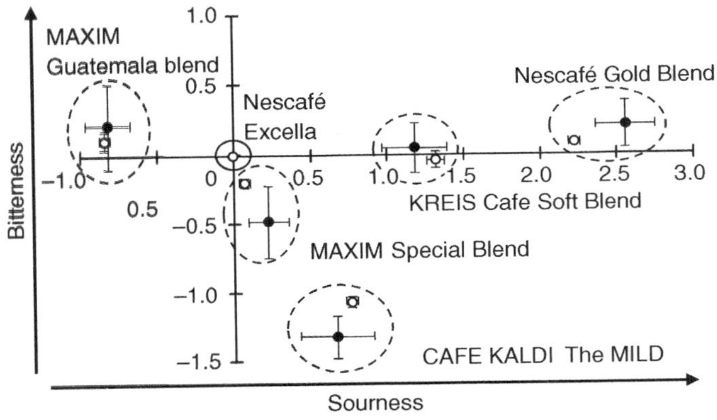

Figure 4.23 Taste map of coffee. All taste information for the control samples, Nescafe Excella, were set to zero. Data are expressed as mean ± SD (n=4). ○, sensor probe; ●, sensor chip. Republished with permission Ref. [1]. Copyright 2013, Pan Stanford Publishing

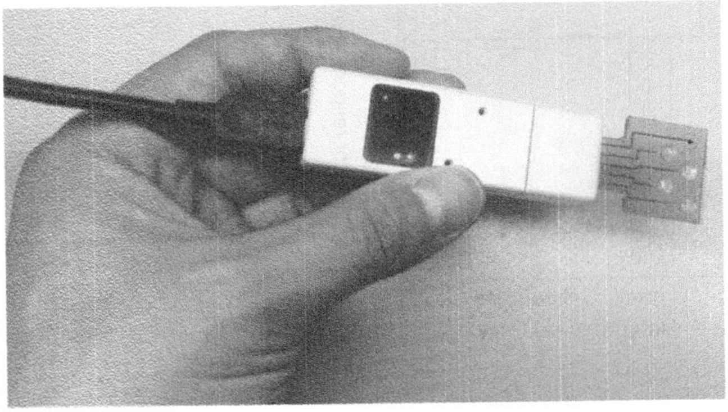

Figure 4.24 Portable taste sensor with sensor chip

sensor chip to six basic taste samples. The SDs of the sensor responses to the basic taste samples demonstrate that the fabricated sensor chips are highly stable and give reproducible results. In addition, the responses of the sensor chip are comparable to those of the conventional taste sensor. Figure 4.23 shows a taste map for coffees that can be bought on the Japanese market. The higher SDs of the sensor chip than those of the sensor probe may be an effect of the pollution of the sensing site by the coffee samples. However, these coffee samples could be distinguished from one another visually. Also, the results of sensor chips and sensor probes were consistent. Thus, these results indicate that the sensor chip can be used for applications involving taste mapping of various foods.

Figure 4.24 shows a fabricated portable taste sensor device with the taste sensor chip. The sensor device consists of high-impedance (high-Z) buffer amplifiers, level shifters, low-pass filters (LPFs), a PIC microcontroller with integrated 12-bit ADCs and a USB interface, and voltage regulators for internal power supply regulation. The performance of the portable

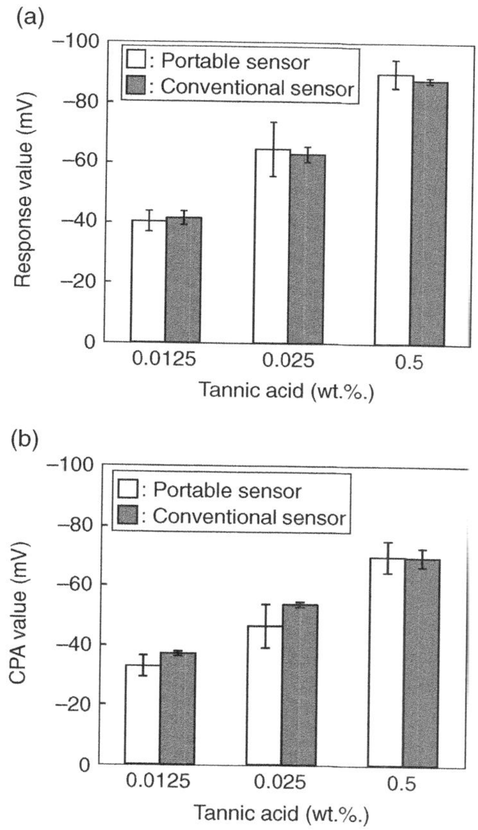

Figure 4.25 Response of the portable sensor to tannic acid. (a) Relative values and (b) CPA values. (From Ref. [185])

taste sensor for astringent substances, tannic acids, was evaluated. The electrical potential was measured at 0.1 s intervals. The average value over a 30 s period was used as the measurement value. The experimental procedure was as follows: first, the electrical potential (V_r) was measured for 30 s after the fabricated sensor chip had been immersed in the reference solution for 3 min. Secondly, the chip was immersed in the astringency sample for 2 min, after which the electrical potential (V_s) was measured for 30 s. Thirdly, the chip was immersed in the reference solution for 1 min, after which the electrical potential (V_r') was measured for 30 s. Finally, the sensor chip was immersed in the washing solution. Here, measurements were made at room temperature and the sensor chip was immersed using a magnetic stirrer and a stirring bar.

Figure 4.25 shows the response potentials (relative and CPA values) in the portable and conventional taste sensors. The result implies that the responses of the fabricated portable sensor show good accuracy and reproducibility comparable to the performance of a commercial taste sensing system, and hence it has the possibility of being used for in-field evaluations and can make a significant contribution in various fields.

4.6 Pungent Sensor

Taste is constructed from five basic taste qualities that are detected at taste receptors in gustatory cells. Additional qualities are pungency and astringency which are related to the sense of pain. In the strict sense of the word, pungency is not a taste quality but a sensation of pain. However, pungent substances which are spices are widely used as food additives. Hence, pungency is one of the important broad taste qualities. Capsaicin, piperine, and allyl isothiocyanate are included in hot red pepper, black pepper, and mustard oil, respectively (Figure 4.26). Transient receptor potential vanilloid type 1 (TRPV 1, capsaicin receptor) [187, 188] and transient receptor potential ankyrin type 1 (TRPA 1) [189, 190], which are members of the transient receptor potential (TRP), are known as the receptors of pungency [191].

There is a variety of capsaicin analytical methods, such as high-performance liquid chromatography [192], liquid chromatography with electrochemical detection [193], gas chromatography [194], and electrochemical impedance spectroscopy [195]. Unfortunately, the lipid/polymer membrane-based taste sensor cannot be applied to pungent sensing because capsaicinoids are nonelectrolytes [93]. We have studied a capsaicin detection technique for a pungent sensor using a surface plasmon resonance (SPR) immunosensor instead of the lipid/polymer membrane-based taste sensor.

We have developed a detection method for capsaicinoids, which are major pungency substances, using an SPR immunosensor [196]. The SPR immunosensors are realized by combining SPR sensors, which can be used to detect changes in the refractive index of the sensor surface with high sensitivity, with an antigen–antibody interaction, resulting in high selectivity. Capsaicinoids detection was performed by SPR using a Biacore J system

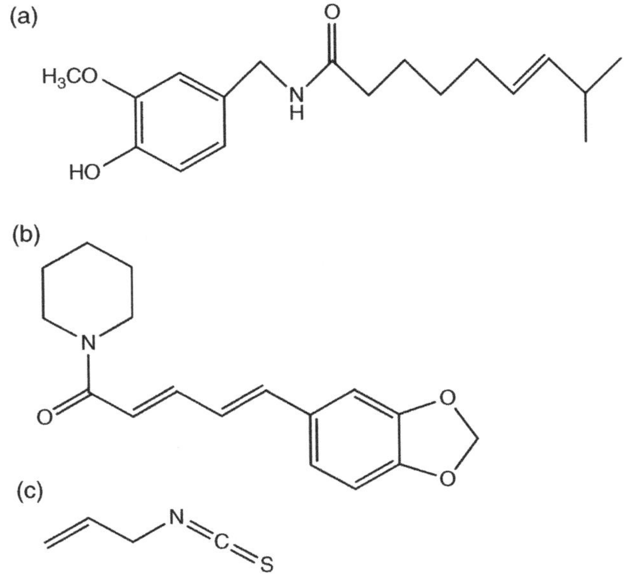

Figure 4.26 Chemical structure of the pungent taste substances: (a) capsaicin, (b) piperine, and (c) allyl isothiocyanate

(GE Healthcare, Japan). A rabbit-derived antihomovanillic acid (HVA) polyclonal antibody was prepared for the measurement of capsaicinoids. Capsaicinoid is a monovalent antigen with a low molecular weight (MW, capsaicin: 307). Therefore, the sandwich immunoassay method cannot be applied. Consequently, we apply an indirect competitive assay using sensor chips with capsaicin analogues immobilized on the SPR sensor surface via an oligo(ethylene glycol)-terminated self-assembled monolayer (SAM).

The procedure of indirect competitive assay is as follows. When the 25 ppm antibody solution is injected onto the chip surface, the antibodies bind to the capsaicinoid analogue (4-hydroxy-3-methoxybenzylamine hydrochloride (HMB)) immobilized onto the chip surface (Figure 4.27), resulting in an increased sensor response. The increase in the sensor response ($\Delta\theta_0$) corresponds to the number of antibodies bound to the chip surface. When a mixture of antibodies and capsaicinoid is injected onto the chip surface for 5 min, the antibodies that have already bound to the capsaicinoid do not bind to the capsaicinoid analogue on the chip surface. Namely, the capsaicinoids in the solution inhibit the antibodies from binding to the chip surface. Therefore, the concentration of antibodies bound to the chip surface decreases from that when only the antibodies are injected ($\Delta\theta_1$). It is considered that $\Delta\theta_1$ decreases as the percentage of capsaicinoids in the mixture increases. Here, the relative change in the concentration of antibodies bound to the chip surface, that is, the bound percentage, is calculated by

$$\left(\frac{\Delta\theta_1}{\Delta\theta_0}\right)\times 100. \tag{4.6}$$

The smaller the bound percentage, the higher the concentration of antibodies bound to the capsaicinoids.

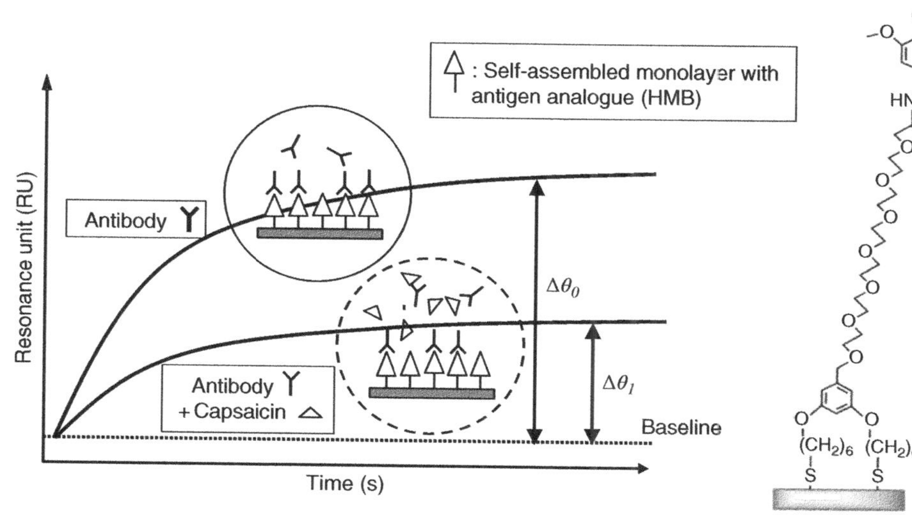

Figure 4.27 Principle of indirect competitive method and sensor chip surface

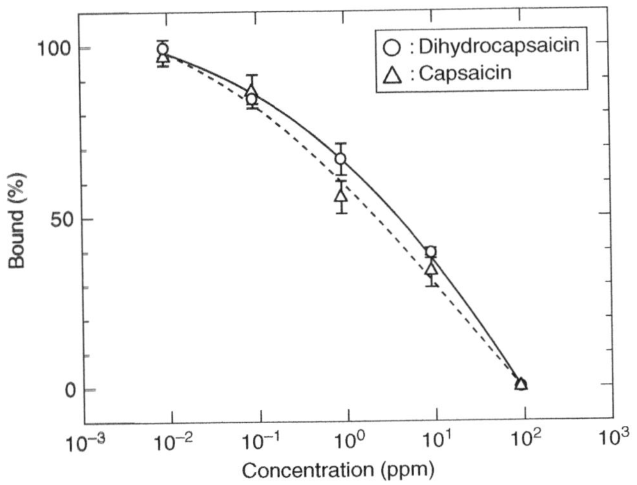

Figure 4.28 Response to capsaicin on HMB sensor surface. Republished with permission Ref. [196]. Copyright 2013, Institute of Electrical Engineers of Japan

Figure 4.28 shows calibration curves for the capsaicinoids using a standard capsaicin and dihydrocapsaicin solution [196]. The limits of detection for capsaicin and dihydrocapsaicin were 25 and 35 ppb, respectively. This finding may indicate that the SPR immunosensor can be used to evaluate pungency. However, the sensor cannot measure other typical pungent substances in foods such as piperine and allyl isothiocyanate. Accordingly, we have developed immunoassays using newly antibodies for measurement of these substances. The next step is to demonstrate the sensor performance using food samples and carry out taste evaluation of foods for not only the five basic tastes and astringency but also pungency.

4.7 Application to Foods and Beverages

4.7.1 Introduction

The taste sensing system equipped with taste sensor electrodes enables taste assessment for a variety of foods and beverages and also allows for effective development of products. This section highlights not only practical examples for foods and beverages but also effective use of the taste sensing system.

4.7.2 Beer

When evaluating the taste attributes of beer, sour and bitter tastes are significant. It is known that the bitterness of beer is mainly derived from alpha acids of hops. The sourness sensor was designed to have a high correlation with pH, and the bitterness sensor was designed mainly to respond to iso-alpha acids corresponding to Bitterness Units (BU), as shown in Figures 4.29 and 4.30, respectively, similarly to the evaluation conducted at Technische Universität München [134, 135]. The basic concept of this application to beer has been previously reported [91, 107, 197].

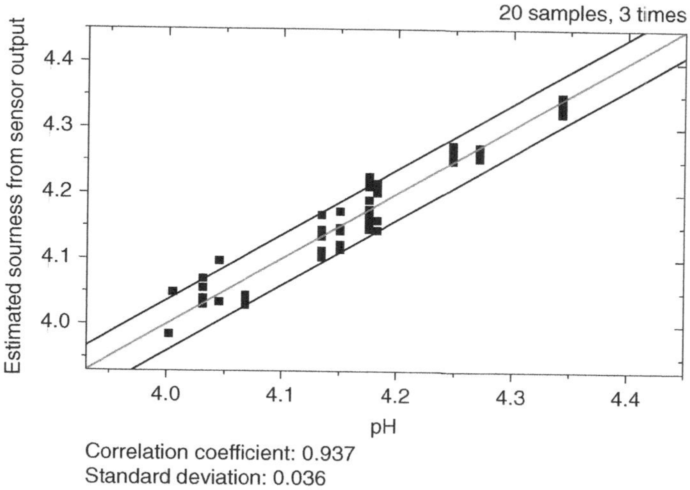

Figure 4.29 Correlation between sourness sensor output and pH

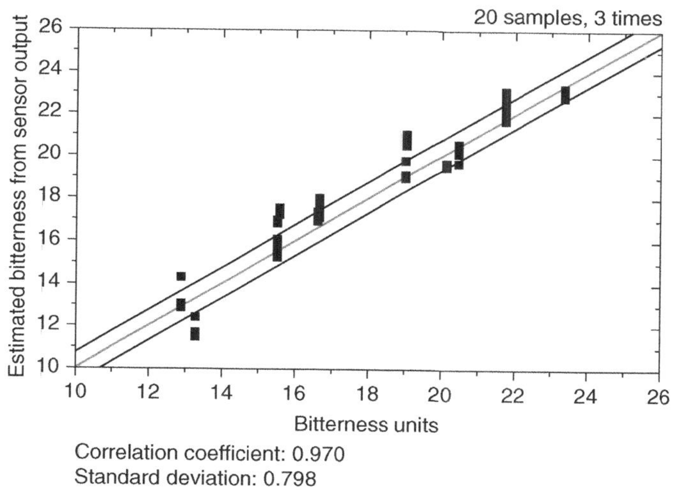

Figure 4.30 Correlation between bitterness sensor output and bitterness units

The hops bittering calculation to obtain BU is commonly used in breweries. Then, what benefit(s) can a bitterness sensor provide? The taste sensing system is able to assess not only taste quality but also its time-dependent information, such as initial taste and aftertaste, as a function of time. Bitterness can be evaluated as both initial and aftertaste caused by the adsorption of bittering compounds on taste receptors. Similar physiochemical phenomena will be observed with the taste sensor. Figure 4.31 shows time-dependent information of bittering components. V_s is the sensor output for the initial bitter taste and contains information on both ions and bittering compounds, and V_{rn} ($n=1, 2, \ldots$) is for aftertaste and indicates how the bitter

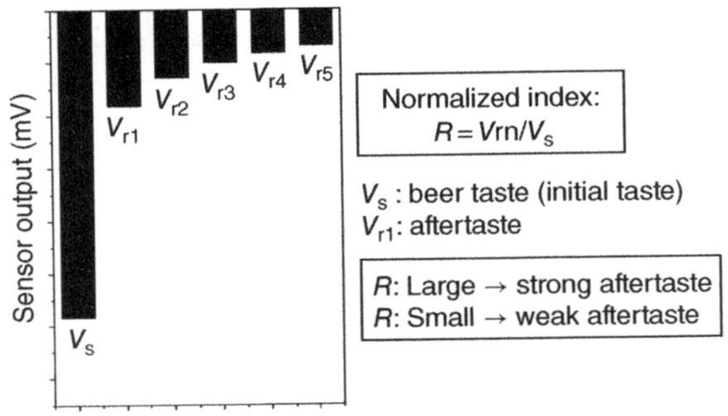

Figure 4.31 Estimated bitter aftertaste

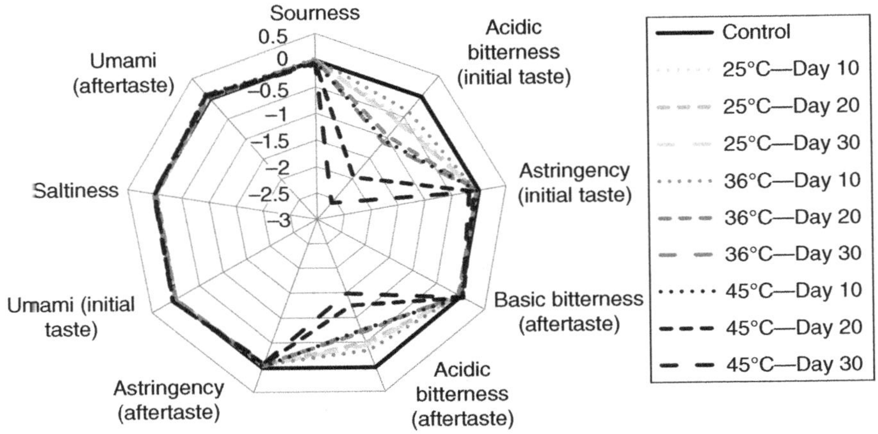

Figure 4.32 Acceleration test of a draft beer kept at 25, 36, and 45°C for 10, 20, and 30 days

taste changes its intensity with time [107]. By using the sensor output, time-sensitive assessment will be possible. For example, sharpness may be defined as the aftertaste divided by initial taste.

Another time-dependent taste quality is taste deterioration with time. Figure 4.32 shows an acceleration test of bottled beers stored at different temperatures for a duration of 30 days. The sample was an original equipment-manufactured product, within one month after its production date. Figures 4.33 and 4.34 illustrate that the deterioration of this beer sample primarily occurs in the initial taste and the aftertaste of bitterness. The heavy dashed line is one taste scale unit from the original bitter intensity and corresponds to a change of approximately 20% in concentration. This change represents the most one can perceive as a minimum difference in taste quality, as described in Section 4.3.6.4. At room temperature, 25°C (77°F), quality deterioration was relatively slow. After 10 days at 45°C (113°F), however, the initial taste of bitterness reached a threshold change in taste, which approximately corresponds to the shelf life of a product from the point of view of taste. In general, quality deterioration is expressed by an Arrhenius equation. By measuring some samples in acceleration tests, it is possible to derive coefficients of the equation and to estimate shelf life.

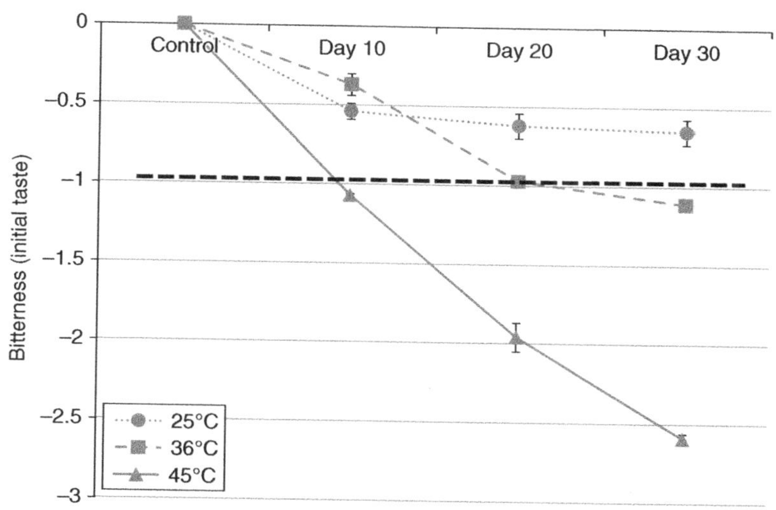

Figure 4.33 Change in initial bitterness taste of a draft beer kept at 25, 36, and 45°C over 10, 20, and 30 days

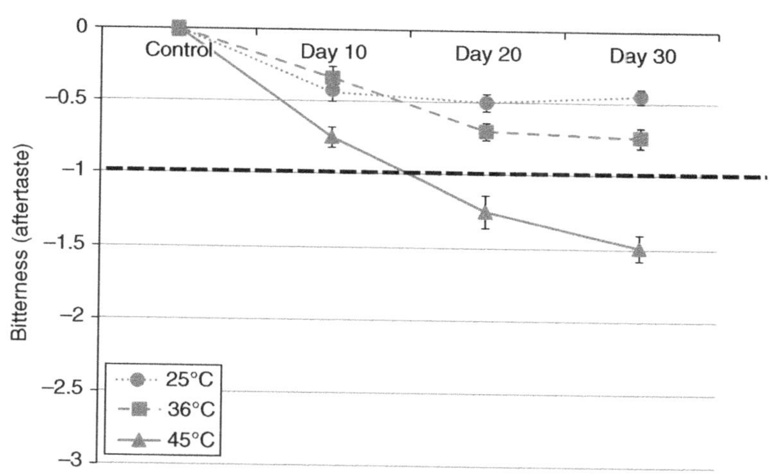

Figure 4.34 Change in bitterness aftertaste of a draft beer kept at 25, 36, and 45°C over 10, 20, and 30 days

4.7.3 Coffee

Recently, coffee (especially regular coffee) has become more popular in Japan than ever. Coffee shops, such as Starbucks®, Doutor®, and Tully's Coffee® are major players, and Japanese-style chain coffee shops such as Komeda Coffee and Hoshino Coffee are holding their own. Fast-food shops such as hamburger shop chains, self-service drink areas in a restaurant, and even conveyor-belt sushi restaurants offer freshly brewed coffee. Other notable examples are coffee at convenience stores, such as Seven-Eleven®, where freshly brewed coffee made by coffee machines is sold over the counter, and the coffee taste is changed every couple of months to prevent consumers from losing interest in the taste.

It is said that the coffee industry in Japan is in its third wave. The first wave was from the late nineteenth century to the 1960s. During this period, Nestle® and MJB® provided coffee to the Japanese market at low cost through mass production and coffee became a popular beverage in Japan. In the second wave, from the 1960s to 2000s, Seattle-type coffee shops, such as Starbucks Coffee®, which offered consistency of taste with small quality dispersion, became popular not only in Japan but throughout the world. In these first and second waves, coffee was heavily consumed and underwent commoditization. As a result, social responsibility and appropriate pricing by places of production arose. Accordingly, in the third wave after 2000, a new style of enjoying coffee has come popular; traceability from production to consumption, sustainable good relationships of trust with growers, and sharing one's sense of value in coffee taste with others. In the third wave, consumers are enjoying the taste of the coffee beans themselves without deep roasting depending on the production area and the species of bean. Furthermore, new machines for coffee extraction have appeared, supporting these new tastes of coffee. The coffee experience has become one that can be linked to that of wine. The enjoyment is tied to the type of bean and the particulars of production, as well as the method of brewing.

Coffee presents a great variety of tastes; the species of beans; place of production; threshing and cleaning techniques, such as dry and wet methods, from the coffee cherry; degree of roasting; and brewing method all affect the taste of coffee [198, 199]. In brewing coffee, water hardness and a temperature also act on the taste of coffee. Hard water contains calcium and magnesium ions that interact with coffee components. A lower water temperature when brewing coffee causes a more sour taste. Additionally, the fineness of ground coffee also affects the taste. An excessively fine grind results in an imbalance between sour and bitter tastes and causes an undesirable sour taste or a stronger taste than expected. The coffee extraction method and type of machine, such as Nel drip, French press coffee, siphon coffee-makers, and espresso machines, can all affect the coffee taste. The most common type of coffee beverage machine is the paper dripper. This type of coffee dripper has a strong effect on the taste. As described, when making coffee at home, the taste quality depends on the brewing method.

A coffee bean is a natural product and thus its quality is susceptible to the environment. Blended coffee is produced by blending beans, whereby care must be taken to maintain the same taste quality. This requirement is not so easy to achieve. Furthermore, expressing the coffee taste and communicating it to others are not so simple. Therefore, studies of coffee taste evaluation using the taste sensing system have been reported by researchers [108, 200, 201].

Ishimitsu & Co., Ltd., a long-established Japanese coffee company, imports coffee beans and sells them to coffee manufacturers. Subsequently, beans are roasted and the taste qualities are assessed. Tomohiro Ishiwaki of Ishimitsu & Co., Ltd. has used the taste sensing system to build a database of coffee beans and coffee tastes under different roasting conditions. An example of his work involves blending beans at different ratios to achieve targeted coffee taste. He said that "the results of the sensor have a high correlation with those of sensory evaluation [155]." The results are calculated using the results of the taste sensing system. Subsequently, the information from the database is shown to enable target taste to be reproduced [155].

A similar method developed by Intelligent Sensor Technology, Inc. using a GA was applied to reproduce the target coffee taste. The details of the method will be discussed in Section 4.7.5.2. Extraction of coffee is as follows:

1. Grinding coffee beans in a coffee mill.
2. Put 7 g of ground coffee into a beaker.
3. Pour 130 g of hot water into the beaker.
4. Wait 10 s.
5. Stir gently with a spoon for five rotations.
6. Wait 4 min.
7. Filter with filter paper and allow to cool in an iced-water bath.
8. Measure the sample at room temperature.

Coffee samples obtained by the above procedure (Brazilian, Colombian, Guatemalan, Ethiopian mocha coffees, a Tanzanian Kilimanjaro coffee, and an Indonesian Mandheling coffee) were measured utilizing the taste sensing system. Figure 4.35 shows that the resulting taste characteristics of each coffee are well assessed. It should be noted that the coffee beans tested were roasted beans available on the market. Next, the mathematical optimization for blending ratios to provide a similar taste of a target coffee was applied. First, the target product was measured by using the taste sensor to obtain its taste qualities as digitized data. After that, the "best combination" was calculated by mathematical optimization; 38.5% Indonesian Mandheling coffee, 33.8% Brazilian mocha, 24.4% Tanzanian Kilimanjaro coffee, and 3.4% mocha were required to reproduce the target taste, as shown in Figure 4.35. The coffee blended in accordance with the results showed almost the same sensor output. In addition, this blend was tasted by coffee specialists. The judgment was that the target taste and the taste of the blend were almost the same. The target coffee mainly contained Brazilian and Colombian beans but the calculated blending ratio did not include Colombian beans. This suggests that the same taste can be reproduced by using alternative beans. The mathematical optimization provides not only combination ratio to a target product but also taste information calculatedly produced from database.

Coffee beans are a natural product, and they sometimes can become unavailable owing to unseasonable weather or disease of coffee plants. Blending of available beans can be a suitable answer to this problem. Historically, blending beans depended on an individual's experience. To become a blender, experience and time are required. Presently, it can be shown that alternative methods help to find the best combination for the target taste. For instance, Brazilian and Mandheling coffee beans were blended as a 50 : 50 blend, and the resultant taste was estimated by mathematical optimization. A coffee sample was made on the basis of the calculated blending ratio and measured by the taste sensor. The taste estimated by calculation and the actual measured value were shown to be in good agreement, indicating that even mathematical calculation can estimate taste qualities of coffee. Other combinations of different coffee beans also gave comparable results as well.

Marketing research from the point of view of taste is possible. In the beginning, the prevalence of regular coffees in Japan, especially in convenience stores, was mentioned.

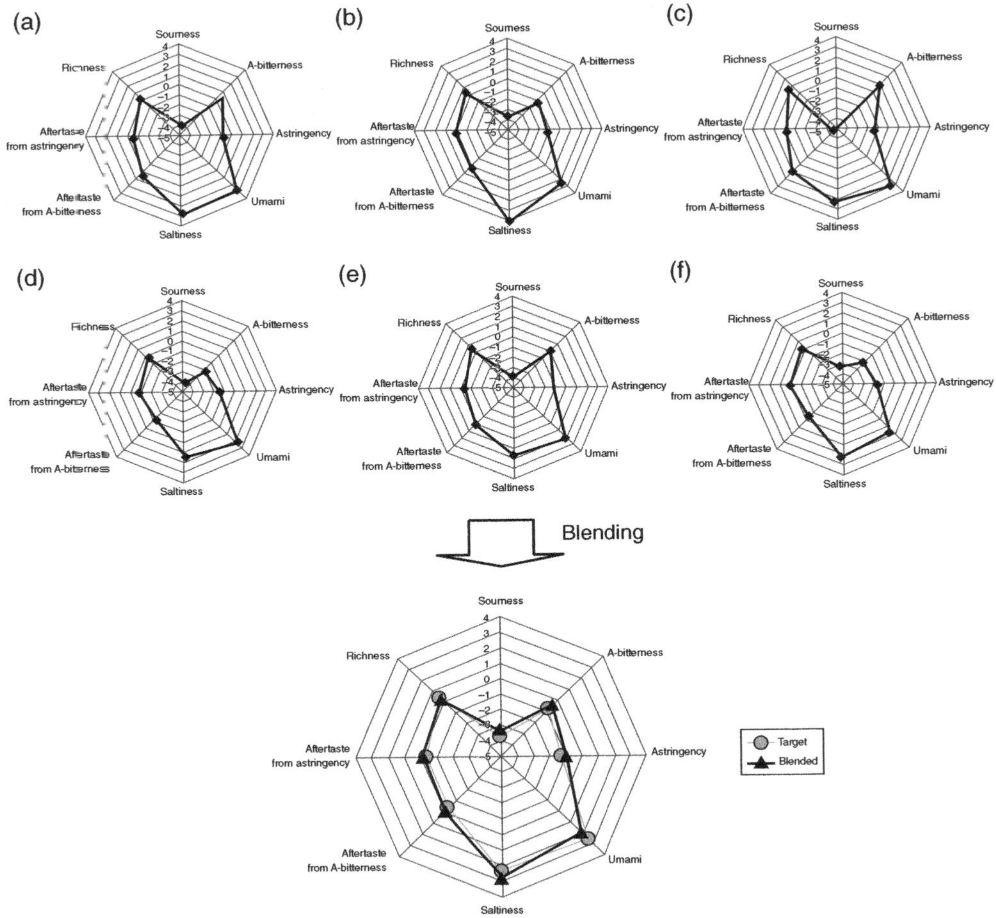

Figure 4.35 Taste patterns of six single-bean coffees, a target coffee, and a blended coffee with a blend ratio based on the result of a genetic algorithm calculation. (a) Mandheling; (b) Kilimanjaro; (c) Mocha; (d) Brazil; (e) Colombia; (f) Guatemala

Figure 4.36 shows taste differences in regular coffee available in convenience stores all over Japan. Each chain convenience store has unique coffee products. This is clearly shown in Figure 4.36. The coffee available at Seven-Eleven has an average taste of all coffees and is the most popular in Japan. On average, 90 cups of coffee are sold at each of the approximately 15 000 Seven-Eleven shops per day.

The grading of coffee beans is different in each production area. In general, larger beans, low-defects beans, and beans from plantations at high altitudes have a tendency to be of higher grades. In Colombia, the grade is defined by the screen size of beans. Supreme beans are larger and of higher grade than Excelso beans. Comparing the taste, Excelso tends to have a slightly stronger taste than that of Supreme. In the case of Brazilian beans, bean grades are defined by the number of defects and screen size. The #2 beans have a smaller number of

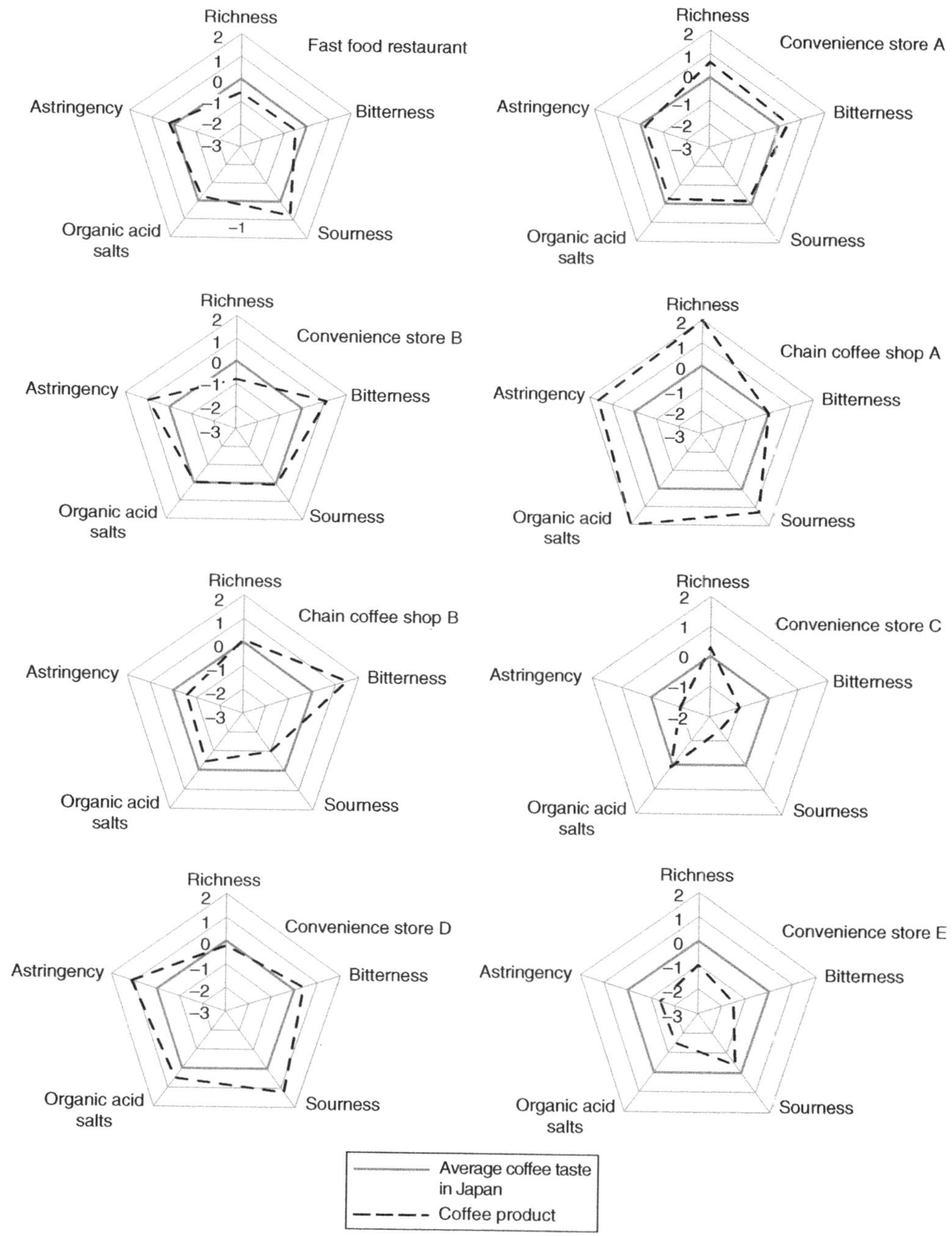

Figure 4.36 Tastes of coffee from a fast-food restaurant, convenience stores, and coffee shop chains. Data was obtained by Intelligent Sensor Technology, Inc.

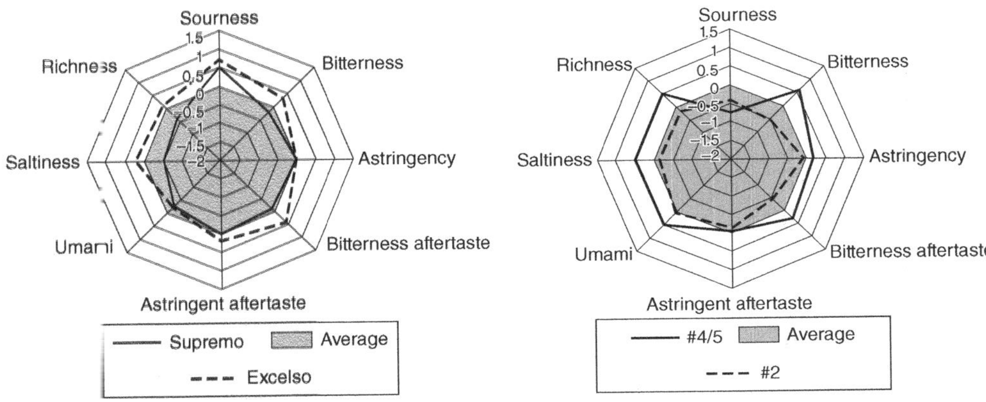

Figure 4.37 Evaluation of Colombian (left) and Brazilian (right) coffee beans. Data was obtained by Intelligent Sensor Technology, Inc.

defects and larger screen size, while the #4/5 beans have a larger number of defects and smaller screen size. Concerning the taste, #4/5 beans tend to have stronger bitterness and astringency than do #2 beans, as shown in Figure 4.37.

In summary, the taste sensing system can be useful in reproducing the taste of a target coffee. To accomplish this, one must first build a database by utilizing the taste sensor. Select beans and roasting conditions and then measure the target product. Finally, mathematically optimize the data to derive the optimum blending ratio. In addition, advanced estimations, which are possible only by using the database information, can demonstrate how blending different beans affects the final result. Furthermore, the taste information can be utilized in preparing a proposal document inside or outside a company.

As described in this section, the taste sensing system is becoming a useful tool for coffee companies, importers, roasters, and affiliated companies (such as makers of paper filters). This application has many possibilities, such as in marketing research, in product development, in inspection when importing beans, as sales tools in proposal document preparation, and in complaint handling.

4.7.4 Meat

Meat evaluation using the taste sensing system and its usefulness has been reported by researchers. There are three typical preparation methods for meat samples: raw, stewing, and barbequing. From the viewpoint of homogeneity of the sample and also the rarity of eating raw meat, stewing meat is considered to be preferable in many cases. Chikuni *et al.* reported the effects of cooking on the taste- and flavor-related compounds in pork [202]. A pork sample was vacuum packed and then incubated at 70°C for 1 h. They suggested that the heating condition did not affect the total free amino acid contents, and thiobarbituric acid reactive substances were major components for flavor and taste created by cooking.

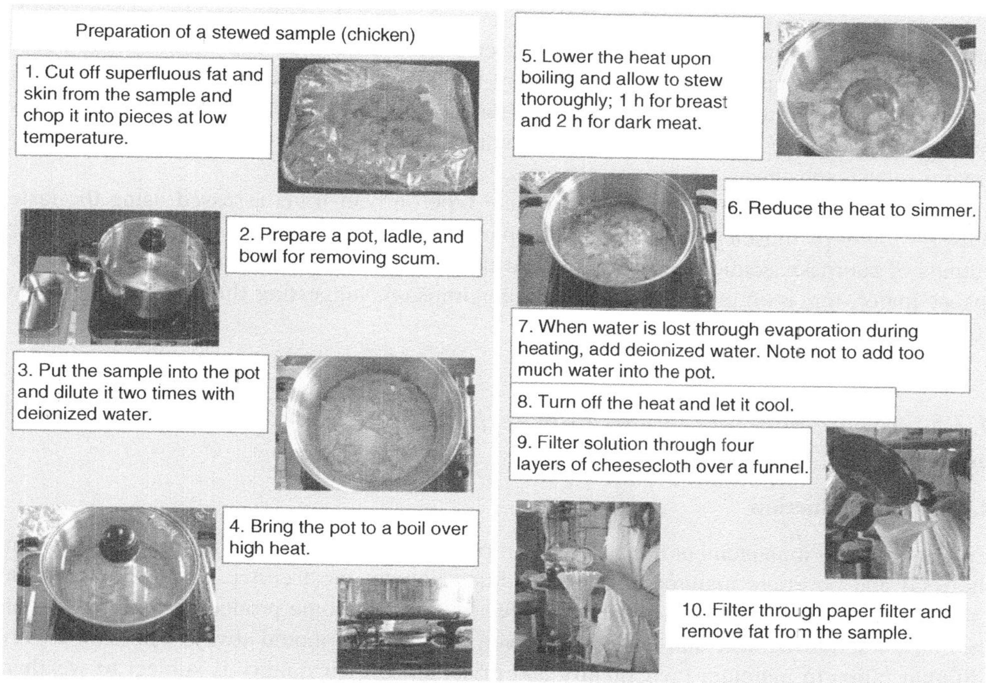

Figure 4.38 Preparation method of stewed chicken sample. Republished with permission from Ref. [203]. Copyright 2008, Japan Poultry Breeders Hatcheries Association

The following is the preparation procedure of a stewed sample (chicken) proposed by Fujimura of Niigata University and his group (Figure 4.38) [203]:

1. Remove superfluous fat and skin from a sample and cut the meat into pieces at low temperature.
2. Prepare a pot, ladle, and bowl for removing scum.
3. Put cut meat into the pot and dilute it two times of meat weight with deionized water.
4. Bring the pot of water to a boil over high heat.
5. Lower the heat upon boiling and allow to stew thoroughly; 1 h for breast and 2 h for dark meat.
6. Reduce the heat and simmer.
7. When water is lost through evaporation during heating, add deionized water. Take care not to add too much water to the pot.
8. Turn off the heat and let cool.
9. Filter solution through four layers of cheesecloth over a funnel.
10. Filter through filter paper and remove fat from the sample.
11. Add salt to the final concentration of 0.3% using a measuring cylinder.
12. Put the prepared sample into a container for testing.

Sasaki *et al.* reported the analysis of pork extract using the taste sensing system [164]. Longissimus muscles from different pig breeds were examined for discrimination and correlation with umami-related substances by applying PCA. The scores were used in a correlation analysis with umami-related substances and revealed that the taste sensor was capable of analyzing the taste of pork; its output was related to the concentration of the umami-related substances [204].

Chikuni *et al.* reported the effects of muscle type on beef traits assessed using the taste sensor. Slow-type muscles (masseter and diaphragm) showed taste differences in the sensor outputs of sourness, acidic bitterness, and astringent tastes compared with fast-type muscles (psoas major, longissimus thoracis, and semitendinosus), suggesting that muscle fiber type affects the beef taste characteristics [166].

4.7.5 Combinatorial Optimization Technique for Ingredients and Qualities Using a GA

4.7.5.1 Introduction

Cost and quality management has been key in providing a certain level of product quality in the food and beverage manufacturing field. For example, most coffee products have been produced by blending different types of coffee beans from some producing regions so that they have the desired taste and flavor. Moreover, coffee beans should always be blended with different ratios to maintain their quality and price, since their quality is subject to weather conditions, soil, and other factors.

Many experts in the field first select the appropriate types of raw ingredients and then decide the blend ratio on the basis of a sensory test. However, these processes require much time. In addition, not all combinations of blending, in practice, can be evaluated as there are a huge number of combinations. For example, if there are 10 types of coffee beans, we can choose any 3 of them in 120 different ways. In mathematics, this is called combinatory optimization and should be treated and computed mathematically [205].

We treated this as a constrained nonlinear optimization problem, and solved it by applying a GA.

4.7.5.2 GA

As an example, we assume that we have measured 10 types of coffee beans (here, referred to as "materials") using the taste sensor, and evaluated them on the basis of taste information, such as sourness, saltiness, sweetness, and bitterness. Then we aim to make a product with the targeted taste qualities by blending three types of coffee beans in an appropriate blend ratio. Figure 4.39 illustrates a matrix of taste information on 10 types of beans. We defined the symbols as follows:

$\hat{m}_{max}$: number of all materials; 10 in this case.

$\hat{M}$: set consisting of 10 types of beans, also called "set of all materials."

m_{max}: number of materials selected; 3 in this case.

M_i: subset of set $\hat{M}$, consisting of any three types of beans, also called "subset of selected materials (Figure 4.40)." As mentioned previously, there are 120 combinations.

Matrix of all taste information ($= F_B$)

	y_1	y_2	y_3	y_4	$\hat{x}$
Material	Sourness	Saltiness	Sweetness	Bitterness	Blend ratio
Bean 1	5	0	1.5	2	0–100%
Bean 2	2	2	2.5	3	0–100%
Bean 3	3	4	1	0	0–100%
Bean 4	5	2.5	0	2	0–100%
Bean 5	0	1.5	2.5	5	0–100%
Bean 6	2	4	4	2	0–100%
Bean 7	1	1	1	5	0–100%
Bean 8	0	0	2	4.5	0–100%
Bean 9	2	1	0	1	0–100%
Bean 10	4.5	1.5	2.5	3	0–100%

$m_{max} = 10$ $d_y = 4$ $\hat{x}_{min} = 0$ $\hat{x}_{max} = 100$

Target $\bar{y}$

	$\bar{y}_1$	$\bar{y}_2$	$\bar{y}_3$	$\bar{y}_4$
Target	Sourness	Saltiness	Sweetness	Bitterness
$\bar{y}$	2.5	2	1	3

First candidate M_1

	y_1	y_2	y_3	y_4	$\hat{x}$
Material	Sourness	Saltiness	Sweetness	Bitterness	Blend ratio
Bean 1	5	0	1.5	2	20%
Bean 3	3	4	1	0	30%
Bean 7	1	1	1	5	50%
y	**2.4**	**1.7**	**1.1**	**2.9**	

$g_1 = \begin{pmatrix} 1 \\ 3 \\ 7 \end{pmatrix}$ $x_1 = \begin{pmatrix} 20 \\ 30 \\ 50 \end{pmatrix}$

Error ε ($= y - \bar{y}$)

	Sourness	Saltiness	Sweetness	Bitterness
ε	−0.1	−0.3	0.1	−0.1

Figure 4.39 Example of matrices for the genetic algorithm

g_i: vector of the serial numbers of three beans selected for set M_i. If coffee beans with the serial numbers 1, 3, and 7 are selected, it is expressed as $(1, 3, 7)^T$. Generally, in GAs, g_i is called an "individual," while a set of several g_i is called "population."

i: serial number of subset M_i; defined as $\{1, 2, \cdots, b_{max}\}$.

b_{max}: number of M_i in a population.

x_i: vector of blend ratio for M_i. If blend proportions of three coffee beans selected in M_i are 20, 30, and 50%, it is expressed as $(20, 30, 50)^T$.

$\hat{x}$: vector of the blend ratio for $\hat{M}$. If $g_1 = (1,3,7)^T$ and $x_1 = (20, 30, 50)^T$ are selected, it is expressed as $(20, 0, 30, 0, 0, 0, 50, 0, 0, 0)^T$. $\hat{x}$ can be expressed as a function of x and g, so it also can be given by $\hat{x} = h_{tr}(x, g)$.

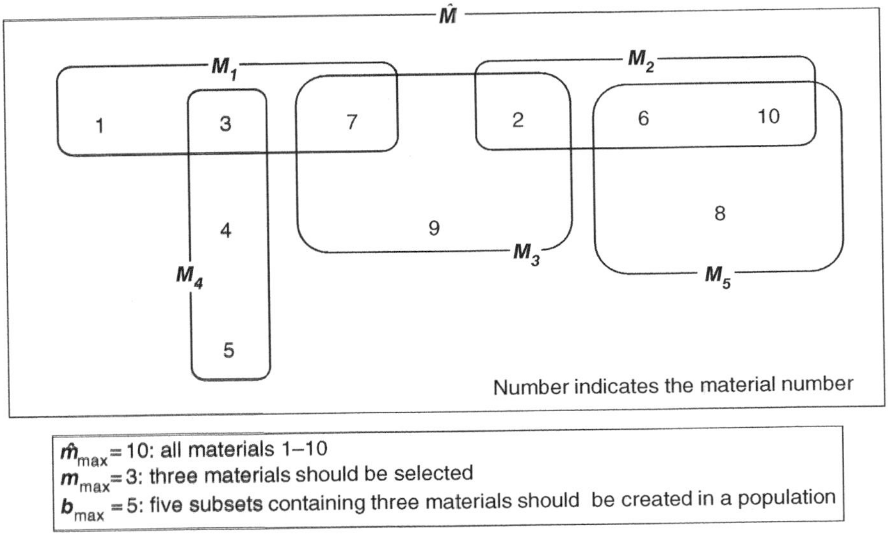

Figure 4.40 Diagram of relationship between $\hat{M}$ and M_i

$\bar{y}$: vector of taste information for the target product, expressed as $\left(\bar{y}_1, \bar{y}_2, \dots \bar{y}_{d_y}\right)^T$.
d_y: number of types of taste information, four in this case; sourness, saltiness, sweetness, and bitterness.
y: vector of taste information calculated for M_i as an estimate, expressed by $\hat{F}_B \hat{x}$.
$\hat{F}_B$: matrix obtained by taste measurement.
ε: vector of difference between $\bar{y}$ and y, expressed as $(y-\bar{y})^T W_y (y-\bar{y})$.
W_y: positive definite of symmetric matrix $d_y \times d_y$ showing the weights for each item of taste information. If the weight for the taste information is 1, W_y is equal to an identity matrix.
* $()^T$ means a transposed matrix.

To search for appropriate materials with appropriate blend ratios, we applied a GA to this problem. GAs are mathematical techniques to solve optimization problems by mimicking natural evolution [206, 207]. Evolutional processes, such as selection, crossover, and mutation, repeatedly take place, and the current population is transformed to a new one with higher fitness to a better solution. Our customized procedure for the GA is as follows:

1. Selection: All individuals $g_1, g_2, \dots g_{b\max}$ are sorted in descending order of fitness, which is evaluated using the objective function J_i. If x_i^* is a solution for g_i, the objective function J_i can be expressed as $J_i = f(h_{tr}(x_i^*, g_i))$. According to the elite strategy, 50% of individuals with higher fitness are recognized as elite and left in a population, while the others are terminated from the population.
2. Crossover: Two individuals g_i and g_j are randomly selected as "parents," and one-point crossover is carried out. One crosspoint is chosen at random in the number range from 2 to $(m_{max} - 1)$. Crossover continues until the number of g_i reaches b_{max}. All the individuals are recognized as "offspring in the next generation."

3. Mutation: Mutation occurs at any probability for only new offspring generated by crossover. All the numbers, each of which represents the serial number of a material, in a new offspring are changed to any value, generally with a low mutation rate.
4. Termination: A series of processes from selection to mutation is repeated until a termination condition is reached.

4.7.5.3 Constrained Nonlinear Optimization

If the maximum and minimum of a blend ratio are $\hat{x}_{min}$ and $\hat{x}_{max}$, respectively, a constraint can be defined as $\hat{x}_{min} \le \hat{x} \le \hat{x}_{max}$. Therefore, x_i^*, a vector of appropriate blend ratios for M_i, can be obtained by solving the following constrained nonlinear optimization problems [208]:

$$f\left(h_{\mathrm{tr}}\left(x_i, g_i\right)\right) = -\omega_\varepsilon \times \varepsilon \qquad (4.7)$$

$$\begin{cases} \max\limits_{x_i} f\left(h_{\mathrm{tr}}\left(x_i, g_i\right)\right) & & (4.8) \\ \text{subject to} & \varepsilon_{\max} - \varepsilon \ge 0 & (4.9) \\ & h_{\mathrm{tr}}\left(x_i, g_i\right) - \hat{x}_{\min} \ge 0 & (4.10) \\ & \hat{x}_{\max} - h_{\mathrm{tr}}\left(x_i, g_i\right) \ge 0 & (4.11) \end{cases}$$

where ω_ε is a vector of weights for each error in ε, and ε_{max} is the maximum of each error in ε. The appropriate solution x_i^* for the aforementioned constrained nonlinear optimization problems can be obtained by minimizing the objective function J_i by the downhill simplex method [209]. Then, by applying the values of J_i to the GA, $g_1, g_2, g_3, \ldots$ and $x_1, x_2, x_3, \ldots$ are obtained in descending order of fitness. In this way, we can select the appropriate types of raw ingredients and decide an appropriate blend ratio [210].

4.7.6 *For More Effective Use of "Taste Information"*

When taking a picture of a backlit subject with a digital camera, it is well known the picture becomes overexposed and the resulting picture is not what was seen. This is a kind of monofunctionalization of the sense of sight. Nevertheless, it has a wide array of uses because both its limitation and usability are already known. A taste sensor is a relatively new concept and device compared with other human sense alternatives. One can easily imagine alternative sensors or devices exhibiting the human senses of sight and sound, such as an optical sensor and a microphone. In contrast, the senses of smell and taste are derived from many chemical substances and alternative devices that are not readily familiar. This section is focused on a basic concept of the taste sensing system and its effective use of taste information.

4.7.6.1 Key Concept

The key to utilizing the taste sensor is to understand that the concept of taste need not be complex. The principle of operation of the sensor unit approximates and translates molecular information into "taste qualities." In general, chemical analysis is detail oriented, and therefore, sometimes, one "can't see the forest for the trees," since detailed information cannot show an overall view. Taste substances are huge molecules that typically interact with each other. Researchers, in many instances, tend to portray taste characteristics as complicated and composed of large varieties of taste factors. As a result, it is conventionally felt that chemical analysis is necessary to adequately analyze taste attributes. Chemical analysis of such compounds as free amino acids cannot provide the most important information on what a material "tastes like." The taste sensor is not a chemical analysis tool but can approximate and translate molecular information into taste qualities or the five basic tastes.

4.7.6.2 Taste Attributes or Qualities become Understandable and Translatable When They Are Simplified

Peak wavelengths or absorbance spectra obtained from chemical analysis are required to interpret the results into terms that can be understood by anyone. Without this interpretation, they are a meaningless row of numbers (except to experts). For example, no intelligent odor sensor is available at present. Coffee aroma contains about 900 odor molecules. Usually, 20–30 of the 900 molecules are analyzed with instruments such as GC/MS. These systems produce information that is too complex for the layman. Therefore, a translation of the molecular information into understandable and simple terms, such as, "caramel" and "fruity" aromas, is required to communicate with consumers.

Once data of a large number of molecules are translated into taste qualities, taste information can be translated into common terms. In architecture, design drawing is based on common metric measurements and therefore the client, design engineer, and contractor can communicate through that common perception. In the food and beverage industry, R&D, sales, and marketing have their own concepts and images for a product. Additionally, there are different and ambiguous ways of expressing taste. Imagine trying to describe the taste of Kentucky Fried Chicken® to someone to whom English is not the mother tongue. If R&D shows free amino acids of chicken, such as 0.86 mg P-Ser, 15 mg Tau, 0.1 mg Trp, 7.1 mg Urea, 3.1 mg Asp, 1.1 mg Thr, 1.2 mg Ser, 3.3 mg Phe, 13.1 mg Gly, 2.4 mg Glu, 3.4 mg Sar…, only a limited number of experts can decode this as meaningful information. Figure 4.41 illustrates the KFC taste data in simple, understandable terms. The solid line indicates the average taste quality of fried chicken available in Japanese retail establishments, and the dotted line is the taste quality of Kentucky Fried Chicken (chicken drumstick).

Once a common image is obtained from a common taste measurement, product design can move on to the next objective, such as "produce a similar-tasting or unique tasting product" or "what are your requirements?"; the decisions can now be made on the basis of accurate information. The results of the taste sensing system can allow discussions among R&D, sales, and marketing members using the same language.

This "taste information" is valuable if it can be communicated to those who need the information. The first step should be to grasp the overall picture. The measurement of samples as much as possible in a food group, such as coffee and tea from national brands to private-label

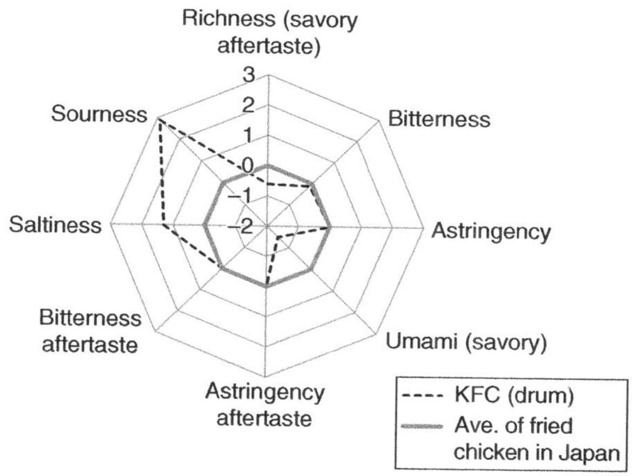

Figure 4.41 Taste information of average fried chicken in Japan and of Kentucky Fried Chicken.

brands (PLB) and from chain shops to famous private cafés is the first step to visualize consumer needs. Once taste information is obtained using the taste sensing system, it can then be combined with other information, such as amount of sales, age, gender, and nationality (from point of sale system), to reveal regional taste preferences. Generational differences such as young and aged, market trends, and high-end choice products verses low-cost products may produce significant information that was not previously recognized.

It is taste information that is finally narrowed down by market values. Are the five basic tastes with 11 different taste qualities as shown in Table 4.4 always required in evaluation? The answer is yes and no. For example, a taste combination of −4.5 of sourness, 1.7 of bitterness, 0.8 of bitterness aftertaste, 1.9 of saltiness, 1.5 of richness, 2.2 of umami, −0.9 of astringency, and −0.4 of astringent aftertaste describes Ethiopian mocha coffee. From the example, coffee taste gets approximated from thousands of molecules to taste qualities by the taste sensor, but it is still an eight-dimensional information quantity. From the viewpoint of R&D, further details may be needed depending on the result. In contrast, from the point of view of sales and marketing, to find more simple and valuable taste information, one must prioritize which taste qualities closely correlates with market value.

For instance, as shown in Figure 4.42, if one or a couple of taste information items, such as bitterness and sourness, correlate with values such as price, market value, product quality, but for another reason can be differentiated from one another or have correlation, a new criterion of values is set, thereby creating a new market value. This is a critical issue to be considered when utilizing taste information. It can be summarized as the well-known 80–20 rule [211]. Applying the Pareto principle allows simplification. It is commonly accepted that a small number of causes are responsible for a large percentage of the effect in a ratio of about 20 : 80. Imagine how to describe the taste of coffee. Coffee contains thousands of components that affect its taste, but bitter and sour taste expressions are perhaps the most important. Utilizing the 80–20 rule for this product, for example, where a couple of coffee tastes such as bitterness and sourness are seen as the overriding qualities for coffee, allows simplification of the taste evaluation.

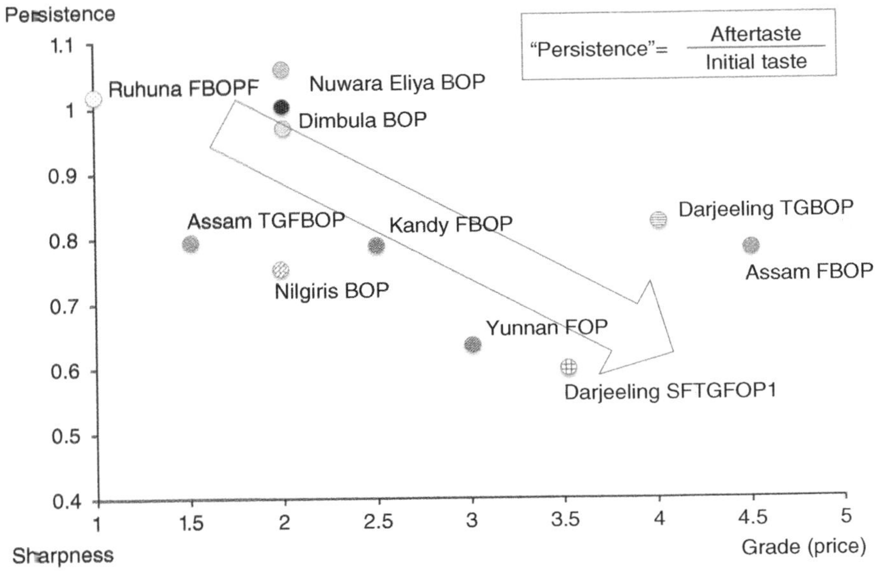

Figure 4.42 Taste information and market value. Tea with a sharper taste has a higher value on the market

4.7.6.3 Simplification of Large Numbers of Molecules into a Couple of Taste Qualities Allows Mathematical Optimization

Without simplification, combinations of factors for statistics lead to an astronomic number of evaluation items; thus it is impossible to apply mathematical optimization. As explained for coffee, a couple of taste indexes are significant to obtain a good approximation of product quality with cost reduction. The optimization method enables the optimization of a large variety of factors such as cost limitations and taste qualities of materials (resulting from blending and preprocessing).

Olfactory information is also essential in evaluating taste quality, but so far, no odor sensor or objective assessment method other than human sensory tests is available to correlate both taste and olfactory information effectively. However, once olfactory information can be digitized, even by using human sensory evaluation, it will be possible to apply this information to some degree of mathematical optimization.

The optimization method sets a target value and calculates the best combination from the known variables of cost limitation and taste qualities. The resulting data becomes suitable particularly for developing PLBs. This optimization method has great potential for application not only to cost reduction but also to a large variety of product developments, such as healthy products without the deterioration of taste.

Simplification allows another mathematical prediction for quality control and quality assurance. As shown in examples of beer samples, comparing beer with taste to a tasteless sample (such as saliva) will reveal deterioration in taste qualities. In the case of beer, acidic bitterness in its initial taste and aftertaste showed changes in taste intensity. In general, quality deterioration is a result of three factors: temperature, light, and oxidization. It is known that the Arrhenius equation [212] empirically describes quality deterioration in food and beverages.

4.7.6.4 Summary

1. Taste characteristics of foods and beverages are a result of combinations of thousands of molecules.
2. The taste sensor can approximate and translate molecular information into taste qualities.
3. Taste qualities become understandable and communicable to most people only if they are simplified.
4. Taste qualities can become market values if they are simplified.
5. Simplification allows mathematical optimization.

4.8 Application to Medicines

4.8.1 Introduction

Palatability is an important factor in determining medication adherence because patients often refuse to take medication because of the unpleasant taste of the API. This could lead to low compliance, adherence, and acceptability in children [213–215]. Therefore, the development of pediatric formulations with better palatability is one of the main issues in the pharmaceutical industry. In taste assessment, the taste sensor has the advantage of allowing the determination of an unpleasant taste of the API or pharmaceutical formulations without the need for sensory tests that may impose risk and burden on panelists. This section concentrates on the taste assessment techniques of APIs and pharmaceutical formulations using the taste sensor.

4.8.2 Bitterness Evaluation of APIs and Suppression Effect of Formulations

In 1998, Toko of Kyushu University and Koichi Wada of Nippon Boehringer Ingelheim Co., Ltd. applied the earliest type of taste sensor electrodes to the evaluation of the bitterness of quinine hydrochloride and an API (not disclosed) [216]. In 2000, Takahiro Uchida and his group of Mukogawa Women's University, Japan, began research and development on the taste assessment techniques of APIs using the next generation of taste sensor electrodes [217]. They measured 10 types of basic APIs, such as quinine hydrochloride and amitriptyline hydrochloride, with the taste sensor and applied measurement data to multiple regression analysis, where three types of information obtained from a taste sensor electrode, including relative value (S), CPA value (C), and their ratio (C/S) were used as explanatory variables [218]. Figure 4.43 shows the correlation between the bitterness sensory score determined by gustatory sensation and the predicted bitterness score obtained using the taste sensor. The predicted bitterness score has a high correlation with the human sensory score (0.824), indicating the effectiveness of taste assessment of medicines by the taste sensor.

Clarithromycin (CAM), one of the macrolide antibiotics, is used extensively for the treatment of certain bacterial infections. This antibiotic, however, has an extremely bitter taste and therefore its bitterness must be suppressed in pharmaceutical formulations. Clarithromycin dry syrup (CAMD) formulation effectively suppresses the bitterness by two taste-masking techniques. One is to mix in a substantial amount of sweet syrup, and the other is to coat particles of CAM with a copolymer, aminoalkyl methacrylate, to prevent the particles from being released [219, 220]. The bitterness of CAM and CAMD was evaluated with the taste sensor [136]. Figure 4.44 shows the correlation between the bitterness sensory score and

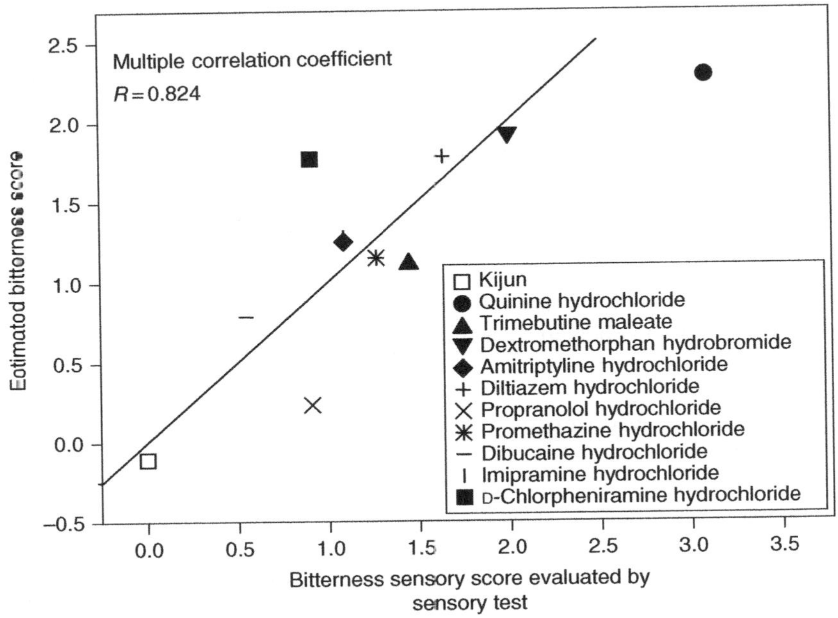

Figure 4.43 Correlation between bitterness sensory score evaluated by gustatory sensation and estimated bitterness score obtained from taste sensors. Reproduced with permission from Ref. [218]. Copyright 2001, The Pharmaceutical Society of Japan

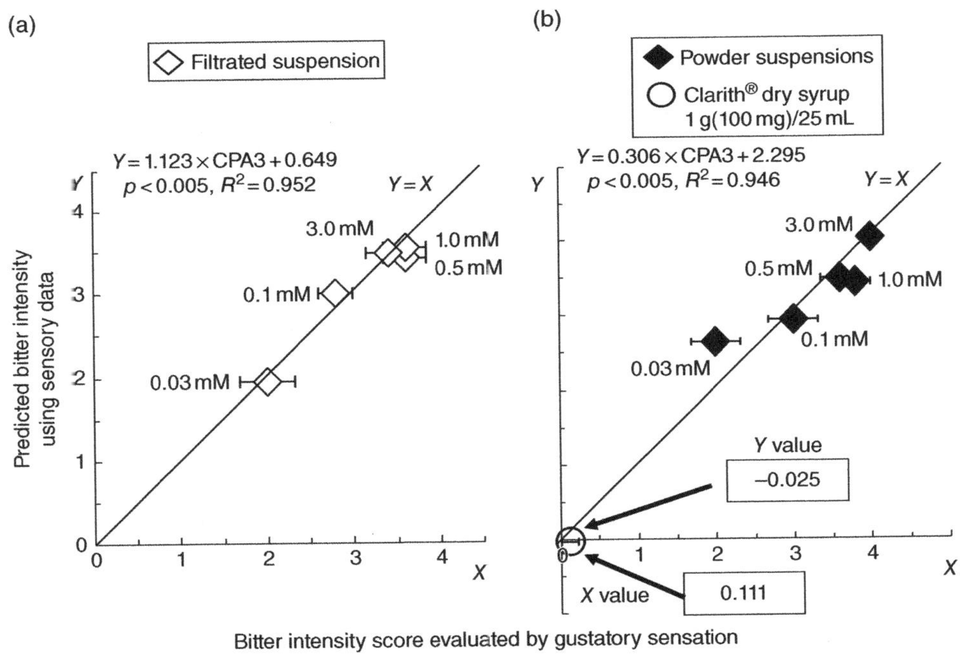

Figure 4.44 Correlation between bitterness sensory score and predicted bitter intensity using bitterness sensor AC0 for clarithromycin dry syrup at different concentrations. (a) data from the filtered solutions, (b) data from the clarithromycin suspensions and Clarith® dry syrup. For further explanation, see text. Error bars represents the mean ± standard deviation (n = 59). Reproduced with permission from Ref. [136]. Copyright 2003, The Pharmaceutical Society of Japan

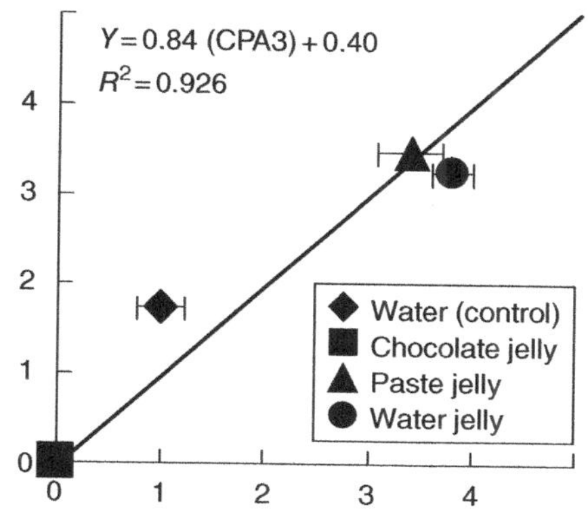

Figure 4.45 Correlation between bitterness sensory score and predicted bitter intensity using bitterness sensor AC0 for mixtures of clarithromycin dry syrup and jelly products. Reproduced with permission from Ref. [137]. Copyright 2006, The Pharmaceutical Society of Japan

bitterness intensity predicted by the bitterness sensor AC0 connected into channel 3. The sensor response not only has high correlation with the bitterness of CAM but also evaluates the suppression effect of CAMD. This demonstrates the effectiveness of taste assessment using the taste sensor.

Jelly products not only help children and the elderly swallow medicines but also suppress their bitterness. To investigate the suppression effect of jelly products, the bitterness of solutions extracted from mixtures of CAMD and some types of jelly products was evaluated with the taste sensor [137]. Jelly products include water, paste, and chocolate types. Figure 4.45 shows the correlation between the bitterness sensory score and bitterness intensity predicted by the bitterness sensor AC0 for mixtures of CAMD and jelly products. The mixtures of CAMD and paste jelly or water jelly have a considerably bitter taste. This is due to their acidic property. They have the pH of almost 4, which promotes the dissolution of CAM and aminoalkyl methacrylate. In contrast, the mixture with chocolate jelly has an extremely low bitter taste compared with that of water jelly, although both mixtures have alkaline pH. This may be because the high viscosity of chocolate jelly suppresses the dissolution of CAM, as well as the sweetness of the jelly suppressing the bitterness.

The taste assessment for other APIs and formulations, such as branched-chain amino acids, which are used as enteral nutrients, and some types of antibiotics including tetracyclines and cephems, also have been reported by Uchida [221–241] and other researchers [7, 8, 138, 242–255].

4.8.3 Development of Bitterness Sensor for Pharmaceutical Formulations

4.8.3.1 Sensor Design

As mentioned in Section 4.8.2, although the taste sensor has been researched for the taste assessment of APIs and pharmaceutical formulations and have shown to be fully beneficial in the pharmaceutical field, there are still some problems with respect to the sensor properties. One problem is the sensitivity and selectivity to the bitterness of APIs. Bitterness sensors AC0 and

ANO can respond to some additives and excipients, such as sweeteners. Since pharmaceutical formulations generally include a variety of such chemicals, higher sensitivity and selectivity are necessary. Another problem is the low correlation with human sensory scores. If a bitterness sensor has low correlation with the bitterness sensory score, multivariate analyses, including PCA or multiple regression analysis, must be used to interpret measurement data. However, such analyses require time and expertise. In the face of such challenges, we developed bitterness sensor BT0 for the evaluation of pharmaceutical formulations.

As discussed in Section 4.3.5, plasticizers are greatly responsible for the sensor properties [123]. As the amount of lipid in the membrane increases, some membranes selectively respond to the bitter substance at the range of CPA values of 10–100 mV. The difference in the CPA value may be explained partly by the LogD value of plasticizers, which is widely used to indicate hydrophobicity. When the LogD value is below 2, it is insufficient for adsorbing bitter substances onto the membrane, thus leading to a zero CPA value, like the TMSPM- and DGDE-based sensors. In contrast, when the LogD value is above 10, electric charge density does not change because excessively high hydrophobicity prevents the dissociation of the lipid in the membrane. Therefore, the TOTM-based sensor exhibits no response. Following these results, BBPA and TBAC were selected as appropriate plasticizers. Next, the amounts of the plasticizer and the lipid PADE incorporated in the membrane were determined by the D-optimal design method [256, 257] to obtain higher sensitivity and selectivity. Among the resulting candidate sensors, the sensor with the highest correlation with the human sensory score for basic APIs was selected, as shown in Figure 4.13a.

4.8.3.2 Prediction of Bitterness Intensity and Threshold

Bitterness sensor BT0 effectively enables the quantitative evaluation of bitterness and the bitterness threshold because of its high correlation with the human sensory score. Figure 4.46 shows the dependence of the CPA value of sensor BT0 on the quinine hydrochloride concentration [258]. In the concentration range of 0.01–1 mM, slight nonlinearity was observed. Therefore, the figure was divided into two for high (0.03–1 mM) and low (0.01–0.1 mM) concentrations to enable more

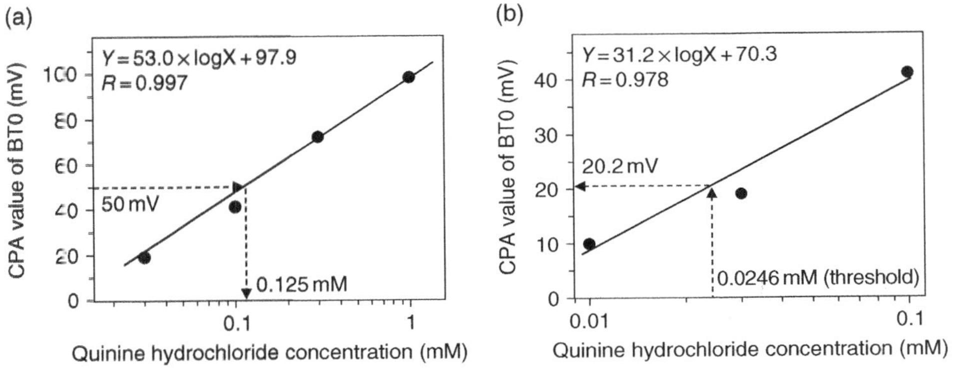

Figure 4.46 Dependence of the CPA value of sensor BT0 on the quinine hydrochloride concentration. (a) High concentrations. (b) Low concentrations. Republished with permission from Ref. [258]. Copyright 2013 CMC publishing

accurate evaluation. When an API with unknown taste exhibits a CPA value of 50 mV, it can be predicted that it has the same bitterness as that of 0.125 mM of quinine hydrochloride.

To predict the bitterness threshold for an API, first the bitterness threshold of quinine hydrochloride must be determined by a sensory test. The sensory test was conducted in accordance with the procedure below [259]:

1. Eight panelists were asked to determine whether a solution of 0.01 mM quinine hydrochloride is bitter compared with a solution of 10 mM KCl, which is tasteless and used as the control sample (a paired-comparison test).
2. When the panelist would detect a difference, for more reliable results, the panelist was asked to identify the odd sample among three unknowns consisting of the control and the bitter solution, two of which are the same (a triangle test). When the correct sample could be identified, the concentration was defined as the bitterness threshold for the panelist.
3. If the panelist could not detect the difference in Step 1, the panelist was asked to determine whether a solution of 0.012 mM quinine hydrochloride, 1.2 times the concentration of the initial sample, was bitter compared with the control. If the panelist could detect the difference, the triangle test for that concentration was conducted to determine the bitterness threshold, the same way as in Step 2. If no difference was yet detected, the concentration was increased to a further 1.2 times to 0.0144 mM, and the bitterness sample was evaluated again.

With the result of the sensory test, the bitterness threshold for quinine hydrochloride was found to be 0.0246 mM, which is the logarithmic mean value for the eight panelists. As shown in Figure 4.46b, the predicted CPA value for 0.0246 mM quinine hydrochloride is 20.2 mV. This also means that an API with the CPA value less than 20.2 mV is expected to exhibit no bitterness. Moreover, the bitterness threshold of an API can be estimated. Figure 4.47 shows

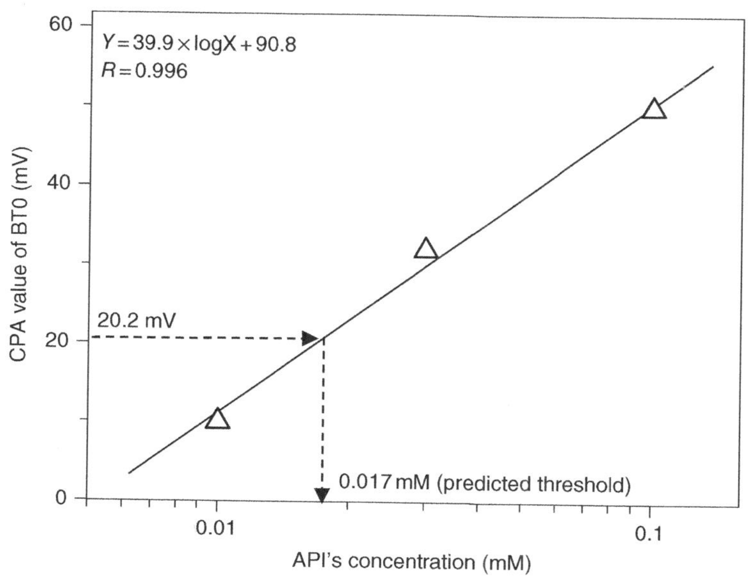

Figure 4.47 Example of dependence of the CPA value of sensor BT0 on API concentration. Republished with permission from Ref. [258]. Copyright 2013, CMC publishing

an example of the API concentration dependence of the sensor BT0 CPA value. The regression equation for the CPA value of 20.2 mV gives a concentration of 0.017 mM, which denotes the predicted bitterness threshold for the API.

4.8.3.3 Applications to Orally Disintegrating Tablets

In 2000, Tsutomu Harada of Eisai Co., Ltd. evaluated some orally disintegrating tablets, or ODTs, of donepezil hydrochloride using the taste sensor [260]. It was shown that the PCA using taste sensor electrodes, except sensor BT0, enables the evaluation of the change in the bitterness with increasing disintegrating time in the same manner as a sensory test.

After the development of sensor BT0, Harada and Uchida evaluated the bitterness of propiverine ODTs with sensor BT0 in combination with a disintegration testing apparatus ODT-101 [261]. Figure 4.48 shows the relationship between the CPA value of bitterness sensor BT0 and the disintegration time of propiverine ODTs. Each ODT contains one type of polysaccharide as a bitterness-masking agent, including carrageenans. Carrageenan has three sulfate groups in its molecule, which help it bind with propiverine hydrochloride, and thus suppress the bitterness with the binding after dissolved fully in water [262]. Fifteen seconds from the start of disintegration, the bitterness of an ODT with λ-carrageenan was suppressed most effectively. This is because λ-carrageenan in the ODT dissolves more rapidly in water than other carrageenans. In contrast, the ODTs with pectin and agar exhibit relatively suppressed bitterness. This is because these polysaccharides cause long disintegration time of the ODTs. With complete disintegration at 30 s, the bitterness of an ODT with ι-carrageenan was suppressed the most effectively, followed by λ-carrageenan and κ-carrageenan. This may be due partly to the number of sulfate groups and the difference in the structure between a random coil and a helix [263]. In contrast, the bitterness of ODTs with pectin and agar was not suppressed completely. This is because these polysaccharides have no sulfate group that can bind with the API, and therefore they cannot suppress the bitterness upon complete disintegration. These results suggest that the bitterness of ODTs containing polysaccharides depends on the balance between their disintegration and dissolution times.

Effective development of ODTs using the taste sensing system was reported by M. Kitamura, T. Nakagawa, A. Harada, and M. Hizaki of Sawai Pharmaceutical Co., Ltd. [264–266].

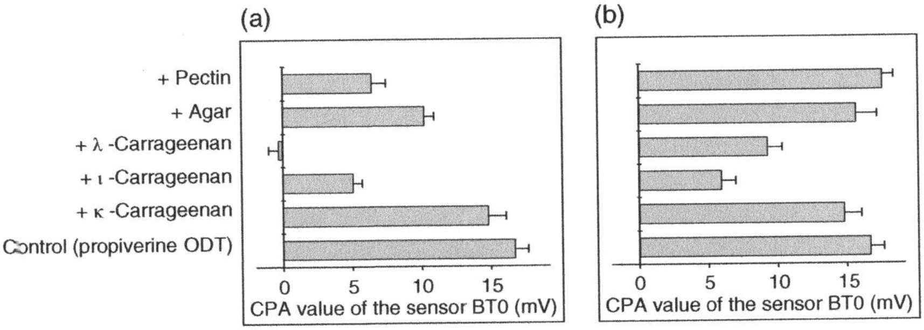

Figure 4.48 CPA value of bitterness sensor BT0 for propiverine ODTs at different disintegration times. (a) 15 s. (b) 30 s (full disintegration). Reproduced with permission from Ref. [218]. Copyright 2010, The Pharmaceutical Society of Japan

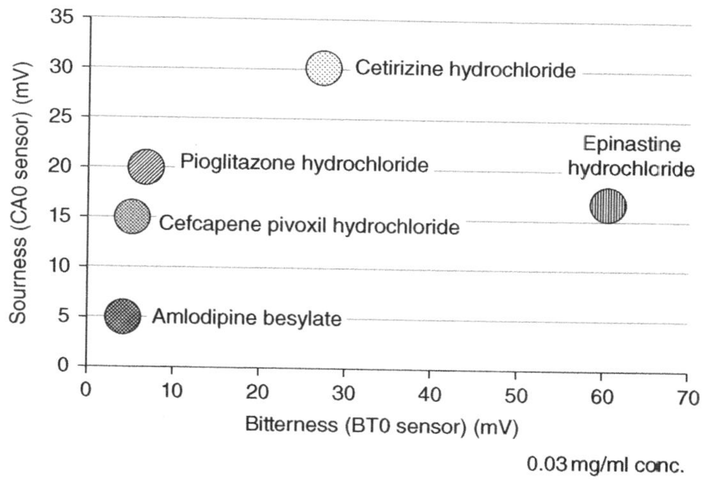

Figure 4.49 Tastes of active pharmaceutical ingredients. Republished with permission from Ref. [265]. Copyright 2012, Jiho Inc

The formulation design becomes very effective if it is based on these taste characteristics. It has become possible to set a design principle corresponding to the taste characteristics of APIs as well as to determine which kinds of masking methods would be effective depending on the degree of difficulty of designing a formulation. Many APIs have an unpleasant taste. Figure 4.49 is a taste map of APIs prepared using the taste sensing system. The vertical axis is sourness intensity and the lateral axis is bitterness intensity. From the figure, it is possible to roughly understand the taste characteristics of each drug. For example, pioglitazone hydrochloride has high sourness but moderate bitterness intensities. Epinastine hydrochloride has the strongest bitterness. Cetirizine hydrochloride has both strong bitter and sour tastes. In fact, both bitter and sour masking were required when designing its ODT.

As shown in Figure 4.50, about 20–50% of bitterness still remained after rinsing a sensor electrode with water. It is known from experience that drug bitterness is often sustained after mouth washing. A bitterness sensor reproduces these kinds of phenomena that are experienced in the oral cavity. Actually, epinastine hydrochloride has strong and sustained bitterness even after rinsing the mouth. On the other hand, sourness is almost completely diminished, as shown in Figure 4.51. Only 1–1.5% of residual sourness remained, suggesting that the sour component is easily removed from the oral cavity. Cetirizine hydrochloride has strong acridity just after its administration but no lasting aftertaste, indicating that acridity results from strong sourness.

Cetirizine hydrochloride possesses very strong bitterness and historically is thought to be unsuitable for ODT development. The primary issues in formulation design were the strong bitterness and the severe acridity of cetirizine. The taste sensing system was used for objectively studying the cause of the harsh tastes. It was found that cetirizine has very strong bitter and sour taste characteristics. The acute acridity suggested that the cause of stimulus was due to its strong sourness. To prepare cetirizine as an ODT, it is necessary to incorporate masking.

"Chemical masking" using cyclodextrin was applied for the above task. A mixture of cetirizine hydrochloride (CTZ) and beta-cyclodextrin was combined with a small amount of

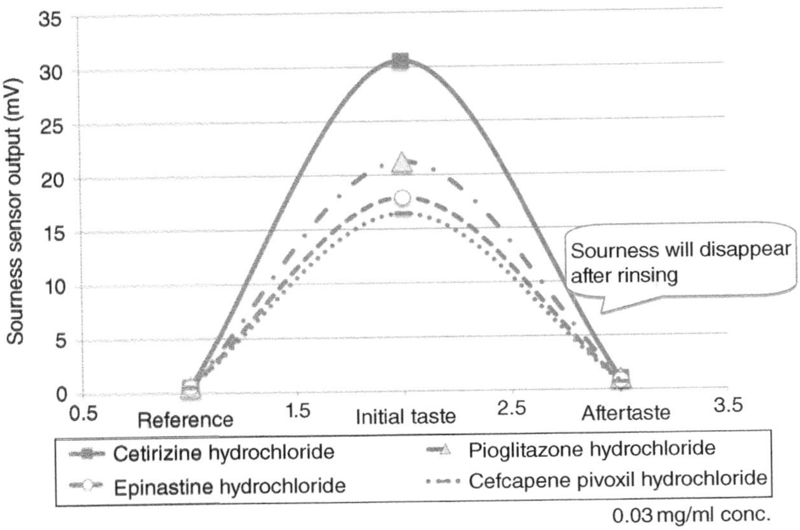

Figure 4.50 Reduction of sourness intensity (sensor CA0). Republished with permission from Ref. [265]. Copyright 2012, Jiho Inc

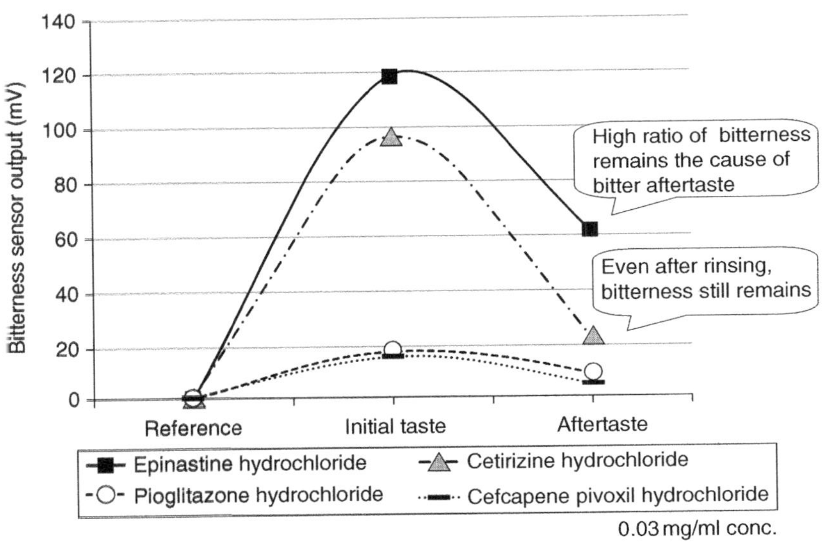

Figure 4.51 Reduction of bitter intensity (the sensor BT0). Republished with permission from Ref. [265]. Copyright 2012, Jiho Inc

purified water and then dehydrated. Figure 4.52 shows the effect of taste masking with and without β-CD. Bitterness was selectively masked by β-CD. As shown in Figure 4.53, β-CD suppressed the bitterness of cetirizine as a function of dosage. The processed mixtures with the molar ratios (CTZ : β-CD) of 1 : 1 to 1 : 4 were tested by a human sensory test (Table 4.6).

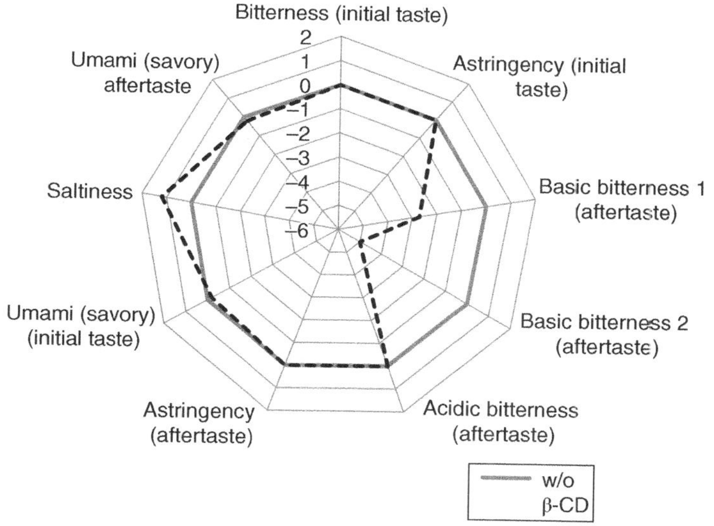

Figure 4.52 Bitterness masking effect of β-CD. Republished with permission from Ref. [265]. Copyright 2012, Jiho Inc

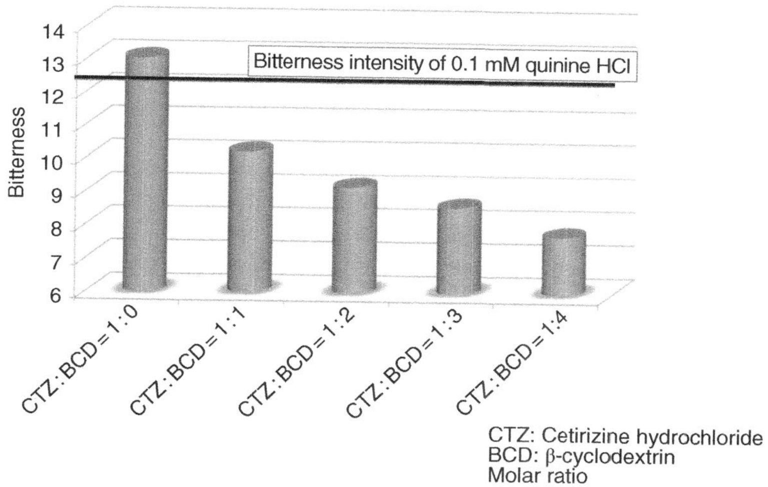

Figure 4.53 Bitterness masking effect and the amount of added β-CD. Republished with permission from Ref. [265]. Copyright 2012, Jiho Inc

Table 4.6 Bitterness masking effect of β-cyclodextrin (human sensory test)

Cetirizine hydrochloride (pts. wt.)	10	10	10	10
β-cyclodextrin (pts. wt.)	24.6	49.2	73.8	98.4
Molar ratios	1 : 1	1 : 2	1 : 3	1 : 4
Human sensory test (bitterness)	Strong bitterness	Weak bitterness	Almost no bitterness	Almost no bitterness

Source: Republished with permission from Ref. [264]. Copyright 2012, Jiho Inc

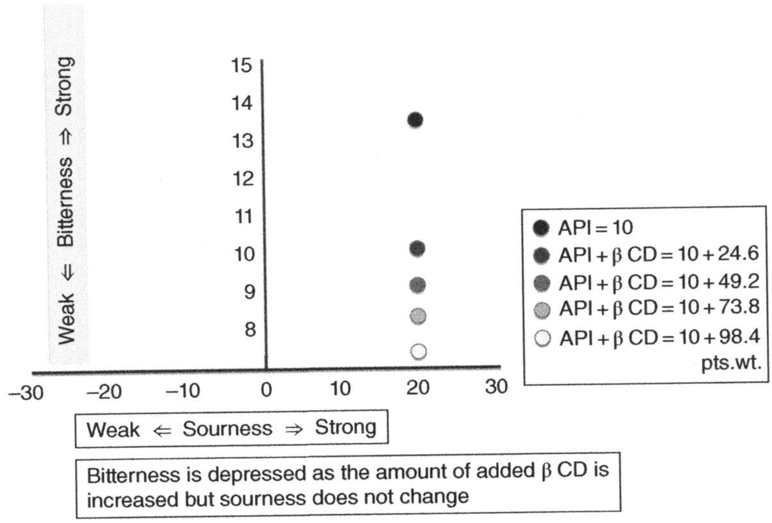

Figure 4.54 Bitterness masking effect of β-cyclodextrin (measured with a taste sensing system). Republished with permission from Ref. [264]. Copyright 2011, Jiho Inc

The test results exhibited dosage dependence. At ratios greater than 1 : 3, almost all bitterness was masked, as shown in Table 4.6. The bitterness evaluation using the taste sensing system also clearly showed the masking effect of adding β-CD.

On the other hand, the sourness of CTZ (cause of acridity) was not masked by adding β-CD, as shown in Figure 4.54, and hence, a way to mask its sourness was needed. Sodium citrate (SC) was used to adjust the pH and to mask the resulting acridity, as shown in Figures 4.55 and 4.56. Increasing the amount of CS decreased the sourness, but excessive dosage caused a salty taste. As the salty taste increases, the ease of consumption is reduced. Therefore, in this case, the formulation was designed to be CTZ : CS = 10 : 10, that is, slightly sour and not excessively salty. The human sensory test was performed for mixtures of 0–20 wt.% SC by adding 10 wt.% CTZ (Table 4.7). Given these findings, masking the bitterness and sourness of the tablet was accomplished by preprocessing with β-CD and SC.

Donepezil hydrochloride is a drug that has bitterness and numbing qualities. Dissociated donepezil in aqueous solution interacts strongly with anionic polymers through electrostatic interaction. Firstly, human sensory tests were performed on samples of donepezil hydrochloride added with equal amounts of polymer additives and D-mannitol as a diluting agent. The results of sensory tests, which indicate a notable masking effect, confirmed the efficacies of methacrylic acid copolymer-LD (EUDRAGIT® L-100-55), carmellose calcium (ECG® -505), and xanthan gum. L-100-55 and ECG were also confirmed to reduce the numbing effect.

ODT samples formulated with these masking materials were evaluated for bitterness and numbness assessment using the taste sensing system. Donepezil hydrochloride has a lasting bitter aftertaste and also a numbing effect, and therefore the test interval and the order of sample solutions easily affected the test results.

Figure 4.57 shows sensor outputs for ODTs composed of donepezil hydrochloride and masking polymers and D-mannitol diluent base. Sample solutions were made by dissolving

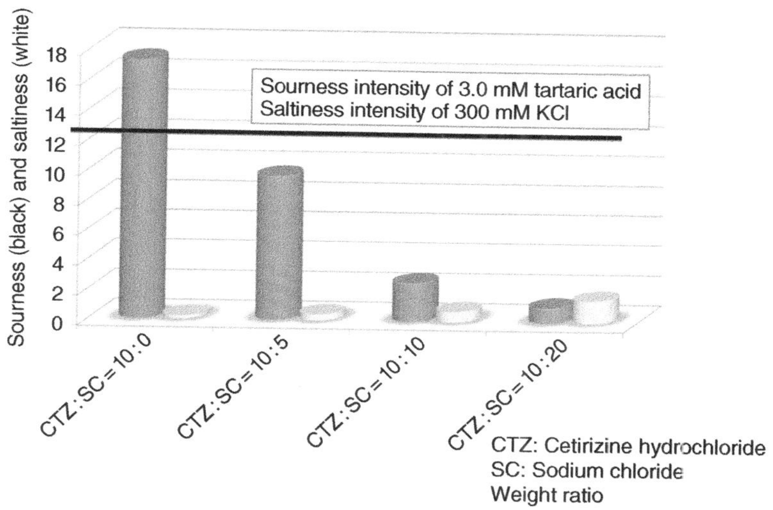

Figure 4.55 Sourness masking effect and the amount of added sodium citrate. Republished with permission from Ref. [265]. Copyright 2012, Jiho Inc

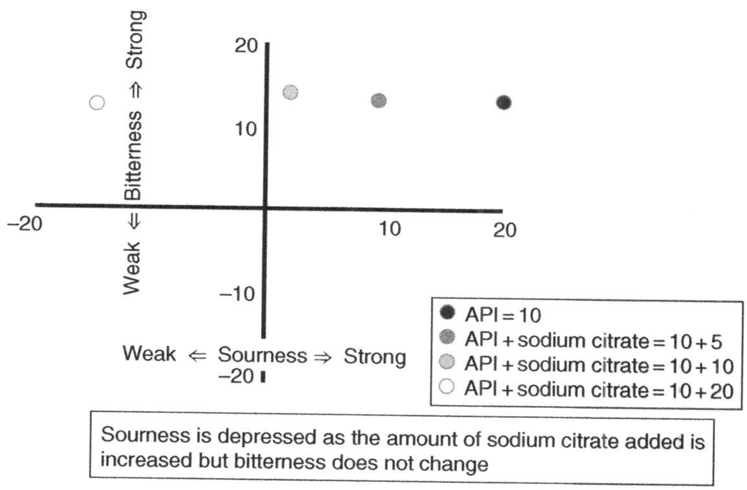

Figure 4.56 Sourness masking effect of β-cyclodextrin (measured with taste sensing system). Republished with permission from Ref. [264]. Copyright 2011, Jiho Inc

Table 4.7 Sourness masking effect of sodium citrate (human sensory test)

Cetirizine hydrochloride (pts. wt.)	10	10	10	10
Sodium citrate (pts. wt.)	0	5	10	20
Ratio by weight	1:0	2:1	1:1	1:2
Human sensory test (sourness)	Strong bitterness	Weak bitterness	Almost no bitterness	Almost no bitterness

Source: Republished with permission from Ref. [264]. Copyright 2012, Jiho Inc

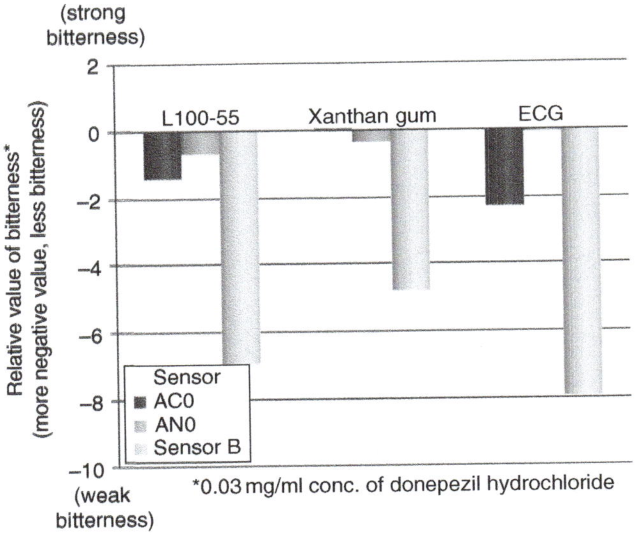

Figure 4.57 Bitterness masking effect of adding polymer additives (measured with taste sensing system). The Sensor B in the figure was a BT0 sensor. Republished with permission from Ref. [266]. Copyright 2012, Jihc Inc

0.03 mg/ml of donepezil hydrochloride. The resulting graph was normalized with the CPA value of 0.03 mg/ml donepezil hydrochloride alone in order to compare the masking effect. From the result, it was confirmed that all materials had masking effects compared with the effect of donepezil hydrochloride alone.

In contrast, the numbness caused by donepezil is recognized as an unpleasant sensation after the diminishing bitter taste, suggesting that adsorbed and residual donepezil manifests as numbness. Hence, persisting aftertaste was used as an index of numbness. In this case, the numbness caused by donepezil was defined as follows:

$$\text{Persistency of numbness} = \frac{\text{After taste}\left(\text{CPA value}\right)}{\text{Initial taste}\left(\text{relative value}\right)}$$

Figure 4.58 shows the persistency of aftertaste for each polymer additive relative to 0.03 mg/ml donepezil output. Samples with L100-55 and ECG exhibited lower outputs than those without masking materials, indicating reduction of numbness. In contrast, a sample with xanthan gum had increased persistence of aftertaste, suggesting that numbness remained. It is indicated that xanthan gum masks the bitterness of donepezil hydrochloride by inhibiting adsorption onto the lipid/polymer membrane of sensor surface. Once adsorbed, however, the viscosity of xanthan gum causes longer adsorption on the sensor surface. The degree of numbness masking by L-100-55 and ECG had a good correlation with the results of the human sensory evaluation.

Next, the masking mechanism of L-100-55 and ECG was investigated. It is said that the masking effect on donepezil is caused by the interaction between dissolved cationic donepezil and counter anionic polymers. Therefore, it is investigated whether determining the quantity

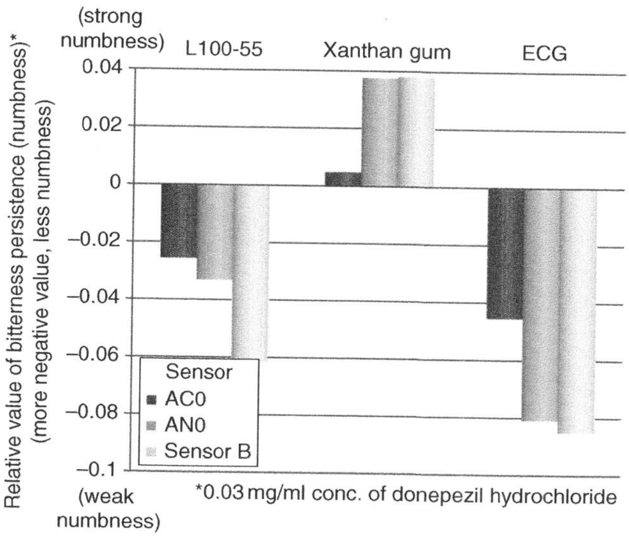

Figure 4.58 Numbness assessment with polymer additives (measured with taste sensing system). The Sensor B in the figure was a BT0 sensor. Republished with permission from Ref. [266]. Copyright 2012, Jiho Inc

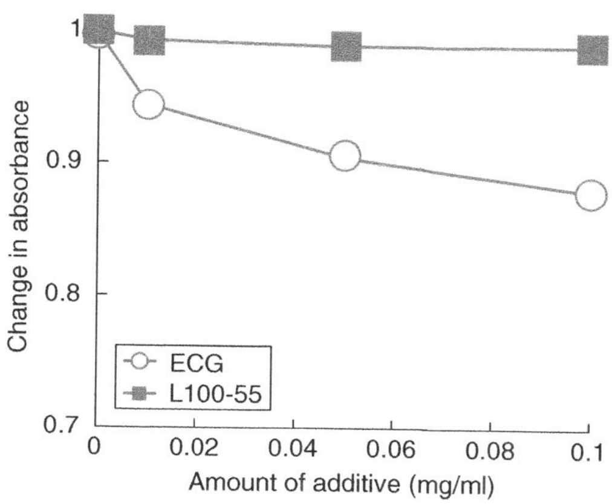

Figure 4.59 Change in absorbance with added polymers. Republished with permission from Ref. [266]. Copyright 2012, Jiho Inc

of dissolved donepezil provides a simple evaluation of the masking effect. The amount of adsorbed donepezil onto the membrane surface was measured from the change in absorbance. In Figure 4.59, the absorbance of 0.01 mg/ml donepezil is designated as 1. Absorbance increases with the amounts of L-100-55 and ECG. As shown in the result, the change in absorbance, which corresponds to the amount of dissociated donepezil, decreased as the amount of ECG increased, which indicates interaction between donepezil and ECG. L100-50 caused no change in absorbance, and therefore, no reduction in dissolved donepezil was confirmed.

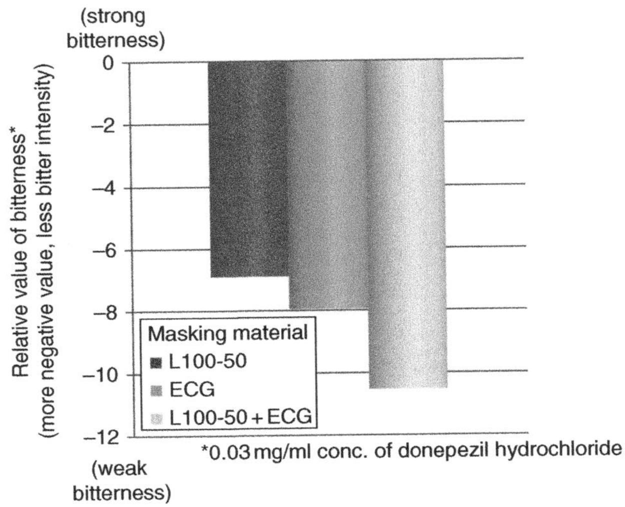

Figure 4.50 Bitterness masking effect of adding both L-100 and ECG (the sensor used in this experiment selectively responds to basic bitter substances). Republished with permission from Ref. [266]. Copyright 2012, Jiho Inc

The results showed that ECG has a masking effect on donepezil owing to the interaction between them. Additionally, L100-55 is considered to have only a small effect on masking donepezil itself but rather acts on the sensor surface composed of the lipid/polymer membrane to inhibit the adsorption of donepezil. From the result, a synergistic masking effect can be expected even with combined use because L100-55 and ECG have different masking mechanisms. Figure 4.60 shows the masking effect of the combination of L100-55 and ECG. As expected, combined use showed better bitterness masking than when using L100-50 or ECG alone.

The evaluation of the bitterness of other APIs and their formulations using sensor BT0 has been reported in other literature [267–270].

4.8.3.4 Response Mechanism to APIs

In this section, the response mechanism of bitterness sensor BT0 to APIs is discussed in detail. The LogD of plasticizers is considered to be one of the parameters associated with the sensor characteristics. However, taste sensor membranes based on the plasticizers BBPA and DOPP, which have almost the same LogD value, exhibit different sensitivities and selectivities to bitter substances (Figure 4.9a and b). This implies that other parameters may be involved in the taste sensor membrane characteristics. To search for such parameters, sixteen hydrochloride salts were examined with respect to the bitterness sensory score, the CPA value, and physicochemical parameters, such as LogD and LogP (Table 4.8) [271]. In the correlation between the CPA value and bitterness sensory score, fursultiamine hydrochloride and papaverine hydrochloride are plotted as outliers. This result suggests that these two APIs cause the accuracy of bitterness evaluation to be low for specific reasons.

Table 4.8 Taste information converted from taste sensor outputs and physicochemical parameters

	Bitterness sensory score	CPA value (mv)	LogD	PSA (Å^2)	MSA (Å^2)	ASA (Å^2)	pK_a	Ionization potential (eV)	Electron density (%)	Polarizability (a.u.)
Quinine	3.93	42.15	−0.55	46.79	488.20	484.24	9.05	8.81	5.18	195.61
Cetirizine	1.64	26.39	0.26	57.04	575.01	615.90	2.12	9.22	5.25	207.55
Hydroxyzine	1.64	32.31	0.08	37.14	582.66	620.94	2.09	9.12	5.25	206.17
Azebestine	6.10	45.60	0.85	37.11	556.25	579.07	8.88	8.92	5.28	226.13
Eperisone	1.14	41.67	1.00	21.51	468.88	502.50	8.77	9.02	5.27	150.69
Bromhexine	−0.14	20.44	1.09	30.46	434.53	424.34	9.32	8.67	5.26	153.75
Ambroxol	0.43	3.39	−0.46	62.86	406.40	430.32	9.30	8.54	5.32	148.80
bperam id	7.20	64.41	1.48	44.98	728.78	640.54	9.41	8.96	5.24	268.85
Theanine	−0.21	0.31	−4.10	105.40	393.63	440.11	5.54	12.69	5.05	154.72
Fursultiamine	3.62	1.21	−0.80	153.42	595.85	524.65	6.26	9.08	5.21	218.43
Papaverine	0.43	46.23	2.12	51.06	518.59	666.35	6.28	8.91	5.14	208.35
Ranitidine	0.43	1.13	−1.46	112.76	490.05	602.19	8.08	8.91	5.26	171.41
Hydralazine	−0.14	4.01	−0.18	65.08	221.41	297.64	6.38	9.06	5.13	100.33
Promethazine	1.36	20.95	−0.40	32.98	416.84	372.54	9.40	7.63	5.26	180.05
Cyproheptadine	1.00	29.13	0.81	16.61	397.65	406.22	9.73	8.54	5.26	199.34
Dicyclomine	2.92	32.01	1.86	30.74	592.81	531.34	8.96	9.09	5.27	164.38

Source: Republished with permission from Ref. [271]. Copyright 2009, The Japanese Association for the Study of Taste and Smell.

The type of standard sample used to calculate the conversion factor depends on the type of taste sensor.

In a sensory test with eight panelists, quinine hydrochloride concentrations of 0.01, 0.021, 0.047, 0.1, 0.21, and 0.47 mM were used as standards of bitterness and were assigned scores of 1, 2, 3, 4, 5, and 6, respectively. 0.1 mM of all APIs was evaluated to determine the bitterness sensory score and the CPA value.

LogD, polar surface area (PSA), molecular surface area (MSA), accessible surface area (ASA), and acid dissociation constant (pK_a) were calculated using the web-based software MarvinSketch 5.2.2. Ionization potential, electron density, and polarizability were calculated with the software WinMOPAC 3.0.

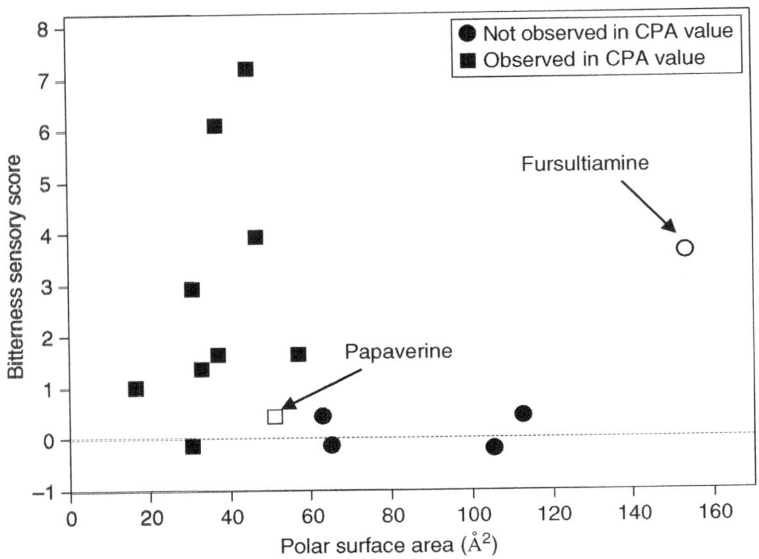

Figure 4.61 Relationship between the polar surface area and bitterness sensory score. Circles: CPA value was not observed. Squares: CPA value was observed. Republished with permission from Ref. [271]. Copyright 2009, The Japanese Association for the Study of Taste and Smell

After the analysis of the parameters focusing on the two APIs, they were found to have unique values of two parameters: the polar surface area (PSA) and LogD. Figure 4.61 shows the relationship between the PSA and the bitterness sensory score. Circles and squares indicate no or some response of sensor BT0 to the corresponding API. Sensor BT0 has almost no response to APIs with the PSA of more than 60. The PSA denotes the sum of the surfaces of polar atoms, such as amino or hydroxyl groups, in one molecule of an API, and has been widely used in the estimation of molecular transport properties, including bioavailability, in drug discovery [272]. This result suggests that APIs with high PSA values cannot penetrate on the sensor membrane, subsequently leading to no sensor response. However, fursultiamine hydrochloride exhibits an extremely bitter taste with the bitterness sensory score of 3.62 although it has the highest PSA value among all the APIs evaluated. This means that sensor BT0 cannot evaluate the bitterness of fursultiamine hydrochloride, and, in practice, does not respond to it. Figure 4.62 shows the relationship between the LogD and CPA values of bitterness sensor BT0. Papaverine hydrochloride has relatively high values of LogD and CPA value although the panelists did not sense any bitter taste at a concentration of 0.1 mM. These results demonstrate that the bitterness sensor response depends partly on the LogD of APIs. With a few exceptions, since most bitter APIs have a certain level of hydrophobicity, they may be detected by bitter taste receptor T2Rs through hydrophobic interaction. To enable the evaluation of the bitterness of the exceptional APIs, a new bitterness sensor must be developed in the near future.

4.8.4 *Evaluation of Poorly Water-Soluble Drugs*

A new measurement methodology of poorly water-soluble drugs was reported by Asako Takakura of Shionogi & Co., Ltd [273]. Dipyridamole, anhydrous quinine, isopropylantipyrine,

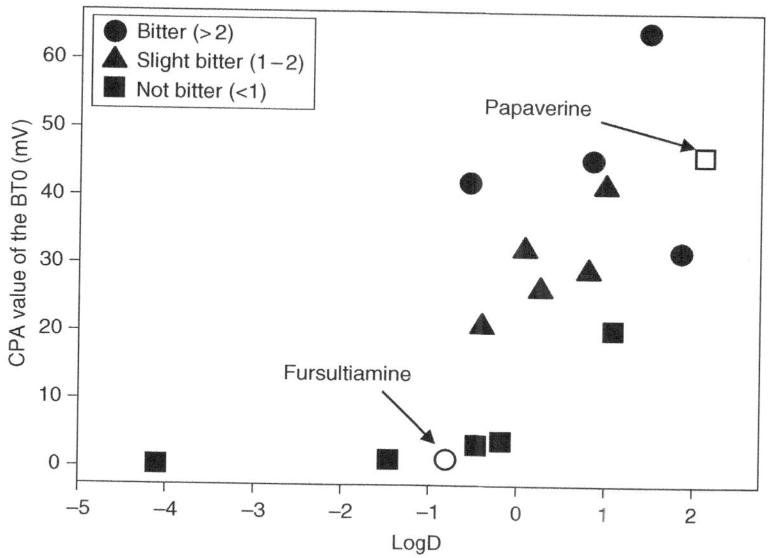

Figure 4.62 Relationship between LogD and CPA value of the bitterness sensor BT0. Circles: bitter (over 2 in sensory score), Triangles: slightly bitter (1–2), Squares: not bitter (<1). Republished with permission from Ref. [271]. Copyright 2009, The Japanese Association for the Study of Taste and Smell

and egg yolk lecithin biochemical agent were purchased from Wako Pure Chemical Industries, Ltd. Other excipients were of Japanese Pharmacopeia (JP) or Japanese Pharmaceutical Excipients (JPE) grade. Dipyridamole was used as a model of poorly water-soluble drugs. It is a basic compound with $pK_a1 = 0.8$ and $pK_a2 = 6.3$, and the literature-based value of its solubility is 5–10 μg/ml at pH 6–7:

1. Saturated dipyridamole solution: An excess amount of dipyridamole was added to 10 mM KCl solution. Next, the mixed solution was ultrasonicated for 20 min and centrifuged for 10 min at 3000 rpm. Supernatant solution was obtained as a sample.
2. Suspension of dipyridamole: Dipyridamole was added into 10 mM KCl solution to a concentration of 0.3 mM (151 mg/ml), and the mixed solution was ultrasonicated for 20 min.
3. Physical mixture with dipyridamole: Dipyridamole and crystalline cellulose at weight ratios of 1 : 9, 1 : 99, and 1 : 999 were triturated finely in a mortar using a pestle.

The bitterness of the drugs was measured with taste sensing system SA402B equipped with sensor BT0:

1. Conventional solution method: In accordance with the standard procedure, the measurement was performed twice and the average was used as the sensor output.
2. Directly Pasting Method: Dipyridamole was directly pasted onto the taste sensor membrane's surface with a finger protected by dry latex examination gloves.

Table 4.9 Measurement procedure of new methodology for poorly soluble APIs

	Measuring time (s)	Cleaning time (s)	Cutoff times	Return step	Stability criterion							
					1ch	2ch	3ch	4ch	5ch	6ch	7ch	8ch
Cleaning solution 1	—	90	—	—	—	—	—	—	—	—	—	—
Cleaning solution 2	—	120	—	—	—	—	—	—	—	—	—	—
Cleaning solution 3	—	120	—	—	—	—	—	—	—	—	—	—
Conditioning solution	30	—	20	—	0.5	0.5	0.5	0.5	0.5	0.5	0.5	0.5
Reference solution	120	—	—	—	—	—	—	—	—	—	—	—
Sample	30	—	—	—	—	—	—	—	—	—	—	—
Cleaning solution 4	—	3	—	—	—	—	—	—	—	—	—	—
Cleaning solution 5	—	3	—	—	—	—	—	—	—	—	—	—
CPA solution	30	—	—	—	—	—	—	—	—	—	—	—

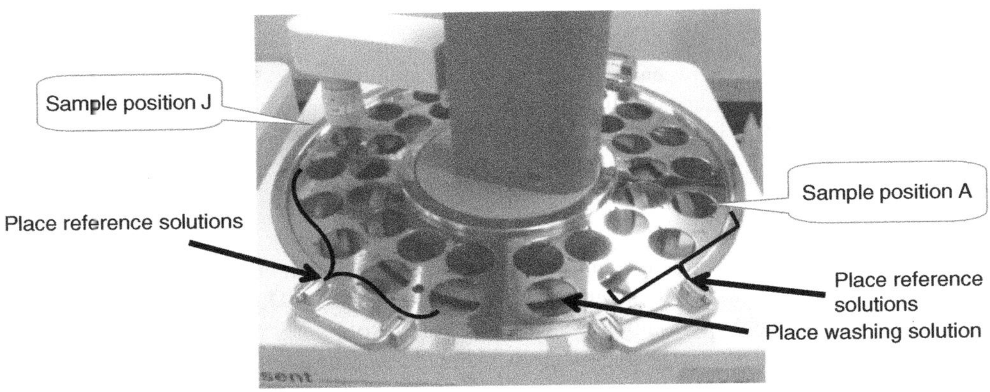

Figure 4.63 Pasting position and placement of solutions

Table 4.9 shows the measurement procedure for insoluble APIs. In this case, all solutions including "Cleaning solution 1" were reference solutions composed of 30 mM KCl and 0.3 mM tartaric acid:

1. The setup for the measurement is shown in Table 4.9. "Reference solution" must be placed on the fifth line between "Conditioning solution" and "Sample." Set "Measuring time" to be as long as necessary to directly paste the drug onto the taste sensor membrane's surface with your finger. In this case, it is set to be 60 s.
2. Leave a vacancy at sample position J and directly paste the drug on the surface of the sensor membrane while it is under the resting condition in accordance with the "measuring time" of the "reference solution" you have set. Place the reference solution composed of 30 mM KCl and 0.3 mM tartaric acid at sample position A. Other than that, the rest of the solutions and their positions should be exactly the same as in the sample measurement procedure given in the handling tutorial (Figure 4.63).

Figure 4.64 Pasting the API on the surface of the taste sensor membrane

Table 4.10 Bitterness evaluated by various methods

Sample	Conventional solution method (mV)		Directly pasted method (mV)	
	Initial taste BT0	Aftertaste CPA(BT0)	Initial taste BT0	Aftertaste CPA(BT0)
Saturated solution	58.1	39.6	9.2	4.4
Suspension	56.8	43.9	150.1	151.3

Source: Republished with permission from Ref. [273]. Copyright 2011, The Society of Powder Technology, Japan.

3. Once measurement starts, the sensor electrode moves to the first, second, and third cleaning solutions, the conditioning solution, and then sample position J. At this position, the sensor head descends and hold its position for 60 s. During the resting position, the drug must be directly pasted onto the sensor membrane's surface by a finger protected by dry latex examination gloves (Figure 4.64). After 60 s, the sensor head moves to sample position A and measures the electric potential. This potential becomes the sensor output of the dipyridamole sample.

In the following experiment, the CPA value was measured. The CPA value of the taste sensing system is obtained, as described in Section 4.3.4. The results are listed in Table 4.10. In the conventional solution method, there was almost no difference in bitterness between saturated solution and a suspension of dipyridamole. In the directly pasted method, however, the suspension showed a higher sensor output for the bitterness of dipyridamole. In the conventional solution method, particles in the suspension could not come into direct contact with the membrane surface, resulting in no potential change in both samples.

Three types of poorly water-soluble drugs and nine excipients assessed by the directly pasted method are listed in Table 4.11. Tasteless powders or with tastes other than bitterness showed slight responses, and powders with bitterness exhibited high outputs. Bitterness evaluation of physical mixtures by the directly pasted method was carried out. The composition ratio of dipyridamole was changed. As listed in Table 4.12, higher outputs were obtained with increasing amounts of dipyridamole.

Table 4.11 Specificity for samples determined by pasting method

Sample	Taste	Initial taste (mV) BT0	Aftertaste (mv) CPA (BT0)
Light anhydrous silicic acid	Tasteless	−1.0	−1.0
Oxidized titanium	Tasteless	3.7	3.0
Crystalline cellulose	Tasteless	2.4	1.4
Stearic acid	Tasteless	1.4	1.1
Sucrose	Sweet	0.9	0.9
Potassium chloride	Salty	0.2	0.0
Sodium chloride	Salty	5.1	1.4
Anhydrous citric acid	Sour	11.4	3.7
Egg yolk lecithin	Nutsy	7.8	5.9
Dipyridamole	Bitter	149.4	149.1
Isopropylantipyrine	Bitter	175.9	178.9
Anhydrous quinine	Bitter	191.1	189.9

Source: Republished with permission from Ref. [273]. Copyright 2011, The Society of Powder Technology, Japan.

Table 4.12 Bitterness assessment with various concentrations of dipyridamole

Composition ratio of physical mixture		Initial taste (mV) BT0	Aftertaste (mV) CPA (BT0)
Dipyridamole	Crystalline cellulose		
0.0	100.0	2.4	1.4
0.1	99.9	82.0	46.0
1.0	99.0	117.8	78.0
10.0	90.0	157.7	154.7

Source: Republished with permission from Ref. [273]. Copyright 2011, The Society of Powder Technology, Japan.

4.9 Perspectives

In this section, we reviewed the state-of-the-art technology of electronic tongues, mainly by focusing on a taste sensor, that is, an electronic tongue with global selectivity. The taste sensor enabled the quantification of the five basic tastes and astringency detected by the human tongue and successfully provided sensor outputs in good agreement with the results of sensory evaluation by panelists. For pungency, a type of taste in a broad sense, the development of a SPR immunosensor that can detect capsaicin, a typical pungent substance, at ultrahigh sensitivity is under way [196, 274]. In the future, it will become possible to quantify the intensities of all types of taste of foods, including the five basic tastes, astringency, and pungency.

Let us conclude by again taking a look at an example of objectivity and subjectivity. In Figure 4.65, the impressions of a family are shown for a girl, her father, and her grandmother drinking green tea. The girl says, "this green tea is bitter," whereas her father calls it "mild," but her grandmother says it is "not bitter." These different responses can be understood in terms of the acceptable range (capacity) of bitterness of each person, shown in Figure 4.65. The evaluation of the bitterness of the tea is made by each person; that is, the girl evaluated it

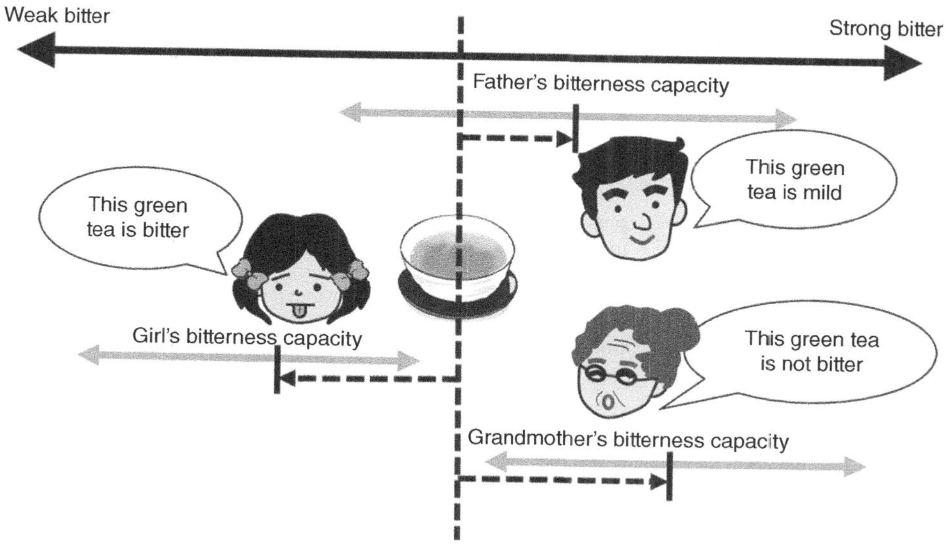

Figure 4.65 Tasting of each person when drinking green tea

to be beyond her acceptable range of bitterness, whereas her father evaluated the bitterness to be in the middle of his own acceptable range, but her grandmother evaluated it to be at the weak end of her range. This means that the criterion of each person's taste is based on a different scale. In addition, the taste is determined by a number of factors, including the five senses (sight, hearing, touch, taste, and odor) as well as food habit and dietary culture. As explained also in Figure 4.1, therefore, "taste" perceived in the brain eventually becomes a subjective concept.

For comparison with the other senses, we select "weight" as another example. The feel and evaluation of weight belong to the sense of touch and are said to be a higher-order function of tactical information. One of the authors (K. T.) usually does not hold anything heavier than a pen, works only with a keyboard at his desk, or carries a laptop computer weighting about 1 kg on business trips. A 3 kg computer will be heavy for the author but will not be heavy and may even be light for other muscular persons. However, the weight of 1 kg or 3 kg is an objective fact. The feeling of "heavy" or "light" is subjective, while the value of "1 kg" or "3 kg" is objective. The same is true for "vision." One example is the shape of a ballpoint pen. It has objective information, such as the thickness, length, color, and shape of the grip, as well as has subjective impressions such as "nifty," "unrefined," or "is one step away from capturing my heart." In this way, both the senses of touch and sight have subjective and objective aspects. If we again focus on taste, the seven basic tastes are subjective perceptions. One of the devices that provide an objective measure ("scale of taste") to the ambiguous sense of taste is a "taste sensor" or "electronic tongue with global selectivity."

In our sense of hearing, we have musical scores, and hence we can reproduce the music of, for example, Bach or Beethoven even now in these modern times. If we can invent "food scores" in the sense of taste, analogous to musical scores, we would be able to preserve and transmit food scores and then reproduce the taste of the desired foods anywhere and at any

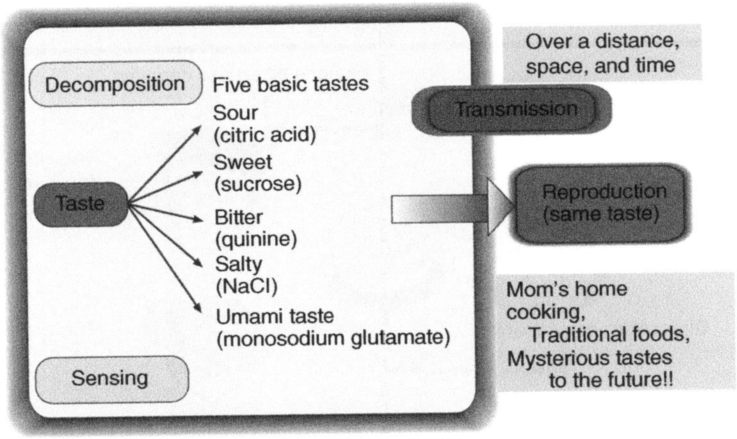

Figure 4.66 Transmission and reproduction of foods using food scores

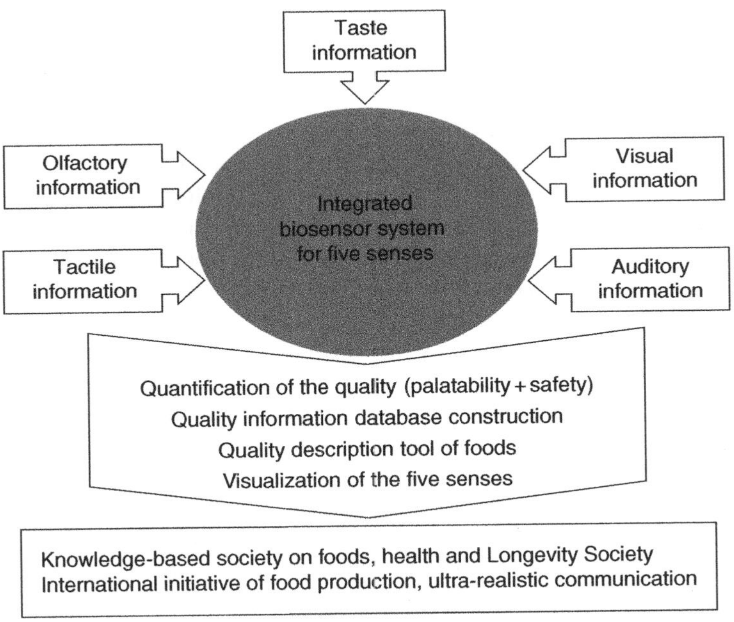

Figure 4.67 Expression of palatability by integrating sensor outputs corresponding to five senses

time, as shown in Figure 4.66. It will become possible, because the taste sensor can quantitatively express tastes experienced by humans, as explained in this section. However, the same situation might not be so easy in the case of smell, because there are no basic smells.

As demonstrated by the examples of application presented in this section, we can record the taste of foods in a computer as digital data. Suppose that we measure a food that does not taste very good using the taste sensor. We can then improve the taste on the basis of data in the archive. Of course, it is desirable for the palatability to be evaluated by also taking into account the history, food culture, and the ethnicity concerned. Therefore, the food score must be expressed mathematically as a function of these factors.

Humans perceive tastes by the tongue and also systematically sense tastes on the basis of information on the five senses, including odor, texture, visual appearance, and sound. In the future, as shown in Figure 4.67, an integrated biosensor system for the five senses will be developed to enable us to quantify the quality (palatability and safety), construct a quality information database, develop a quality description tool of foods, and visualize the five senses. It will lead to a knowledge-based society on foods, and also health and longevity society; international initiative on food production; and ultrarealistic communication.

References

[1] K. Toko, ed. *Biochemical Sensors: Mimicking Gustatory and Olfactory Senses*, Singapore: Pan Stanford Publishing, 2013.

[2] Y. Tahara and K. Toko, Electronic tongues: A review, *IEEE Sensors J.*, vol. 13, no. 8, pp. 3001–3011, 2013.

[3] L. Cassiday, Making sense of e-sensing. 2013. AOCS. Available from http://www.aocs.org/Membership/informArticleDetail.cfm?itemnumber=18699 (accessed on August 27, 2015).

[4] K. Toko, T. Onodera, and Y. Tahara, Nano-biosensors for mimicking gustatory and olfactory senses, *Bio-Nanotechnology: A Revolution in Food, Biomedical and Health Sciences*, D. Bagchi, M. Bagchi, H.Moriyama, and F. Shahidi, eds., pp. 270–291, New York: John Wiley & Sons, Inc., 2013.

[5] R. S.Latha and P. K.Lakshmi, Electronic tongue: An analytical gustatory tool, *J. Adv. Pharm. Technol. Res.*, vol. 3, no. 1, pp. 3–8, 2012.

[6] N. Savage, Technology: The taste of things to come, *Nature*, vol. 486, no. 7403, pp. S18–S19, 2012.

[7] K. Woertz, C. Tissen, P. Kleinebudde, and J. Breitkreutz, A comparative study on two electronic tongues for pharmaceutical formulation development, *J. Pharm. Biomed. Anal.*, vol. 55, no. 2, pp. 272–281, 2011.

[8] K. Woertz, C. Tissen, P. Kleinebudde, and J. Breitkreutz, Taste sensing systems (electronic tongues) for pharmaceutical applications, *Int. J. Pharm.*, vol. 417, no. 1–2, pp. 256–271, 2011.

[9] M.del Valle, Electronic tongues employing electrochemical sensors, *Electroanalysis*, vol. 22, no. 14, pp. 1539–1555, 2010.

[10] L. Escuder-Gilabert and M. Peris, Review: Highlights in recent applications of electronic tongues in food analysis, *Anal. Chim. Acta*, vol. 665, no. 1, pp. 15–25, 2010.

[11] A. Riul, Jr., C. A. R. Dantas, C. M. Miyazaki, and O. N. Oliveira, Jr., Recent advances in electronic tongues, *Analyst*, vol. 135, no. 10, pp. 2481–2495, 2010.

[12] M. Ghasemi-Varnamkhasti, S. S. Mohtasebi, and M. Siadat, Biomimetic-based odor and taste sensing systems to food quality and safety characterization: An overview on basic principles and recent achievements, *J. Food Eng.*, vol. 100, no. 3, pp. 377–387, 2010.

[13] Y. Kobayashi, M. Habara, H. Ikezazki, R. Chen, Y. Naito, and K. Toko, Advanced taste sensors based on artificial lipids with global selectivity to basic taste qualities and high correlation to sensory scores, *Sensors*, vol. 10, no. 4, pp. 3411–3443, 2010.

[14] D. Citterio and K. Suzuki, Smart taste sensors, *Anal. Chem.*, vol. 80, no. 11, pp. 3965–3972, 2008.

[15] F. Winquist, Voltammetric electronic tongues: Basic principles and applications, *Microchim. Acta*, vol. 163, no. 1–2, pp. 3–10, 2008.

[16] V. Anand, M. Kataria, V.Kukkar, V. Saharan, and P. K. Choudhury, The latest trends in the taste assessment of pharmaceuticals, *Drug Discov. Today*, vol. 12, no. 5–6, pp. 257–265, 2007.

[17] P. Ciosek and W. Wroblewski, Sensor arrays for liquid sensing: Electronic tongue systems, *Analyst*, vol. 132, no. 10, pp. 963–978, 2007.

[18] M. Habara and K. Toko, Taste sensor, *Encyclopedia of Sensors*, C.A.Grimes, E. C. Dickey, and M. V. Pishko, eds., pp. 107–119, Stevenson Ranch: American Scientific Publishers, 2006.

[19] K. Toko and M. Habara, Taste sensor, *Chem. Senses*, vol. 30, no. Suppl 1, pp. i256–i257, 2005.

[20] Y. Vlasov, A. Legin, A. Rudnitskaya, C. D. Natale, and A. D'Amico, Nonspecific sensor arrays ("electronic tongue") for chemical analysis of liquids, *Pure Appl. Chem.*, vol. 77, no. 11, pp. 1965–1983, 2005.

[21] F. Winquist, S. Holmin, C. Krantz-Rülcker, P. Wide, and I. Lundström, A hybrid electronic tongue, *Anal. Chim. Acta*, vol. 406, no. 2, pp. 147–157, 2000.
[22] K. Toko, Taste sensor, *Sens. Actuators B*, vol. 64, no. 1–3, pp. 205–215, 2000.
[23] K. Toko, *Biomimetic Sensor Technology*, Cambridge, MA: Cambridge University Press, 2000.
[24] A. V. Legin, A. M. Rudnitskaya, Y. G. Vlasov, C. Di Natale, and A. D'Amico, The features of the electronic tongue in comparison with the characteristics of the discrete ion-selective sensors, *Sens. Actuators B*, vol. 58, no. 1–3, pp. 464–468, 1999.
[25] K. Toko, Taste sensor with global selectivity, *Mater. Sci. Eng. C*, vol. 4, no. 2, pp. 69–82, 1996.
[26] K. Hayashi, M. Yamanaka, K. Toko, and K. Yamafuji, Multichannel taste sensor using lipid membranes, *Sens. Actuators B*, vol. 2, no. 3, pp. 205–213, 1990.
[27] C. Di Natale, A. D'Amico, Y. G. Vlasov, A. V. Legin, and A. M. Rudnitskaya, Multicomponent analysis of heavy metal cations and in organic anions in liquids by a non-selective chalcogenide glass sensor array, Proceedings of the International Conference EUROSENSORS IX, Stockholm, Sweden, June 25–29, 1995. pp. 36–37.
[28] Y. Kawamura and M. R. Kare, *Umami: A Basic Taste. Physiology, Biochemistry, Nutrition, Food Science*, New York Marcel Dekker, 1987.
[29] C. Pfaffmann, The sense of taste, *Handbook of Physiology: Neurophysiology*, J. Field, ed., Washington, DC: American Physiological Society, 1959.
[30] L. M. Bartoshuk, Taste mixtures: Is mixture suppression related to compression?, *Physiol. Behav.*, vol. 14, no. 5, pp. 643–649, 1975.
[31] E. Adler, M. A. Hoon, K. L. Mueller, J. Chandrashekar, N. J. P. Ryba, and C. S. Zuker, A novel family of mammalian taste receptors, *Cell*, vol. 100, no. 6, pp. 693–702, 2000.
[32] J. Chandrashekar, K. L. Mueller, M. A. Hoon, E. Adler, L. Feng, W. Guo, C. S. Zuker, and N. J. P. Ryba, T2Rs function as bitter taste receptors, *Cell*, vol. 100, no. 6, pp. 703–711, 2000.
[33] W. Meyerhof, C. Batram, C. Kuhn, A. Brockhoff, E. Chudoba, B. Bufe, G. Appendino, and M. Behrens, The molecular receptive ranges of human TAS2R bitter taste receptors, *Chem. Senses*, vol. 35, no. 2, pp. 157–170, 2010.
[34] G. Nelson, M. A. Hoon, J. Chandrashekar, Y. Zhang, N. J. P. Ryba, and C. S. Zuker, Mammalian sweet taste receptors, *Cell*, vol. 106, no. 3, pp. 381–390, 2001.
[35] G. Nelson, J. Chandrashekar, M. A. Hoon, L. Feng, G. Zhao, N. J. P. Ryba, and C. S. Zuker, An amino-acid taste receptor, *Nature*, vol. 416, no. 6877, pp. 199–202, 2002.
[36] Y. Ishimaru, H. Inada, M. Kubota, H. Zhuang, M. Tominaga, and H. Matsunami, Transient receptor potential family members PKD1L3 and PKD2L1 form a candidate sour taste receptor, *Proc. Natl. Acad. Sci. U. S. A.*, vol. 103, no. 33, pp. 12569–12574, 2006.
[37] A. L. Huang, X. Chen, M. A. Hoon, J. Chandrashekar, W. Guo, D. Trankner, N. J. P. Ryba, and C. S. Zuker, The cells and logic for mammalian sour taste detection, *Nature*, vol. 442, no. 7105, pp. 934–938, 2006.
[38] J. Chandrashekar, C. Kuhn, Y. Oka, D. A. Yarmolinsky, E. Hummler, N. J. P. Ryba, and C. S. Zuker, The cells and peripheral representation of sodium taste in mice, *Nature*, vol. 464, no. 7286, pp. 297–301, 2010.
[39] P. M. Schweizer-Berberich, S. Vaihinger, and W. Göpel, Characterisation of food freshness with sensor arrays, *Sens. Actuators B*, vol. 18, no. 1–3, pp. 282–290, 1994.
[40] H.-K. Hong, H. W. Shin, H. S. Park, D. H. Yun, C. H. Kwon, K. Lee, S.-T. Kim, and T. Moriizumi, Gas identification using micro gas sensor array and neural-network pattern recognition, *Sens. Actuators B*, vol. 33, no. 1–3, pp. 68–71, 1996.
[41] T. C. Pearce, S. S. Schiffman, H. T. Nagle, and J. W. Gardner, *Handbook of Machine Olfaction*, Weinheim: Wiley-VCH, 2003.
[42] F. Röck, N. Barsan, and U. Weimar, Electronic nose: Current status and future trends, *Chem. Rev.*, vol. 108, no. 2, pp. 705–725, 2008.
[43] T. Nakamoto, *Human Olfactory Displays and Interfaces: Odor Sensing and Presentation*, Hershey, PA: IGI Global, 2012.
[44] C. Di Natale, A. Macagnano, F. Davide, A. D'Amico, A. Legin, Y. Vlasov, A. Rudnitskaya, and B. Selezenev, Multicomponent analysis on polluted waters by means of an electronic tongue, *Sens. Actuators B*, vol. 44, no. 1–3, pp. 423–428, 1997.
[45] L. Lvova, C. Di Natale, and R. Paolesse, Porphyrin-based chemical sensors and multisensor arrays operating in the liquid phase, *Sens. Actuators B*, vol. 179, pp. 21–31, 2013.
[46] Y. Vlasov, A. Legin, and A. Rudnitskaya, Cross-sensitivity evaluation of chemical sensors for electronic tongue: Determination of heavy metal ions, *Sens. Actuators B*, vol. 44, no. 1–3, pp. 532–537, 1997.

[47] P. Ivarsson, C. Krantz-Rülcker, F. Winquist, and I. Lundström, A voltammetric electronic tongue, *Chem. Senses*, vol. 30, no. Suppl 1, pp. i258–i259, 2005.
[48] G. Verrelli, L. Francioso, R. Paolesse, P. Siciliano, C. Di Natale, A. D'Amico, and A. Logrieco, Development of silicon-based potentiometric sensors: Towards a miniaturized electronic tongue, *Sens. Actuators B*, vol. 123, no. 1, pp. 191–197, 2007.
[49] G. Verrelli, L. Lvova, R. Paolesse, C. Di Natale, and A. D'Amico, Metalloporphyrin: Based electronic tongue—An application for the analysis of Italian white wines, *Sensors*, vol. 7, no. 11, pp. 2750–2762, 2007.
[50] A. Legin, A. Rudnitskaya, L. Lvova, Y. Vlasov, C. Di Natale, and A. D'Amico, Evaluation of Italian wine by the electronic tongue: Recognition, quantitative analysis and correlation with human sensory perception, *Anal. Chim. Acta*, vol. 484, no. 1, pp. 33–44, 2003.
[51] M. Ghasemi-Varnamkhasti, M. L. Rodríguez-Méndez, S. S. Mohtasebi, C. Apetrei, J. Lozano, H. Ahmadi, S. H. Razavi, and J. Antonio de Saja, Monitoring the aging of beers using a bioelectronic tongue, *Food Control*, vol. 25, no. 1, pp. 216–224, 2012.
[52] F. Winquist, P. Wide, and I. Lundström, An electronic tongue based on voltammetry, *Anal. Chim. Acta*, vol. 357, no. 1–2, pp. 21–31, 1997.
[53] F. Winquist, I. Lundström, and P. Wide, The combination of an electronic tongue and an electronic nose, *Sens. Actuators B*, vol. 58, no. 1–3, pp. 512–517, 1999.
[54] A. Rudnitskaya, D. Kirsanov, Y. Blinova, E. Legin, B. Seleznev, D. Clapham, R. S. Ives, K. A. Saunders, and A. Legin, Assessment of bitter taste of pharmaceuticals with multisensor system employing 3 way PLS regression, *Anal. Chim. Acta*, vol. 770, pp. 45–52, 2013.
[55] G. Sehra, M. Cole, and J. W. Gardner, Miniature taste sensing system based on dual SH-SAW sensor device: An electronic tongue, *Sens. Actuators B*, vol. 103, no. 1–2, pp. 233–239, 2004.
[56] Q. Liu, F. Zhang, D. Zhang, N. Hu, K. J. Hsia, and P. Wang, Extracellular potentials recording in intact taste epithelium by microelectrode array for a taste sensor, *Biosens. Bioelectron.*, vol. 43, pp. 186–192, 2013.
[57] Q. Liu, D. Zhang, F. Zhang, Y. Zhao, K. J. Hsia, and P. Wang, Biosensor recording of extracellular potentials in the taste epithelium for bitter detection, *Sens. Actuators B*, vol. 176, pp. 497–504, 2013.
[58] Q. Liu, F. Zhang, D. Zhang, N. Hu, H. Wang, K. Jimmy Hsia, and P. Wang, Bioelectronic tongue of taste buds on microelectrode array for salt sensing, *Biosens. Bioelectron.*, vol. 40, no. 1, pp. 115–120, 2013.
[59] C. Zhang and K. S. Suslick, A colorimetric sensor array for organics in water, *J. Am. Chem. Soc.*, vol. 127, no. 33, pp. 11548–11549, 2005.
[60] C. Zhang, D. P. Bailey, and K. S. Suslick, Colorimetric sensor arrays for the analysis of beers: A feasibility study, *J. Agric. Food Chem.*, vol. 54, no. 14, pp. 4925–4931, 2006.
[61] C. Zhang and K. S. Suslick, Colorimetric sensor array for soft drink analysis, *J. Agric. Food Chem.*, vol. 55, no. 2, pp. 237–242, 2006.
[62] F. Winquist, C. Krantz-Rülcker, and I. Lundström, Electronic tongues and combinations of artificial senses, *Sensors Update*, vol. 11, no. 1, pp. 279–306, 2002.
[63] N. Prieto, P. Oliveri, R. Leardi, M. Gay, C. Apetrei, M. L. Rodriguez-Méndez, and J. A. de Saja, Application of a GA–PLS strategy for variable reduction of electronic tongue signals, *Sens. Actuators B*, vol. 183, pp. 52–57, 2013.
[64] Y. G. Vlasov, E. A. Bychkov, and A. V. Legin, Chalcogenide glass chemical sensors: Research and analytical applications, *Talanta*, vol. 41, no. 6, pp. 1059–1063, 1994.
[65] D. Itoh, F. Sassa, T. Nishi, Y. Kani, M. Murata, and H. Suzuki, Droplet-based microfluidic sensing system for rapid fish freshness determination, *Sens. Actuators B*, vol. 171–172, pp. 619–626, 2012.
[66] P. J. Mazzone, X. F. Wang, Y. Xu, T. Mekhail, M. C. Beukemann, J. Na, J. W. Kemling, K. S. Suslick, and M.Sasidhar, Exhaled breath analysis with a colorimetric sensor array for the identification and characterization of lung cancer, *J. Thorac. Oncol.*, vol. 7, no. 1, pp. 137–42, 2012.
[67] Aissy. Inc. Available from http://aissy.co.jp/en/index.html (accessed on September 14, 2015).
[68] J. K. Lorenz, J. P. Reo, O. Hendl, J. H. Worthington, and V. D. Petrossian, Evaluation of a taste sensor instrument (electronic tongue) for use in formulation development, *Int. J. Pharm.*, vol. 367, no. 1–2, pp. 65–72, 2009.
[69] K. Toko and K. Yamafuji, Influence of monovalent and divalent cations on the surface area of phosphatidylglycerol monolayers, *Chem. Phys. Lipids*, vol. 26, no. 1, pp. 79–99, 1980.
[70] K. Toko, J. Nitta, and K. Yamafuji, Dynamic aspect of a phase transition in DOPH-millipore membranes, *J. Phys. Soc. Jpn*, vol. 50, no. 4, pp. 1343–1350, 1981.
[71] K. Toko and K. Yamafuji, Stabilization effect of protons and divalent cations on membrane structures of lipids, *Biophys. Chem.*, vol. 14, no. 1, pp. 11–23, 1981.

[72] K. Toko, K. Ryu, S. Ezaki, and K. Yamafuji, Self-sustained oscillations of membrane potential in DOPH-millipore membranes, *J. Phys. Soc. Jpn*, vol. 51, no. 10, pp. 3398–3405, 1982.
[73] K. Toko, M. Tsukiji, S. Ezaki, and K. Yamafuji, Current-voltage characteristics and self-sustained oscillations in dioleyl phosphate-millipore membranes, *Biophys. Chem.*, vol. 20, no. 1–2, pp. 39–59, 1984.
[74] K. Toko, M. Nosaka, M. Tsukiji, and K. Yamafuji, Dynamic property of membrane formation in a protoplasmic droplet of Nitella, *Biophys. Chem.*, vol. 21, no. 3–4, pp. 295–313, 1985.
[75] K. Toko, K. Yoshikawa, M. Tsukiji, M. Nosaka, and K. Yamafuji, On the oscillatory phenomenon in an oil/water interface, *Biophys. Chem.*, vol. 22, no. 3, pp. 151–158, 1985.
[76] K. Toko, M. Tsukiji, S. Iiyama, and K. Yamafuji, Self-sustained oscillations of electric potential in a model membrane, *Biophys. Chem.*, vol. 23, no. 3–4, pp. 201–210, 1986.
[77] K. Toko, N. Nakashima, S. Iiyama, K. Yamafuji, and T. Kunitake, Self-oscillation of electric potential of a porous membrane impregnated with polymer multi-bilayer complexes, *Chem. Lett.*, vol. 15, no. 8, pp. 1375–1378, 1986.
[78] S. Iiyama, K. Toko, and K. Yamafuji, Effect of bitter substances on a model membrane system of taste reception, *Agric. Biol. Chem.*, vol. 50, no. 11, pp. 2709–2714, 1986.
[79] S. Iiyama, K. Toko, and K. Yamafuji, Electric oscillation in an excitable model membrane impregnated with lipid analogues, *Biophys. Chem.*, vol. 28, no. 2, pp. 129–135, 1987.
[80] K. Hayashi, K. Yamafuji, K. Toko, N. Ozaki, T. Yoshida, S. Iiyama, and N. Nakashima, Effect of taste substances on electric characteristics of a lipid cast membrane with a single pore, *Sensors Actuators*, vol. 16, no. 1–2, pp. 25–42, 1989.
[81] K. Toko, K. Hayashi, T. Fujiyoshi, and K. Yamafuji, Self-organized electric structure in uni- and multicellular biological systems, in *Cooperative Dynamics in Complex Physical Systems*, H. Takayama, ed., pp. 326–327, New York: Springer, 1989.
[82] S. Iiyama, K. Toko, K. Hayashi, and K. Yamafuji, Effect of several sweet substances on the electrical characteristics of a dioleyl phosphate-millipore membrane, *Agric. Biol. Chem.*, vol. 53, no. 3, pp. 675–681, 1989.
[83] K. Hayashi, K. Toko, and K. Yamafuji, Effect of taste substances on aperiodic oscillation of an electric potential in a synthetic lipid membrane, *Jpn. J. Appl. Phys.*, vol. 28, pp. 1507–1512, 1989.
[84] K. Toko, T. Yamanaka, K. Hayashi, and K. Yamafuji, Multi-channel taste sensor with lipid membranes, Technical Digest of the Ninth Sensor Symposium, Tokyo, Japan, May 30–31, 1990, pp. 193–196.
[85] H. Ikezaki, K. Hayashi, M. Yamanaka, R. Tatsukawa, K. Toko, and K. Yamafuji, Multichannel taste sensor with artificial lipid membrane, *Trans. JEICE Jpn*, vol. J74-C-II, no. 5, pp. 434–442, 1991 (in Japanese).
[86] K. Hayashi, H. Ikezaki, K. Toko, and K. Yamafuji, Testing a food taste with taste sensor, *J. Brew. Soc. Jpn.*, vol. 86, pp. 633–639, 1991 (in Japanese).
[87] H. Ikezaki, K. Toko, K. Hayashi, R. Toukubo, T. Yamanaka, K. Sato, and K. Yamafuji, Intelligent multi-channel taste sensor with lipid membranes, *Technical Digest of the 10th Sensor Symposium*, Tokyo, Japan, May 30–31, 1991 pp. 173–176 (in Japanese).
[88] S. Iiyama, Y. Miyazaki, K. Hayashi, K. Toko, K. Yamafuji, H. Ikezaki, and K. Sato, Highly sensitive detection of taste substances using monolayer lipid membranes, *Sensors Mater.*, vol. 4, no. 1, pp. 21–27, 1992.
[89] T. Murata, K. Hayashi, K. Toko, H. Ikezaki, K. Sato, R. Toukubo, and K. Yamafuji, Quantification of sourness and saltiness using a multichannel sensor with lipid membranes, *Sensors Mater.*, vol. 4, no. 2, pp. 81–88, 1992.
[90] S. Iiyama, K. Toko, T. Matsuno, and K. Yamafuji, Responses of lipid membranes of taste sensor to astringent and pungent substances, *Chem. Senses*, vol. 19, no. 1, pp. 87–96, 1994.
[91] K. Toko, T. Matsuno, K. Yamafuji, K. Hayashi, H. Ikezaki, K. Sato, R. Toukubo, and S. Kawarai, Multichannel taste sensor using electric potential changes in lipid membranes, *Biosens. Bioelectron.*, vol. 9, no. 4–5, pp. 359–364, 1994.
[92] K. Hayashi, K. Toko, M. Yamanaka, H. Yoshihara, K. Yamafuji, H. Ikezaki, R. Toukubo, and K. Sato, Electric characteristics of lipid-modified monolayer membranes for taste sensors, *Sens. Actuators B*, vol. 23, no. 1, pp. 55–61, 1995.
[93] S. Iiyama, S. Ezaki, K. Toko, T. Matsuno, and K. Yamafuji, Study of astringency and pungency with multichannel taste sensor made of lipid membranes, *Sens. Actuators B*, vol. 24, no. 1–3, pp. 75–79, 1995.
[94] K. Oohira, K. Toko, H. Akiyama, H. Yoshihara, and K. Yamafuji, Electric characteristics of hybrid polymer membranes composed of two lipid species, *J. Phys. Soc. Jpn*, vol. 64, pp. 3554–3561, 1995.
[95] K. Oohira and K. Toko, Theory of electric characteristics of the lipid/PVC/DOPP membrane and PVC/DOPP membrane in response to taste stimuli, *Biophys. Chem.*, vol. 61, no. 1, pp. 29–35, 1996.
[96] K. Toko, R. Yasuda, S. Ezaki, and T. Fujiyoshi, Taste sensing FET (TSFET), Technical Digest of the 15th Sensor Symposium, Kawasaki, Japan, June 3–4, 1997, pp. 243–246.

[97] K. Oohira and K. Toko, Electrical characteristics of lipid/PVC/DOPP membrane and PVC/DOPP membrane used as transducers in chemical sensors, *Sensors Mater.*, vol. 9, no. 1, pp. 57–68, 1997.
[98] K. Toko, Y. Nakagawa, M. Obata, and T. Yahiro, Discrimination of taste qualities using static and dynamic responses of multichannel taste sensor, *Sensors Mater.*, vol. 9, no. 5, pp. 297–306, 1997.
[99] T. Tsukatani and K. Toko, Response of a lipid membrane to ethanol, *IEEJ Trans. Sensors Micromachines*, vol. 118-E, no. 5, pp. 266–271, 1998 (in Japanese).
[100] K. Hayashi, H. Shimoda, S. Matsufuji, and K. Toko, Adsorption of taste substances on lipid membranes of taste sensor, *IEEJ Trans. Sensors Micromachines*, vol. 119, no. 7, pp. 374–382, 1999.
[101] H. Ikezaki, A. Taniguchi, and K. Toko, Detection of hot taste by new measurement method using a multichannel taste sensor, Technical Digest of the 17th Sensor Symposium, Kawasaki, Japan, May 30–31, 2000, pp. 275–278.
[102] R. Takamatsu, K. Toko, H. Takeguchi, and A. Kawabata, Development of technology for separating and identifying bitter substances, *Sensors Mater.*, vol. 13, no. 3, pp. 179–187, 2001.
[103] T. Arimura, S. Takagi, M. Iwakura, K. Toko, and T. Imoto, Research on the taste change phenomenon using a taste sensor, *Jpn. J. Taste Smell Res.*, vol. 8, no. 3, pp. 453–456, 2001 (in Japanese).
[104] S. Iiyama, H. Kuga, S. Ezaki, K. Hayashi, and K. Toko, Peculiar change in membrane potential of taste sensor caused by umami substances, *Sens. Actuators B*, vol. 91, no. 1–3, pp. 191–194, 2003.
[105] R. Takamatsu, K. Toko, and H. Takeguchi, Development of technology for extracting bitter substance in mixed solution, *Jpn. J. Taste Smell Res.*, vol. 10, no. 3, pp. 435–438, 2003 (in Japanese).
[106] M. Iwakura, T. Adachi, and K. Toko, Discrimination of beer and estimation of beer foam with the taste sensor, *Tech. Rep. IEICE. OME*, vol. 100, no. 252, pp. 65–69, 2000 (in Japanese).
[107] Y. Naito, K. Sato, Y. Kobayashi, H. Ikezaki, A. Taniguchi, and K. Toko, Evaluation of aftertaste of beer using taste sensing system, *Tech. Rep. IEICE. OME*, vol. 100, pp. 157–160, 2000 (in Japanese).
[108] H. Komai, Y. Naito, K. Sato, and H. Ikezaki, Measurement of coffee taste using lipid membrane taste sensors, Proceedings of the ASIC 16th International Scientific Colloquium on Coffee, Kyoto, Japan, April 9–14, 1995, pp. 300–308.
[109] S. Iiyama, Y. Suzuki, S. Ezaki, Y. Arikawa, and K. Toko, Objective scaling of taste of sake using taste sensor and glucose sensor, *Mater. Sci. Eng. C*, vol. 4, no. 1, pp. 45–49, 1996.
[110] Y. Arikawa, K. Toko, H. Ikezaki, Y. Shinha, T. Ito, I. Oguri, and S. Baba, Analysis of sake mash using multichannel taste sensor, *J. Ferment. Bioeng.*, vol. 82, no. 4, pp. 371–376, 1996.
[111] T. Imamura, K. Toko, S. Yanagisawa, and T. Kume, Monitoring of fermentation process of miso (soybean paste) using multichannel taste sensor, *Sens. Actuators B*, vol. 37, no. 3, pp. 179–185, 1996.
[112] H. Yamada, Y. Mizota, K. Toko, and T. Doi, Highly sensitive discrimination of taste of milk with homogenization treatment using a taste sensor, *Mater. Sci. Eng. C*, vol. 5, no. 1, pp. 41–45, 1997.
[113] H. Ikezaki, A. Taniguchi, and K. Toko, Quantification of taste of green tea with taste sensor, *IEEJ Trans. Sensors Micromachines*, vol. 117, no. 9, pp. 465–470, 1997 (in Japanese).
[114] K. Toko and T. Fukusaka, Measurement of hydrophobicity of amino acids using a multichannel taste sensor, *Sensors Mater.* , vol. 9, no. 3, pp. 171–176, 1997.
[115] K. Toko and T. Nagamori, Quantitative expression of mixed taste of amino acids using multichannel taste sensor, *IEEJ Trans. Sensors Micromachines*, vol. 119, no. 11, pp. 528–531, 1999 (in Japanese).
[116] H. Chibvongodze, K. Hayashi, and K. Toko, Discrimination of D-amino acids from L-amino acids using membrane potential change, *Sensors Mater.*, vol. 13, no. 2, pp. 99–106, 2001.
[117] S. Baldacci, T. Matsuno, K. Toko, R. Stella, and D. D. Rossi, Discrimination of wine using taste and smell sensor, *Sensors Mater.*, vol. 10, no. 3, pp. 185–200, 1998.
[118] S. Iiyama, M. Yahiro, and K. Toko, Measurements of soy sauce using taste sensor, *Sens. Actuators B*, vol. 66, no. 1–3, pp. 205–206, 2000.
[119] S. Iiyama, S. Ezaki, and K. Toko, Taste of inorganic salts analyzed with multichannel electrode, *Tech. Rep. IEICE. OME*, vol. 100, no. 252, pp. 1–6, 2000 (in Japanese).
[120] M. Habara and K. Toko, Discrimination of saltiness with coexisting components using multichannel taste sensor with lipid membranes, *IEICE Trans. Electron.*, vol. E83-C, no. 7, pp. 1040–1045, 2000.
[121] S. Iiyama, S. Ezaki, and K. Toko, Analysis of saltiness and bitterness of inorganic salts using taste sensors, *Sensors Mater.*, vol. 13, no. 3, pp. 137–144, 2001.
[122] K. Sato, K. Sasaki, H. Ikezaki, A. Taniguchi, M. Mitsumoto, S. Yamaguchi, H. Nakai, and K. Toko, Taste sensing system and its application to beef evaluation, *Tech. Rep. IEICE. OME*, vol. 100, no. 253, pp. 13–17, 2000 (in Japanese).

[123] Y. Kobayashi, H. Hamada, Y. Yamaguchi, H. Ikezaki, and K. Toko, Development of an artificial lipid-based membrane sensor with high selectivity and sensitivity to the bitterness of drugs and with high correlation with sensory score, *IEEJ Trans. Electr. Electron. Eng.*, vol. 4, no. 6, pp. 710–719, 2009.

[124] Y. Kobayashi and H. Ikezaki, Advanced taste sensors based on artificial lipid membrane, in *Biochemical Sensors: Mimicking Gustatory and Olfactory Senses*, K. Toko, ed., pp. 5–44, Singapore: Pan Stanford Publishing, 2013.

[125] M. Gouy, Sur la constitution de la charge électrique à la surface d'un électrolyte, *J. Phys. Theor. Appl.*, vol. 9, no. 1, pp. 457–468, 1910 (in French).

[126] D. L. Chapman, A contribution to the theory of electrocapillarity, *Philos. Mag.*, vol. 25, no. 148, pp. 475–481, 1913.

[127] T. A. J. Payens, Ionized monolayers, *Philips Res. Rep.*, vol. 10, pp. 425–481, 1955.

[128] H. Träuble, M. Teubner, P. Woolley, and H. Eibl, Electrostatic interactions at charged lipid membranes: I. Effects of pH and univalent cations on membrane structure, *Biophys. Chem.*, vol. 4, no. 4, pp. 319–342, 1976.

[129] H. Ikezaki, Y. Naito, Y. Kobayashi, R. Toukubo, A. Taniguchi, and K. Toko, Improvement of selectivity of taste sensor by control of charge density and hydrophobicity of lipid membrane, *Tech. Rep. IEICE. OME*, vol. 100, no. 253, pp. 19–24, 2000 (in Japanese).

[130] H. Ikezaki, Y. Kobayashi, R. Toukubo, Y. Naito, A. Taniguchi, and K. Toko, Techniques to control sensitivity and selectivity of multichannel taste sensor using lipid membranes, Proceedings of the 10th International Conference on Solid-State Sensors and Actuators, Sendai, Japan, June 7–10, 1999, pp. 1634–1637.

[131] G. Wypych, *Handbook of Plasticizers*, Toronto: ChemTec Publishing, 2003.

[132] H. Ikezaki, Bitterness suppression techniques for foods, in *Taste Modification Technology of Food and Medicine*, K. Toko and T. Uchida, eds., pp. 131–141, Tokyo: CMC Publishing, 2008 (in Japanese).

[133] Y. Ueda, M. Yonemitsu, T. Tsubuku, M. Sakaguchi, and R. Miyajima, Flavor characteristics of glutathione in raw and cooked foodstuffs, *Biosci. Biotechnol. Biochem.*, vol. 61, no. 12, pp. 1977–1980, 1997.

[134] M. Gastl, S. Hanke, and W. Back, Analytical investigations to evaluate bitter sensation using a taste sensing system, *Brew. Sci.*, vol. 60, pp. 48–54, 2008.

[135] M. Gastl, S. Hanke, and W. Back, "Drinkability"—Balance and harmony of components as well as an incentive for continuing to drink, *Brauwelt Int.*, vol. 26, no. 3, pp. 148–153, 2008.

[136] A. Tanigake, Y. Miyanaga, T. Nakamura, E. Tsuji, K. Matsuyama, M. Kunitomo, and T. Uchida, The bitterness intensity of clarithromycin evaluated by a taste sensor, *Chem. Pharm. Bull.*, vol. 51, no. 11, pp. 1241–1245, 2003.

[137] E. Tsuji, T. Uchida, A. Fukui, R. Fujii, and H. Sunada, Evaluation of bitterness suppression of macrolide dry syrups by jellies, *Chem. Pharm. Bull.*, vol. 54, no. 3, pp. 310–314, 2006.

[138] M. Okamoto, H. Sunada, M. Nakano, and R. Nishiyama, Bitterness evaluation of orally disintegrating famotidine tablets using a taste sensor, *Asian J. Pharm. Sci.*, vol. 4, no. 1, pp. 1–7, 2009.

[139] R. Chen, H. Ikezaki, N. Hayashi, K. Kohata, Y. Kugimiya, Y. Kobayashi, A. Taniguchi, and K. Toko, Study on evaluating jimi-taste of green tea using multichannel taste sensor, Proceedings of the International Conference on O-CHA(tea) Culture and Science, Shizuoka, Japan, November 4–6, 2004, pp. 736–740.

[140] N. Hayashi, R. Chen, H. Ikezaki, S. Yamaguchi, D. Maruyama, Y. Yamaguchi, T. Ujihara, and K. Kohata, Techniques for universal evaluation of astringency of green tea infusion by the use of a taste sensor system, *Biosci. Biotechnol. Biochem.*, vol. 70, no. 3, pp. 626–631, 2006.

[141] N. Hayashi, R. Chen, H. Ikezaki, T. Ujihara, H. Kitajima, and Y. Mizukami, Evaluation of the astringency of black tea by a taste sensor system: Scope and limitation, *Biosci. Biotechnol. Biochem.*, vol. 71, no. 2, pp. 587–589, 2007.

[142] N. Hayashi, R. Chen, H. Ikezaki, and T. Ujihara, Evaluation of the umami taste intensity of green tea by a taste sensor, *J. Agric. Food Chem.*, vol. 56, no. 16, pp. 7384–7387, 2008.

[143] N. Hayashi, R. Chen, M. Hiraoka, T. Ujihara, and H. Ikezaki, β-Cyclodextrin/surface plasmon resonance detection system for sensing bitter-astringent taste intensity of green tea catechins, *J. Agric. Food Chem.*, vol. 58, no. 14, pp. 8351–8356, 2010.

[144] Y. Uchiyama, M. Yamashita, M. Kato, T. Suzuki, M. Omori, and R. Chen, Evaluation of the taste of tea with different degrees of fermentation using a taste sensing system, *Sensors Mater.*, vol. 23, no. 8, pp. 501–506, 2011.

[145] N. Hayashi, T. Ujihara, R. Chen, K. Irie, and H. Ikezaki, Objective evaluation methods for the bitter and astringent taste intensities of black and oolong teas by a taste sensor, *Food Res. Int.*, vol. 53, no. 2, pp. 816–821, 2013.

[146] M. Habara, H. Ikezaki, A. Taniguchi, and K. Toko, Electric response of taste sensor with lipid/polymer membranes to sweet taste substances, *Jpn. J. Taste Smell Res.*, vol. 8, no. 3, pp. 313–316, 2001 (in Japanse).

[147] M. Habara, H. Ikezaki, A. Taniguchi, and K. Toko, Improvement of sensitivity to sweet taste substances using taste sensor with lipid membranes, *IEEJ Trans. Sensors Micromachines*, vol. 121, no. 12, pp. 641–646, 2001 (in Japanese).
[148] M. Habara, H. Ikezaki, and K. Toko, Study of sweet taste evaluation using taste sensor with lipid/polymer membranes, *Biosens. Bioelectron.*, vol. 19, no. 12, pp. 1559–1563, 2004.
[149] H. Cui, M. Habara, H. Ikezaki, and K. Toko, Selectivity control in a sweetness sensor using lipid polymer membranes, *Sensors Mater.*, vol. 17, no. 7, pp. 385–390, 2005.
[150] M. Habara, D. Beppu, H. Cui, H. Ikezaki, and K. Toko, Detection of sugars using lipid/polymer membranes, *Sensors Mater.*, vol. 19, no. 5, pp. 325–331, 2007.
[151] H. Cui, M. Habara, H. Ikezaki, and K. Toko, Development of lipid/polymer membrane for detecting sweet taste substances, *Jpn. J. Taste Smell Res.*, vol. 14, no. 3, pp. 629–632, 2007 (in Japanese).
[152] H. Cui, M. Habara, H. Ikezaki, and K. Toko, Study of membrane potential change in surface-modified lipid/polymer membrane using phenol compound, *Jpn. J. Taste Smell Res.*, vol. 15, no. 3, pp. 401–404, 2008 (in Japanese).
[153] K. Toyota, H. Cui, K. Abe, M. Habara, K. Toko, and H. Ikezaki, Sweetness sensor with lipid/polymer membranes: Sweet-responsive substances, *Sensors Mater.*, vol. 23, no. 8, pp. 465–474, 2011.
[154] K. Toyota, H. Cui, K. Abe, M. Habara, K. Toko, and H. Ikezaki, Sweetness sensor with lipid/polymer membranes: Response to various sugars, *Sensors Mater.*, vol. 23, no. 8, pp. 475–482, 2011.
[155] T. Ishiwaki, Application of taste sensor to blending of coffee, in *Biochemical Sensors: Mimicking Gustatory and Olfactory Senses*, K. Toko, ed., pp. 83–90, Singapore: Pan Stanford Publishing, 2013.
[156] H. Matsuo, N. Hayashi, T. Ujihara, S. Fujita, T. Tatsuno, M. Mitarai, Y. Gejima, Y. Toyomitsu, O. Kinoshita, and T. Taniguchi, Astringency of kamairi-cha and sen-cha, *J. Jpn. Soc. Food Sci. Technol.*, vol. 59, no. 1, pp. 6–16, 2012 (in Japanese).
[157] T. Ujihara, N. Hayashi, and H. Ikezaki, Objective evaluation of astringent and umami taste intensities of matcha using a taste sensor system, *Food Sci. Technol. Res.*, vol. 19, no. 6, pp. 1099–1105, 2013.
[158] A. Totsuka, Application of multichannel taste sensor (MCTS) for winemaking, in *Biochemical Sensors: Mimicking Gustatory and Olfactory Senses*, K. Toko, ed., Singapore: Pan Stanford Publishing, 2013.
[159] N. Nakahara, H. Sakaida, T. Kai, Y. Sakakibara, K. Nishiyama, N. Fukuda, and M. Suiko, Taste evaluation of Honkakeshochu with taste sensor (research on the taste of Honkakeshochu Part I), *J. Jpn. Soc. Food Sci. Technol.*, vol. 52, no. 4, pp. 145–153, 2005 (in Japanese).
[160] N. Nakahara, S. Furta, H. Sakaida, T. Kai, Y. Sakakibara, K. Nishiyama, and M. Suiko, Lipid membrane sensors that identifies basic taste of Honkakeshochu (research on the taste of Honkakeshochu Part II), *J. Jpn. Soc. Food Sci. Technol.*, vol. 52, no. 8, pp. 355–365, 2005 (in Japanese).
[161] M. Doi, Evaluation of kokumi taste of Japanese soup stock materials using taste sensor, *Sensors Mater.*, vol. 23, no. 8, pp. 493–499, 2011.
[162] Y. Mizota, H. Matsui, M. Ikeda, K. Iwatsuki, and K. Toko, Study on evaluation of UHT milk processed with indirect and direct heating methods by sensor analysis, *Jpn. J. Taste Smell Res.*, vol. 13, no. 3, pp. 517–520, 2006 (in Japanese).
[163] Y. Mizota, H. Matsui, M. Ikeda, N. Ichihashi, K. Iwatsuki, and K. Toko, Flavor evaluation using taste sensor for UHT processed milk stored in cartons having different light permeabilities, *Milchwissenschaft*, vol. 64, no. 2, pp. 143–146, 2009.
[164] K. Sasaki, F. Tani, K. Sato, H. Ikezaki, A. Taniguchi, T. Emori, F. Iwaki, K. Chikuni, and M. Mitsumoto, Analysis of pork extracts by taste sensing system and the relationship between umami substances and sensor output, *Sensors Mater.*, vol. 17, no. 7, pp. 397–404, 2005.
[165] T. Okumura, R. Yamada, and T. Nishimura, Sourness-suppressing peptides in cooked pork loins, *Biosci. Biotechnol. Biochem.*, vol. 68, no. 8, pp. 1657–1662, 2004.
[166] K. Chikuni, M. Oe, K. Sasaki, M. Shibata, I. Nakajima, K. Ojima, and S. Muroya, Effects of muscle type on beef taste-traits assessed by an electric sensing system, *Anim. Sci. J.*, vol. 81, no. 5, pp. 600–605, 2010.
[167] K. Tokubo, K. Kamijo, Y. Naito, and H. Ikezaki, Measurement of solid food using multichannel taste sensor, *Jpn. J. Taste Smell Res.*, vol. 16, no. 3, pp. 505–508, 2009 (in Japanese).
[168] T. U. Tran, K. Suzuki, H. Okadome, S. Homma, and K. Ohtsubo, Analysis of the tastes of brown rice and milled rice with different milling yields using a taste sensing system, *Food Chem.*, vol. 88, no. 4, pp. 557–566, 2004.
[169] T. U. Tran, K. Suzuki, H. Okadome, H. Ikezaki, S. Homma, and K. Ohtsubo, Detection of changes in taste of japonica and indica brown and milled rice (*Oryza sativa* L.) during storage using physicochemical analyses and a taste sensing system, *J. Agric. Food Chem.*, vol. 53, no. 4, pp. 1108–1118, 2005.

[170] K. Toko, R. Chen, and M. Habara, Evaluation of saltiness using taste sensor, *Bull. Soc. Sea Water Sci. Jpn.*, vol. 58, no. 1, pp. 57–63, 2005 (in Japanese).

[171] R. Chen, H. Ikezaki, and K. Toko, Development of sensor with high selectivity for saltiness and its application in taste evaluation of table salt, *Sensors Mater.*, vol. 22, no. 6, pp. 313–325, 2010.

[172] A. Fujita, A. Isogai, M. Endo, H. Utsunomiya, S. Nakano, and H. Iwata, Effects of sulfur dioxide on formation of fishy off-odor and undesirable taste in wine consumed with seafood, *J. Agric. Food Chem.*, vol. 58, no. 7, pp. 4414–4420, 2010.

[173] M. Nakamura, F. Sato, S. Yoshida, M. Kumagai, and Y. Suzuki, Evaluation of taste change resulting from thickening of food with instant food thickeners, *J. Jpn. Soc. Food Sci. Technol.*, vol. 57, no. 9, pp. 380–388, 2010 (in Japanese).

[174] H. Kurose, M. Hasebe, H. Ikezaki, A. Taniguchi, Y. Naito, and K. Toko, Application study of drinking water using taste sensing system, *J. Jpn. Water Works Assoc.*, vol. 75, no. 8, pp. 2–8, 2005 (in Japanese).

[175] H. Stone, Gustatory responses to the L-amino acids in man, in *Olfaction and Taste II*, T. Hayashi, ed., pp. 289–306, Oxford: Pergamon Press, 1967.

[176] S. Miura, S. Sato, M. Yoshida, T. Kaneko, S. Namba, and M. Kainosho, Taste and other sensations in mouth, *Sensory Evaluation Handbook*, K. Ichikawa, T. Indow, S. Sato et al., eds., pp. 156–175, Tokyo: JUSE Press, 1973 (in Japanese).

[177] L. M. Beider, Part 2: Taste, in *Handbook of Sensory Physiology IV: Chemical Senses*, L. M. Beider, ed., pp. 200–220, New York: Springer-Verlag, 1971.

[178] H. G. Schutz, and F. J. Pilgrim, Differential sensitivity in gustation, *J. Exp. Psychol.*, vol. 54, no. 1, pp. 41–48, 1957.

[179] T. Fukagawa, Y. Tahara, M. Yasuura, M. Habara, H. Ikezaki, and K. Toko, Relationship between taste sensor response and amount of quinine adsorbed on lipid/polymer membrane, *J. Innov. Electron. Commun.*, vol. 2, no. 1, pp. 1–6, 2012.

[180] D. Hara, T. Fukagawa, Y. Tahara, M. Yasuura, and K. Toko, Examination of amount of astringent substances adsorbed onto lipid/polymer membrane used in taste sensor, *Sens. Lett.*, vol. 12, pp. 1172–1176, 2014.

[181] K. Twomey, E. A. de Eulate, J. Alderman, and D. W. M. Arrigan, Fabrication and characterization of a miniaturized planar voltammetric sensor array for use in an electronic tongue, *Sens. Actuators B*, vol. 140, no. 2, pp. 532–541, 2009.

[182] F. Winquist, C. Krantz-Rulcker, and I. Lundstrom, A miniaturized voltammetric electronic tongue, *Anal. Lett.*, vol. 41, no. 5, pp. 917–924, 2008.

[183] K. Twomey, A. Truemper, and K. Murphy, A portable sensing system for electronic tongue operations, *Sensors*, vol. 6, no. 11, pp. 1679–1696, 2006.

[184] Y. Tahara, A. Ikeda, Y. Maehara, M. Habara, and K. Toko, Development and evaluation of a miniaturized taste sensor chip, *Sensors*, vol. 11, no. 10, pp. 9878–9886, 2011.

[185] Y. Tahara, K. Nakashi, K. Ji, A. Ikeda, and K. Toko, Development of a portable taste sensor with a lipid/polymer membrane, *Sensors*, vol. 13, no. 1, pp. 1076–1084, 2013.

[186] Y. Tahara, Y. Maehara, K. Ji, A. Ikeda, and K. Toko, Development of a multichannel taste sensor chip for a portable taste sensor, Proceedings of the IEEE Sensors 2012, Taipei, Taiwan, October 28–31, 2012, pp. 1–4.

[187] M. J. Caterina, M. A. Schumacher, M.Tominaga, T. A. Rosen, J. D. Levine, and D.Julius, The capsaicin receptor: A heat-activated ion channel in the pain pathway, *Nature*, vol. 389, no. 6653, pp. 816–824, 1997.

[188] M. Tominaga, M. J. Caterina, A. B. Malmberg, T. A. Rosen, H. Gilbert, K. Skinner, B. E. Raumann, A. I. Basbaum, and D.Julius, The cloned capsaicin receptor integrates multiple pain-producing stimuli, *Neuron*, vol. 21, no. 3, pp. 531–543, 1998.

[189] M. Bandell, G. M. Story, S. W. Hwang, V. Viswanath, S. R. Eid, M. J. Petrus, T. J. Earley, and A. Patapoutian, Noxious cold ion channel TRPA1 is activated by pungent compounds and bradykinin, *Neuron*, vol. 41, no. 6, pp. 849–857, 2004.

[190] S.E. Jordt, D. M. Bautista, H. Chuang, D. D. McKemy, P. M. Zygmunt, E. D. Hogestatt, I. D. Meng, and D.Julius, Mustard oils and cannabinoids excite sensory nerve fibres through the TRP channel ANKTM1, *Nature*, vol. 427, no. 6971, pp. 260–265, 2004.

[191] C. T. Simons and E. Carstens, Oral chemesthesis and taste, in *The Senses: A Comprehensive Reference*, A. Basbaum, M. Bushnell, D. Smith, G. Beauchamp, S. Firestein, P. Dallas, D. Oertel, R. Masland, T. Albright, J. Kaas and E. Gardner, eds., pp. 345–369, New York: Academic Press, 2008.

[192] I. Perucka and W. Oleszek, Extraction and determination of capsaicinoids in fruit of hot pepper *Capsicum annuum* L. by spectrophotometry and high-performance liquid chromatography, *Food Chem.*, vol. 71, no. 2, pp. 287–291, 2000.

[193] V. Supalkova, H. Stavelikova, S. Krizkova, V. Adam, A. Horna, L. Havel, P. Ryant, P. Babula, and R. Kizek, Study of capsaicin content in various parts of pepper fruit by liquid chromatography with electrochemical detection, *Acta Chim. Slov.*, vol. 54, no. 1, pp. 55–59, 2007.

[194] B. V. Thomas, A. A. Schreiber, and C. P. Weisskopf, Simple method for quantitation of capsaicinoids in peppers using capillary gas chromatography, *J. Agric. Food Chem.*, vol. 46, no. 7, pp. 2655–2663, 1998.

[195] E. P. Randviir, J. P. Metters, J. Stainton, and C. E. Banks, Electrochemical impedance spectroscopy versus cyclic voltammetry for the electroanalytical sensing of capsaicin utilising screen printed carbon nanotube electrodes, *Analyst*, vol. 138, no. 10, pp. 2970–2981, 2013.

[196] S. Tian, M. Ma, M. Yasuura, R. Yatabe, T. Onodera, and K. Toko, Detection of capsaicinoids using surface plasmon resonance immunosensor, Papers of Technical Meeting on Chemical Sensor 2013, IEE, Japan, pp. 65–69, 2013 (in Japanese).

[197] S. Ezaki, T. Yuki, K. Toko, Y. Tsuda, and K. Nakatani, Analysis of taste qualities and ingredients of beer by taste sensing system, *IEEJ Trans. Sensors Micromachines*, vol. 117, no. 9, pp. 449–455, 1997 (in Japanese).

[198] International Trade Centre, *The Coffee Exporter's Guide*, third edition, Geneva: United Nations Publications, 2011.

[199] W. H. Ukers, *All about Coffee: A History of Coffee from the Classic Tribute to the World's Most Beloved Beverage*, Avon: Adams Media, 2012.

[200] T. Fukunaga, K. Toko, S. Mori, Y. Nakabayashi, and M. Kanda, Quantification of taste of coffee using sensor with global selectivity, *Sensors Mater.*, vol. 8, no. 1, pp. 47–56, 1996.

[201] M. Sakata, I. Tago, Y. Ueki, T. Tamura, and K. Fukushima, Investigation of taste for infusion coffee and correlation between taste and ingredients, *Jpn. J. Taste Smell Res.*, vol. 6, no. 3, pp. 669–672, 1999 (in Japanese).

[202] K. Chikuni, K. Sasaki, T. Emori, F. Iwaki, F. Tani, I. Nakajima, S. Muroya, and M. Mitsumoto, Effect of cooking on the taste- and flavor-related compounds in pork, *Jpn. J. Swine Sci.*, vol. 39, no. 3, pp. 191–199, 2002.

[203] PR publication of Japan Poultry Breeders Hatcheries Association. Jidori oyobi broiler niku no sikibetsu hyouka hou (in Japanese). Available from http://www.syukeifuran.or.jp/official/books/ (accessed on September 14, 2015) (in Japanese).

[204] K. Sasaki, M. Motoyama, and M. Mitsumoto, Changes in the amounts of water-soluble umami-related substances in porcine longissimus and biceps femoris muscles during moist heat cooking, *Meat Sci.*, vol. 77, no. 2, pp. 167–172, 2007.

[205] C. H. Papadimitriou and K. Steiglitz, *Combinatorial Optimization: Algorithms and Complexity*, Englewood Cliffs: Prentice-Hall, Inc., 1982.

[206] D. E. Goldberg, *Genetic Algorithms in Search, Optimization and Machine Learning*, Reading: Addison-Wesley Longman Publishing Co., Inc., 1989.

[207] J. H. Holland, Genetic algorithms, *Sci. Am.*, vol. 267, pp. 66–72, 1992.

[208] D. G. Luenberger, *Linear and Nonlinear Programming*, Reading: Addison-Wesley, 1984.

[209] J. A. Nelder and R. Mead, A simplex method for function minimization, *Comput. J.*, vol. 7, no. 4, pp. 308–313, 1965.

[210] H. Sato, and M. Sato, Nonlinear sparse optimization based on genetic algorithm–optimization for food ingredients and their quantities, *IEICE Tech. Report*, vol. 113, no. 271, pp. 47–52, 2013 (in Japanese).

[211] R. Koch, *The 80/20 Principle: The Secret to Achieving More with Less*, London: Nicholas Brealey Publishing, 2010.

[212] R. Chang, *Physical Chemistry for the Biosciences*, London: Nicholas Brealey Publishing, 2010.

[213] D. Bagger-Sjoback and G. Bondesson, Taste evaluation and compliance of two paediatric formulations of phenoxymethylpenicillin in children, *Scand. J. Prim. Health Care*, vol. 7, no. 2, pp. 87–92, 1989.

[214] J. L. Powers, Properties of azithromycin that enhance the potential for compliance in children with upper respiratory tract infections, *Pediatr. Infect. Dis. J.*, vol. 15, no. Suppl 9, pp. 30–37, 1996.

[215] D. Matsui, Assessing the palatability of medications in children, *Paediatr. Perinat. Drug Ther.*, vol. 8, no. 2, pp. 55–60, 2007.

[216] S. Takagi, K. Toko, K. Wada, H. Yamada, and K. Toyoshima, Detection of suppression of bitterness by sweet substance using a multichannel taste sensor, *J. Pharm. Sci.*, vol. 87, no. 5, pp. 552–555, 1998.

[217] T. Uchida, Y. Miyanaga, H. Tanaka, K. Wada, S. Kurosaki, T. Ohki, M. Yoshida, and K. Matsuyama, Quantitative evaluation of the bitterness of commercial medicines using a taste sensor, *Chem. Pharm. Bull.*, vol. 48, no. 11, pp. 1843–1845, 2000.

[218] T. Uchida, Y. Kobayashi, Y. Miyanaga, R. Toukubo, H. Ikezaki, A. Taniguchi, M. Nishikata, and K. Matsuyama, A new method for evaluating the bitterness of medicines by semi-continuous measurement of adsorption using a taste sensor, *Chem. Pharm. Bull.*, vol. 49, no. 10, pp. 1336–1339, 2001.

[219] T. Yajima, N. Umeki, and S. Itai, Optimum spray congealing conditions for masking the bitter taste of clarithromycin in wax matrix, *Chem. Pharm. Bull.*, vol. 47, no. 2, pp. 220–225, 1999.

[220] T. Yajima, Y. Fukushima, S. Itai, and Y. Kawashima, Method of evaluation of the bitterness of clarithromycin dry syrup, *Chem. Pharm. Bull.*, vol. 50, no. 2, pp. 147–152, 2002.

[221] T. Nakamura, A. Tanigake, Y. Miyanaga, T. Ogawa, T. Akiyoshi, K. Matsuyama, and T. Uchida, The effect of various substances on the suppression of the bitterness of quinine-human gustatory sensation, binding, and taste sensor studies, *Chem. Pharm. Bull.*, vol. 50, no. 12, pp. 1589–1593, 2002.

[222] Y. Miyanagai, Y. Kobayashi, H. Ikezaki, A. Taniguchi, and T. Uchida, Bitterness prediction or bitterness suppression in human medicines using a taste sensor, *Sensors Mater.*, vol. 14, no. 8, pp. 455–465, 2002.

[223] Y. Miyanaga, A. Tanigake, T. Nakamura, Y. Kobayashi, H. Ikezaki, A. Taniguchi, K. Matsuyama, and T. Uchida, Prediction of the bitterness of single, binary- and multiple-component amino acid solutions using a taste sensor, *Int. J. Pharm.*, vol. 248, no. 1–2, pp. 207–218, 2002.

[224] T. Uchida, A. Tanigake, Y. Miyanaga, K. Matsuyama, M. Kunitomo, Y. Kobayashi, H. Ikezaki, and A. Taniguchi, Evaluation of the bitterness of antibiotics using a taste sensor, *J. Pharm. Pharmacol.*, vol. 55, no. 11, pp. 1479–1485, 2003.

[225] Y. Miyanaga, N. Inoue, A. Ohnishi, E. Fujisawa, M. Yamaguchi, and T. Uchida, Quantitative prediction of the bitterness suppression of elemental diets by various flavors using a taste sensor, *Pharm. Res.*, vol. 20, no. 12, pp. 1932–1938, 2003.

[226] Y. Miyanaga, J. Mukai, T. Mukai, M. Odomi, and T. Uchida, Suppression of the bitterness of enteral nutrients using increased particle sizes of branched-chain amino acids (BCAAs) and various flavours: A taste sensor study, *Chem. Pharm. Bull.*, vol. 52, no. 4, pp. 490–493, 2004.

[227] T. Ishizaka, Y. Miyanaga, J. Mukai, K. Asaka, Y. Nakai, E. Tsuji, and T. Uchida, Bitterness evaluation of medicines for pediatric use by a taste sensor, *Chem. Pharm. Bull.*, vol. 52, no. 8, pp. 943–948, 2004.

[228] J. Mukai, Y. Miyanaga, T. Ishizaka, K. Asaka, Y. Nakai, E. Tsuji, and T. Uchida, Quantitative taste evaluation of total enteral nutrients, *Chem. Pharm. Bull.*, vol. 52, no. 12, pp. 1416–1421, 2004.

[229] M. Kataoka, Y. Miyanaga, E. Tsuji, and T. Uchida, Evaluation of bottled nutritive drinks using a taste sensor, *Int. J. Pharm.*, vol. 279, no. 1–2, pp. 107–114, 2004.

[230] E. Tsuji, M. Takada, and T. Uchida, Effective bitterness evaluation of macrolide dry syrup formulations by a taste sensor, *Jpn. J. Pharm. Health Care Sci.*, vol. 31, no. 3, pp. 186–193, 2005 (in Japanese).

[231] H. Tachiki, H. Uchiyama, Y. Okuda, R. Uchida, Y. Kobayashi, and T. Uchida, Bitterness evaluation of famotidine orally disintegrating tablets using a taste sensor, *Jpn. J. Med. Pharm. Sci.*, vol. 54, no. 3, pp. 321–327, 2005 (in Japanese).

[232] T. Ishizaka, E. Tsuji, J. Mukai, K. Asaka, and T. Uchida, Evaluation of bitterness due to breaking up and crushing tablets, *Jpn. J. Pharm. Health Care Sci.*, vol. 32, no. 3, pp. 259–265, 2006 (in Japanese).

[233] J. Mukai, T. Ishizaka, E. Tokuyama, E. Tsuji, and T. Uchida, Quantitative taste evaluation of semi elemental diets, *Jpn. J. Pharm. Health Care Sci.*, vol. 32, no. 5, pp. 383–391, 2006 (in Japanese).

[234] E. Tokuyama, T. Shibasaki, H. Kawabe, J. Mukai, S. Okada, and T. Uchida, Bitterness suppression of BCAA solutions by L-ornithine, *Chem. Pharm. Bull.*, vol. 54, no. 9, pp. 1288–1292, 2006.

[235] Y. Hashimoto, C. Matsunaga, E. Tokuyama, E. Tsuji, T. Uchida, and H. Okada, The quantitative prediction of bitterness-suppressing effect of sweeteners on the bitterness of famotidine by sweetness-responsive sensor, *Chem. Pharm. Bull.*, vol. 55, no. 5, pp. 739–746, 2007.

[236] T. Ishizaka, S. Okada, E. Takemoto, E. Tokuyama, E. Tsuji, J. Mukai, and T. Uchida, The suppression of enhanced bitterness intensity of macrolide dry syrup mixed with an acidic powder, *Chem. Pharm. Bull.*, vol. 55, no. 10, pp. 1452–1457, 2007.

[237] S. Okada, E. Takemoto, T. Ishizaka, and T. Uchida, Quality evaluation of original product and generic versions of clarithromycin dry syrup, *Jpn. J. Med. Pharm. Sci.*, vol. 33, no. 11, pp. 905–912, 2007 (in Japanese).

[238] M. Kataoka, E. Tokuyama, Y. Miyanaga, and T. Uchida, The taste sensory evaluation of medicinal plants and Chinese medicines, *Int. J. Pharm.*, vol. 351, no. 1–2, pp. 36–44, 2008.

[239] S. Okada, E. Takemoto, T. Ishizaka, K. Taniuchi, A. Tanoue, E. Tsuji, and T. Uchida, Study on palatability of antibacterial formulations for pediatric use: Evaluation of suspension/dispersion capabilities and bitterness intensity, *Jpn. J. Med. Pharm. Sci.*, vol. 34, no. 1, pp. 32–39, 2008 (in Japanese).

[240] T. Haraguchi, A. Miyazaki, M. Yoshida, and T. Uchida, Bitterness evaluation of intact and crushed Vesicare orally disintegrating tablets using taste sensors, *J. Pharm. Pharmacol.*, vol. 65, no. 7, pp. 980–987, 2013.

[241] T. Uchida, M. Yoshida, M. Hazekawa, T. Haraguchi, H. Furuno, M. Teraoka, and H. Ikezaki, Evaluation of palatability of 10 commercial amlodipine orally disintegrating tablets by gustatory sensation testing, OD-mate as a new disintegration apparatus and the artificial taste sensor, *J. Pharm. Pharmacol.*, vol. 65, no. 9, pp. 1312–1320, 2013.

[242] N. Anjiki, N. Kawahara, and Y. Goda, Evaluation of the taste of Kampo formulae by taste-sensing system (1), *Nat. Med.*, vol. 59, no. 4, pp. 164–170, 2005 (in Japanese).

[243] N. Anjiki, A. Suzuki, N. Kawahara, and Y. Goda, Evaluation of the taste of Kampo formulae by taste-sensing system (2), *Jpn. J. Pharmacol.*, vol. 60, no. 1, pp. 21–27, 2006 (in Japanese).

[244] K. Woertz, C. Tissen, P. Kleinebudde, and J. Breitkreutz, Performance qualification of an electronic tongue based on ICH guideline Q2, *J. Pharm. Biomed. Anal.*, vol. 51, no. 3, pp. 497–506, 2010.

[245] K. Woertz, C. Tissen, P. Kleinebudde, and J. Breitkreutz, Rational development of taste masked oral liquids guided by an electronic tongue, *Int. J. Pharm.*, vol. 400, no. 1–2, pp. 114–123, 2010.

[246] F. M. Mady, A. E. Abou-Taleb, K. A. Khaled, K. Yamasaki, D. Iohara, T. Ishiguro, F. Hirayama, K. Uekama, and M. Otagiri, Enhancement of the aqueous solubility and masking the bitter taste of famotidine using drug/SBE-beta-CyD/povidone K30 complexation approach, *J. Pharm. Sci.*, vol. 99, no. 10, pp. 4285–4294, 2010.

[247] T. Kagaya, G. Inouea, M. Aya, K. Matsumoto, T. Hasegawa, H. Hamada, M. Akimoto, K. Atsuda, and K. Sugibayashi, Masking effect of deglutition aid jelly on the bitterness of acetaminophen as a bitter substance, *Food Funct.*, vol. 6, no. 1, pp. 8–15, 2010 (in Japanese).

[248] N. Ono, Y. Miyamoto, T. Ishiguro, K. Motoyama, F. Hirayama, D. Iohara, H. Seo, S. Tsuruta, H. Arima, and K. Uekama, Reduction of bitterness of antihistaminic drugs by complexation with β-cyclodextrins, *J. Pharm. Sci.*, vol. 100, no. 5, pp. 1935–1943, 2011.

[249] N. Anjiki, J. Hosoe, H. Fuchino, F. Kiuchi, S. Sekita, H. Ikezaki, M. Mikage, N. Kawahara, and Y. Goda, Evaluation of the taste of crude drug and Kampo formula by a taste-sensing system (4): Taste of processed aconite root, *J. Nat. Med.*, vol. 65, no. 2, pp. 293–300, 2011.

[250] K. Woertz, C. Tissen, P. Kleinebudde, and J. Breitkreutz, Development of a taste-masked generic ibuprofen suspension: Top-down approach guided by electronic tongue measurements, *J. Pharm. Sci.*, vol. 100, no. 10, pp. 4460–4470, 2011.

[251] C. Eckert, C. Lutz, J. Breitkreutz, and K. Woertz, Quality control of oral herbal products by an electronic tongue—Case study on sage lozenges, *Sens. Actuators B*, vol. 156, no. 1, pp. 204–212, 2011.

[252] T. Ogata, A. Koide, M. Kinoshita, and T. Ozeki, Taste masking of propiverine hydrochloride by conversion to its free base, *Chem. Pharm. Bull.*, vol. 60, no. 8, pp. 976–984, 2012.

[253] M. Preis, M. Pein, and J. Breitkreutz, Development of a taste-masked orodispersible film containing dimenhydrinate, *Pharmaceutics*, vol. 4, no. 4, pp. 551–562, 2012.

[254] N. Anjiki, H. Fushimi, J. Hosoe, N. Fushimi, K. Komatsu, S. Q. Cai, H. Ikezaki, M. Mikage, N. Kawahara, and Y. Goda, Use of a taste-sensing system to discriminate kasseki (aluminum silicate hydrate with silicon dioxide) in the Japanese pharmacopoeia from Huashi (Talc) in pharmacopoeia of the People's Republic of China, *J. Tradit. Med.*, vol. 30, no. 1, pp. 34–40, 2013.

[255] C. Eckert, M. Pein, J. Reimann, and J. Breitkreutz, Taste evaluation of multicomponent mixtures using a human taste panel, electronic taste sensing systems and HPLC, *Sens. Actuators B*, vol. 182, pp. 294–299, 2013.

[256] W. F. Kuhfeld, R. D. Tobias, and M. Garratt, Efficient experimental design with marketing research applications, *J. Mark. Res.*, vol. 31, no. 4, pp. 545–555, 1994.

[257] A. Todoroki and T. Ishikawa, Design of experiments for stacking sequence optimizations with genetic algorithm using response surface approximation, *Compos. Struct.*, vol. 64, no. 3–4, pp. 349–357, 2004.

[258] M. Habara, Y. Kobayashi, and H. Ikezaki, Design of taste sensors and applications to taste-masked formulations, in *Strategy and Novel Technology on Pharmaceutical Preparation II*, H. Takeuchi, ed., pp. 239–248, Tokyo: CMC Publishing, 2013 (in Japanese).

[259] Y. Kobayashi, Y. Yamaguchi, T. Yamaji, M. Hiraoka, and H. Ikezaki, Prediction of bitterness threshold of drugs using artificial lipid-based bitterness sensor, Proceedings of the 26th Symposium on Particulate Preparations and Designs, 2009, pp. 106–107 (in Japanese) November 26, 27, Hiroshima, Japan.

[260] T. Harada, T. Kato, and K. Iwamoto, Quantification of taste of medicines with a taste sensor, *Tech. Rep. IEICE. OME*, vol. 100, no. 252, pp. 125–130, 2000 (in Japanese).

[261] T. Harada, T. Uchida, M. Yoshida, Y. Kobayashi, R. Narazaki, and T. Ohwaki, A new method for evaluating the bitterness of medicines in development using a taste sensor and a disintegration testing apparatus, *Chem. Pharm. Bull.*, vol. 58, no. 8, pp. 1009–1014, 2010.

[262] N. Caram-Lelham and L.-O.Sundelöf, Some aspects on characterization and properties of charged polysaccharides. An investigation of the system carrageenan/amitriptyline/water with relation to amphiphile adsorption and charge density, *Int. J. Pharm.*, vol. 115, no. 1, pp. 103–111, 1995.

[263] V. L. Campo, D. F. Kawano, D. B. da Silva, Jr., and I.Carvalho, Carrageenans: Biological properties, chemical modifications and structural analysis—A review, *Carbohydr. Polym.*, vol. 77, no. 2, pp. 167–180, 2009.

[264] M. Kitamura, Development concept and formulation design of cetirizine hydrochloride OD tablet, *Pharm. Technol. Jpn.*, vol. 27 no. 4, pp. 91–98, 2011 (in Japanese).

[265] M. Kitamura, T. Nakagawa, A. Harada, and M. Hizaki, Drug development research with taste sensor. Part 1, *Pharm. Technol. Jpn*, vol. 28, no. 1, pp. 43–48, 2012 (in Japanese).

[266] T. Nakagawa, M. Kitamura, A. Harada, and M. Hizaki, Drug development research with taste sensor. Part 2, Taste masking and formulation development of donepezil hydrochloride OD tablets, *Pharm. Technol. Jpn.*, vol. 28, no. 3, pp. 31–37, 2012 (in Japanese).

[267] A. Harada, M. Kitamura, T. Nakagawa, and M. Hizaki, Drug development research with taste sensor. Part 3, Using something together with cefcapene pivoxil hydrochloride fine granules, *Pharm. Technol. Jpn.*, vol. 28, no. 4, pp. 97–101, 2012 (in Japanese).

[268] K. Ohsugi, Y. Tahara, H. Akitomi, M. Yasuura, Y. Kobayashi, M. Nakano, H. Ikezaki, T. Haraguchi, T. Uchida, and K. Toko, Consideration of an evaluation method for bitterness and suppression of bitterness of branched chain amino acids by taste sensor, *Jpn. J. Taste Smell Res.*, vol. 19, no. 3, pp. 421–424, 2012 (in Japanese).

[269] M. Ito, M. Yoshida, Y. Kobayashi, M. Hiraoka, H. Ikezaki, and T. Uchida, Bitterness evaluation of H1-receptor antagonists using a taste sensor, *Sensors Mater.*, vol. 23, no. 8, pp. 483–492, 2011.

[270] H. Akitomi, Y. Tahara, M. Yasuura, Y. Kobayashi, H. Ikezaki, and K. Toko, Quantification of tastes of amino acids using taste sensors, *Sens. Actuators B*, vol. 179, pp. 276–281, 2013.

[271] Y. Kobayashi, Y. Yamaguchi, H. Hamada, H. Ikezaki, and K. Toko, Study on drug bitterness evaluation using artificial lipid-based membrane bitterness sensor and physicochemical parameters, *Jpn. J. Taste Smell Res.*, vol. 16, no. 3, pp. 497–500, 2009 (in Japanese).

[272] P. Ertl, B. Rohde, and P. Selzer, Fast calculation of molecular polar surface area as a sum of fragment-based contributions and its application to the prediction of drug transport properties, *J. Med. Chem.*, vol. 43, no. 20, pp. 3714–3717, 2000.

[273] A. Takakura, H. Ikezaki, and M. Hiraoka, Development of evaluation techniques for poorly-water soluble drugs using the taste sensing system, Proceedings of the 28th Symposium on Particulate Preparations and Designs, 2011, pp. 98–99 (in Japanese) October 27, 28, Osaka, Japan.

[274] S. Nakamura, R. Yatabe, T. Onodera, and K. Toko, Sensitive detection of capsaicinoids using a surface plasmon resonance sensor with anti-homovanillic acid polyclonal antibodies, *Biosensors*, vol. 3, no. 4, pp. 374–384, 2013.

5

Pattern Recognition

Saverio De Vito[1], Matteo Falasconi[2] and Matteo Pardo[3]
[1] *ENEA C.R. Portici, P.le E. Fermi, Naples, Italy*
[2] *Dipartimento di Ingegneria dell'Informazione, Università degli Studi di Brescia, Brescia, Italy*
[3] *Institute for Applied Mathematics and Information Technologies, National Research Council (IMATI CNR), Genoa, Italy*

5.1 Introduction

Since its first appearance in the scientific literature, the so-called electronic nose (e-nose) has coupled the olfactive data-gathering capability of chemical sensor arrays with the high-value information extraction capability of multivariate data analysis or pattern recognition algorithms.

Machine learning tools have been typically used to extract information from sensor readings, specifically for chemical identification (i.e., detection of a particular analyte in complex mixtures, possibly at different concentrations) or classification (i.e., discrimination among different complex mixtures) and for quantification of gas concentrations in complex mixtures. Unpredictable fluctuations of environmental conditions, together with the limitations still affecting chemical sensing devices—in the first place their lack of selectivity and stability—call for the adoption of state-of-the-art statistical data analysis methods.

In the following chapters we will analyze various pattern recognition challenges in artificial olfaction (AO) (we will use interchangeably the terms "statistical data analysis," "pattern recognition," and "machine learning"). Major application frameworks will be revised in Section 5.2. Two consecutive sections will present supervised and unsupervised machine learning frameworks with applicative examples. The last section will analyze advanced computational intelligence approaches to e-noses.

Essentials of Machine Olfaction and Taste, First Edition. Edited by Takamichi Nakamoto.

Companion website: www.wiley.com/go/nakamoto/olfaction

5.2 Application Frameworks and Their Challenges

5.2.1 Common Challenges

The major limitations of chemical sensor-based electronic noses (ENs) remain their lack of selectivity (or specificity) and stability.

Lack of selectivity means that chemical sensors can show a significant response toward gases they are not designed to sense, the so-called interferent gases. Sensors with poor selectivity adversely affect the discriminating power of the array if no information is available about the interferent gas. Advancements in sensor development and fabrication technology have led in the past years to a steady increase of sensor sensitivity, but a trade-off with specificity and stability seems always to apply, constantly shifting the line of the horizon in the search for a specific, sensitive, and stable chemical sensor.

In many cases, lack of specificity can be overcome by multivariate data analysis. Each sensor then provides some piece of information to build a knowledge base about interferents. This capability, however, is limited by the number of interferents and the dimension of the sensor array.

As an example, most solid-state gas sensors show very high cross-sensitivity to humidity [57]. Although a number of signal processing approaches for humidity compensation have been proposed, ENs do not yet provide satisfactory performance. The implementation of a dedicated, compact humidity control subsystem, together with advanced humidity compensation algorithms, into portable devices could constitute an important advancement, especially for in-field deployment (e.g., for environmental monitoring purposes).

The problem of chemical sensor instability over time is known as "sensor drift"; it manifests itself through small and nondeterministic temporal variations of the sensor response under identical conditions. This is generally attributed to sensor aging or thermomechanical degradation, which depends on the sensor (sensor material, fabrication process, etc.) and on its operational environment. In practice, it is actually unimportant whether the sensors or the e-nose as a whole (e.g., tubing, sampling system) is drifting. This phenomenon eventually hampers the initial capability of the e-nose to detect, identify, or quantify gas mixtures.

Another source of loss of calibration quality over time is the often neglected "concept" drift, which is the change of the distribution of target odorant concentrations. This may set the operating point of a calibration algorithm outside of its "known" envelope represented by the previous recordings on which it has been trained. Up to now, drift correction remains one of the most relevant issues in machine olfaction data processing [24].

From a practical point of view, the EN training still remains the major bottleneck in terms of time resources and costs. Very little effort has been devoted to the development of transferable calibration models that can be applied to multiple equivalent devices. The majority of published results are limited to datasets gathered with a single device which are never transferred, hampering large-scale EN deployment. Actually, individual calibration of a multitude of devices for a specific application area is generally unrealistic due to the requirement of actual samples to be measured. The development of calibration transfer models (master-to-slave or project-to-project) would be of critical importance.

A peculiar instance of the transferability problem is commonly encountered when a sensor needs to be replaced due to sensor breakage or malfunctioning. Sensor replacement often jeopardizes the use of the previously collected databases. The underlying cause is always the poor reproducibility of chemical sensors, also inside the same production batch—a fact that is rarely discussed and that still critically hinders the commercial diffusion of e-noses. This problem has been addressed from different points of view, either improving sensors or using multivariate calibration methods; however, it is not fully solved yet [71].

5.2.2 *Static In-Lab Applications*

Laboratory applications often rely on desktop instruments, permitting the usage of several add-ons, such as preconcentration units, humidity and temperature regulators, baseline adjustments, air filters, and so on. As a reference, Alpha MOS (www.alpha-mos.com) commercializes a series of laboratory e-nose analyzers: FOX and GEMINI series based on metal oxide (MOX) sensor detectors, HERACLES and ULYS based on gas chromatography, and AIRSENSE using soft ionization mass spectrometry technology.

To automate sample incubation, headspace generation, collection, and injection, laboratory e-nose analyzers can be coupled to autosamplers instead of using manual injection. Headspace sampling can be performed in either a static or a dynamic way, the former being more reproducible and the latter being more sensitive due to the possibility of extracting larger sample volumes (attention has to be paid to the possible confounding extraction of low-volatility compounds). Additionally, these autosamplers can be further equipped with a variety of devices for extracting or concentrating volatile organic compounds before analysis, such as solid phase microextraction (SPME), solid phase dynamic extraction (SPDE), in-tube extraction (ITEX), thermal desorption autosampler (TDAS), and flow cell dynamic on-line analysis.

Moreover, the absence of power consumption constraints also allows for off-line and high-performance computing systems.

Experimental procedure design [96] is usually critical for pattern recognition performances in lab application where conditions can be controlled accurately. Generally, both for training and testing phases, different sample classes should be measured with an interdispersion sufficient for reproducing real working conditions. Measuring different sample classes in a per-class order must be avoided, since this may record the difference of experimental condition, resulting in a fake interclass variability.

ENs as laboratory equipments mainly focus on two areas: food analysis and medical diagnostics.

Food analysis has been, and still seems to be, one of the most promising application areas for ENs because of their simplicity of use, low cost, and rapidity with regard to traditional analytical methodologies while still allowing a good correlation with sensory panels [123]. In fact, aroma is one of the key food parameters; the characteristic bouquet of volatile organic compounds may provide information and act as an indicator for food safety and quality. ENs have been applied in various food contexts, such as process monitoring, freshness evaluation, shelf-life investigation, authenticity determination, and product traceability [88].

The use of olfaction for medical diagnosis dates back to 400 BC during the time of the Hippocratic school. The introduction of the e-nose technology in medicine is envisaged to overcome the traditional shortcomings of olfactive diagnosis. As such, EN's noninvasive and rapid diagnostic technology has been applied to the detection of several pathologies, including pulmonary diseases, microbial infections, type 2 diabetes, and urinary system diseases [13].

The application of e-nose medical diagnosis has largely developed in the last two decades, and many experiments performed with a number of different instruments have been reported in literature [111]. Among others, in vitro measurements were mainly oriented to the study of bacterial cultures, which have also been the most investigated matter in this field. In vivo studies have focused on body fluids (urine), skin odors, and breath exhalate. Some approaches involve the analysis of off-odors captured by swabs as in the case of searching for specific bacterial growth [32]. Direct breath analysis is instead usually carried out for early diagnosis of lung cancer or for detection of other pulmonary or gastroesophageal pathologies such as gastroesophageal reflux disease, COPD, and asthma (see [90, 107]). Recently application of e-nose to cancer detection has been reviewed by D'Amico *et al.* [13].

Most of the in-lab application scenarios focus on classification problems. Discriminant power is a first rank concern. In medical applications the focus is often on two-class classification problems: one class associated with healthy patients, or "negatives," and the other one associated with affected patients, also labeled as "positive" (see Ref. [124]). Performance assessment is then traditionally conducted in terms of false-positives and false-negatives, with false-negatives being generally associated with a high misclassification cost. A significant challenge is the class imbalance and the correct assessment of priors, which is normally done with the help of historical records. However, the most significant challenge is to obtain a significant dataset, capable of adequately covering real world variability both in the pathology and in the patients, which is generally very high. These aspects are also common to some industry settings in which a go/no-go decision is to be made (quality assessment or nondestructive tests).

5.2.3 On-Field Applications

On-field EN deployment typically focuses on detecting and quantifying chemicals in the surrounding uncontrolled atmosphere. Gas detection and quantification represent valuable information in many safety and security applications, ranging from pollution monitoring to detection of inflammable or dangerous gases, explosives, or drug factories. In indoor environments, volatile organic compounds (e.g., formaldehyde) released as off-gas by furniture adhesives or cleaning agents or by smoking can reach concentration levels that can cause cancer [27]. Outdoor city pollution is basically made up of inorganic gases and particulate matters. Olfactive nuisance can greatly impact populations living near to solid wastes or waste water treatment sites as well as to some industrial manufacturing plants. Exposure to olfactive nuisance may in fact severely affect working capabilities (inability to concentrate) and make living in such sites extremely unpleasant, being a source of nausea, headaches, fatigue, throat irritation, and sleep disturbance, which also affects social skills [22]. Since relevant events occur intermittently, are unpredictable in both space and time, and require prompt problem identification, in-lab e-noses are not suitable [44]. Indeed, the outdoor deployment of multiple electronics is required.

The main peculiarities of on-field applications concern sampling (controlled vs. uncontrolled), calibration, and concept drift.

As far as sampling is concerned, the choice is whether to provide an active pneumatic sampling of the ENs' surrounding atmosphere or to passively let the atmosphere reach the sensor array. Portable e-nose systems, especially those targeted at punctual, local quali-quantitative measurements, can usefully exploit the first approach to control the flow toward the sensor chambers, mitigating environmental variations with a filtering subsystem. Air quality monitoring systems may exploit pneumatic sampling to provide a steady flow toward an open or sealed sensor array, avoiding fluid dynamic stagnation issues. This is often the case for long-term on-demand or cyclic olfactive nuisance monitoring applications. The system can then in principle operate in a cyclic or on-demand manner, producing sensor array responses resembling those encountered in lab operations, with the possibility of relying on base response acquisition (for drift counteraction) and sensor response dynamic behavior as a source of valuable information.

However, whenever an energy consumption is an issue (pervasive or personal air pollution monitors), uncontrolled passive sampling may be dictated. Sensors are then in near equilibrium with the environment and their response is constantly available (real-time monitoring). However base response is no longer available and drift counteraction has to rely on more complex approaches.

When it comes to complex mixtures and dynamic environments, like the ones involved in city air pollution monitoring, it is lengthy and costly to synthetically (i.e., with in-lab measurement) generate representative datasets [61, 91]. In facts, this approach usually fail in modeling adequately the range of environmental and process value distribution strongly limiting the application of solid-state multisensory devices in this framework. As a way out, ENs can make use of ground truth data from by colocated conventional stations for calibration [15, 74]. Environmental conditions and target variables are affected by seasonal variations and this may hamper the representativeness of a limited duration recording. On the other hand, it is impracticable to extend the duration of the calibration recording over an entire year; adapting the obtained calibration to the changing environment and sensor performances has been proposed [108]. Furthermore, there is no guarantee that the conditions recorded for a particular place will be representative of conditions in other geographical locations.

Whenever the sensor system has to be used for real-time monitoring (such as in the case of environmental nuisance measurement) or for rapid routine controls during industrial production (like in food applications), the system response time, and hence data processing time, is a crucial factor. While, generally, chemosensors deliver their steady state response in minutes, fast available transient information may be considered. On the other hand, transient information is less stable and may require some sophisticated data analysis. However, data processing time becomes then the limiting factor due to the fact that, given certain hardware specifications (e.g., hardware processing speed and accuracy), there are intrinsic constraints on the methods that may be implemented in real time [19, 20].

A special case of on-field EN deployment is distributed and mobile chemical sensing, which is becoming increasingly relevant. Indeed, gas propagation requires a dense network of measuring points in order to correctly assess its distribution [130]. Chemical propagation is affected by several phenomena including the 3D structure of the environment, diffusion, transport, and turbulence, with each one playing a different role in different indoor or outdoor settings. Correlation of chemical signal with wind (speed, direction) has to be analyzed to localize and characterize the source of olfactive information. Dense measurements can be obtained with a network of distributed chemical multisensor devices [21, 76]. Alternatively, a single mobile robot equipped with chemical sensors can obtain similar results. In both cases, 3D continuous assessments of chemicals are needed for source identification and declaration. Generally speaking, the distributed approach has greater time accuracy, while a mobile robot

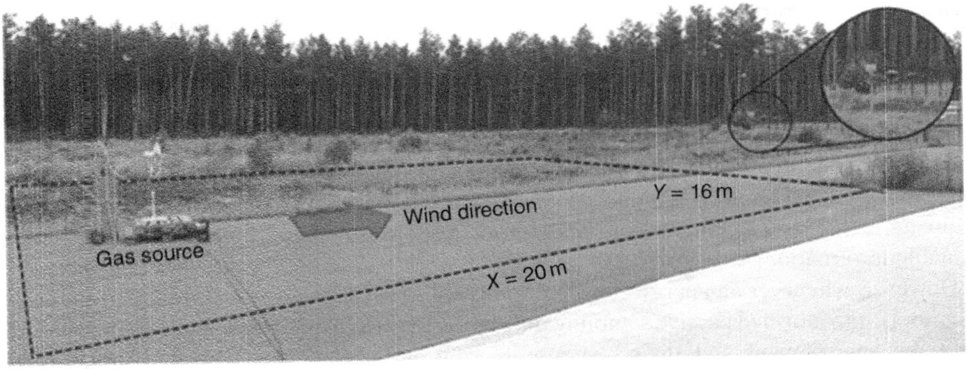

Figure 5.1 A chemical sensing microdrone while performing gas source identification (Source: From Ref. [78]. Reprinted with permission from Taylor & Francis)

is more space accurate. Mobile systems opened the path for biomimetic approaches to source localization by using steering algorithms that allow the robot to autonomously track the plume (see Figure 5.1) after being hit by it, eventually locating the source [5, 78].

Of course, the two approaches can be combined in a network of fixed and mobile analyzers with potential application to citizen-based air pollution monitoring.

In the mobile and personal air pollution monitoring framework, the deployment of hundreds of multisensory devices is also foreseen (e.g., EveryAware project). Eventually these devices can generate an incoming data stream that, together with appropriate dispersion models, will give a picture of a city's air pollution state. The analysis of the rich data stream that is produced may give birth to novel distributed olfaction applications such as remote characterization of mobile pollutant emission sources [93]. It is clear, though, that novel, high-throughput or distributed pattern analysis tools should be devised to cope with these emerging scenarios.

In wireless chemical sensor networks, as mentioned earlier, energy management is a primary concern. Allowing a node to become self-aware of the informative level of the data it is sensing can allow for sensor censoring strategies, avoiding the transmission of uninformative data and thereby saving transmission-related power. It is mandatory to also explore low-computational-load pattern recognition approaches, such as feedforward neural networks, with a memory footprint that can be beard by most microcontroller architectures.

5.3 Unsupervised Learning and Data Exploration

5.3.1 Feature Extraction: Static and Dynamic Characteristics

Data acquisition is the first step of the data analysis cycle: sensor response curves (raw data), for example, electric conductivity variations versus time of MOX gas sensors, are collected and saved into a format suitable for computer-based analysis. The main subsequent step, feature extraction, is perhaps the most critical stage of a pattern recognition system. Its responsibility is to produce new salient and informative features, resulting in a concise pattern also known as an olfactive fingerprint. Feature extraction is a dimensionality reduction process:

data is converted from the original space to the feature space typically characterized by a reduced dimensionality. The choice of feature extraction processes can significantly affect the performance of the subsequent modules in the pattern analysis system.

In duty cycled measurements, mostly applied in lab e-nose settings, one can recognize a typical response cycle made by four time segments. In the first one, the so-called base, the sensor response—to a reference gas or to filtered outer air—is captured. In the second step, the sensor array is abruptly exposed to the gas mixture, revealing its dynamic characteristics. In the third step the sensor array response is supposed to reach a steady state while in the last phase the sensor array is flushed revealing its desorption dynamic characteristics (see Figure 5.2). In continuous monitoring settings, instead, transient characteristics may be triggered by abrupt changes in gas mixtures concentrations.

The standard feature extraction procedure is to select "local" features such as the steady-state response of the raw sensor, normalized or not by the baseline response [10]. Baseline response level is usually used for drift effect counteraction, weighting the above mentioned steady state response that has historically been considered the most informative response segment [39]. For unmanned settings when continuous monitoring is performed, the baseline is typically unavailable. In these cases, the instantaneous sensor array response may be the only feature considered.

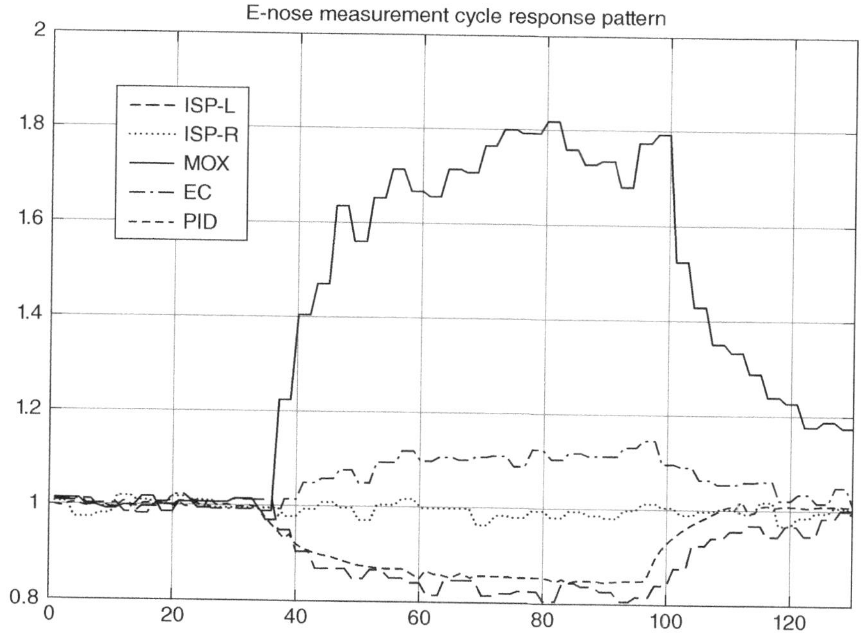

Figure 5.2 Typical normalized response of a 5 × hybrid sensor array. Different phases of the basic normalized response of a duty cycled e-nose sensor array are clearly distinguishable. After the baseline acquisition phase, the sensor array is abruptly exposed to the gas mixture, producing an absorption transient after which the responses settle to a pseudo steady-state level ($t = 80$). Steady state response level was the primary feature during the first decade of artificial olfaction development. At $t = 95$ the flushing mixture set in, causing the array to exhibit the desorption transient pattern

During the last decade we can observe a path from a few local features toward complex and more descriptive features pattern including global waveform features and transient features. Several algorithms have been proposed to extract (generate) features from the whole sensors' response [26, 79, 101]. Many papers have consistently shown that absorption and desorption transients possess an informative content that complements the steady state response. Synthetic local indexes, for example, sensor response max derivatives, may be used to capture the informative content of sensor dynamics. Extracting synthetic indexes of the overall characteristic of the entire response waveform (e.g., the area under the curve, peak response value, peak first or second derivative, etc.) has also consistently shown positive results [49].

The sensors' phase space has also been investigated. Works referring to this approach analyze the trajectory of the instantaneous response in a bidimensional space characterized by the actual response and its derivatives (see Figure 5.3). Typical examples of phase space features are phase space integral and dynamic moments [73, 83, 116].

AO researchers have also considered functional basis decomposition, for example, using FFT (see Figure 5.5), Bessel Functions and Wavelet transform approaches (e.g., Haar DWT). Coefficients of the obtained decomposition captured discriminative information very efficiently [30, 55, 114].

For real-time or continuous concentration estimation, several authors have proposed to capture the sensor array transient behavior by coupling the instantaneous sensor response with a set of past sensor responses in a tapped delay line fashion (see Figures 5.4 and 5.5).

De Vito *et al.* [16] have shown that this approach obtained robust and almost instantaneous concentration estimation during abrupt exposition to gas mixtures.

More complex sensor operation, like temperature modulation of MOX sensors and voltage scanning for voltammetric sensors, or the use of ion mobility spectrometers (IMSs) as an e-nose sensor, increase data dimensionality and enlarged the range of possible feature extraction methods [115, 131]. In this framework, wavelet transform [28] and transient analysis [51]

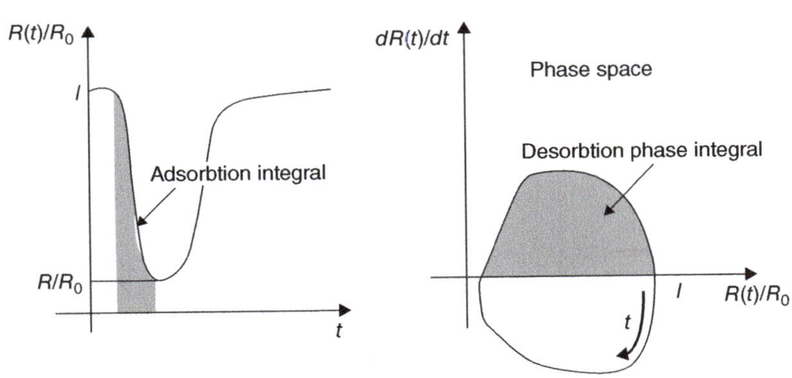

Figure 5.3 Examples of response features and phase space features. The normalized $R(t)/R_0$ response curve over time (left) and the corresponding response curve in the first-order phase space (right) are shown. Note that, in the phase space, $t=0$ corresponds to the point with coordinates (1,0); indeed $R_0=T(t=0)$ by definition. The integral of the area under the curve (AuC) for adsorption phase (left) and the so-called desorption phase integral (right) are highlighted. (Source: From Ref. [83]. Reprinted with permission from Elsevier Science Ltd.)

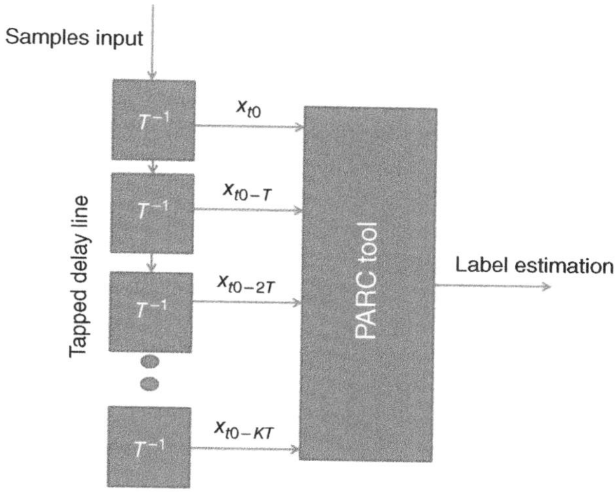

Figure 5.4 Tapped delay line block diagram. Consecutive T-delay blocks allow for the construction of a feature vector embedding information on sensors' dynamic behavior in response to input changes

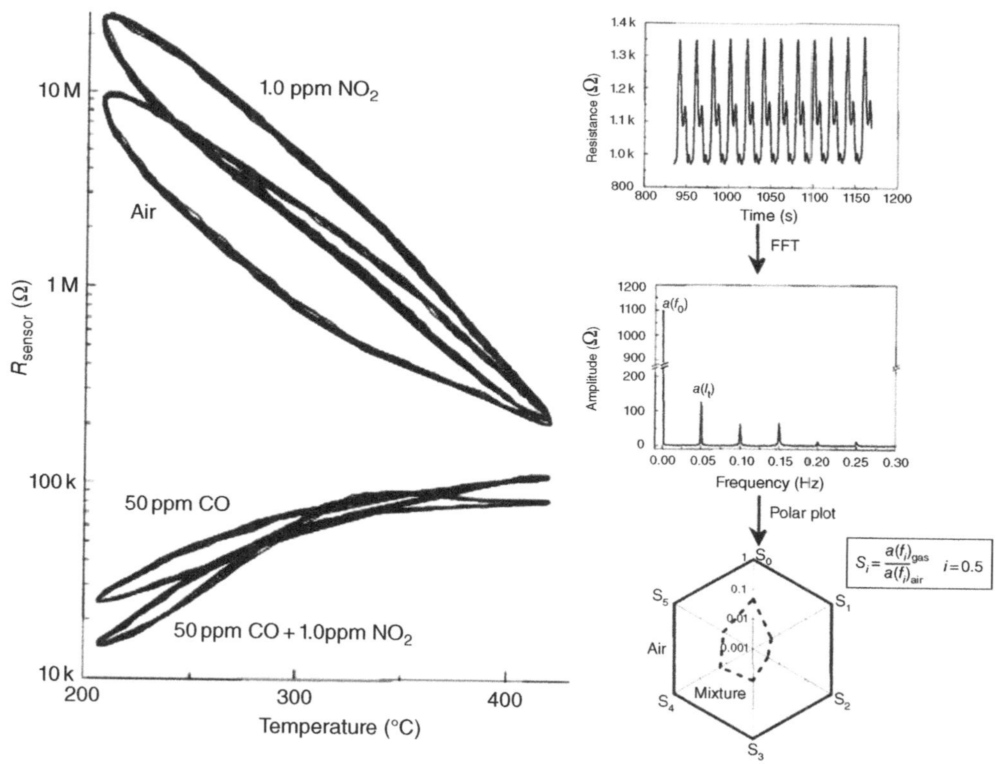

Figure 5.5 Response curve of a single temperature-modulated MOX sensor to the exposition of synthetic mixtures of CO and NO_2 in a temperature modulation framework (left). On the right, the proposed FFT-based fingerprint extraction process is depicted (Source: From Ref. [55]. Reprinted with permission from Elsevier Science Ltd.)

for temperature-modulated MOX response profiles and spectra bucketing, originally developed for GC–MS, for IMS have been experimented [54, 94]. In the last 10 years, bioinspired ad-hoc techniques have also been developed to deal with these responses complexities specifically tackling dimensionality reduction [89].

As one can expect, there is no general definition or consensus about the informativeness of a particular response feature. Surprisingly, feature performance comparison has been very rarely tackled: Pardo and Sberveglieri [83] reported a consistent advantage of phase space integral features, in particular desorption phase features, over classical steady state features. More recently, Vergara *et al.* [114] reported an informational advantage of Bessel functions' decomposition factors with regard to steady state and moving average features.

In general, and especially when high-dimensional data are concerned, feature selection is to be applied both to avoid the curse of dimensionality, and uninformative, noisy or even detrimental features.

5.3.2 Exploratory Data Analysis

Initial data examination, also known as exploratory data analysis (EDA) [110], is often the most important part of the data analysis cycle [122] and may be implemented on raw data as well as after a feature extraction step. In fact, EDA is a necessary step in which the user interacts with the machine to check the quality of experimental results before embarking in successive, more automated steps, thus saving lot of unnecessary efforts and sometime preventing gross errors.

The aims of EDA are manifold: maximize insight into a dataset, uncover underlying structure, extract important features, and detect outliers. The most valuable outcome of EDA is to check for prior assumptions, understand how they affect the EN response, and determine the optimal experimental settings.

EDA includes three relevant aspects:

1. *Checking the quality of the data.* A first look at the sensor responses can help to evaluate the instrument operative state identifying and preventing problems. For example, in the case of the responses of chemical sensors to controlled mixtures in ENs, the expected structure of the response is usually known. Malfunctioning of the equipment (sampling system, sensors, electronics) can be spotted by plotting the sensor time response.
2. *Calculating summary statistics.* Summary statistics can be used to characterize the data: few numbers can convey fundamental properties. For example, by calculating (sample) mean and variance for each feature and for each class, it is possible to detect the (more obvious) outliers and to get clues about the important variables for discriminating particular classes.
3. And finally, *producing plots of the data in order to get a feel of their structure.* This aspect of data investigation should ideally be performed iteratively together with data collection in order to adjust the experimental conditions for maximizing the system performance (e.g., sample classification by the EN).

Although a number of classical approach are universally recognized, EDA is perhaps one of the most uncodified data analysis phases, mostly relying on the experience of the data analyst.

There are various types of plots:

1. Histograms, box plots, and bar charts are a simple yet effective (see Figure 5.6) way of visualizing the univariate sensor features. Single response values or the mean of several samples can be plotted for each sensor individually [45, 120].
   Simple feature plots can be used, for instance, to display the individual chemical sensor signals and evidence whether the given sensor/feature discriminates among different classes or provides a response proportional to a given gas concentration. [118] used single feature plots for EDA of industrial safety application.
   Instead of feature plots, box plots can be used; these convey meaningful statistical information more synthetically. The box plot summarizes different properties of a data distribution: (i) the box has lines at the lower or first quartile (bottom line in Figure 5.6), median or second quartile (line in the middle), and upper or third quartile (top line) values; (ii) whiskers are lines extending from each end of the box showing the extent of the tails of the sample distribution. Whiskers extend from the box out to the most extreme data value within 1.5*IQR, where IQR is the inter-quartile range (i.e., difference between 3rd and 1st quartile values) of the sample; (iii) outliers are data with values beyond the ends of the whiskers and they are marked with a red cross. If there is no data outside the whisker, a dot is placed at the bottom whisker.
2. Scatter plots or joint distribution plot of pairwise combinations of variables can be a valid aid to evidence of (changing) cross-correlation between selected input (Figure 5.7a) or target variables (Figure 5.7c) but are useful only if the dimensionality of the data is not too high. Pearson correlation matrix (Figure 5.7b) is another graphical tool used to visualize similarities and differences between sensor responses; correlation values span from 0 to 1, where 0 means no correlation (completely different behavior between the sensors,

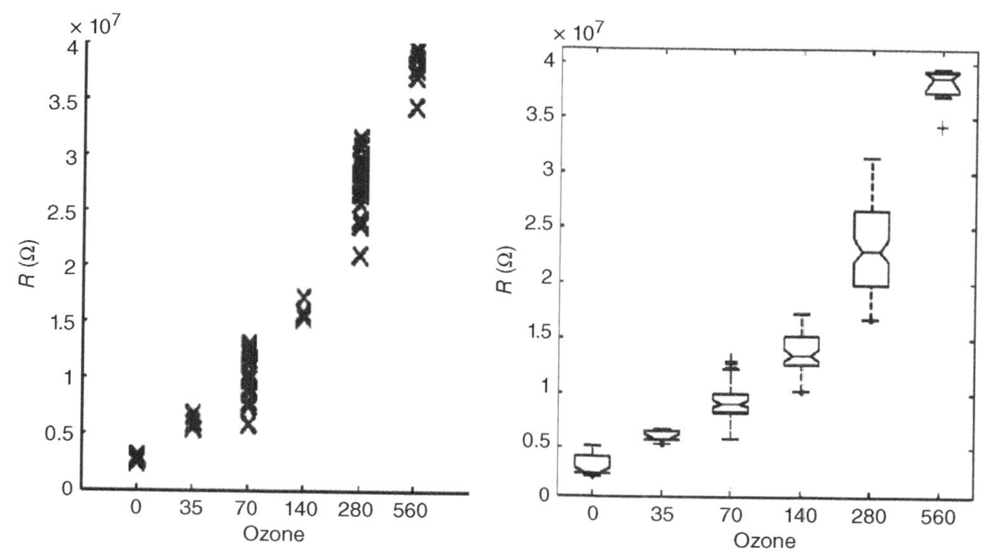

Figure 5.6 Feature plot versus box plot (Source: Modified from Ref. [118])

i.e., completely different information) and 1 indicates the maximum correlation degree (identical behavior); for this reason, uncorrelated features are preferred.

3. Polar plots and radar plots (see Figure 5.8) display multivariate dimensional data in two dimensions in form of a map (often referred to as "fingerprint"). Axes for each feature radiate from the origin at equal angles with the magnitudes of the features joined by straight lines. These plots have been used frequently in the past and many examples can be found in the literature. The problem with radar plots is that they become confusing if too many data points, for example, replicated sample measurements, are plotted. Authors are therefore used to plot only the average response values (the average being taken over measurement replicas), which makes sense only provided that intraclass variance is smaller than interclass variability, feature-per-feature. The latter information is rarely provided though, making radar plots more a way for showing the supposed machine "fingerprint" than a robust analysis tool.
4. Plots of data projections (linear or nonlinear). Generally two-dimensional projection has been most commonly used.

There are various kinds of projections; a major distinction is that between supervised and unsupervised projections.

In exploratory analysis, the elective method is principal component analysis (PCA). Geometrically, PCA can be thought of as a rotation of the axes of the original coordinate system to a new set of orthogonal axes that are ordered in terms of the amount of variation (variance) of the original data they account for. PCA is a linear transformation: this means that the new variables are linear combinations of the original variables and are (due to their construction) uncorrelated. In determining the PCA projection no class information or clustering assumption is used, the technique is hence termed unsupervised.

PCA should be systematically adopted for detecting clusters in the data, indicating important variables and detecting outliers. PCA plots are best used in a (iterative) sequence either by using diverse data labeling (where the labels correspond to the variables of interest) and by taking data subsets which correspond to well defined target values. This approach leads to discover the best operative conditions (e.g., the best sampling parameters) to solve the specific problem. In AO, first principal components can be associated with sensor or experimental condition (environmental) variations; useful discriminant information may be well embedded in low-variance directions in the principal component space [84].

In general, the projections at EDA stage are unsupervised, as in PCA. If the projections are supervised, they make use of easy to calculate error criteria, for example, general separability measures like Fisher criterion (DFA), based on within-class and between-class covariance.

Linear discriminant analysis (LDA) is a supervised method that assumes unimodal Gaussian classes with different mean and equal class covariance. Although quite often used for data visualization, as any supervised technique, LDA runs the risk of overfitting, providing false class separability on the training set for small sample number to dimensionality ratio. Thus, authors using LDA approach should pay much attention on deriving conclusions about electronic olfaction system performance or operate cross validation schemes. LDA has the additional limitation of having a maximum dimensionality given by $C-1$, with C being the number of classes.

PLS in conjunction with a binary class-coding (partial least squares (PLS) discriminant analysis (PLS-DA)) can be found in the sensor literature as a mean for feature extraction and data

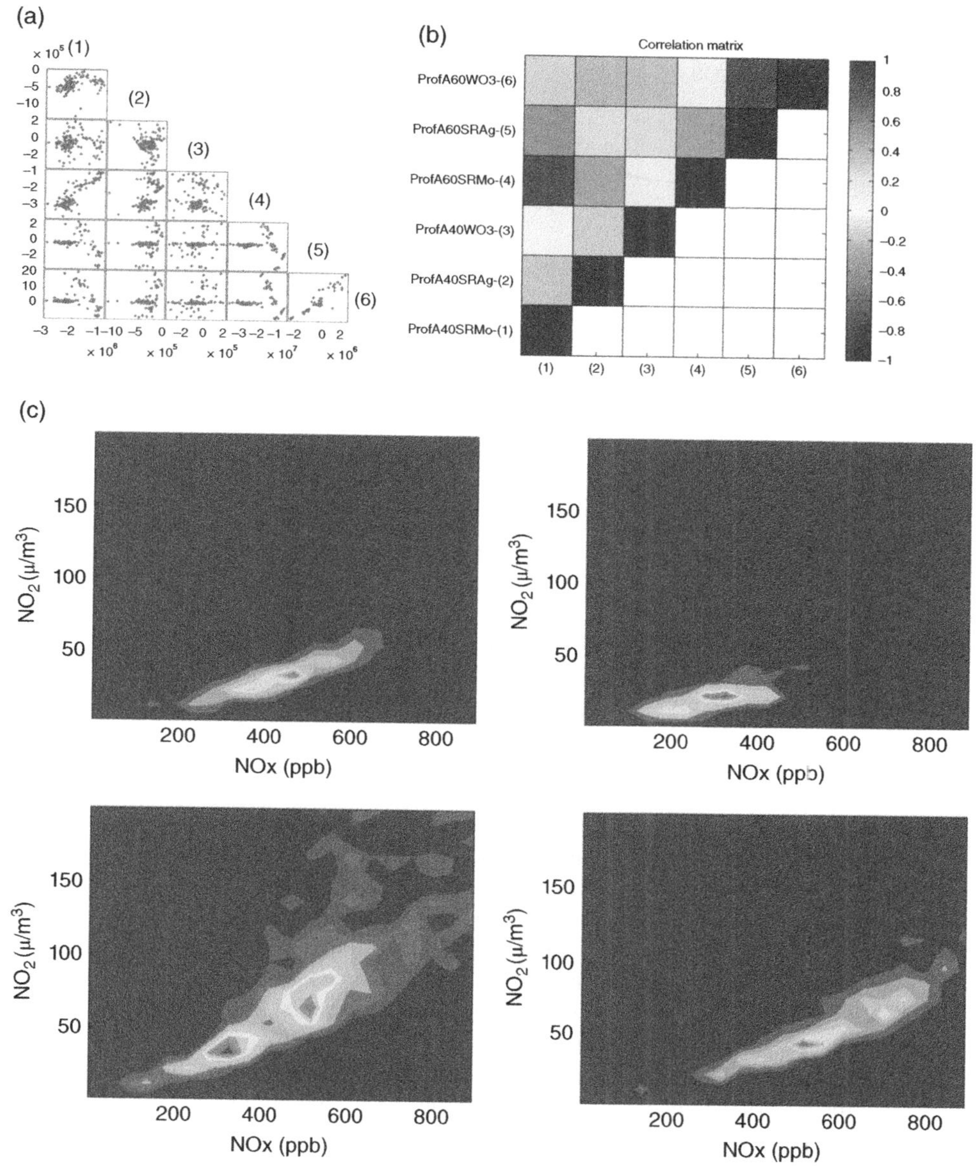

Figure 5.7 (a) Scatter plot of electronic nose data (Source: From Ref. [41]. Reprinted with permission from Elsevier Science Ltd.) for six different features (three sensors, two diverse extracted features), (b) correlation matrix evaluated on the same samples, all values are positive (Source: From Ref. [41]. Reprinted with permission from Elsevier Science Ltd.), (c) combined distribution estimation plot of two pollutants in different seasons highlighting seasonal concept drift in air pollution scenarios) (Source: From Ref. [17])

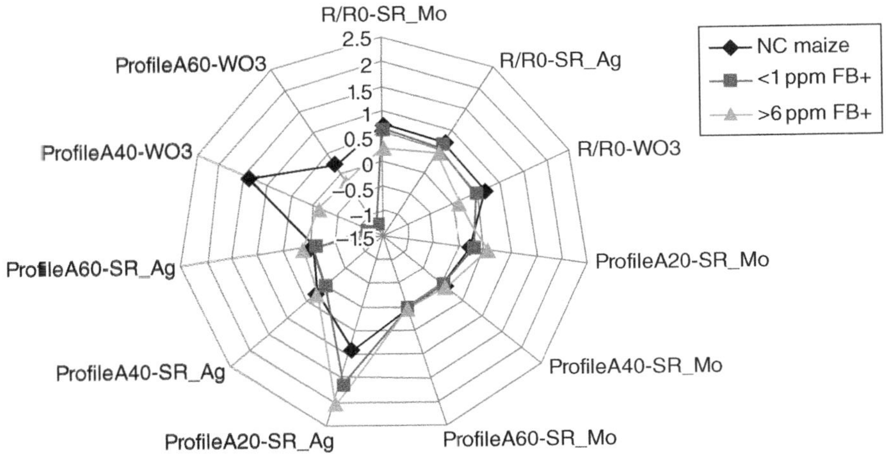

Figure 5.8 Radar plot of data (Source: From Ref. [41]) of a MOX sensor array response for three different sample classes (noncontaminated maize (NC maize), low-level fumonisin-contaminated maize—<1 ppm FB+, high-level fumonisin-contaminated maize—>6 ppm FB+)

visualization with direct class prediction abilities [11, 26]. PLS provides a decomposition of the sensor data in latent variables that can be plotted to provide a visual representation of the classification properties. The use of PLS helps avoiding the drawbacks due to sensor correlation.

Independent component analysis (ICA) [12, 62] is a statistical and computational technique for revealing hidden factors that underlie sets of random variables or measurements. As the name implies, the goal is to find the transformation of variables by which the new components are statistically as independent from each other as possible. ICA defines a general model of the observed multivariate data. In this model, the data variables are assumed to be linear or nonlinear mixtures of some unknown latent variables. The objective of ICA algorithm is to determine the mixing matrix, which is also unknown, with minimum loss of information. ICA can be seen as an extension of the PCA for the extraction of independent features [65].

ICA is capable of finding the underlying factors or sources when classic methods fail. Indeed, ICA was proposed to identify which part of the observed signal was due to environmental disturbances, such as temperature or humidity variations [26].

One main limitation of LDA is the assumption of symmetric classes with separated centroids. When the covariance matrices of the classes are equal, the optimal decision boundary to separate two normal distributions (minimal Bayes error) is given by the distribution's symmetry plane, and it is not difficult to calculate the Bayes error nor to find subspaces in which the Bayes error is minimal. Two distributions with unequal covariance matrices do not in general have a symmetry plane.

A linear feature extraction technique called asymmetric class projection (ACP) was especially thought to deal with asymmetric distributions of data [66]. The ACP is similar to LDA in the respect that both aim at extracting discriminating features (linear combinations or projections) from many variables. Yet, ACP maximizes a measure of variance while the LDA maximizes a measure of mean difference; LDA is not even defined when there is no mean difference. The ACP, in contrast, works on (two) possibly concentric distributions with unequal covariance matrices. For instance, we may have two sets of samples (good and bad) and the measurements in the "good samples" are well clustered around one point while the "bad samples" are scattered around. The technique was successfully tested with synthetic data but provides comparable results to LDA for an experimental dataset [66].

5.3.3 Cluster Analysis

Cluster analysis (CA) is an unsupervised learning method frequently used to assess data structure, to discover hidden structures in the datasets, and to extract (or compress) the information by identifying cluster representative samples also known as *prototypes* [60, 132].

In CA essentially patterns are divided into groups (clusters) so that the objects belonging to the same group are more similar to each other than those within different groups. Similarity is usually measured by a distance function defined on pairs of patterns. The most popular metric for continuous (numerical) features is the Euclidean distance.

CA has been introduced in the chemical sensor field more than two decades ago [38] and then has been often adopted in combination with PCA for data mining and visualization [33].

A widely agreed taxonomy distinguishes between *hierarchical clustering* (HC) and *partitional clustering* (PC), on the basis of the properties of the generated clusters [60].

HC organizes the data into a hierarchical structure, that is, a nested set of subsequent non-overlapping partitions.

Algorithms for HC are mainly classified as divisive and agglomerative methods. Divisive HC begins with all objects in one cluster and then splits it in subsequent nested sub-clusters until all object belong to individual clusters; this is not very used in practice due to exponentially diverging computational cost with the number of objects to be clustered. Agglomerative HC starts with clusters that include exactly one object, then a series of merge operations are followed out to bring all objects into the same group. The most popular methods are single link (SL), average link (AL), complete link (CL), and Ward strategy (Ward).

The results of HC are graphically depicted by a binary tree or dendrogram (see Figure 5.9a). The root node of the dendrogram represents the whole dataset and each leaf node is regarded as a data object. The height of the dendrogram usually expresses the distance between each pair of objects or clusters, or an object and a cluster. A data partition can be obtained by cutting the tree at different distance levels. In the experiments reported in Figure 5.9b [41], some maize samples contaminated by fumonisin B1 (FB) were screened by using a MOX sensor-based e-nose. The three clusters corresponding to noncontaminated samples and to the samples contaminated with high and low FB levels are clearly visible.

PC aims at obtaining a single partition of the data and it is advantageous for large datasets when the use of HC is computationally prohibitive. One of the most important factors of PC is the criterion (or cost) function that must be minimized. The sum of squared error function, for instance, led to the well known *k*-means algorithm [70]:

$$\sum_{k=1}^{C}\sum_{i=1}^{n} xi - ck^2 \tag{5.1}$$

where C is the number of clusters, xi is the ith pattern belonging to the kth cluster, and ck is the center of cluster k. K-means algorithm is very popular in the sensor literature because it is easy to implement. A major drawback is that it is sensitive to the selection of the initial partition

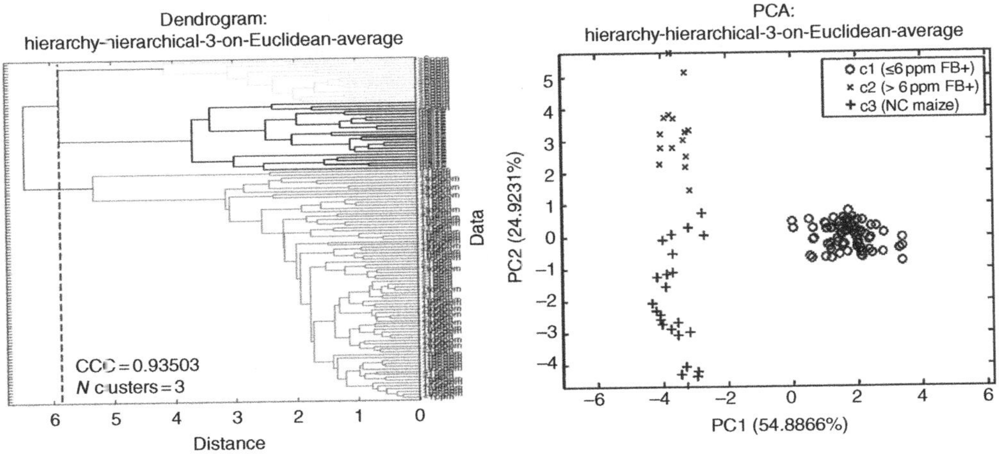

Figure 5.9 Graphical visualization of EN data by hierarchical clustering (Euclidean distance, average link): (a) dendrogram plot and (b) PCA score plot of the same clusters (noncontaminated maize (NC maize), low-level fumonisin-contaminated maize—<1 ppm FB+, high-level fumonisin-contaminated maize—>5 ppm FB+) (Source: From Ref. [41]. Reprinted with permission from Elsevier Science Ltd.)

and may converge to a local minimum of the criterion function if the initial partition is not well chosen. Besides, it works very well only for compact hyper-spherical clusters and it is very sensitive to outliers and noise. Once the clustering results are achieved it may be visualized using, for example, a PCA plot where data labels are represented by the cluster labels, as done for in Figure 5.9.

5.4 Supervised Learning

Supervised learning is the branch of machine learning that focuses on the exploitation of a set of (previously) recorded data samples and a reference knowledge source in a process that is termed *learning by examples*. The initial knowledge about the relationship between data samples and their reference *target values* is actually generalized to cope with real-world samples, that is, with previously "unseen" samples.

Most of the problems that e-noses are meant to solve can be taxonomized as either classification or regression problems. In the first case, the e-nose platform is requested to identify a chemical (or a chemical mixture) usually discriminating it from several others. Practically it has to assign a recorded sample to a class (sometimes referred to as a *label*) among *k* others (with *k*<+inf), by recognizing its olfactive fingerprint. Detection problems, in which the presence of a substance is to be determined, can be considered as a subset of a classification problem and, specifically, a two-class classification problem.

In regression problems, the e-nose is requested to estimate the concentration of a particular analyte either alone or as an element in a mixture, that is, assigning a sample to a peculiar value in a continuous range in R^n with $n \geq 1$. Regression can also be obtained by approximation, by discretizing the range of concentrations, and by assigning a class label to each discrete interval. In this case, we obtain the so-called regression by classification [47].

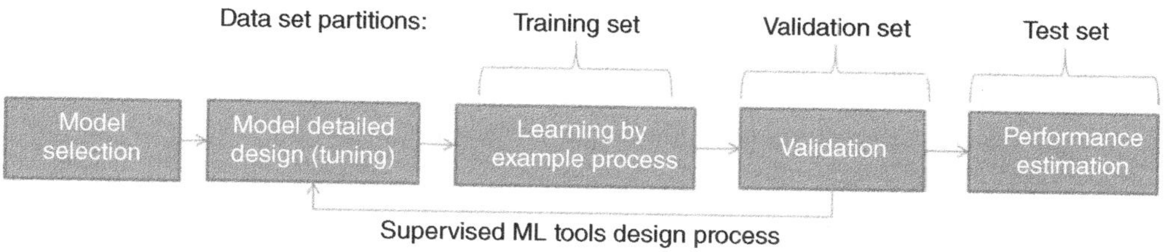

Figure 5.10 Design process of a supervised learning tool (either a classifier or a regressor)

Supervised learning can be pursued by either *generative* or *discriminative* approaches. With y being the target variables and x the observed olfactive pattern, generative approaches aim to learn a model for $p(x,y)$, that is, the joint statistical distribution of input and output variables, while discriminative approaches focus on practical modeling of the $p(y|x)$ distribution, the distribution of output values (labels or concentration estimations), given the realization of the input variable x. The latter approach dominates the AO literature.

The development of a supervised learning model encompasses multiple design steps (see Figure 5.10). At first, model selection concerns the choice of an appropriate modeling tool capable of performing the fingerprint classification or the concentration estimation task. Multilayer perceptrons (MLPs) or support vector machine (SVM) are state-of-the-art examples in the machine olfaction field. The next step concerns the detailed design of the selected tool and requires the tuning of structural parameters, also known as hyperparameters. Examples of hyperparameters are the number of hidden neurons in MLPs and the complexity parameters in SVM. For example, the optimal hyperparameters of an SVM regressor can be selected by a brute force search aiming at obtaining the best performance [81]. The actual learning then takes place by fine-tuning the relevant learning parameters (i.e., network weights for an MLP) relying on the training dataset. Afterward, performance of the trained model can be estimated while a validation dataset may be used for estimating the suitability of the chosen architecture, providing feedback to the hyperparameters tuning or to the learning process itself. At the end of these, often iterative, steps, the identified model has been developed in a fully trained classification or regression tool. In order to correctly assess the generalized performances of the chosen approach, model design, development, and final performance estimation must rely on strictly separate datasets.

In order to capture the accuracy performance of supervised classification techniques, performance estimation is implemented by computing the percentage of correct classified samples for each class. As prior probability can be significantly different from a class to another, the confusion matrix is even more informative, as it shows the percentages of attribution to the different classes in columns for each of the correct classes (rows).

In order to express the performance of regressor systems, a useful measure is mean absolute error (MAE), that is, the average absolute value of the difference between the estimated concentration and the real one. For comparison reasons among different gases' concentration estimation, the MAE value can be divided by the range of concentrations, obtaining an effective relative measure of the error. Mean relative error (MRE) can be simply defined as the average of the relative errors and is usually used for providing a local unbiased estimation of the error in each estimation attempt.

Fundamental limitations of the supervised approach include the so-called overtraining. This term describes the lack of generalization performance of a trained classifier or regressor due to the design of an overcomplex solution. Pursuing the minimization of the sole training error may in fact lead only to perform well on training set samples.

Supervised learning techniques may be designed to deal with several specific issues of machine olfaction field. As previously stated, interferents induce sensor response changes that are indistinguishable from the one induced by the target gas, hampering detection and quantification capabilities. Supervised learning algorithm may offer a solution by exploiting the partial overlapping responses of an array of non-specific sensors to build knowledge on interferents and correct for their presence. Of course this may only be obtained by using a training set that include interferents affected samples. This capability can be extended even to recognize or quantify gases toward which none of the sensors have been targeted during design and fabrication. In this case, the response of a sensor subset can provide information on them by describing partial specificities toward them or, more rarely, constant relationships between measured and unmeasured chemical species described in the training set data—the latter is often referred to as proxy sensing.

5.4.1 Classification: Detection and Discrimination of Analytes and Mixtures of Volatiles

In discriminative approaches, the olfactive fingerprints space is partitioned in domains each of which is representative of a peculiar class. Multiple classification tools have been applied to AO settings. Assuming Gaussian distributions, LDA has been a frequent choice. In a two class problem, LDA searches the optimal vector w and scalar w_0 defining the separating hyperplane:

$$y(x) = w^T x + w_0 \tag{5.2}$$

where x is the sample to analyze.

If $y(x_k) > 0$ the pattern x_k is assigned to a class C_1; otherwise to class C_2. The distance between x_k and the hyperplane $w^T x + w_0$ is called margin.

LDA has been used for wine characterization [8], tomato shelf-life evaluation [42], analyte discrimination [103], water pollution recognition [43], and medical diagnostic [64].

LDA, as a classification tool, has some relevant limitations: its generalization performance can be affected by the presence of outliers, non-Gaussian class sample distributions, and significant difference among covariance matrices. It is also known to be severely affected by overtraining issues in case of high-dimensional feature spaces coupled with limited dataset numerosity. The biggest limitation, quite naturally, is the imposition of a linear separation surface.

A nonlinear data learning technique for supervised classification, which has also been extensively applied in AO, is the MLP neural network [84]. Combined with informative features it has been shown to be one of the most versatile tools for analyzing e-nose data. The MLP is composed of multiple interconnected layers (typically three: input, hidden, and output layers) of sigmoidal functional units (neurons) with weighted interconnections, with its response depending on the value of both the inputs (sensor responses) and the connection weights. It requires the design of the network architecture and in particular the choice of the number of layers and hidden neurons, which are generally designed on the base of the number of inputs and outputs. Expanded binary coding of the classification output is usually adopted, with each of the output neurons being associated with a specific class. The network is hence trained to respond to a vector, with only a single saturation output in the output neuron being associated with the training sample real class and null output for the other ones. When dealing with unseen samples, the output neurons' level rarely reaches 1.0, ranging from 0 to the max level for each output neuron. The neuron scoring the maximum output level is designated as the winning output neuron and its associated classes will be selected as estimated output label. Several iterative learning algorithms have been proposed to train MLPs, ranging from gradient descent to automatic Bayesian regularization [15, 16, 35]. Generally, training samples are presented for network classification in successive epochs. The computation of the empirical error induces a modification in the network weights aimed to reduce the error itself. Similar to most supervised learning approaches, neural networks are prone to overtraining because of the minimization of the sole empirical error. Several strategies have been devised to limit its effect. Early stopping of the learning phase can be used to limit the optimization level of the empirical error while keeping sufficient generalization properties. This implies setting apart a significant number of highly valued e-nose samples as an intermediate validation set, for deciding when to stop learning on the base of the validation set error value. Automatic Bayesian regularization instead balances the contribution of empirical error and model complexity (number and value of neuron connection weights), relieving the designer from the necessity of carefully designing the number of hidden layers at the cost of introducing less sensitive parameters.

Radial basis function (RBF) networks and probabilistic neural networks are specialized implementation of the neural network paradigm relying on training instances to build a set of representative radial functions in the feature space:

$$\varphi_{ij}\left(x,\alpha_{ij}\right)=\frac{\exp\left(-x-\alpha_{ij}\right)^{2}}{2\sigma_{ij}^{2}} \tag{5.3}$$

where centroids α_{ij} are representative prototypes for each class. Functions act as a middle layer in a three-layer network architecture with the output layer computing a weighted sum of functional invocations over the presented pattern x:

$$y=\sum_{i=1}^{m}\sum_{j=1}^{k}w_{ij}\varphi_{ij}\left(x-\alpha_{ij}\right)^{2}+w_{\text{bias}} \tag{5.4}$$

RBF network have principally been selected in AO settings for their enhanced generalization capabilities when a limited number of training sets are available [6]. Another interesting capability, shared with most instance-based classifiers, is the possibility of applying an incremental training procedure that could be exploited for continuous learning when the knowledge base is not entirely available at first or to adapt the embedded knowledge to counteract drift effects on classification performances during the network operative life [63, 109].

The ability to deal with uncertainty, that is, to qualitatively describe the complexity of odors has led to consider fuzzy logic [4]. Singh *et al.* has exploited multi-value logic in AO by using fuzzy neural networks attempting to reduce dependence to random initial weights choice in

MLP-based olfactive patterns classification architectures [99]. Similarly Ping *et al.* have tried to improve RBF network performances by the use of Fuzzy c-means [92].

Developed by Vapnik in the early 1990s [112], SVMs have been introduced in AO community by Distante *et al.* [30]. A review on their application to e-nose data analysis can be found in Ref. [81].

Based on structural risk minimization theory, SVMs are designed to reduce overtraining by a regularization approach balancing empirical error minimization with maximization of the class margin (see Figure 5.11). In fact, the SVM classification algorithm tries to explicitly find an hyperplane $w^T x + b = 0$ that minimizes

$$\frac{1}{2}w^2 + C\sum_{i=1}^{n}\xi_i \tag{5.5}$$

with n being the number of training set samples, ξ_i being a slack variable accounting for the empirical classification error, and C a constant trading off the cost of empirical errors with the maximization of the margin (that is equivalent to the minimization of the norm of w).

Strongly nonlinearly separable problems are tackled by projecting data into higher-dimensional spaces by generally nonlinear functions. Distance among samples, essential to the optimization procedure, can be however directly computed in the original space by means of the so-called kernel functions avoiding the actual projection of the original data. Most common kernels used by practitioners are the polynomial kernel and the Gaussian (RBF) kernel:

$$K(x,y) = e^{-\frac{x-y_2^2}{2\sigma^2}}$$

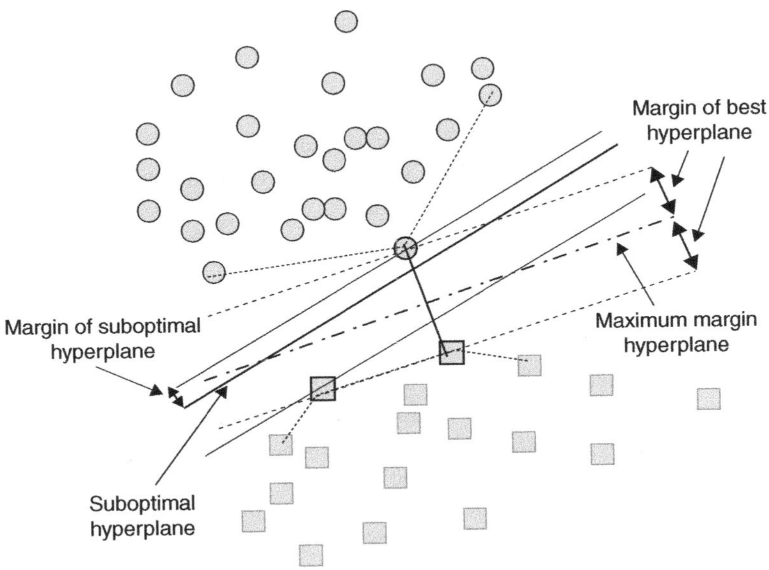

Figure 5.11 Representation of margin regularization problem in SVMs (Source: From Ref. [85]. Reprinted with permission from Elsevier Science Ltd.)

whose orders of complexity are controlled by the order of the polynomial function and the σ parameter of the Gaussian kernel function, respectively.

It can also be shown that the SVM framework contains, by an appropriate choice of the kernel functions, several other classification schemes as special cases, for example, MLPs and RBF networks. It has been frequently shown that often SVMs outperform those approaches [1, 31].

C and the variables controlling the kernel complexity are the most significant free parameters to be optimized (Pardo *et al.*[86], see Figure 5.12): C controls the relative cost of training set empirical errors in the search for the maximum margin classifier, while the complexity parameter effectively represents the complexity of the structure of the separating curve in the original sample representation space. Although this is not always the case, knowledge representation in a trained SVM can be very compact, especially with low kernel complexity levels. In fact, the computation of the separating hyperplane in the transformed space depends only on the support vectors, determining a low memory footprint, and a limited number of invocations of the kernel function.

Research in SVM tackles the development and testing of new ad-hoc kernels for dealing with complex, nonstandard data representations. A rather interesting work in AO settings has been performed by Vembu *et al.* [113]. They show the application of kernels specific for time series representations, capable of analyzing the time dependence of the chemical sensor response. Practically, the kernel allows performance of a relevance-based dynamic feature extraction within the classification step, outperforming a classical filter approach in which features are selected before designing the classification model (see Section 4.3).

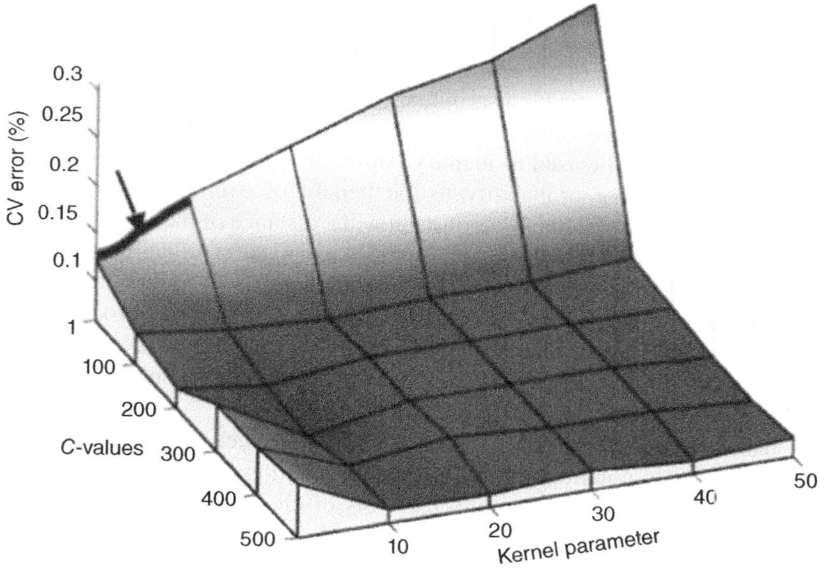

Figure 5.12 Relationships among the values of C, kernel complexity parameters σ, and cross-validated estimated generalization error. In this case small values of kernel parameter produce more complex separation surfaces; that is the case, for example, in RBF kernels. Actually limited overfitting effects are visible for low kernel parameter values at $C>300$ (Source: From Ref. [85]. Reprinted with permission from Elsevier Science Ltd.)

Instance-based classifiers base their classification algorithms on a set of class representative instances called prototypes. Their success is partly due to the human-friendly representation of their knowledge base: prototypes can be used to efficiently represent the olfactive fingerprint of an olfactive class. *k*NN classifiers, for example, rely on distance measures, assigning an unknown sample to the class to which the majority of its *k* nearest training samples belong to. Artificial immune systems (AIS) are modern representatives of this class of tools based on analogies with the mammalian immune system that has been shown to have significant memory and adaptivity properties [14]. The working principle of their artificial counterparts is based on evolving a compact set of representative instances (antibodies) via evolutionary computing strategies (mutation and selection). Recently, AIS have been tested in AO in different settings [18, 75].

Classifier ensembles are collections of simple classifiers that cooperatively perform a complex classification task by means of combination strategies (such as majority voting). A notable and modern example for classifiers ensembles is Random forests (RF), that is, ensembles of decision or classification trees. In a RF, each decision tree is trained over a partition of the original training dataset. Each node of the tree is generated by choosing the best feature predictor among a subset of the available features that have been randomly extracted. Because of the significant amount of randomness involved in their training, RF both attain a high performance and are resilient to overtraining. RF have been introduced to AO by Pardo and de Vito [81]. When applied to food industry classification problems (contamination and defect detection), they have been shown to be equivalent to mainstream classifiers such as SVMs. Their results have been recently confirmed in an artificial tongue setting [133].

The combination of two class classifiers with different strategies is also adopted to solve multiclass problems. For example, SVMs can be designed to discriminate one class against all the others and then combined (e.g., with majority voting) to determine the final cooperative estimation [7, 31]. In classification problems, a measure of posterior probability of classification, that is, the probability with which a sample belongs to a class, is sometimes desirable. The measure can be used to identify "difficult" samples, discarding those that are classified with low posterior probability to the benefit of overall performance measure. Brudzewski *et al.* have shown in the AO setting how the distance of the sample under analysis from a fingerprint of the winning class can be used to estimate the reliability of classification of SVMs. Alternatively, Acevedo proposed to compute a posterior probability exploiting the voting procedure across $k(k-1)/2$ SVMs employed with a 1 versus 1 approach to multiclass problems. Relevance vector machines can more generally solve this issue while providing a reduction in computational needs at the cost of the loss of the unique solution property of the SVM-based approach (see Ref. [121]).

From an historical perspective, we have seen a rather steep evolution from the linear classification schemes to nonlinear ones. Despite the additional efforts needed to cope with multiclass scenarios, SVMs are becoming the tools of choice thanks to the availability of implementations and performance advantage compared with classic tools.

5.4.2 Regression: Machine Olfaction Quantification Problems and Solutions

In AO, multivariate regression problems require the estimation of actual concentrations of odorant or, more generally, chemical analytes even in complex mixtures, by relying on the combined response of multiple chemical sensors. Typical applications focus on pollutant concentration estimation in indoor or outdoor samples. More generally, regression can be aimed to determine the value of quality estimators such fruit ripeness or sensorial indicators [69]. Recently, regression has been employed for medical diagnostics applications, for example to quantify Blood Glucose levels starting from exhaled breath analysis [97].

During the last decades, several statistical regression tools have been used to perform quantitative estimations in AO. In one of the first attempts, Sundgren *et al.* have proposed PLS technique for concentration estimation of single analytes when in a mixture [102]. PLS finds the multidimensional direction in the input X space that explains the maximum multidimensional variance direction in the output Y space. Differently from PCA, PLS is also suited in case of multicollinearity among observations, overcoming the limitation of standard regression approaches. PLS continues to be proposed as a regressor model in AO and remains a tool of choice, for its simplicity, in food industry applications [69].

The generally nonlinear, unstable, and nonspecific response of chemical sensors may require the use of multivariate nonlinear functional estimators. MLP, with its universal function estimator property has offered a valid theoretical and practical framework to AO practitioners. A regression-oriented MLP differs from a pattern recognition MLP basically for being trained to estimate continuous varying target variables. Usually its architecture features sigmoidal transfer functions in the hidden layer while relying on linear transfer functions in the single output layer neuron. A single MLP may actually be designed and trained to simultaneously estimate the concentration of several analytes in a mixture. The network design used by Szcuzureka *et al.* is an example of such an architecture using steady state response of an array of TGS sensors as input to perform simultaneous estimation of butanol and xylene within a single tool (Szcuzurek *et al.*[104], see Figure 5.13).

Alternatively, one can aim to quantify each analyte with a dedicated regressor [48].

The influence of environmental variability as well as interferent gases induced the researchers to include, within the relevant observable sets, the response of sensors specifically targeted to interferents like RH (see Ref. [59]). Generalizing this approach, De Vito *et al.* [17] have exploited the universal functional estimation and noise insensitivity properties of MLPs for quantitative multivariate calibration using on-field recordings for distributed outdoor pollution monitoring. As highlighted by these authors, the number of training samples and hence the cost of the calibration procedure are linked by a nonlinear relationship (see

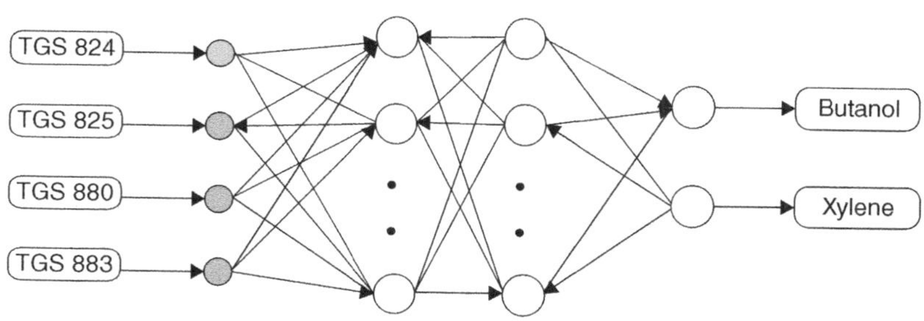

Figure 5.13 Regression NN architecture for simultaneous estimation of butanol and xylene based on MOX chemical sensor response (Source: From Ref. [104]. Reprinted with permission from Elsevier Science Ltd.)

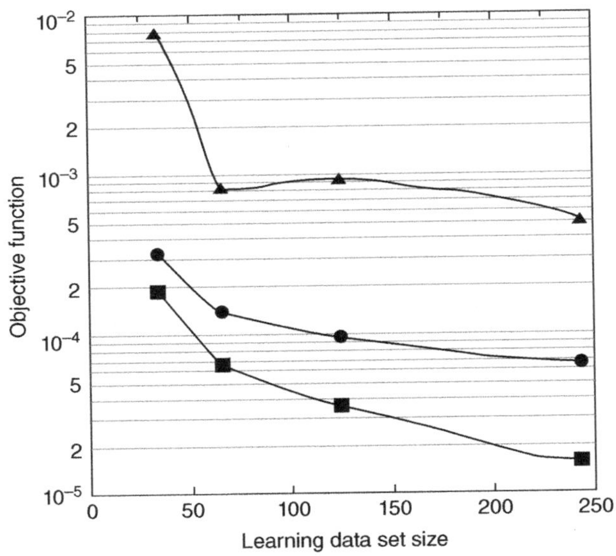

Figure 5.14 Relationship between performance of multiple NN architectures and the samples' numerosity in its training set for a generic problem. From the figure, it is apparent how the dataset numerosity has a significant—though nonlinear and usually not predictable—influence in increasing the generalization performances of the neural networks by building a more complete knowledge about the distribution of samples and their relationships with desired labels in the generic problem under analysis (Source: From Ref. [59]. Reprinted with permission from Elsevier Science Ltd.)

Figure 5.14). Initially, performance improve significantly while enlarging the database until a plateau is reached where sample addition don't hold for any improvement. A careful analysis should be conducted in order to locate the optimal trade-off set point.

Other neural architectures, such as regression-targeted RBF neural networks, have shown their suitability for quantitative olfactive analysis in highly sensitive medical applications [97].

In many applications, the slow response of the chemical sensors is reflected in the dynamic performance of the regressor unless the dynamic behavior of the sensors is captured by the feature extraction method or by a suitable regression architecture. Pardo *et al.* [82] have proposed time delay neural networks to improve the precision of the network estimation during steady state phase.

Similarly, De Vito *et al.* have exploited the dynamic response by coupling a tapped delay line (see Section 3.1) to a regression SVM (rSVM). They obtained a fast concentration estimation, overcoming the slow and somewhat complex sensor array response to abrupt changes of analyte concentrations and environmental conditions [16]. In that paper, the sensitivity of rSVM to the *C* and kernel complexity parameter is also highlighted (see Figure 5.15). Although employed in a limited number of studies, rSVM can rival and sometimes outperform the MLP in regression tasks.

Fuzzy logic approaches can be pursued for regression application in AO usually by encompassing the typical fuzzification, inference, and defuzzification steps. Practically,

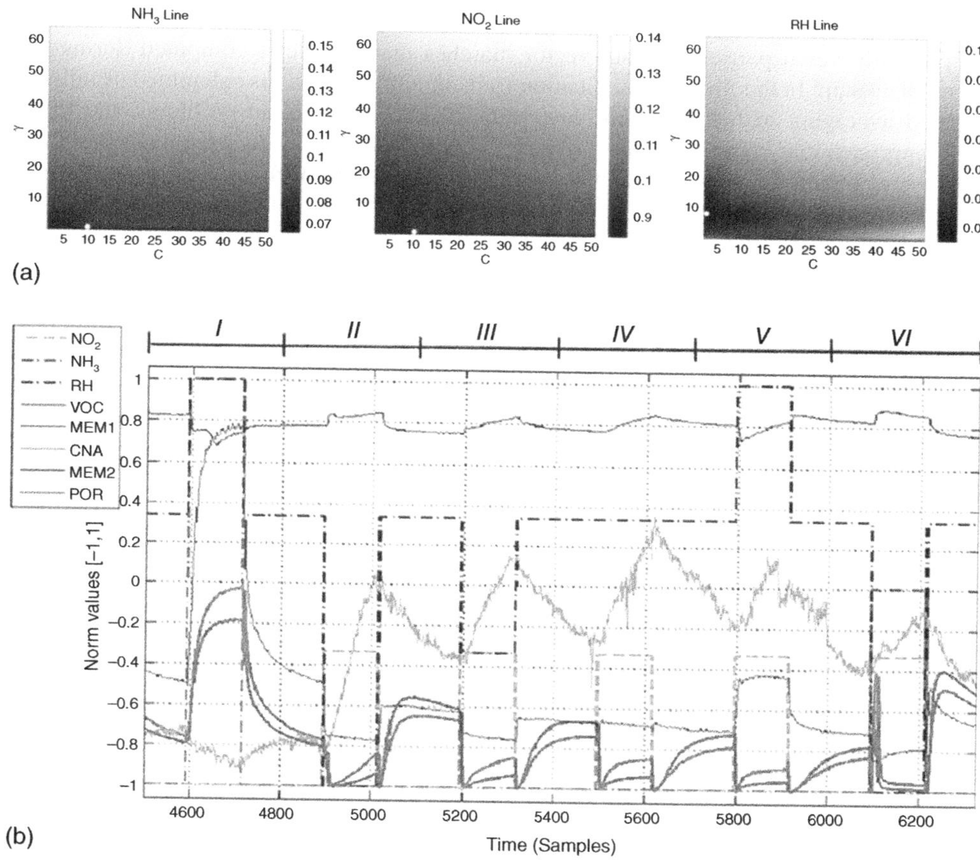

Figure 5.15 (a) Influence of C and RBF kernel parameter ($\gamma = 1/2\sigma^2$) in determining the performance level of trained rSVM estimations. Gray scale indicates mean relative regression error in estimating $NH_3$and NO_2 concentrations and RH in a gas ternary mixture (Source: From Ref. [16]. Reprinted with permission from Elsevier Science Ltd.). (b) Fast response of a tapped delay rSVM in a concentration estimation problem. Lines 1–3 depict normalized network output (solid) versus actual concentration (dashed), while line 4 provides a subset of adopted sensor normalized response readings. It is possible to spot the slow original sensor response and compare it with the near real-time responsiveness of the network output) (Source: From Ref. [16]. Reprinted with permission from Elsevier Science Ltd.)

a set of if-then rules can be set for assigning a fuzzy set membership value to regions of combined response of the sensor array by means of a fuzzy membership function. The fuzzy sets effectively cover the range of concentrations under analysis. RBF is an interesting example of these class of functions:

$$F_i^{(r)}(x_i) = \exp - \left(\frac{x_i - c_i^{(r)}}{\sigma_i^{(r)}} \right)^2 \quad (5.6)$$

where $c_i^{(r)}$ and $\sigma_i^{(r)}$ are the center and the width of the ith fuzzy set from the rth rule. The strength with which particular input vector matches the rth rule is computed during the intermediate step. In the final defuzzification step, an output value y is calculated usually by weighted averaging of the activated fuzzy sets. As in classification problems, the fuzzy neural approach can effectively make use of fuzzy logic structure to strengthen the noise resilience of neural network architectures and generally the performance of the regression system [126].

In a recent paper, Badura *et al.* provide a performance comparison of some of the most used regression approaches in AO. Although limited to the use of single dataset, the study highlights the performance advantage of rSVM and a specific fuzzy neural approach over conventional MLP neural networks [3]. The study also highlights a significant performance drop when rSVM work outside the response range defined by training samples, an issue shared with MLPs (see Ref. [17]). In contrast the issue seems to have a more limited influence on the analyzed fuzzy neural approach. This study confirms a trend that can be observed in literature, with the once gold standard MLP-based solutions being replaced by rSVM-based solutions.

5.4.3 Feature Selection

Large feature sets may reduce generalization capabilities and interpretability. Moreover some features may be irrelevant, uninformative, or even detrimental. The primary objective of feature selection (FS) is to find an "optimal" subset of features that maximizes information content or predictive accuracy of sensor readings. More generally, in AO, feature selection schemes may be used to select sensors, sensor material, or sensor tuning characteristics (e.g., MOX heating temperature level or profile), all regarded as descriptive features. The simplest FS approach consists of evaluating the information content of each feature individually, selecting those features which rank firsts with respect to a selected score function. Unfortunately, this simple approach fails to consider relationships among features (e.g., correlation) and will rarely find an optimal subset, so that feature combinations should be considered.

Any procedure for feature selection must rely on two basic components:

1. A scoring criterion must be defined to rank subsets of features.
2. A systematic procedure for searching through the set of candidate subsets of features.

The search procedure could simply consist of an exhaustive search of all possible subsets of features since this is in general the only approach which is guaranteed to find the optimal subset. This brute force approach of course bears a significant computational footprint, as the number of possible combinations among the basic features set to be evaluated grows exponentially with its dimension d (2^d-1). In principle the selection criterion could make use of the performance assessment score of the complete pattern recognition (e.g., misclassification rate for a classification problem or MAE for a regression problem) giving raise to what are called wrapper approaches. Often, in practical applications, the researcher may be forced to consider simplified selection criteria (termed as filter approach) as well as non-exhaustive search procedures in order to limit the computational complexity of the search process. Targeting a defined dimension k of the optimal dataset may help to reduce the search set dimension to $\binom{d}{k}$. To avoid the exponential explosion of an exhaustive search, several methods have been devised that explore the feature space in a more efficient fashion. These search strategies can be grouped into three categories: exponential, sequential, and randomized.

Exponential techniques perform a search in which the complexity grows exponentially with the number of states. Belonging to this class, *branch and bound* (BB) is particularly effective when the selection criterion J satisfies a monotonicity relation, that is, $J(X^{+}) > J(X)$ where X denotes a set of features and X^{+} is a larger set that contains X as a subset. Let us consider an original set of d features and suppose we would find the best set of $m < d$ features. The starting point of BB procedure is the construction of a search tree structure. At the top of the structure, the measure $J(D)$ of the selected criterion is evaluated, and then the features are discarded once at time with the constraint to maximize the measure $J(M)$. Each possible subset of m features over d is represented by one node at the bottom of the tree. To avoid evaluating all the possible subsets, the tree threshold is fixed—for example, the value of criterion at a certain node of the tree. If at any point in the search structure an intermediate node—corresponding to $m < x < d$ features—is encountered for which the value $J(X)$ is smaller than the threshold, there is no need to evaluate any of the sets which lie below this node on the tree, since, as a consequence of the monotonicity relation, such nodes necessarily have values of the criterion smaller than $J(X)$ and then smaller than the threshold. The algorithm terminates when every final-layer node either has been evaluated or excluded using the monotonicity relation. Unfortunately the monotonicity condition on the criterion J is not always verified.

Sequential search algorithms are greedy strategies that reduce the number of states to be visited during the search by applying local search. The basic methods are sequential *forward selection* and sequential *backward elimination*. The former approach starts by considering each of the variables individually and selecting the one which gives the largest value for the devised selection criterion. Successive stage of the algorithm are characterized by the addition of a feature to the candidate subset set, again chosen on the basis of which of the possible candidates at that stage gives rise to the largest increase in the value of the selection criterion of the candidate subset. However, should two feature variables provides little discrimination considered singularly, but being very effective when considered together, then the forward selection procedure may never consider this combination since either feature alone would never be selected. Sequential *backward elimination* starts with the full set of d features and then eliminate them one at a time. At each stage of the algorithm, one feature is deleted from the set, chosen among all available candidates as the one which gives the smallest reduction in the value of the selection criterion.

Finally, randomized search algorithms attempt to overcome the computational cost of exponential methods and the tendency of sequential methods to become trapped in local minima. Among these techniques, simulated annealing and genetic algorithms are most widely used.

AO practitioners have employed several different criteria and selection schemes for feature selection, so exploring both filter and wrapper approaches, with the latter being more common. Examples of scoring criteria are Mahalanobis distance [77] and univariate [68] and multivariate Fisher discriminant power [46, 95].

Brute force search has been considered by Pardo *et al.*, who obtained a remarkable shrinking of the initial 19-dimension feature space to a bidimensional space while also simultaneously obtaining better results [86]. In that paper, authors used *k*NN as a performance estimator in a wrapper approach. A similar approach had been previously followed by Ref. [9], who selected the best sensor subset within a 20-polymer sensor array using a high-performance computing parallel architecture.

Forward selection has been employed by Ref. [87] to reduce a 140-dimension feature space with an artificial neural network (ANN) as a performance estimator in a wrapper approach. The space was generated by the use of five features in 14 sensors and two different normalizations.

Genetic algorithms have been also applied to feature selection, trying to compare and integrate readings by multiple e-noses for fruit solution discrimination [6]. Comparisons among results obtained by algorithmic and stochastic selection methods were reported in Ref. [86] and in Ref. [40]; they showed the advantages of evolutionary computing, with the genetic algorithm providing a computationally effective method (with regard to brute force) to stochastically reduce the number of subset evaluations retaining global minimization capabilities. A similar comparison on a different dataset was performed by Gualdron *et al.* in 2006 who in turn showed that while a genetic algorithm provides the best performance, forward (and backward) selection provides a good trade-off between computational load and performance.

FS methods have recently shown their relevance for the identification of optimal operative chemical sensor parameters—in particular temperature levels and profiles for MOX sensors. In particular, Vergara and coauthors [115, 117] have investigated these methods by using Kullback–Leibler divergence and Fisher-like multivariate discrimination as scoring criteria.

5.5 Advanced Topics

5.5.1 System Instability Compensation

Sensor drift consists of small and nondeterministic temporal variations of the sensor response when it is exposed to the same analytes under identical conditions [56]. Today it is still impossible to fabricate chemical sensors that are immune to drift. As we strive to put e-nose to work outside laboratories, we face the problem that input and target variables distribution change over the time; more awkwardly, their relationship is also changing.

One of the main consequence is that the EN's selectivity and sensitivity decrease. Actually, the gas sensor drift changes the way samples distribute in the data space, thus limiting the ability to operate over long periods. In facts, pattern recognition models become useless after a period of time, in some cases being weeks or a few months. We are hence forced to adapt our knowledge, either supervised or not, to the changes occurring during time. Sensor drift must be therefore compensated or corrected to achieve reliable measurements from a sensor array.

Algorithms to mitigate the negative effect of gas sensor drift are known since the early 1990s. Nevertheless, the study of sensor drift is still a challenging task for the chemical sensor community. Solutions proposed in the literature can be grouped into five main categories (see Figure 5.16): (i) sensor signal processing (baseline manipulation, filtering), (ii) periodic sensor calibration, (iii) attuning methods, (iv) adaptive correction models, and (v) semisupervised learning approach.

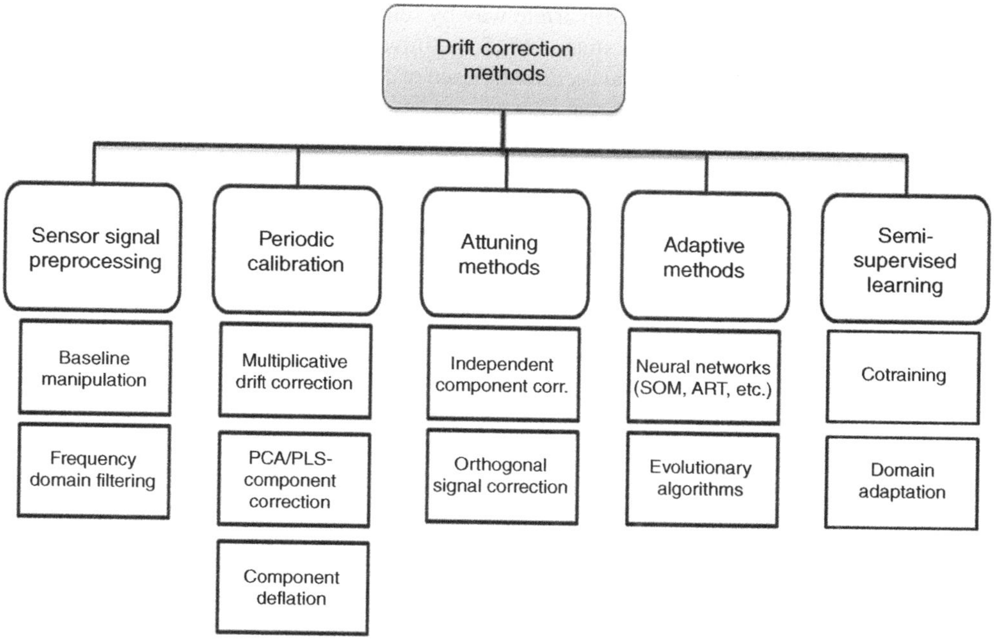

Figure 5.16 Taxonomy of different drift correction methods emerging from the electronic nose literature for the past 20 years (Source: Modified from Ref. [24])

One of the simplest methods for drift compensation that has been proposed in the literature, which is also widely used as a preprocessing method, is the normalization of individual sensor signals with the value of the sensor response before exposition to the target gas (the so-called baseline) [39]. This is mostly used for the rejection of common mode perturbations, such as additive or multiplicative effects, and of pattern scattering due to changes in concentration, instabilities in sampling, and changes in environmental parameters. The most commonly approaches are:

1. *Differential*: subtracts the baseline of each sensor. It helps compensating additive drift effects which are present both in the baseline and in the gas response.
2. *Relative*: divide by the baseline. It corrects for (constant and uniform) multiplicative drift effects.
3. *Fractional*: a combination of the previous two that works for multiplicative drift and has the advantage of providing dimensionless measurements and normalized sensor responses.

As observed by Llobet [67], drift typically occurs in a different frequency domain with respect to the interesting signals, being in general a slower process. Therefore, transformations of sensor signals from time to frequency domain and removal of the lowest-frequency components can filter drift out; for example, discrete wavelet transform has been proposed by several authors [58, 67, 129].

A single calibrant, or a set of calibrants, is perhaps the most robust method to mitigate drift effects over a long period of time. This strategy can be applied in a univariate way (sensor-by-sensor) [100] or in a multivariate way by removing the directions of dispersion of the reference data in the feature space [37, 53]. Univariate calibration indeed combines simplicity with good performance and is currently used in commercial e-noses (e.g., EOS507C, Sacmi Imola SC, Imola, Italy).

One of the first attempts of performing robust drift correction by multivariate methods, called component correction (CC), was proposed by Artursson *et al.* [2]. CC uses PCA in conjunction with the reference gas technique. If the sensor responses to a certain reference gas contain a significant amount of drift, the first component identified by the PCA analysis on these measurements, which is the one that describes the maximum variability, will likely define the direction of the drift. Drift corrected data can be then obtained by subtracting from the original data matrix the score values projection along the first PC. Removing the first component is usually enough whenever we are facing drift effects caused by sensors' aging. However, if non-linear drift effects are present (affecting also higher-order PCs) more than one component can be subtracted. A similar correction strategy can be also obtained with a PLS regression model instead of PCA. The main problem of both (unsupervised) approaches is that also part of relevant information contained in these components can be subtracted, thus reducing the EN classification performance.

A different approach to perform multivariate drift correction was proposed by Gutierrez-Osuna [50]. The overall idea is to find a set of variables whose variance can be attributed to drift or interferents. One way is to search these variables in the sensor response during sensor washing or reference gas measurement cycles. Since washing/reference gases are virtually constant over time, variations in their transient response can be used to estimate the amount of sensor drift present in each experiment. Canonical correlation analysis (CCA) and PLS are then performed to find out a subset of "latent variables" that summarizes the linear dependencies between odor and washing/reference responses. After that Ordinary Least Squares Regression (OLSR) is used to subtract these "latent variables" from the sensor response to odor samples.

Attuning methods try to perform CC without resorting to the use of calibration samples and try to deduce drift components directly from the training data.

Di Natale *et al.* [26] proposed the exploit ICA. As seen before, in PCA or PLS correction methods the computed components are mutually uncorrelated but usually they contain information on both signal and disturbances; thus subtracting one component means losing information. ICA is applied to EN data in a supervised way, to preserve only those components correlated with the sample features relevant to the application. The ICs that mostly correlate with the objective of the measurement are chosen, while those more correlated with the disturbances are discarded. This solution provides good results especially in removing drift effects due to environmental variables changes (e.g., temperature, pressure, and humidity) that can be measured during experiments.

Given the sensor array data matrix X, representing a set of independent variables, and the class label vector C, representing a set of dependent variables, orthogonal signal correction (OSC) removes the variance of X which is not directly correlated to the variables in C. The approach is to find the data subspace best correlated with the C vector and remove from X its projection on the direction orthogonal to this subspace. The condition of orthogonality assures that the signal correction process removes as little information as possible from the original data. Although very effective, one of the main issues (assumptions) of OSC is the existence of a training set containing a significant amount of drift, allowing the correct identification of the orthogonal (drift) direction to be rejected. Padilla *et al.* have applied OSC to a dataset of 17 conductive polymer sensors operating for 10 months, with a total of 3415 measurements for three analytes at different concentration levels [80]. OSC boosts classification performance of about 15% (on average).

Adaptive methods for drift correction try to follow the odor samples' trajectory evolution in feature space and hence continuously "adapt" the pattern recognition model to new data.

The first adaptive approaches have considered unsupervised ANN, such as Kohonen self-organizing maps (SOMs). Some authors [25, 72, 128] used a single SOM common to all classes, while others [29] proposed to use a separate SOM for each odor. In the same framework, neural networks applying the adaptive resonance theory (ART) allow for new classes to be created [119].

The basic idea to allow for ANN drift compensation is to maintain a certain learning rate during the normal use of the network in order to learn changes of the input patterns due to drift effects. The learning rate must be kept to a low level in order to avoid overfitting of the model in a plasticity–stability trade-off.

Selecting the appropriate learning rate to keep during normal operation is complex and may strongly impact the correction capability. To the best of our knowledge, no automatic method has been proposed so far to efficiently tune this parameter. Moreover, adaptive models typically require a high number of training samples.

Recently, a new adaptive drift correction method based on evolutionary algorithms has been presented [23]. The overall idea is to exploit the learning capabilities of evolutionary algorithms to compute a simple linear correction model correcting (shifting) the incoming test samples toward the most probable classification region. This model works very well under the (generally true) hypothesis that short-term drift is linear in time. The coefficients of the correction matrix adapt automatically using an optimization method called Covariance Matrix Adaptation Evolution Strategy (CMA-ES) [52].

This method introduces a number of important improvements: (i) it can work in cooperation with any pattern recognition algorithm (either for classification or for regression); (ii) it is robust to discontinuity in the data; and (iii) it gives better performance in the long term.

Adaptive drift corrections still present limitations that prevent their widespread application: they require equiprobable and frequent sampling of all classes to avoid a single class from drifting too much, making it unrecognizable, and they rely on the correct identification of the pattern recognition model to track how different classes change. Local errors in the classification may easily reduce the adaptation capability. Still, adaptive methods represent an important step forward in tackling the problem of sensor drift in AO. Until now they have not been sufficiently investigated and compared with other methods in order to definitively assess their superior capabilities.

Very recent newcomers are the earlier introduced AIS-based methods. Martinelli *et al.* have exploited their intrinsic adaptivity properties to address and limit drift effects on classification performances with interesting results for real-world applications. Their results indicate that AISs, if properly designed, seem to overcome the need for a balanced class presentation order.

Drift is in many ways very similar to incomplete or nonrepresentative description of the data distribution. Furthermore, procuring the supervised samples capable of correctly describing the emergence of drift has a high cost, requiring skilled personnel for a significant time interval. Cost is therefore another reason for incomplete description.

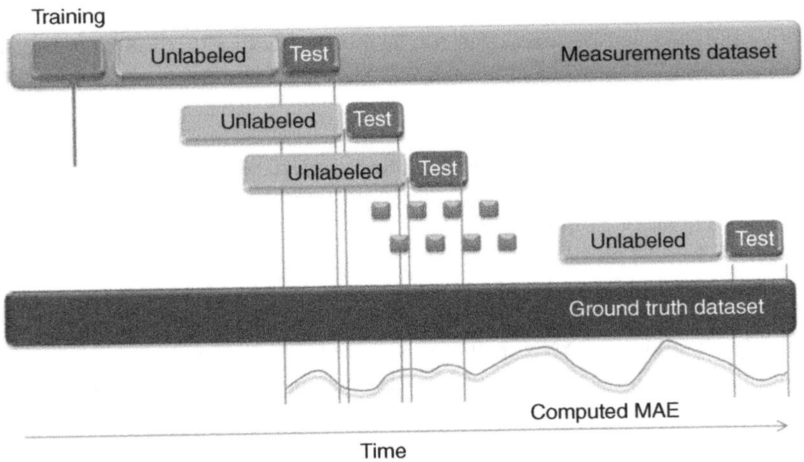

Figure 5.17 Sliding window partitioning method for SSL-based drift counteraction. A fixed length and position training set is used together with a sliding window set of unlabeled (and so unsupervised) samples to estimate gas concentration in a sliding fixed length test set. Estimations are then compared with the ground truth to provide MAE performance. In the proposed test, training set size l was 24 samples, while unlabeled reservoir size was set as 100 samples. Test set size was 24 (Source: From Ref [21])

On the other hand, unsupervised samples are relatively easy to obtain. We could then, in principle design data analysis algorithms such that they enlarge their knowledge after a brief supervised training phase, exploiting the unlabeled samples they will encounter during operative life. This is what semisupervised learning (SSL) tries to accomplish.

SSL has just begun being applied to AO; a recent paper considers SSL for drift counteraction in long term air quality monitoring application [21]. Their challenging scenario is affected by both sensor drift and a multiscale recurring concept drift on a daily, weekly, and season basis induced by human behavior (working hours and days induce increased car mobility; weather phenomena (temperature, atmospheric pressure, and RH variations) or both (use of gas- or fuel-based house heating in winter time). As such an initial calibration computed with 360 on-field recorded sample become unreliable after a few months. Recalibration may be helpful but rapidly become unfeasible with the growing number of deployed e-nose devices.

The author's approach instead relies on the use of cooperating semisupervised learning regressors. Practically, after a brief supervised learning phase, each regressor knowledge base is continuously updated by extending its training set with incoming unknown samples whose labels are predicted by the other cooperating regressor (see Figure 5.17). In this framework, regressor diversity is a key enabler for new knowledge learning. Furthermore, meaningful unlabeled sample selection is crucial in order to avoid the embedding of wrongly labeled samples in the knowledge base.

The process is iterated over a one-year-long period in a sliding window fashion providing the regressors with a moving reservoir of unlabeled samples within which the

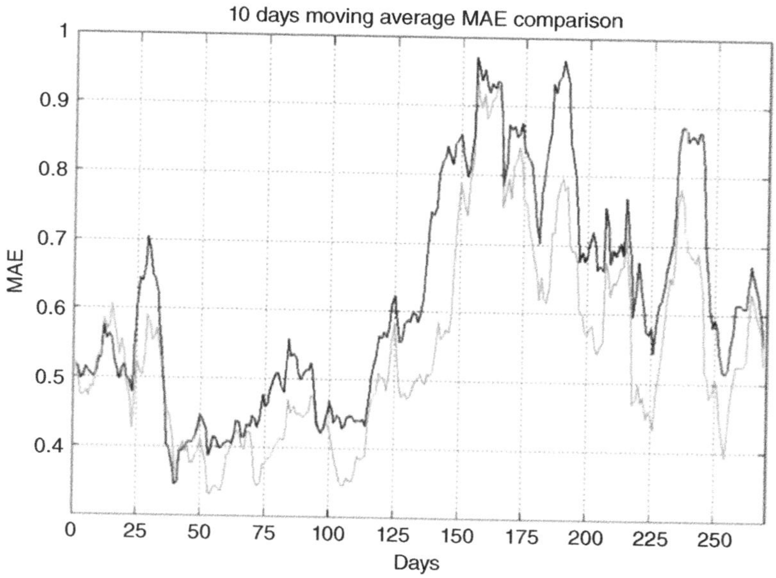

Figure 5.18 Drift counteraction experiment. CO estimation comparison with integrated SSL algorithm (lighter color) and standard MLP algorithm (black). The SSL approach achieved a 11.5% performance gain with respect to the 1-year-long averaged MAE score (Source: From Ref. [21])

meaningful samples for continuous learning are chosen. Performance was estimated over the contiguous test sets in a pure inductive way.

Figure 5.18 depicts the results of applying semisupervised cotraining in this setting, comparing the 10-day moving average MAE between a base MLP and the SSL-based strategy for the CO concentration estimation using 1 day of hourly training samples. Authors shown that cotraining algorithm obtained an overall performance gain of 11.5% on the MAE when applied along the entire one-year-long dataset. Hence, the moving window SSL-based strategy has allowed the regressor to adapt for concept and sensor drift effects by using unlabeled samples to modify hypothesis learnt over a very limited initial supervised time segment, strongly reducing the number of samples needed for correction-based approaches. This adaptation eventually reduced the drift effects on CO concentration estimation performances.

SSL has been proposed also in classification problems: the work from Kim *et al.*, for example, explicitly tries to adapt initial knowledge, embedded in a RBF network classifier, using unlabeled samples [63]. However the work heavily rely on the strong assumption of nonexisting concept drift (i.e., $p(y)$ is assumed constant) that is rarely verified in real-world applications.

Very recently Qihe Liu *et al.* have shown the use of semisupervised manifold regularization for drift adaptation (Qihe Liu *et al.*, 2014). Unlabeled data have been used to incrementally upgrade the kernel-based classifier knowledge obtaining the improvement of the needed recalibration rate. Authors also show that the semisupervised approach outperformed conventional baseline correction method.

5.5.2 *Calibration Transfer*

Most of the problems for practical uses of EN in real-world environment arise because the conditions in which the instrument is used differ from the conditions in which the multivariate calibration model was built. Apart from sensor drift which was discussed previously, other problems may come from: (a) change in odor sampling conditions (different baseline conditions between laboratory and in-field context); (b) sample background change (measured samples are somewhat different from the ones used during the training); (c) the calibration model needs to be transferred to another instrument or project; (d) sensors' malfunctioning and replacement (with nominally identical ones). In general it is difficult, or even impossible, to overcome problems (a) and (b); the issue may only be solved by retraining the instrument in the new baseline conditions and with the new samples. Conversely, in the literature some solutions have been proposed for tackling the last two issues.

Actually, with e-noses being around for over 30 years, it is surprising how little effort has been devoted to the development of transferable calibration models, which are models that can be applied to multiple equivalent devices with minor or without adjustments. The majority of published results is limited to datasets gathered with a single device. This lack of insight in transferable models hampers large scale implementations of ENs, as individual calibration of a multitude of devices for a specific application area is generally unrealistic due to the requirement of actual samples to be measured.

Calibration transfer (case c) covers two different aspects:

(c1) Transfer of the calibration set collected by one instrument (master) to another one (slave) or to a whole series of nominally identical instruments. This facilitates the training phase, which is often long and laborious, and drastically reduces its costs.

(c2) After having completed one project (e.g., the setup of an EN for screening of microbial contamination of tomato pulp), one desires to transfer the knowledge (training parameters) into a different project (for instance, the screening of microbial contamination of vegetable soups) with a minimal amount of effort and with a minimum number of recalibration measurements.

Calibration transfer can be accomplished by transforming the secondary instrument readings to resemble the master target values or labels used in the calibration. In the electronic nose literature this is the most common approach, either using univariate or multivariate approaches.

Sensor replacement (case d) is necessary whenever one sensor of the array brakes or starts providing erratic results. Due to the inherent technological tolerances in the fabrication of chemical sensors of diverse technologies, the new sensor will be different from the initial one. Much work has been done at industrial level to improve sensor fabrication reproducibility; however, changes in baseline and in sensor profile sensitivity to the compounds of interest are always observed after sensor replacement. This leads to a distortion of the original probability density functions along the feature space dimension of the replaced sensor.

In fact this situation is nothing but a special case of the previous case (c1) where the EN after sensor replacement can be considered as a different instrument indeed. Therefore similar strategies and algorithms can be used.

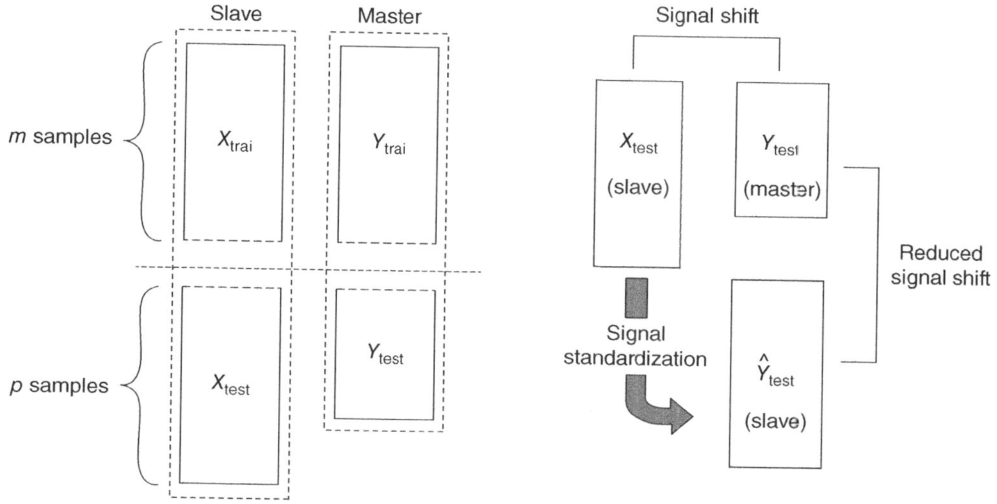

Figure 5.19 The matrices of the slave instrument and the master instrument divided into the training and test sets (left); procedure to obtain shift corrected values of the slave instrument according to the master instrument response (right) (Source: From Ref. [105]. Reprinted with permission from Elsevier Science Ltd.)

There is a literature on spectrophotometer instrument standardization [34] based on transferring near-infrared (NIR) calibration through methods such as the classical direct standardization (DS) and piece direct standardization (PDS), OSC, wavelet techniques, and principal components regression (PCR). All these methods can also be applied to (off-line) multivariate calibration transfer of EN systems. The main limitation experimented until now concerns the quality and amount of data necessary to ensure robustness. Tomic *et al.* [105] describe two approaches to map the responses of slave instruments to a master instrument: univariate direct standardization (UDS) and a multivariate method based on PLS (see Figure 5.19). The former method, based on simple, computationally inexpensive univariate linear regression, calculates a unique shift compensation function for each sensor. The second method uses the data from all sensors for the design of one shift model for the complete gas-sensor array. These methods were applied for calibration transfer of five commercial EN systems configured with eight quartz crystal microbalance (QMB) sensors and provided similar calibration performances.

UDS and PLS require two or more different types of reference samples to be measured at regular intervals. If only one type of reference sample is available, other recalibration methods have to be considered, such as multiplicative drift correction (MDC) and CC [106].

Carmel *et al.* [98] considered the task of finding a mapping between two commercial ENs that employ two different sensor technologies: QMB (MOSES-EN) and conducting polymers (Cyranose 320). Such a mapping is a model that predicts the response of one EN (slave) based on the response of the other (master). They investigated a number of methods, including PCR, PLSs, neural networks, and tessellation-based linear interpolation. A major results was that the mapping direction was a key factor: in mappings from CP to QMB, prediction accuracy was much higher than in the opposite mapping, using the same classifier. This was attributed to QMB possessing a high power of odor discrimination. The selection of the EN master is thus crucial. It was also observed that in PCR and PLS, the optimal prediction accuracy was achieved using a dimensionality that was lower than the total number of sensors: dimensionality reduction should be performed before calibration transfer.

Gas sensor array calibration transfer is particularly relevant for distributed applications of sensors. In indoor air monitoring application, for instance, it is not practical to perform instrument calibrations for each sensor array at each field installation. In addition, it would also be too costly to perform separate calibrations on each array in the factory before field installation. Instead, it would be desirable to perform calibrations on identical arrays located in the controlled environment of a laboratory and then transfer the calibrations to the field-deployed instruments. In this context, PLS calibration models were developed using steady-state tin oxide sensor array responses based both on single analytes (toluene) and compounds- [125]. The authors were able to successfully transfer single-component calibrations by sorting the sensors in each array by sensitivity prior to transfer and have demonstrated a limited compound-class calibration transfer by successfully predicting *o*-xylene concentrations via a calibration model for toluene. While the accuracy for the *o*-xylene prediction was not as good as with the *o*-xylene model, it did have some success. The strategy of recognizing that similar VOCs tend to have similar array signatures may also be applicable to arrays containing other types of sensors. However, subsequent tests for calibration transfer of ternary mixture calibrations were not successful.

More recently, Zhang and coworkers [127] have proposed a high-performance on-line calibration transfer based on global affine transformation (GAT) and Kennard–Stone sequential algorithm (KSS). GAT indicated a general linear transformation between the master and slave instruments. Robust Weighted Least Square (RWLS) is applied for mapping the slave instrument onto the master in order to realize instruments standardization. Basically, this method minimizes a weighted sum of squares, in which the weight assigned to each sample point depends on how far the point is from the fitted line. In the calibration process, the selection of the most representative samples is performed through the KSS algorithm. In prediction step, three ANNs for concentration prediction of three analytes were trained based on an error backpropagation algorithm.

As a result, the instrumental related signal shifts were significantly reduced.

5.6 Conclusions

At its first appearances, the e-nose owed to the capability of fingerprint analysis a significant amount of its possibility to mimic the capability of its biologic counterpart. Since then and over three decades, researchers in AO have tried to face e-nose limitations and emerging applications, developing ad-hoc pattern recognition algorithms. From the early approaches to classification, olfactive pattern recognition techniques have evolved to encompass modern data mining and signal processing tools, classifier ensembles, and advanced regression techniques supporting application in industry, medical diagnostics, environmental monitoring, and so on. Many works, though, focus on the validation step to a single applicative scenario and mostly on a single dataset. A limited effort has however been spent in comparing the different developed approaches; as such, there is no general consensus about their suitability or optimality in different applications. In this sense, the lack of shared datasets and algorithm repository represents a limitation to the organic development of the field.

In general, pattern recognition has been shown to be an essential enabler to e-nose application to real-life environment, but this relationship has been shown to have more profound implications. The enhanced capabilities to extract and deal with complex and multivariate sensor array responses and to highlight hidden structures in olfactive data have had a significant impact on the way researchers think about chemical sensors themselves as both devices and olfactive information sources. Dynamic feature extraction techniques, drift counteraction strategies, and solid-state sensor thermal modulation are meaningful examples. Interestingly enough, the biomimetic nature of the e-nose architecture has led researchers from the AO field to develop results that may allow a deeper insight in the biologic olfaction data processing path (see Ref. [36]). However, despite these efforts, a limited number of issues, as described in this chapter, are still to be solved, while new challenges arise from emerging real-world, unmanned, or distributed deployment of e-nose tools. These challenges will eventually foster the future development of olfactive pattern recognition field.

References

[1] Acevedo, F. J., Maldonado, S., Domínguez, E., Narváez, A., & López, F. (2007). Probabilistic support vector machines for multi-class alcohol identification, *Sensors and Actuators B: Chemical*, 122 (1): 227–235.

[2] Artursson, T., Eklov, T., Lundström, I., Mårtensson, P., Sjöström, M., & Holmberg, M. (2000). Drift correction for gas sensors using multivariate methods, *Journal of Chemometrics*, 14(5–6): 711–723

[3] Badura, M., Szczurek, A., Szecówka, & P. M. (2013). Statistical assessment of quantification methods used in gas sensor system, *Sensors and Actuators B: Chemical*, 188: 815–823.

[4] Barko, G., Abonyi, J., Hlavay, J. (1999). Application of fuzzy clustering and piezoelectric chemical sensor array for investigation on organic compounds, *Analytica Chimica Acta*, 398(2–3): 219–226.

[5] Hernandez Bennetts, V., Lilienthal, A. J., Neumann, P. P., & Trincavelli, M. (2012). Mobile robots for localizing gas emission sources on landfill sites: Is bio-inspiration the way to go?, *Frontiers in Neuroengineering*, 4: 1–12.

[6] Boilot, P., Hines, E. L., Gongora, M. A., Folland, R. S. (2003). Electronic noses inter-comparison, data fusion and sensor selection in discrimination of standard fruit solutions, *Sensors and Actuators B: Chemical*, 88(1): 80–88.

[7] Brudzewski, K., Osowski, S., & Markiewicz, T. (2004). Classification of milk by means of an electronic nose and SVM neural network, *Sensors and Actuators B: Chemical*, 98(2–3): 291–298.

[8] Buratti, S., Benedetti, S., Scampicchio, M., & Pangerod, E. C. (2004). Characterization and classification of Italian Barbera wines by using an electronic nose and an amperometric electronic tongue, *Analytica Chimica Acta*, 525(1): 133–139.

[9] Burl, M. C., Doleman, B. J., Schaffer, A., & Lewis, N. S. (2001). Assessing the ability to predict human percepts of odor quality from the detector responses of a conducting polymer composite-based electronic nose, *Sensors and Actuators B: Chemical*, 72(2): 149–159.

[10] Clifford, P. K. & Tuma, D. T. (1982–1983). Characteristics of semiconductor gas sensors. I. Steady state gas response, *Sensors and Actuators*, 3: 233–254.

[11] Ciosek, P., Brzózka, Z., Wróblewski, W., Martinelli, E., Di Natale, C., & D'Amico, A. (2005). Direct and two-stage data analysis procedures based on PCA, PLS-DA and ANN for ISE-based electronic tongue—Effect of supervised feature extraction, *Talanta*, 67(3): 590–596.

[12] Comon, P. (1994). Independent component analysis, a new concept? *Signal Process*, 36: 287–314.

[13] D'Amico, A., Di Natale, C., Paolesse, R., Macagnano, A., Martinelli, E., Pennazza, G., Santonico, M., Bernabei, M., Roscioni, C., Galluccio, G., Bono, R., Finazzi Agrò, E., & Rullo, S. (2008). Olfactory systems for medical applications, *Sensors and Actuators B: Chemical*, 130(1): 458–465.

[14] de Castro Nunes, L. & Von Zuben, F. J., *Artificial Immune Systems. Part 1, Basic Theory and Applications*, Technical Report, DCA 01/99, Campinas, UNICAMP (1999).

[15] De Vito, S., Massera, E., Piga, M., Martinotto, L., & Di Francia, G. (2007). On-field calibration of an electronic nose for benzene estimation in an urban pollution monitoring scenario. *Sensors and Actuators B: Chemical* 129 (2): 750–757.

[16] De Vito, S., Castaldo, A., Loffredo, F., Massera, E.Polichetti, T., Nasti, I., Vacca, P., Quercia, L., & Di Francia, G. (2007). Gas concentration estimation in ternary mixtures with room temperature operating sensor array using tapped delay architectures, *Sensors and Actuators B: Chemical*, 124(2): 309–316.

[17] De Vito, S., Piga, M., Martinotto, L., & Di Francia, G. (2009). CO, NO_2 and NO_x urban pollution monitoring with on-field calibrated electronic nose by automatic Bayesian regularization, *Sensors and Actuators B: Chemical*, 143(1): 182–191.

[18] De Vito, S., Martinelli, E., Di Fuccio, R., Tortorella, F., Di Francia, G., D'Amico, A., & Di Natale, C., An artificial immune systems for artificial olfaction data analysis: Comparison between AIRS and ANN models. Proceedings of 2010 International Joint Conference on Neural Networks—World Conference on Computational Intelligence, Barcelona, Spain, July 18–23, 2010, IEEE Conference Publications, pp. 1–7, Piscataway, New Jersey.

[19] De Vito, S., Di Palma, P., Ambrosino, C., Massera, E., Burrasca, G., Miglietta, M. L., & Di Francia, G. (2011). Wireless sensor networks for distributed chemical sensing: Addressing power consumption limits with on-board intelligence, *IEEE Sensors Journal*, 11(4): 947–955.

[20] De Vito, S., Fattoruso, G., Liguoro, R., Oliviero, A., Massera, E., Sansone, C., Casola, V., & Di Francia, G. (2011). Cooperative 3D air quality assessment with wireless chemical sensing networks, *Procedia Engineering*, 25: 84–87.

[21] De Vito, S., Fattoruso, G., Pardo, M., Tortorella, F., & Di Francia, G. (2012). Semi-supervised learning techniques in artificial olfaction: A novel approach to classification problems and drift counteraction, *IEEE Sensors Journal*, 12, 3215.

[22] Diaconu, M., *Senses and the City: An Interdisciplinary Approach to Urban Sensescapes*, Munster, LIT Verlag (2011).

[23] Di Carlo, S., Falasconi, M., Sanchez, E., Scionti, A., Squillero, G., & Tonda, A. (2011). Increasing pattern recognition accuracy for chemical sensing by evolutionary based drift compensation, *Pattern Recognition Letters* 32(13): 1594–1603.

[24] Di Carlo, S. & Falasconi, M. (2014). Drift correction methods for gas chemical sensors in artificial olfaction systems: techniques and challenges. In: Wang, W. (Ed.), *Advances in Chemical Sensors*, New York, InTech.

[25] Di Natale, C., Davide, F. A., Damico, A., Gopel, W., & Weimar, U. (1994). Sensor array calibration with enhanced neural networks, *Sensors and Actuators B: Chemical*, 18: 654–657.

[26] Di Natale, C., Martinelli, E., & D'Amico, A. (2002). Counteraction of environmental disturbances of electronic nose data by independent component analysis, *Sensors and Actuators B: Chemical* 82(2–3): 158–165.

[27] Diamond, D., Coyle, S., Scampagnani, S., & Hayes, J. (2008). Wireless sensor networks and chemo-/biosensing, *Chemical Reviews*, 108(2): 652–679.

[28] Ding, H., Ge, H., & Liu, J. (2005). High performance of gas identification by wavelet transform-based fast feature extraction from temperature modulated semiconductor gas sensors, *Sensors and Actuators B: Chemical*, 107(2): 749–755.

[29] Distante, C., Siciliano P., & Persaud, K. C. (2002). Dynamic cluster recognition with multiple self-organising maps, *Pattern Analysis and Applications*, 5: 306–315.

[30] Distante, C., Leo, M., Siciliano, P., & Persaud, K. (2003). On the study of feature extraction methods for an electronic nose, *Sensors and Actuators B*, 87: 274–288.

[31] Distante, C., Ancona, N., & Siciliano, P. (2003). Support vector machines for olfactory signals recognition, *Sensors and Actuators B: Chemical*, 88: 30–39.

[32] Dutta, R., Morgan, D., Baker, N., Gardner, J. W., & Hines, E. L. (2005). Identification of Staphylococcus aureus infections in hospital environment: Electronic nose based approach, *Sensors and Actuators B: Chemical*, 109(2): 355–362.

[33] Falasconi, M., Pardo, M., & Sberveglieri, G. (2013). Methods and graphical tools for exploratory data analysis of artificial olfaction experiments. In: Nakamoto, T. (Ed.), *Human Olfactory Displays and Interfaces: Odor Sensing and Presentation*, Hershey, PA, Information Science Reference, pp. 317–339.

[34] Feudale, R. N., Woody, N. A., Tan, H., Myles, A. J., Brown, S. D., & Ferré, J. (2002). Transfer of multivariate calibration models: A review, *Chemometrics and Intelligent Laboratory Systems*, 64(2): 181–192.

[35] Foresee, F. D. & Hagan, M. T., Gauss–Newton approximation to Bayesian regularization. Proceedings of the 1997 International Joint Conference on Neural Networks, Houston, TX, USA, June 9–12, 1997, pp. 1930–1935. Available at http://hagan.okstate.edu/icnn97a.pdf (accessed on December 4, 2015).

[36] Fonollosa, J., Gutierrez-Galvez, A., & Marco, S. (2012). Quality coding by neural populations in the early olfactory pathway: Analysis using information theory and lessons for artificial olfactory systems, *PLoS One*, 7(6): e37809.

[37] Fryder, M., Holmberg, M., Winquist, F., & Lündstrom, I., A calibration technique for an electronic nose. Proceedings of the Transducers'95 and Eurosensors IX, Stockholm, Sweden, June 25–29, 1995, pp. 683–686, Piscataway, New Jersey.

[38] Gardner, J. W. (1991). Detection of vapours and odours from a multisensor array using pattern recognition. Part 1. Principal component and cluster analysis, *Sensors and Actuators B: Chemical*, 4: 109–115.

[39] Gardner, J. W. & Bartlett, P. N., *Electronic Noses: Principles and Applications*, New York, Oxford University Press (1999).

[40] Gardner, J. W., Boilot, P., & Hines, E. L. (2005). Enhancing electronic nose performance by sensor selection using a new integer-based genetic algorithm approach, *Sensors and Actuators B: Chemical*, 106(1): 114–121.

[41] Gobbi, E., Falasconi, M., Torelli, E., & Sberveglieri, G. (2011). Electronic nose predicts high and low fumonisin contamination in maize cultures, *Food Research International*, 44(4): 992–999.

[42] Gómez, A. H., Wang, J., Hu, G., & Pereira, A. G. (2008). Monitoring storage shelf life of tomato using electronic nose technique, *Journal of Food Engineering*, 85(4): 625–631.

[43] Goschnick, J., Koronczi, I., Frietsch, M., & Kiselev, I. (2005). Water pollution recognition with the electronic nose KAMINA, *Sensors and Actuators B: Chemical*, 106(1): 182–186.

[44] Grimes, C., On, K. G., Varghese, O. K., Yang, X., Mor, G., Paulose, M., Dickey, E. C., & Ruan, C. (2003). A sentinel sensor network for hydrogen sensing, *Sensors*, 3: 69–82.

[45] Groves, W. A., Zellers, E. T., & Frye, G. C. (1998). Analyzing organic vapors in exhaled breath using a surface acoustic wave sensor array with preconcentration: Selection and characterization of the preconcentrator adsorbent, *Analytica Chimica Acta*, 371: 131.

[46] Gualdrón, O., Llobet, E., Brezmes, J., Vilanova, X., & Correig, X. (2006). Coupling fast variable selection methods to neural network-based classifiers: Application to multisensor systems, *Sensors and Actuators B: Chemical*, 114(1): 522–529.

[47] Gulbag, A. & Temurtas, F. (2006). A study on quantitative classification of binary gas mixture using neural networks and adaptive neuro-fuzzy inference systems, *Sensors and Actuators B: Chemical*, 115(1): 252–262.

[48] Gulbag, A., Temurtas, F., Tasaltin, C., & Öztürk, Z. Z. (2007). A study on radial basis function neural network size reduction for quantitative identification of individual gas concentrations in their gas mixtures, *Sensors and Actuators B: Chemical*, 124(2): 383–392.

[49] Gutierrez-Osuna, R., Troy Nagle, H., & Schiffman, S. S. (1999). Transient response analysis of an electronic nose using multi-exponential models, *Sensors and Actuators B*, 61: 170–182.

[50] Gutierrez-Osuna, R., Drift reduction for metal-oxide sensor arrays using canonical correlation regression and partial least squares. Proceedings of the Seventh International Symposium on Olfaction and Electronic Nose, Brighton, UK, July 2000, Institute of Physics Publishing, pp. 147–152, Bristol, UK.

[51] Gutierrez-Osuna, R., Gutierrez Galvez, A., & Powar, N. (2003). Transient response analysis for temperature modulated chemoresistors, *Sensors and Actuators B: Chemical*, 93(1): 57–66.

[52] Hansen, N., Ostermeier, A., & Gawelczyk, A. On the adaptation of arbitrary normal mutation distributions in evolution strategies: The generating set adaptation. In: L.Eshelman (Ed.). Proceedings of the Sixth International Conference on Genetic Algorithms, Pittsburgh, PA, USA, July 15–19, 1995. San Francisco, CA, Morgan Kaufmann, pp. 312–317.

[53] Haugen, J.-E., Tomic, O., & Kvaal, K. (2000). A calibration method for handling the temporal drift of solid state gas-sensors, *Analytica Chimica Acta*, 407(1–2): 23–39.

[54] Hauschild, A.-C., Schneider, T., Pauling, J., Rupp, K., Jang, M., Baumbach, J. I., & Baumbach, J. (2012). Computational methods for metabolomic data analysis of ion mobility spectrometry data—Reviewing the state of the art, *Metabolites*, 2(4): 733–755.

[55] Heilig, A., Bârsan, N., Weimar, U., Schweizer-Berberich, M., Gardner, J. W., & Göpel, W. (1997). Gas identification by modulating temperatures of SnO_2-based thick film sensors, *Sensors and Actuators B: Chemical*, 43(1–3): 45–51.

[56] Holmberg, M., Davide, F., Di Natale, C., D'Amico, A., Winquist, F., & Lundstrom, I. (1997). Drift counteraction in odor recognition applications: Lifelong calibration method, *Sensors and Actuators B: Chemical*, 42(3): 185–194.

[57] Hübnera, M., Simionb, C. E., Tomescu-Stănoiub, A., Pokhrela, S., Bârsana, N., & Weimara, U. (2011). Influence of humidity on CO sensing with p-type CuO thick film gas sensors, *Sensors and Actuators B: Chemical*, 153: 347–353.

[58] Hui, D., Jun-Hua, L., & Zhong-Ru, S. (2003). Drift reduction of gas sensor by wavelet and principal component analysis, *Sensors and Actuators B: Chemical*, 96(1–2): 354–363.

[59] Huybrechts, G., Szecówka, P., Roggen, J., & Licznerski, B. W. (1997). Simultaneous quantification of carbon monoxide and methane in humid air using a sensor array and an artificial neural network, *Sensors and Actuators B: Chemical*, 45(2): 123–130.

[60] Jain, K., Murty, M. N., & Flynn, P. J. (1999). Data clustering: A review, *ACM Computing Surveys*, 31(3): 264–323.

[61] Kamionka, M., Breuil, P., & Pijolat, C. (2006). Calibration of a multivariate gas sensing device for atmospheric pollution measurement, *Sensors & Actuators: B. Chemical*, 18: 323–327.

[62] Kermit, M. & Tomic, O. (2003). Independent component analysis applied on gas sensor array measurement data, *IEEE Sensors Journal*, 3(2): 218–228.

[63] Kim, N., Byun, H. G., Kwon, K. H., Persaud, K. C., & Lim, O. J., P2. 9.33 Unsupervised adjustment of centers in RBF networks for sensor drift compensation. Tagungsbund Meeting on Chemical Sensors (IMCS), AMA Conferences, Nuremberg, Germany, May 20–23, 2012, pp. 1802–1804. AMA Service GmbH, Wunstorf, Germany.

[64] Kolk, A., Hoelscher, M., Maboko, L., Jung, J., Kuijper, S., Cauchi, M., Bessant, C., van Beers, S., Dutta, R., Gibson, T., & Reither, K. (November 2010). Electronic-nose technology using sputum samples in diagnosis of patients with tuberculosis, *Journal of Clinical Microbiology*, 48(11): 4235–4238.

[65] Lathauwer, L. D. & Moor, B. D. (2000). An introduction to independent component analysis, *Journal of Chemometrics*, 14: 123–149.

[66] Lindgren, D. & Spångéus, P. (2004). A novel feature extraction algorithm for asymmetric classification, *IEEE Sensors Journal*, 4(5): 643–650.

[67] Llobet, E., Brezmes, J., Ionescu, R., Vilanova, X., Al-Khalifa, S., Gardner, J. W., Bârsan, N., & Correig, X. (2002). Wavelet transform and fuzzy artmap-based pattern recognition for fast gas identification using a micro-hotplate gas sensor, *Sensors and Actuators B: Chemical*, 83(1–3): 238–244.

[68] Llobet, E., Gualdrón, O., Vinaixa, M., El-Barbri, N., Brezmes, J., Vilanova, X., Bouchikhi, B., Gómez, R., Carrasco, J. A., & Correig, X. (2007). Efficient feature selection for mass spectrometry based electronic nose applications, *Chemometrics and Intelligent Laboratory Systems*, 85(2): 253–261.

[69] Lozano, J., Santos, J. P., Arroyo, T., Aznar, M., Cabellos, J. M., Gil, M., & del Carmen Horrillo, M. (2007). Correlating e-nose responses to wine sensorial descriptors and gas chromatography–mass spectrometry profiles using partial least squares regression analysis, *Sensors and Actuators B: Chemical*, 127(1): 267–276.

[70] MacQueen, J. B. (1967). Some methods for classification and analysis of multivariate observations. In: *Proceedings of Fifth Berkeley Symposium on Mathematical Statistics and Probability*, 1, Le Cam, L. M. and Neyman, J. (Eds). Berkeley, University of California Press, pp. 281–297.

[71] Marco, S. & Gutierrez, A. (2012). Signal and data processing for machine olfaction and chemical sensing: A review, *IEEE Sensors Journal*, 12(11): 3189.

[72] Marco, S., Ortega, A., Pardo, A., & Samitier, J. (1998). Gas identification with tin oxide sensor array and self-organizing maps: Adaptive correction of sensor drifts, *IEEE Transactions on Instrumentation and Measurement*, 47(1): 316–321.

[73] Martinelli, E., Falconi, C., D'Amico, A., & Di Natale, C. (2003). Feature extraction of chemical sensors in phase space, *Sensors and Actuators B: Chemical*, 95: 132.

[74] Martinelli, E., Santonico, M., Pennazza, G., Paolesse, R., D'Amico, A., & Di Natale, C. (2011), Short time gas delivery pattern improves long-term sensor reproducibility, *Sensors and Actuators B: Chemical*, 156: 753–759.

[75] Martinelli, E., Magna, G., De Vito, S., Di Fuccio, R., Di Francia, G., Vergara, A., & Di Natale, C. (2013). An adaptive classification model based on the artificial immune system for chemical sensor drift mitigation, *Sensors and Actuators B: Chemical*, 177: 1017–1026.

[76] Mead, M. I., Popoola, O. A. M., Stewart, G. B., Landshoff, P., Calleja, M., Hayes, M., Baldovi, J. J., McLeod, M. W., Hodgson, T. F., Dicks, J., Lewis, A., Cohen, J., Baron, R., Saffell, J. R., & Jones, R. L. (2013). The use of electrochemical sensors for monitoring urban air quality in low-cost, high-density networks, *Atmospheric Environment*, 70: 186–203.

[77] Muezzinoglu, M. K., Vergara, A., Huerta, R., & Rabinovich, M. I. (2010). A sensor conditioning principle for odour identification, *Sensors and Actuators B: Chemical*, 145(2): 472–476.

[78] Neumann, P., Hernandez Bennetts, V., Lilienthal, A. J., Bartholmai, M., & Schiller, J. H. (2013). Gas source localization with a micro-drone using bio-inspired and particle filter-based algorithms. *Advanced Robotics (AR), Special Issue on Aerial Robots*, 27(9): 725–738.

[79] Padilla, M., Montoliu, I., Pardo, A., Perera, A., & Marco, S. (2006). Feature extraction on three way enose signals, *Sensors and Actuators B: Chemical*, 116(1–2): 145–150.

[80] Padilla, M., Perera, A., Montoliu, I., Chaudry, A., Persaud, K., & Marco, S. (2010). Drift compensation of gas sensor array data by orthogonal signal correction, *Chemometrics and Intelligent Laboratory Systems*, 100(1): 28–35.

[81] Pardo, M. and de Vito, S. (2008). Support vector machines for electronic nose pattern analysis. In: Hines, E., Martínez-Ramón, M., Pardo, M., Llobet, E., Leeson, M., & Iliescu, D. (Eds.), *Intelligent Systems: Techniques and Applications*, Aachen, Shaker Publications, pp. 173–198.

[82] Pardo, M., Faglia, G., Sberveglieri, G., Corte, M., Masulli, F., & Riani, M. (2000). A time delay neural network for estimation of gas concentrations in a mixture, *Sensors and Actuators B: Chemical*, 65(1–3): 30.

[83] Pardo, M. & Sberveglieri, G. (2007). Comparing the performance of different features in sensor arrays, *Sensors and Actuators B: Chemical*, 123(1): 437–443.

[84] Pardo, M. & Sberverglieri, G. (2004). Remarks on the use of multilayer perceptrons for the analysis of chemical sensor arrays, *IEEE Sensors Journal*, 4(3): 355, 363.

[85] Pardo, M. & Sberveglieri, G. (2005). Classification of electronic nose data with support vector machines, *Sensors and Actuators B: Chemical*, 107(2): 730–737.

[86] Pardo, M., Kwong, L. G., Sberveglieri, G., Brubaker, K., Schneider, J. F., Penrose, W. R., & Stetter, J. R. (2005). Data analysis for a hybrid sensor array, *Sensors and Actuators B: Chemical*, 106(1): 136–143.

[87] Paulsson, N., Larsson, E., & Winquist, F. (2000). Extraction and selection of parameters for evaluation of breath alcohol measurement with an electronic nose, *Sensors and Actuators A: Physical*, 84(3): 187–197.

[88] Peris, M. & Escuder-Gilabert, L. (2009). A 21st century technique for food control: Electronic noses, *Analytica Chimica Acta*, 638(1): 1–15.

[89] Perera, A., Yamanaka, T., Gutierrez-Galvez, A., Raman, B., & Gutierrez, O. R. (2006). A dimensionality reduction technique inspired by receptor convergence in the olfactory system, *Sensors and Actuators B: Chemical*, 116(1): 17–22.

[90] Pennazza, D. A., Santonico, M., Martinelli, E., Roscioni, C., & Galluccio, G. (2010). An investigation on electronic nose diagnosis of lung cancer, *Lung Cancer*, 68(2): 170–176.

[91] Pijolat, C., Pupier, C., Sauvan, M., Tournier, G., & Lalauze, R. (1999). Gas detection for automotive pollution control, *Sensors and Actuators B: Chemical*, 59: 195–202.

[92] Ping, W. & Jun, X. (1996). A novel recognition method for electronic nose using artificial neural network and fuzzy recognition, *Sensors and Actuators B: Chemical*, 37(3): 169–174.

[93] Popoola, O., Mead, I., Bright, V., Baron, R., Saffell, J., Stewart, G., Kaye, P., & Jones, R., A portable low-cost high density sensor network for air quality at London Heathrow Airport, EGU General Assembly 2013, Vienna, Austria, April 7–12, 2013, EGU2013-1907.

[94] Pomareda, V., Calvo, D., Pardo, A., & Marco, S. (2010). Hard modeling multivariate curve resolution using LASSO: Application to ion mobility spectra, *Chemometrics and Intelligent Laboratory Systems*, 104(2): 318–332.

[95] Raman, B., Meier, D., Evju, J. K., & Semancik, S. (2009). Designing and optimizing microsensor arrays for recognizing chemical hazards in complex environments, *Sensors and Actuators B: Chemical*, 147(2): 617–629.

[96] Roussel, S., Forsberg, G., Grenier, P., & Bellon-Maurel, V. (1999). Optimisation of electronic nose measurements. Part II: Influence of experimental parameters, *Journal of Food Engineering*, 39(1): 9–15.

[97] Saraoglu, H. M., Selvi, A. O., Ebeoglu, M. A., & Tasaltin, C. (2013). Electronic nose system based on quartz crystal microbalance sensor for blood glucose and HbA1c levels from exhaled breath odor, *IEEE Sensors Journal*, 13(11): 4229, 4235.

[98] Shaham, O., Carmel, L., & Harel, D. (2005). On mappings between electronic noses, *Sensors and Actuators B: Chemical*, 106(1): 76–82.

[99] Singh, S., Hines, E. L., & Gardner, J. W. (1996). Fuzzy neural computing of coffee and tainted water data from an electronic nose, *Sensors and Actuators B: Chemical*, 30: 190–195.

[100] Sisk, B. C. & Lewis, N. S. (2005). Comparison of analytical methods and calibration methods for correction of detector response drift in arrays of carbon black-polymer composite vapor detector, *Sensors and Actuators B: Chemical*, 104(2): 249–268.

[101] Skov, T. & Bro, R. (2005). A new approach for modeling sensor based data, *Sensors and Actuators B: Chemical*, 106(2): 719–729.

[102] Sundgren, H., Winquist, F., Lukkari, I., & Lundström, I. (1991). Artificial neural networks and gas sensor arrays: Quantification of individual components in a gas mixture, *Measurement Science and Technology*, 2: 464.

[103] Sysoev, V. V., Goschnick, J., Schneider, T., Strelcov, E., & Kolmakov, A. (2007). A gradient microarray electronic nose based on percolating SnO_2 nanowire sensing elements, *Nano Letters*, 7(10): 3182–3188.
[104] Szczurek, A., Szecówka, P. M., & Licznerski, B. W. (1999). Application of sensor array and neural networks for quantification of organic solvent vapours in air, *Sensors and Actuators B: Chemical*, 58(1–3): 427–432.
[105] Tomic, O., Ulmer, H., & Haugen, J.-E. (2002). Standardization methods for handling instrument related signal shift in gas-sensor array measurement data, *Analytica Chimica Acta*, 472: 99–111.
[106] Tomic, O., Eklov, T., Kvaal, K., & Haugena, J. (2004). Recalibration of a gas-sensor array system related to sensor replacement. *Analytica Chimica Acta*, 512: 199–206.
[107] Timms, C., Thomas, P. S., & Yates, D. H. (2012). Detection of gastro-oesophageal reflux disease (GORD) in patients with obstructive lung disease using exhaled breath profiling, *Journal of Breath Research*, 6(1): 016003.
[108] Tsujita, W., Yoshino, A., Ishida, H., Moriizumi, T. (2005). Gas sensor network for air pollution monitoring, *Sensors and Actuators B: Chemical*, 110: 304–311.
[109] Tudu, B., Jana, A., Metla, A., Ghosh, D., Bhattacharyya, N., & Bandyopadhyay, R. (2009). Electronic nose for black tea quality evaluation by an incremental RBF network, *Sensors and Actuators B: Chemical*, 138(1): 90–95.
[110] Tukey, J. W., *Exploratory Data Analysis*, Reading, MA, Addison-Wesley (1977).
[111] Turner, A. P. F. & Magan, N. (2004). Electronic noses and disease diagnostic, *Nature Reviews. Microbiology*, 2: 161–166.
[112] Vapnik, V. N., *The Nature of Statistical Learning Theory*, New York, Springer-Verlag (1995).
[113] Vembu, S., Vergara, A., Muezzinoglu, M. K., & Huerta, R. (2012). On time series features and kernels for machine olfaction, *Sensors and Actuators B: Chemical*, 174: 535–546.
[114] Vergara, A., Martinelli, E., Huerta, R., D'Amico, A., & Di Natale, C. (2011). Orthogonal decomposition of chemosensory cues, *Sensors and Actuators B: Chemical*, 159(1): 126–134.
[115] Vergara, A., Llobet, E., Brezmes, J., Ivanov, P., Cane, C., Gracia, I., Vilanova, X., & Correig, X. (2007). Quantitative gas mixture analysis using temperature modulated micro-hotplate gas sensors: Selection and validation of optimal modulating frequencies, *Sensors and Actuators B: Chemical*, 123(2): 1002–10016.
[116] Vergara, A., Llobet, E., Martinelli, E., Di Natale, C., D'Amico, A., & Correig, X. (2007). Feature extraction of metal oxide gas sensors using dynamic moments, *Sensors and Actuators B: Chemical*, 122(1): 219–226.
[117] Vergara, A., Muezinoglu, K., Rulkov, N., & Huerta, R. (2010). Information-theoretic optimization of chemical sensors, *Sensors and Actuators B: Chemical*, 148(1): 298–306.
[118] Vezzoli, M., Ponzoni, A., Pardo, M., Falasconi, M., Faglia, G., & Sberveglieri, G. (2008). Exploratory data analysis for industrial safety application, *Sensors and Actuators B: Chemical*, 131: 100–109.
[119] Vlachos, D., Fragoulis, D., & Avaritsiotis, J. (1997). An adaptive neural network topology for degradation compensation of thin film tin oxide gas sensors, *Sensors and Actuators B: Chemical*, 45(3): 223–228.
[120] Walmsley, A. D., Haswell, S. J., & Metcalfe, E. (1991). Methodology for the selection of suitable sensors for incorporation into a gas sensor array. *Analytica Chimica Acta*, 242: 31.
[121] Wang, X., Ye, M., & Duanmu, C. J. (2009). Classification of data from electronic nose using relevance vector machines, *Sensors and Actuators B: Chemical*, 140(1): 143–148.
[122] Webb, A., *Statistical Pattern Recognition*, New York, John Wiley & Sons, Inc. (2002).
[123] Wilson, A. D. & Baietto, M. (2009). Applications and advances in electronic-nose technologies, *Sensors*, 9(7): 5099–5148.
[124] Wilson, A. D. & Baietto, M. (2011). Advances in electronic-nose technologies developed for biomedical applications, *Sensors*, 11(1): 1105–1176.
[125] Wolfrum, E. J., Meglen, R. M., Peterson, D., & Sluiter, J. (2006). Calibration transfer among sensor arrays designed for monitoring volatile organic compounds in indoor air quality, *IEEE Sensors Journal*, 6(6): 1638–1643.
[126] Zhang, L., Wen, L., Lu, Y., & Yang, P. (2002). Quantitative fuzzy neural network for analytical determination, *Analytica Chimica Acta*, 468: 105–117.
[127] Zhang, L., Tian, F., Kadri, C., Xiao, B., Li, H., Pan, L., & Zhou, H. (2011). On-line sensor calibration transfer among electronic nose instruments for monitoring volatile organic chemicals in indoor air quality, *Sensors and Actuators B: Chemical*, 160: 899–909.
[128] Zuppa, M., Distante, C., Siciliano, P., & Persaud, K. C. (2004). Drift counteraction with multiple self-organising maps for an electronic nose, *Sensors and Actuators B: Chemical*, 98(2–3): 305–317.
[129] Zuppa, M., Distante, C., Persaud, K. C., & Siciliano, P. (2007). Recovery of drifting sensor responses by means of DWT analysis, *Sensors and Actuators B: Chemical*, 120(2): 411–416.

[130] Yamanaka, T., Ishida, H., Nakamoto, T., & Moriizumi, T. (1998). Analysis of gas sensor transient response by visualizing instantaneous gas concentration using smoke, *Sensors and Actuators A: Physical*, 69(1): 77–81.
[131] Montoliu, I., Tauler, R., Padilla, M., Pardo, A., Marco, S. (2010). Multivariate curve resolution applied to temperature-modulated metal oxide gas sensors, *Sensors and Actuators B: Chemical*, 145 (1): 464–473.
[132] Gordon, A. D. (1999). *Classification*. Chapman and Hall/CRC, Boca Raton, FL.
[133] Liu, M., Wang, M., Wang, J., Li, D. (2013). Comparison of random forest, support vector machine and back propagation neural network for electronic tongue data classification: Application to the recognition of orange beverage and Chinese vinegar. *Sensors and Actuators B Chemical*, 177: 70–980.

6

Using Chemical Sensors as "Noses" for Mobile Robots

Hiroshi Ishida[1], Achim J. Lilienthal[2], Haruka Matsukura[1], Victor Hernandez Bennetts[2] and Erik Schaffernicht[2]
[1]*Graduate School of Bio-Applications and Systems Engineering, Tokyo University of Agriculture and Technology, Tokyo, Japan*
[2]*School of Science and Technology, AASS, Mobile Robotics & Olfaction Lab, Örebro University, Örebro, Sweden*

6.1 Introduction

Gas sensors detect the presence of gaseous chemical compounds in air. They are often used in the form of gas alarms for detecting dangerous or hazardous gases. However, a limited number of stationary gas alarms may not be always sufficient to cover a large industrial facility, for example. Human workers having a portable gas detector in their hand need to be sent to thoroughly check gas leaks in the areas not covered by stationary gas alarms. However, making repetitive measurements with a gas detector at a number of different locations is laborious. Moreover, the places where the gas concentration level needs to be checked are often potentially dangerous for human workers. If a portable gas detector is mounted on a mobile robot, the task of patrolling in an industrial facility for checking a gas leak can be automated. Robots are good at doing repetitive tasks and can be sent into harsh environments that might be dangerous for humans.

Mobile robots with a gas sensing capability are also expected to serve as robotic sniffer dogs. If appropriate sensors are provided and effective search algorithms are implemented, robots would be able to search for prohibited narcotics at airports, locate landmines buried underground, and find survivors in the wake of disasters, just as trained dogs. As in the case of robotic sniffer dogs, the gas sensing capability certainly offers new solutions for the problems that cannot be easily solved by other sensing modalities. However, gas sensing (or chemical sensing in general) has not been as popular in robotics as visual sensing or range detection. This is partly because robotics researchers have not been familiar with chemical

Essentials of Machine Olfaction and Taste, First Edition. Edited by Takamichi Nakamoto.

Companion website: www.wiley.com/go/nakamoto/olfaction

sensing technologies. The developers of chemical sensors have also been unfamiliar with robotics. Now the situation is changing. Advances in both technologies have resulted in many off-the-shelf products. Even novice researchers can easily try putting gas sensors on mobile robots by purchasing commercial gas sensors and robot kits.

However, the real research problems toward the realization of mobile robots with olfactory sensing capabilities start here. From the perspective of robotics researchers, chemical sensing is different from other sensing modalities in various aspects. Compared to visual and other popular sensing modalities, chemical sensing inherently involves much larger uncertainties. The response of a gas sensor not only changes with the concentration of the target chemical species but also is affected by temperature and humidity variations in the environment. The sensor output often shows significant drift over time. The sensing layer of a gas sensor is eventually poisoned and degrades. Moreover, all currently available gas sensors more or less have limitations. Dogs can sniff out a trace scent of explosives at ppb levels in a fraction of seconds. The response time, sensitivity, and selectivity of the existing gas sensors do not match those of the dog nose.

With gas sensors, a robot can detect the presence of a remote gas source. It would seem from this fact that gas sensors enable remote sensing of objects as vision does. In reality, most gas sensors are contact-type sensors. A gas sensor shows response only when a chemical substance is actually transported from its source to the sensor surface. This transportation process adds another uncertainty in chemical sensing. A chemical substance evaporated from its source is generally dispersed in the environment by turbulent airflow. Even if a mobile robot would be equipped with ideal, fast, and sensitive gas sensors, the robot searching for a gas leak would still have to be able to cope with a time-varying complicated distribution of the target chemical in the environment.

Over the past two decades, various research efforts have been made to employ chemical sensing capabilities on mobile robot platforms. In the following sections, the tasks that mobile robots are expected to accomplish with the help of chemical sensing capabilities are described together with the technical challenges involved in those tasks. Then, hardware setups that we can use and various algorithms that have been proposed to accomplish the tasks are presented. This chapter deals mostly with the detection of airborne gas-phase chemical substances by mobile robots but briefly mentions underwater chemical sensing as well. By showing the current status of the mobile robot olfaction research, the authors hope to stimulate further research efforts to make advances in this relatively new field in robotics.

6.2 Task Descriptions

6.2.1 Definitions of Tasks

Suppose that a conventional gas alarm is mounted on an autonomous mobile robot. We can program the robot to patrol along a predefined path in an office building and check the gas concentration level at some specified locations. In this case, the robot is utilized just as a means to bring the gas alarm to the points of interest. This way of using a robot is still helpful in making repetitive measurements. However, if the task is simply to make independent measurements at various locations, there is no fundamental difference in the chemical sensing part of the task between using a robot and a human worker to carry gas sensors from one place to another.

One way to make full use of robots in chemical sensing tasks is to use their self-localization capabilities. A robot making gas concentration measurements at a number of locations can record the coordinates of each location together with the gas sensor reading obtained at that location. The recorded coordinate values can be used later to build a gas distribution map in a given environment or to calculate spatial correlation among the measured gas sensor readings. Alternatively, gas sensor readings that a robot obtains can be used to navigate the robot toward areas of higher gas concentration. Such robot-specific chemical sensing tasks that have been addressed in the literature are classified into three categories, that is, chemical trail following, chemical source localization, and chemical distribution mapping [1].

Chemical trail following is to track a trail of a chemical substance laid on the ground or the floor. In animal behavior, chemical substances are often used to mark trails and territories on the ground [2, 3]. For example, a worker ant lays a pheromone trail on her way back home from a food source. Other ants can locate the food simply by tracking the smell of the pheromone. The chemical components of the pheromone laid on the ground gradually evaporate, and the trail eventually disappears. However, the pheromone trail is maintained as long as the line of ants returning from the food to the nest reinforces the trail by laying additional pheromone. The trail disappears when the food is exhausted and the reinforcement of the pheromone trail is stopped.

Russell and coworkers proposed several scenarios in which mobile robots can make use of chemical trails [4, 5]. A possible application is, for example, multiple robots working together to clean the entire floor of a room. As each robot moves, a trail of detergent used for floor cleaning is left on the floor. This detergent trail can be used as a repellent marker to let the other robots easily recognize the areas already cleaned. Multiple robot coordination can thus be accomplished by simply programming each robot to avoid the chemical marker. There is no need to remove the detergent marker after the entire floor is cleaned. It spontaneously disappears as the detergent evaporates. Chemical trails can also be used when transporting large quantities of materials using multiple robots. In this scenario, an intelligent pathfinder robot first explores the environment. Once an appropriate path is found, much simpler trail-following robots can transport the materials by following the chemical trail laid by the pathfinder robot. A chemical trail can also be used as a virtual umbilical cable. If the robot lays a chemical trail during the exploration of an unknown environment, the robot can always come back to its initial position simply by tracking the chemical trail.

In a chemical source localization task, a robot is expected to find the location of a chemical source by tracking the distribution of an airborne chemical substance or a chemical dissolved in water. Many animals can localize food, mates, or nests by tracking their smells [3, 6, 7]. Mobile robots with such a chemical source localization capability can be used to search for leaks of flammable gas [8]. The potential applications of chemical source localization robots also include robotic sniffer dogs as mentioned in Section 6.1. The task of chemical source localization is generally divided into three subtasks: chemical finding, chemical tracking, and chemical source declaration [1, 9, 10]. When a robot is brought into a new environment, it first needs to find the area in which the target chemical substance is distributed. The chemical sensors on the robot show no response until they come in contact with the target chemical. Therefore, the chemical finding task must be executed without using chemical signals. Once the robot is brought into the spatial distribution of the target chemical, the chemical sensors start responding. Then, the robot can navigate toward the location of the chemical source by using the sensor readings to track the chemical distribution. This chemical tracking task is terminated when the robot reaches the chemical source location and declares that the goal has been reached.

Chemical distribution mapping is to build a map showing the distribution of a target chemical substance based on the measurements collected in a given environment. In some applications, it is sufficient to know the distribution of a target chemical even though the exact location of the chemical source cannot be tracked down [1]. In a rescue mission, for example, it is necessary to immediately determine where a highly concentrated hazardous gas is expected and where the rescue workers can safely do their jobs. The gas distribution can be measured if a sufficient number of gas sensors are placed in the given environment. Although various research efforts are being made to develop easy-to-use sensor networks, the deployment of a number of gas sensors is yet a laborious work. It is generally easier to send a mobile robot to the area of interest and make the robot scan the area with gas sensors.

6.2.2 *Characteristics of Turbulent Chemical Plumes*

In order to correctly understand the technical challenges involved in the robotic chemical sensing tasks, one must know how a chemical substance released from its source spreads in the environment. It is often misunderstood that the released chemical vapor spreads isotropically in all directions from the source. However, the isotropic chemical distribution is generated only in special occasions. As shown in Figure 6.1, all gas molecules at temperatures above absolute zero are in random thermal motion. When gas molecules are released into air from their source, they gradually spread via the random thermal motion. This process is known as molecular diffusion and makes the gas molecules spread isotropically. However, this molecular diffusion is an extremely slow process. A typical value of the diffusion constant of a small gas molecule like methane in air is 2×10^{-5} m^2/s [3]. The characteristic diffusion length in 1 h is calculated to be only 0.5 m [3], which means that most gas molecules remain within the 0.5 m radius of their source location even 1 h after being released from the source.

In reality, there exists airflow that overpowers the slow molecular diffusion. In an outdoor environment, we generally feel wind blowing on our face. The velocity of the wind is generally at least 0.2 m/s. Therefore, the gas molecules released from their source are immediately

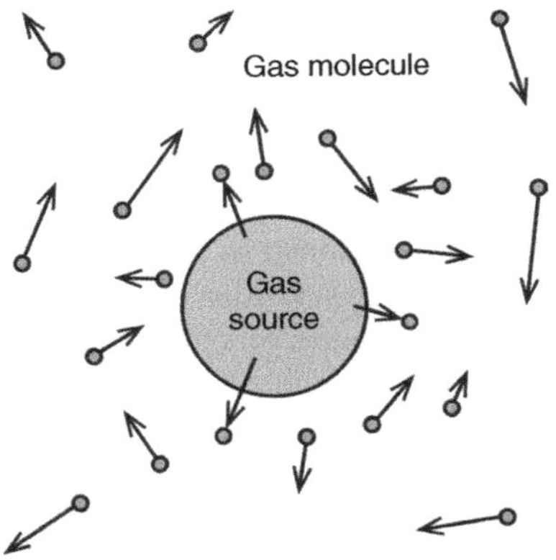

Figure 6.1 Diffusion of gas molecules released from a gas source

carried away by the airflow. The transport of gas molecules by fluidic air motion dominates molecular diffusion also in most indoor environments. Even in a closed room, small temperature variations cause convective air currents at least in the order of a few millimeters per second [11, 12]. Gas molecules released from their source are therefore carried by the airflow and spread mostly in the downwind direction from the source. The resultant aerial trail of gas is called a plume [13].

Moreover, most flows we encounter are almost always turbulent [13]. The speed and the direction of the airflow randomly fluctuate since the turbulent flow contains a number of eddies of different sizes. As shown in Figure 6.2, large eddies contained in the turbulent flow twist and stretch the plume. Eddies smaller than the width of the plume mix the perimeter of the plume with surrounding air. The small eddies also stir the gas concentration distribution inside the plume. As a result, the plume is made to meander randomly and comes to have a complicated filamentous patchy structure like the smoke from a cigarette shown in Figure 6.3.

Technical challenges in chemical source localization mostly come from this complicated structure of turbulent chemical plumes. When a gas sensor is placed in a turbulent gas plume,

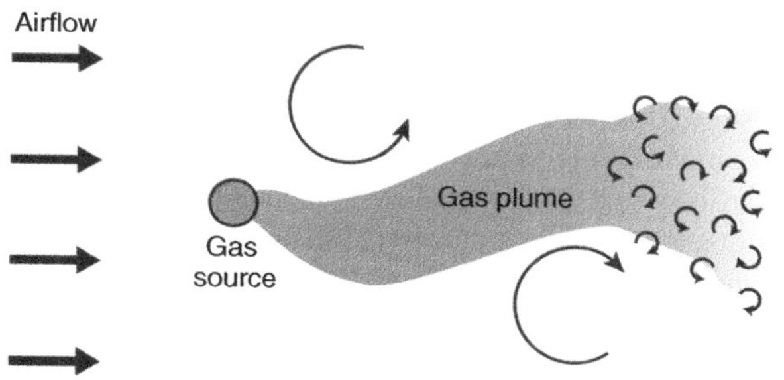

Figure 6.2 Effects of eddies of different sizes in turbulent flow on the structure of a gas plume. Large eddies make the plume meander, while small eddies determine the internal structure of the plume

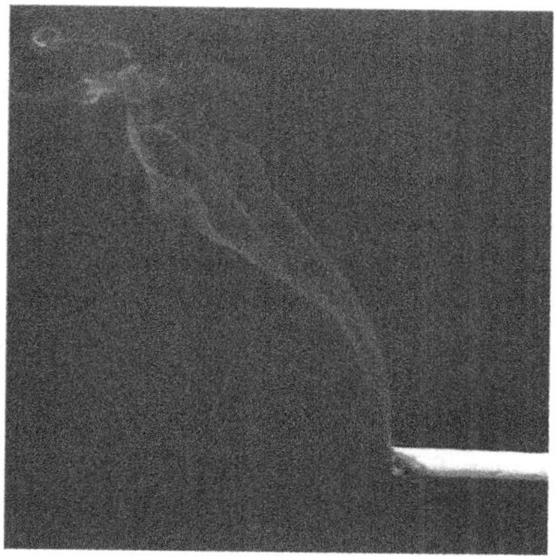

Figure 6.3 Filamentous structure in smoke rising from a cigarette

the sensor is exposed to highly intermittent changes in the gas concentration [13]. As the filamentous patches of the plume pass over the sensor, the response of the gas sensor shows large fluctuations. Moreover, the plume has a shape elongated in the direction of the airflow that carries the gas molecules. Therefore, the chemical concentration gradient along the air-flow direction is inherently small. Detecting a small spatial change in a highly fluctuating signal is a challenging task. The turbulent nature of the plume also poses a problem in chemical distribution mapping. Even if a robot is used to measure the gas concentration at various locations, there is a limitation in the number of measurements that the robot can make in a given time period. If the measurement takes too long time, the environment may change as the time passes. The major technical challenge in chemical distribution mapping is therefore how to obtain a reliable map with a sufficient resolution from fluctuating gas sensor readings measured at coarse grid points. The turbulence of the airflow is not a serious problem in chemical trail following. The gas sensor is generally brought close to the floor to detect a chemical trail that has been laid on the floor. Therefore, the fluctuations in the observed gas sensor response are generally much smaller than those observed in chemical source localization.

6.3 Robots and Sensors

6.3.1 Sensors for Gas Detection

Animals have a keen sense of smell. It is still unclear what kind of chemical substances mine detection dogs use to identify the smell of explosives evaporated from landmines buried underground. However, it is said that trained dogs respond to 2,4-dinitrotoluene (one of the primary impurities contained in trinitrotoluene explosives) at a ppb level [14]. Dogs can also discriminate a faint smell of explosives even under the presence of interfering smells. Male moths search for females by tracking airborne volatile sexual pheromone. It is said that their pheromone detection system has single-molecule sensitivity [15]. Another distinct characteristic of animal olfaction is its fast response. The response time of animals' chemoreceptors is typically 0.1–0.5 s [7, 15]. There is a significant gap between the sensitivity, the selectivity, and the response time of the currently available gas sensors and those of animal olfaction. For example, various types of sensors have been proposed for the detection of explosives, but have not yet replaced the trained dogs [16].

Most of the gas sensing robots reported so far are equipped with commercial metal-oxide gas sensors because of their reasonably high sensitivity and fair response time (typically <5 s) [17–29] They respond to flammable gas at a sub-ppm level. They also have excellent long-term stability and can be used for many years. However, the slow recovery of the response (typically 1 min) after gas is removed is a serious drawback to be used on a mobile robot. The locomotion speed of the robot must be slowed down to wait for the sensor response to follow the change in the gas concentration. There are individual variations in the sensitivity of the sensors. Therefore, it is a good practice to choose a set of sensors with matched sensitivity if multiple sensors are mounted on a mobile robot, and there response values are to be compared. The gas-sensitive mobile robots developed so far are mostly made for testing the ideas in laboratory environments rather than for real applications. Therefore, the detection target was chosen mainly from the ease of handling. For mobile robots equipped with metal-oxide gas sensors, ethanol vapor was most often used as the target chemical. Saturated ethanol vapor was generated by using a pump for bubbling air through liquid ethanol. Alternatively, liquid ethanol was poured in a dish and was placed on the floor in the environment.

Robots equipped with quartz crystal microbalance (QCM) sensors [4, 30] and polymer-based conductometric sensors [9, 31–33] were also reported. Although these sensors generally have better selectivity and show faster response than metal-oxide gas sensors, they tend to lack long-term stability. A live insect antenna can be used as an odor detector since it yields a voltage difference between its tip and base according to the odor intensity [34]. However, its lifetime is rather short. A recovery time of less than a second was achieved using an ion detector for detecting an aerial plume of ionized air [13, 35–37]. Instead of a gas source, an ion generator was used in combination with the ion detector. In Ref. [24], the recovery time of a metal-oxide gas sensor on the robot was reduced to 1 s by selecting a sensor with fast response and using it with a suction pump to quickly replace air samples around the sensor. Mobile robots equipped with an electronic nose system can detect a specific target gas even under the presence of other interfering gases [8, 11, 17, 19, 27, 28, 38–41]. An electronic nose (or e-nose in short) consists of an array of gas sensors and a pattern classifier. An array of gas sensors with slightly different selectivities yields a response pattern unique to each specific gas species. The discrimination between the target gas and others is accomplished by pattern recognition. However, different gas sensors tend to have different time constants. The response pattern is distorted when the sensor array is exposed to a gas with fluctuating concentration as in a turbulent gas plume [38]. Therefore, the pattern recognition becomes challenging.

6.3.2 *Airflow Sensing*

As described in Section 6.2.2, gas molecules released from their source are carried by airflow and trail in the downwind direction. It means that if a gas is detected at a certain location, the gas source is likely to be in the upwind direction. Thus, the direction of the airflow can be used as a clue for estimating the gas source location. Therefore, some gas sensing robots are equipped with airflow sensors or an anemometer. In most cases, especially in early work, thermal thermistor-type airflow sensors are used [9, 20–22, 24, 35, 36]. When airflow strikes a heated thermistor element, the heat dissipation from the thermistor increases. Therefore, the resistance of the thermistor changes accordingly. Generally, a thermistor sensor cannot tell the direction of the airflow because the amount of the heat dissipation is the same for the airflow striking the thermistor element from the opposite direction. Therefore, in a typical setup, a pillar-like object is placed in the center, and four thermistors are aligned around the pillar (one in the front, one in the back, one on the left, and one on the right). The thermistor on the downwind side of the pillar shows the smallest resistance change since the pillar obstructs the airflow. Thus, the direction of the airflow can be determined from the response pattern of the four thermistors. The thermistor sensors are compact and inexpensive. However, when the airflow velocity goes below 0.05 m/s, the accuracy in measuring the airflow velocity and determining its direction degrades significantly. In such a low-velocity flow field, the effect of the upward convective air currents rising from heated thermistor elements cannot be neglected. Some robots have a mechanical vane to determine the airflow direction [30, 33].

Recently, ultrasonic anemometers are becoming popular in gas sensing robots [8, 23, 25, 26, 28]. The anemometer has pairs of ultrasonic transducers. The time of flight of ultrasonic bursts when sending from one transducer to another is measured for each pair and compared to the time for ultrasonic bursts to travel in the opposite direction. The speed and direction of the airflow are calculated from the collected time-of-flight measurements. A two-axis anemometer has two pairs of ultrasonic transducers to measure two orthogonal components of the airflow velocity vector. A three-axis anemometer has three pairs of transducers and can measure all three components of the three-dimensional airflow velocity. Ultrasonic anemometers are expensive and bulky since flight paths of a certain length (typically 0.1 m) are required to measure the small change in the flight time of ultrasonic bursts between the transducers. However, ultrasonic anemometers outperform thermistor sensors especially in the measurement of low flow velocity in indoor environments. With a typical ultrasonic anemometer, the direction of airflow as low as 0.01 m/s can be accurately measured.

6.3.3 Robot Platforms

Most of the mobile gas sensing robots reported so far have used small wheeled robots for their platforms to test the proposed gas sensing algorithms in laboratory environments. Only recently, robots equipped with a laser scanner or GPS devices have started to be used [8, 25, 26, 28, 38–41]. Self-localization, path-planning, and obstacle avoidance capabilities of these robots enable fully autonomous operation even in cluttered environments. A six-legged robot that mimics pheromone trail following of ants was reported in Ref. [5]. Legged robots generally offer better maneuverability but add complexity in controlling the locomotion of the robots. RoboLobster in Refs. [42–44] and the silkworm moth robot in Ref. [34] are specifically designed after their model animals. The size and locomotion speed of the robots are matched with those of the model animals so that various biological hypotheses on olfactory-guided behavior of the model animals can be tested on the robot platforms.

Although terrestrial mobile robots have been used in the majority of the work on mobile robot olfaction, there are some exceptions. Submersible wheeled robots [42–45] and swimming robots [46, 47] were used to find underwater chemical sources. Flying robots were developed to try three-dimensional gas source localization and gas distribution mapping [12, 48, 49]. Research efforts were made also to develop robotic systems to search for underground gas leaks [50, 51].

6.4 Characterization of Environments

Experiments to show the performance of gas sensing robots have been conducted in various environments. In this section, results of experiments conducted to characterize different types of environments are presented to show the difficulties involved in the gas sensing tasks. As described in Section 6.2.2, the gas distribution in a real environment can be extremely complex. Suppose that you are planning to do some experiments to see if your mobile robot can find a gas source placed in your laboratory. Unless you are an expert in mobile robot olfaction, you generally have no prior knowledge on the flow field generated in your laboratory. Therefore, you cannot predict how the released gas spreads from the source. Since the airflow field in the laboratory may change over time, reproducibility of the experiments is generally poor. For this reason, the experiments on gas sensing mobile robots were often conducted in wind tunnels [20, 24, 34–36]. When a gas source is placed in a strong uniform flow field generated in a wind tunnel, the gas plume trails in the straight downwind direction from the source. You can easily check the response of your robot to this gas plume. Alternatively, you can generate a unidirectional airflow field in your laboratory by placing an electric fan or an array of electric fans in your room [9, 22, 27, 32, 33]. Since the same airflow conditions can always be prepared by turning on the fans, a sequence of experiments can be conducted under similar conditions to statistically evaluate the performance of the robots. In some cases, experiments were conducted in places where steady airflow fields were generated by air-conditioning systems [21, 30].

Even in a closed room with all air-conditioning systems turned off, small temperature variations in the room cause airflow in the order of 0.01–0.1 m/s [12]. Under some conditions, a fairly steady convective flow field is generated in a room. We set up a room shown in Figure 6.4 for conducting experiments with mobile gas sensing robots [52]. Arrays of DC fans were placed on the floor and on one of the walls of a closed room. The right wall of the room was on the north side of the building, and the temperature there was almost always lower than at the other walls. When the fans were turned off, the air cooled down by the right wall descended along the wall and spread over the floor. The air then rose on the other side of the room and came back to the right wall along the ceiling. Thus, a weak circulating flow field was generated. The airflow velocity measured by an ultrasonic anemometer was in the range from 0.01 to 0.03 m/s. When the fans were turned on, a similar circulating flow field with higher velocities can be generated. The mobile robot shown in Figure 6.5 was used in the experiments to see the gas dispersion in this room [52]. The robot is equipped with e-nose systems consisting of arrays of metal-oxide gas sensors and a photoionization gas detector (ppbRAE 3000, RAE) on its bottom part to detect a gas hanging along the floor. A two-dimensional ultrasonic anemometer (WindSonic, Gill) is mounted on top of the robot. The ideal location of the anemometer is near the gas sensors so that the gas concentration and the airflow can be measured at the same position. However, a compromise had to be made because the anemometer also needs to be at a place where the airflow coming to the anemometer is not blocked by the body of the robot. A laser range scanner (LMS200, SICK) is also mounted on the robot for autonomous navigation and obstacle avoidance. The gas distribution map measured by the robot is shown in Figure 6.6 [52]. Saturated ethanol vapor was released from a tube placed on the floor at a constant flow rate (200 ml/min). The gas distribution map was drawn from the response of

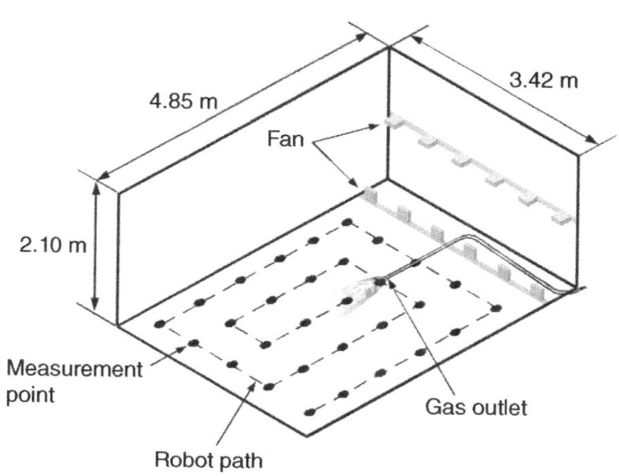

Figure 6.4 Setup of a room for generating a circulating airflow field that mimics natural convection (Source: From Ref. [52], figure 3, p. 185. Reproduced with permission from the Japan Society of Mechanical Engineers)

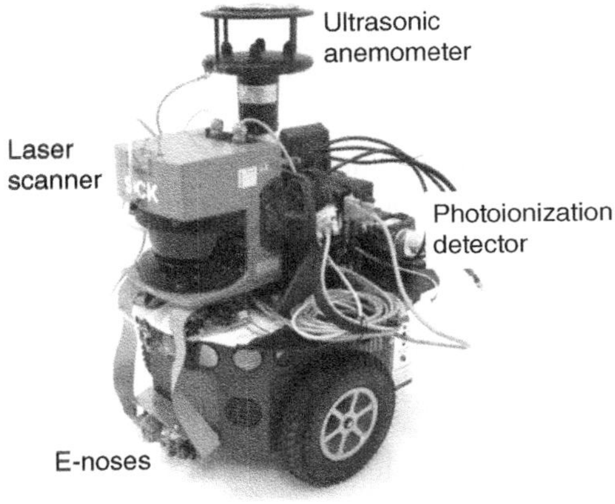

Figure 6.5 Mobile robot used for measuring the airflow vectors and gas distribution in various environments (Source: From Ref. [52], figure 1, p. 184. Reproduced with permission from the Japan Society of Mechanical Engineers)

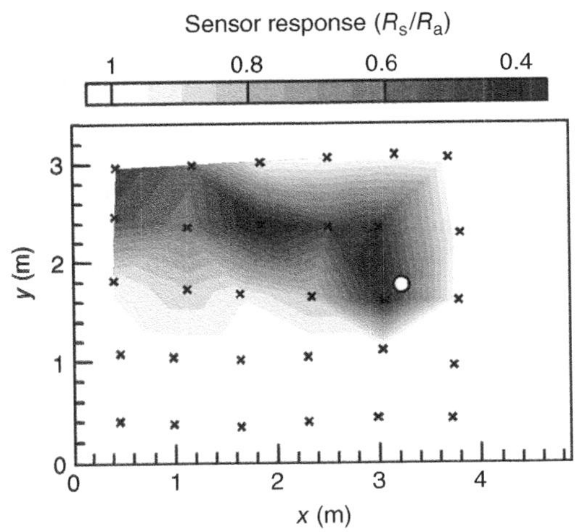

Figure 6.6 Gas distribution map in the closed experimental room shown in Figure 6.4. The open circle shows the location of the gas source. The crosses indicate the points at which the robot made stops for data collection (Source: From Ref. [52], figure 5, p. 185. Reproduced with permission from the Japan Society of Mechanical Engineers)

Figaro TGS2620 gas sensor in the e-nose. The sensor response is defined as the ratio of the sensor resistance in gas (R_s) to that in air (R_a). As the gas concentration increases, the resistance of the metal-oxide gas sensor decreases. Therefore, the value of the sensor response (R_s/R_a) also decreases. The robot was programmed to scan the environment along a predefined spiral path shown in Figure 6.4. The robot was also programmed to stop at specified measurement points for data collection. At each measurement point, the sensor data were recorded for 30 s. Figure 6.6 was drawn using the time average of the gas sensor response value for this 30 s time period. A gas plume trailing from the gas source location along the floor can be clearly seen in Figure 6.6. All DC fans were turned off during the gas distribution measurement, and therefore, the gas plume was formed by the natural convection in the room. The convective airflow along the floor was less than 0.03 m/s and was not perceivable to humans but enough to generate an airborne gas plume. The direction of the plume was always to the left (away from the north wall) in all repeated measurements.

Similar gas distribution mapping experiments were also done in a meeting room (14 m × 6 m) [52]. All desks and chairs were put aside as shown in Figure 6.7. A gas distribution map obtained in this environment is shown in Figure 6.8 together with the airflow velocity vectors measured at each measurement point [52]. The robot stopped for 30 s at each measurement point for data collection. All windows and doors were closed, and the air-conditioning system was turned off. A glass dish filled with liquid ethanol was used as the gas source. As shown in Figure 6.8, the airflow field in this meeting room was not uniform. Therefore, the ethanol vapor was widely but irregularly distributed in the room. As the size of the room increases, more complicated temperature distributions are generally developed. Therefore, the airflow field and the gas distribution also tend to become more complicated. Gas distribution mapping experiments were also done in an outdoor tennis court shown in Figure 6.9 [52]. In this case, the robot stopped at each measurement point for 10 s. The direction of the measured airflow vectors appears quite random, as shown in Figure 6.10 [52]. Therefore, the gas distribution became sporadic.

Figure 6.7 Meeting room as an example of uncontrolled indoor environment (Source: From Ref. [52], figure 6, p. 185. Reproduced with permission from the Japan Society of Mechanical Engineers)

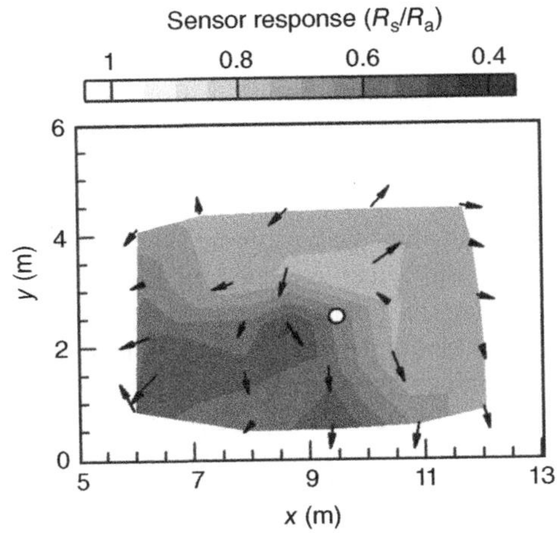

Figure 6.8 Gas distribution map in the meeting room shown in Figure 6.7. The open circle shows the location of the gas source. The arrows indicate the airflow velocity vectors measured at the grid points defined on the path of the robot (Source: From Ref. [52], figure 9, p. 186. Reproduced with permission from the Japan Society of Mechanical Engineers)

Figure 6.9 Outdoor tennis court (Source: From Ref. [52], figure 10, p. 186. Reproduced with permission from the Japan Society of Mechanical Engineers)

6.5 Case Studies

6.5.1 Chemical Trail Following

A worker ant lays a pheromone trail on her way back to the nest from a food source so that other ants can follow the trail. In principle, tracking of a stationary odor trail on the ground is much easier than tracking of an airborne turbulent odor plume. The behavioral strategy of the

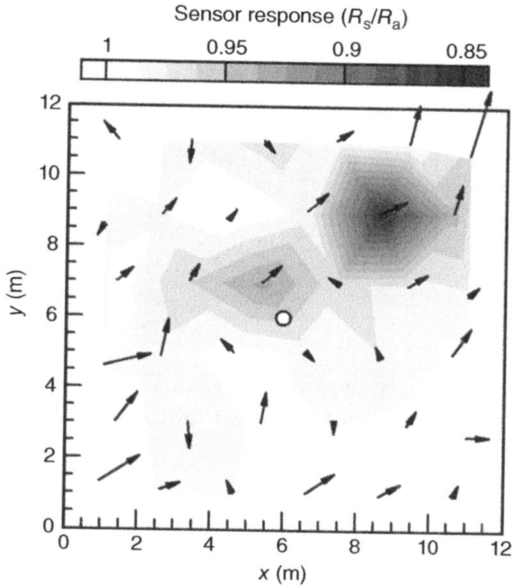

Figure 6.10 Gas distribution map in the outdoor tennis court shown in Figure 6.9. The open circle shows the location of the gas source. The arrows indicate the measured airflow velocity vectors (Source: From Ref. [52], figure 13, p. 187. Reproduced with permission from the Japan Society of Mechanical Engineers)

pheromone trail following by ants seems to be the bilateral comparison of the pheromone intensities detected by the two antennae [2, 3]. The variations in the pheromone concentration along the trail are extremely small, while the pheromone concentration variations across the trail can be easily detected by the bilateral comparison. Nevertheless, trail-following ants appear to be able to determine the direction in which a pheromone trail has been made [53].

It appears to be straightforward to develop a mobile robot with the chemical trail-following capability because the distribution of the chemical vapor evaporated from a stationary chemical trail is much more stable than an airborne gas plume. Indeed, successful demonstrations of chemical trail following for meters were reported [4, 5, 31]. In the early work of Russell and coworkers, a robot equipped with two QCM gas sensors was developed [4, 5]. Solid camphor was dissolved in an organic solvent, and the solution was applied on the floor to lay a chemical trail. The robot successfully followed the camphor trail by bilateral comparison of the sensor responses. Later, several practical issues were pointed out. Figure 6.11 shows the improved version of Russell's chemical trail-following robot. To avoid the problems incurred by mismatches in sensitivities and response times of the left and right gas sensors, the robot shown in Figure 6.11 has only one QCM sensor on its front. It follows a chemical trail by swinging the sensor transversely and measuring the change in the response of the sensor with its location. Moreover, the QCM sensor is now mounted under a DC fan that generates an air curtain. The flow pattern that the fan generates is designed so that the chemical vapor is brought to the sensor only from the trail just beneath the sensor. Thus, reception of chemical vapor from nearby chemical trails is avoided. In the floor-cleaning scenario mentioned in Section 6.2.1, multirobot cooperation was accomplished by programming the robots to avoid chemically

Figure 6.11 Reactive Autonomous Testbed (RAT) carrying a QCM gas sensor for chemical trail following (Source: R. A. Russell, Monash University, Clayton, Victoria, Australia. Reproduced with permission from R. A. Russell)

marked areas. However, the robots driven by this simple algorithm may leave some areas not visited by any robots. Therefore, an advanced algorithm was proposed to guarantee the full area coverage [17]. In the proposed algorithm, navigations using chemical trails and mapping based on cellular decomposition of the area being explored were combined.

6.5.2 Chemotactic Search versus Anemotactic Approach

The most straightforward approach for gas source localization is to move a mobile robot along the gas concentration gradient. In this approach, a robot is generally equipped with a pair of gas sensors and is programmed to steer toward the side with the higher gas concentration [18, 19, 32]. However, there are some issues in applying this approach in real environments. As described in Section 6.2.2, the airflow we encounter is turbulent. Therefore, a gas plume with a patchy intermittent structure is formed owing to a number of eddies in the turbulent flow. Local and instantaneous concentration gradients show large fluctuations. Therefore, the robot is often misled by erroneous gradient measurements.

If the temporal variations in the airflow direction are not so large, the time-averaged gas concentration distribution has a smooth shape as shown in Figure 6.12. The robot is more reliably oriented toward the gas source if the time-averaged values of the left and right gas sensor responses are compared rather than using their instantaneous values. However, the gas concentration gradient along the airflow direction is extremely small except in the vicinity of the gas source. The gas concentration along the plume centerline is generally inversely proportional to the distance from the gas source [54]. Downstream from the source along the plume, the gas concentration decays rapidly at the beginning. However, this rapid decay ceases at some distance, and after that, the gas concentration asymptotically approaches zero. As the plume extends downstream, its width grows and its chemical content becomes more diluted. Therefore, the gas concentration gradient across the plume also becomes smaller at downstream locations. Robot 1 in Figure 6.12 has successfully approached the source location by tracking the gas concentration gradient. However, robot 2 who has started from a distant place may not reach the gas source location. In theory, the gas concentration gradient

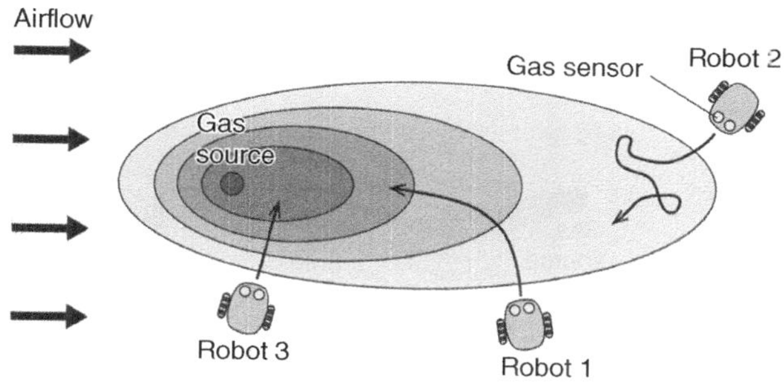

Figure 6.12 Behavior of robots tracking the gas concentration gradient

is always perpendicular to the contour lines of the gas distribution map. Therefore, the gradient vector at the side of the plume near the gas source tends to point toward the downwind direction. Thus, robot 3 in Figure 6.12 does not go directly to the source location but makes a detour in the downwind direction. A slight change in the airflow direction causes large fluctuations in the gas concentration in the vicinity of the gas source. Therefore, the time-course waveform of the gas concentration measured using a gas sensor is peaky. In contrast, a slight shift in the position of a wide diluted plume at a far downstream location causes little concentration fluctuations. Thus, not only the measured gas concentration itself but also the frequency of the peaks in the gas concentration can be used as an index showing the proximity to a gas source [29, 55].

Despite the difficulties in tracking turbulent odor plumes, many animals can localize food, mates, or nests by navigating toward odor sources using olfactory cues [7]. Extensive studies conducted to reveal the behavioral strategy underlying pheromone plume tracking of male moths have shown that their flight path to a pheromone-releasing female is a result of repeated "upwind surges" and "casting" [7, 56]. A male moth has pheromone receptors on its pair of antennae [2]. In principle, he can steer toward the side with the higher pheromone concentration by comparing the stimulus intensities at the left and right antennae. Instead, in reality, when a male moth encounters a pheromone plume and detects the pheromone on his antennae, it turns and proceeds a certain distance in the upwind direction. This odor-gated anemotaxis (or odor-gated rheotaxis in water) is commonly seen in olfactory search behavior of a variety of animal species [7]. As long as the moth is in contact with the pheromone plume, repeated "upwind surges" bring the moth closer to the female. As shown in Figure 6.2, the pheromone plume randomly meanders because of the turbulence of the airflow carrying the pheromone molecules. Therefore, the male moth flying upwind may sometimes leave the plume accidentally. To cope with such situations, the moth flies across the wind with gradually broadened scanning widths when the pheromone signal is lost. This behavior is called "casting." When the contact with the plume is regained, the male resumes upwind surges.

As in the case of successful flights of male moths to females, robots searching for a gas source can use the direction of the airflow as a reliable directional cue. Therefore, mobile robots equipped with airflow sensors in addition to gas sensors have been developed [9, 20–26, 30, 33, 35, 36]. The basic part of the algorithms implemented into these robots is the same although the details of the algorithms are different. As shown in Figure 6.13, a robot equipped with an anemometer goes upwind when the target gas is detected. When the gas is not detected,

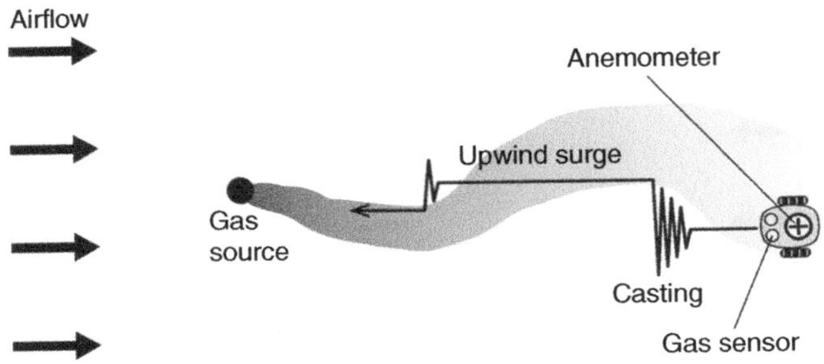

Figure 6.13 Behavior of robots tracking the gas plume with repeated upwind surges and casting

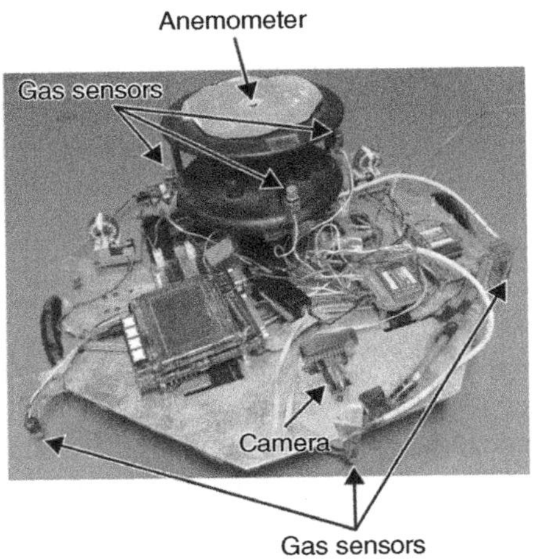

Figure 6.14 TUAT mobile robot equipped with metal-oxide gas sensors, a two-dimensional ultrasonic anemometer, and a CMOS camera

the robot performs casting-like behavior. Some robots go straight upwind as long as the contacts with the plumes are maintained [9, 24, 35, 36]. Some have multiple gas sensors for lateral comparison of gas concentrations in the plume so that they can adjust their headings toward the plume centerline while moving upwind [20–22]. Some robots perform casting exactly like moths [21, 35]. However, some robots move along outward spiral trajectories when the contact with the plume is lost [9, 22, 24]. In theory, casting across the airflow direction maximizes the probability of regaining contact with the plume because the plume has a shape elongated along the airflow direction [57]. Since most of the gas source localization robots reported so far have a differential drive system, moving them along spiral trajectories can be easily done by adjusting the speeds of the left and right wheels.

An example of an anemotactic gas source localization robot is shown in Figure 6.14. The robot has three metal-oxide gas sensors (TGS2620, Figaro Engineering) attached on a

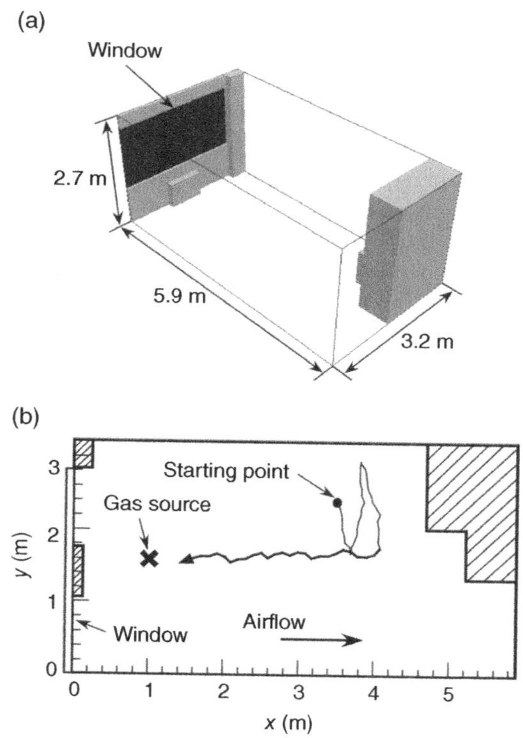

Figure 6.15 Result of upwind plume tracking by TUAT robot. (a) Schematic diagram of the room in which the experiments were performed (Source: From Ref. [10], figure 8(a), p. 3168. Reproduced with permission from IEEE). (b) Typical trajectory of the robot. Thick and thin lines represent trajectories of the robot in upwind surges and casting, respectively) (Source: From Ref. [58], figure 7.5, p. 196. Reproduced with permission from Fragrance Journal Ltd.)

two-dimensional ultrasonic anemometer (WindSonic, Gill) [23]. The trajectory of the robot successfully tracking a gas plume is shown in Figure 6.15 [10, 58]. The experiments were conducted in a closed room with a large window as shown in Figure 6.15a. In wintertime or at night even in summertime, the temperature of the window facing outside generally becomes lower than at the other places in the room. Therefore, circulating airflow field is generated as in the room shown in Figure 6.4. The air cooled down at the window descends along the window and spreads over the floor. When the experiments shown in Figure 6.15b were conducted, the airflow velocity on the floor at the center of the room was only 0.08 m/s, which is below the human detection limit of airflow. Saturated ethanol vapor was generated by bubbling air through liquid ethanol and was released at a flow rate of 100 ml/min from a tube placed on the floor. Since the starting position of the robot was off the ethanol plume, the robot first conducted casting across the airflow. When the robot entered the ethanol plume, it tracked the plume in the upwind direction and successfully approached the gas source.

6.5.3 *Attempts to Improve Gas Source Localization Robots*

As described in Section 6.2.1, the task of chemical source localization can be divided into three subtasks, that is, chemical finding, chemical tracking, and chemical source declaration. Most of the work on gas source localization robots was focused on chemical tracking, and the other two subtasks were not often discussed in the literature. It seems that there is not much we can do for chemical finding. As described in Section 6.2.1, when a robot is brought into a new environment, it first needs to find the area in which the target gas is present. However, the gas sensors on the robot show no response until the robot happens to run into that area. Therefore, the initial chemical finding must be accomplished without having a means to predict where the gas is distributed. A robot can systematically scan the neighborhood for a gas distribution by moving along an outward spiral path [9]. Casting across the airflow is effective not only for regaining contact with a gas plume but also for initial plume finding. In theory, if no gas is detected at a certain position, it is likely that there is no gas source in the area upwind from that position. Moving across the airflow direction maximizes the probability of encountering a gas plume that trails along the airflow direction [57]. Alternatively, scanning the environment in parallel using multiple robots can expedite the initial chemical search [59].

In most of the work reported so far, source declaration algorithms were not implemented on the robots to avoid too many complications in the experiments. In most cases, the search was simply terminated when the robot came within a specified radius of the gas source [20–22, 24, 27, 33, 35, 36]. When the robot passes over the source, a sudden change in the gas concentration is observed in the gas sensor response. Almost no chemical exists on the upstream side of the source, while the gas concentration becomes the highest on the immediate downstream side of the source. This feature can be used to declare that the source has been found [9, 30, 47]. However, care should be taken to correctly determine whether the robot has passed over the source or it has accidentally left the plume to the side. If a chemical source is placed directly on the floor, it takes some distance for the height of the chemical plume trailing along the floor to grow to the height of the chemical sensors on the robot. When a robot with multiple chemical sensors at different heights approaches the chemical source by tracking the plume, the sensor at the highest position is the first that stops responding to the chemical, followed by the second, third, and so forth. This change in the sensor response pattern caused by the change in the plume height can be used for source declaration. Male moths appear to use visual cues to terminate the upwind flight toward a pheromone-releasing female. It is known that a male moth initiates mating behavior when he visually recognizes a female [7]. Attempts were made to incorporate vision into the search for a chemical source. When we notice a burning smell, we turn around and look for a fire or smoke. If the appearance of the gas source is known a priori, a robot can find a suspicious object using a camera. Then, the robot can approach the object to check its smell using gas sensors for source declaration [23, 60, 61]. The robot shown in Figure 6.14 has a CMOS camera and metal-oxide gas sensors on its lower front side.

Attempts were also made to find the optimal upwind plume-tracking algorithm by comparing variations of algorithms by theoretical analysis, simulations, and experiments [24, 33, 35, 62, 63]. However, it is difficult at this moment to derive general conclusions from such work since different research groups are working with different robots in different environments. It seems so far that the differences in experimental setups have more effects on the gas source localization performance than the differences of the algorithms. Moreover, the comparison was so far made mostly in simplified environments, for example, in a wind tunnel or in an artificially generated uniform airflow field. To what extent the algorithms can cope with temporal variations and nonuniformity of the airflow fields in real environments is not yet fully investigated. Suppose that a robot is tracking a gas plume in an outdoor environment. If a casting algorithm is implemented in the robot, it can cope with a shift in the airflow direction [36]. When the airflow direction changes, the plume starts to trail in the new direction. Then, the robot starts casting across the new airflow direction with gradually expanding scanning width. This motion eventually brings the robot back into the plume, which is now trailing at a new location.

If the fluctuations in the airflow direction are too large, a different search algorithm may be required [25, 26]. In an open outdoor field, the wind direction frequently changes, but the uniformity of the flow can be assumed for the instantaneous flow field. If the wind is blowing to the north at the position of the robot, it is highly likely that the wind at the position of the gas source is also blowing to the north. If the wind direction at the robot position changes, the wind direction at the gas source location is also likely to change in the same way. A puff of gas released from the gas source travels along a complicated trajectory as a result of the changes in the airflow direction. When a puff of gas is detected at a gas sensor on the robot, the trajectory along which the puff of gas has arrived to the robot can be estimated using the history of the airflow direction measured over a certain period of time by the anemometer on the robot. A particle filter implementing this estimation process was proposed in Ref. [25]. Successful plume tracking over tens of meters in an outdoor field was reported for this particle filter-based algorithm. This approach can also be used for gas source declaration because the estimated trajectories of gas puffs should in theory converge to the location of the gas source [25, 49].

Various algorithms were also proposed to cope with unstable flow fields and to speed up the search for a gas source. Some algorithms use mathematical models of the plume shape to estimate the source location from a limited number of measurements [64–66]. Some use a mathematical formulation of advection of chemical substances to determine the most preferable heading of a robot [67, 68]. Chemical source localization algorithms based on hidden Markov methods, Bayesian inference methods, naïve physics, and particle filters were also reported [25, 69–71]. If multiple robots are employed to search for a gas source in parallel or cooperatively, the speed and the robustness of the search can be improved [9, 18, 59, 72]. In some applications, for example, landmine detection, a robot is required to find all gas sources present in a given environment. Even if a gas source is found by tracking a gas plume, it may not be the only one. A search algorithm based on a Bayesian occupancy grid mapping method was proposed for mapping out all gas sources in the environment [73].

Gas source localization in an environment cluttered with obstacles is also challenging. In some work, obstacle avoidance functions were integrated with gas source localization algorithms [27, 30]. Typically, a limited number of obstacles were placed in an environment. A robot with an obstacle avoidance function successfully approached a gas source by moving through the obstacles. However, it appears that the obstacles placed in the environment were rather small and that their presence did not significantly alter the environment. Real challenges arise from large obstacles that significantly modify the flow field and chemical distribution in the environment. The flow immediately behind an obstacle is generally highly fluctuating because the object disturbs the flow. The size of this wake region is determined by various factors, but it surely depends on the size of the object. If there is an object in the environment whose size is several times larger than the gas source localization robot, a large wake region is generated. If the robot comes completely in the wake of the object, the gas sensors and the anemometer on the robot are exposed to highly fluctuating airflow and gas concentration. The robot may need to escape from such a region before resuming the search for a gas source.

6.5.4 *Flying, Swimming, and Burrowing Robots*

In a typical experiment on robotic gas source localization, a wheeled robot searches for a gas source by tracking a gas plume trailing along the floor. The location to place the gas source was carefully chosen to ensure that the released gas is carried by airflow blowing constantly along the floor. Even electric fans were often used to generate such an airflow. However, gas plumes in reality may not always stay on the floor or the ground. They may extend three-dimensionally. A flying blimp robot was developed aiming at tracking a three-dimensional aerial plume by mimicking the flight of moths [48]. Recently, a quadrotor-type microdrone equipped with a metal-oxide gas sensor was reported [49]. The microdrone succeeded in tracking a methane plume over 10 m in an outdoor environment by using upwind plume-tracking algorithms and a particle filter-based source declaration algorithm. The three-dimensionality of gas plumes may be more prominent in indoor environments. In a closed indoor environment as shown in Figures 6.4 and 6.7, convection caused by the temperature distribution in the room is the main force that spreads gas molecules released from their source. A gas plume generated in such an environment is three-dimensional in nature. A body of air rises if it is warmer than its surroundings and sinks if it is colder. Therefore, the convection always involves vertical fluid motion. A blimp robot was successfully applied to three-dimensional gas distribution mapping in indoor environments [12]. Robo-moth in Ref. [37] is a unique experimental setup developed to study the behavioral mechanisms of pheromone orientation in moths. Three-dimensional flight of a moth in a wind tunnel was simulated by moving a pheromone detector in the wind tunnel by a robotic arm attached at the end of the wind tunnel.

Not only terrestrial animals but also many aquatic animals have a keen sense of smell. Vision is not always useful in aquatic environments since the sunlight intensity decreases with the depth in water. RoboLobster was designed to serve as a test bed to study the olfactory search behavior of lobsters [42–44]. Lobsters can search for food by tracking its smell. Various hypotheses on the underlying behavioral mechanisms can be tested using this wheeled underwater robot by changing the computer program for the robot and its sensor configuration. The original RoboLobster was equipped with a pair of conductivity sensors, and salt water was used as a detection target. The second version of RoboLobster has optical sensors that respond to a fluorescent dye. The performance of purely chemotactic search algorithms and rheotactic algorithms similar to pheromone plume tracking in moths was tested and compared with the olfactory search capabilities of live lobsters. Similar fluorescent sensors were mounted on a torpedo-shaped underwater robot platform called REMUS (remote environmental monitoring units) [46, 47]. A full set of algorithms for chemical source localization, that is, initial plume finding based on zigzag scanning, rheotactic plume tracking, casting, and source declaration, were implemented in the robot. In the experiments reported in Ref. [47], the robot successfully tracked fluorescent dye plumes as long as 100 m in a real marine environment near San Clemente Island, CA. Crayfish are known to generate directed water currents by waving their small appendages with a fanlike shape [74]. A wheeled submersible robot equipped with electrochemical sensors and fanning arms was developed to mimic the olfactory search behavior of crayfish [45]. Water currents generated by using the fanning arms were used to draw water samples from the surroundings to the chemical sensors on the robot. Thus, the chemical signal reception at the sensors mounted on the robot was enhanced. A similar behavior for collecting chemical samples using actively generated flow can also be seen in many other animal species [75].

Leaks from underground gas pipes and gas tanks pose serious hazards [10]. However, there is no effective means to determine the location of the leaks. The distribution of a chemical substance released into the soil is much more stable than that of chemical released into the air or water. Little or no turbulence exists in underground environments since there is almost no fluid flow in the soil. However, the difficulty lies in moving a robot or at least a chemical sensor in soil. Considering the high cost for moving through soil, the length of the path that the robot travels has to be minimized, and therefore, the underground chemical source localization task needs to be accomplished with a minimal number of measurement points [50]. It is also difficult to estimate the location of an underground gas leak using only the gas concentration measurements from above the ground [10]. For example, the path of gas diffusion from the leak point to the ground surface is altered if the soil is covered by pavement. The MOLE I robot reported in Ref. [50] takes readings of underground chemical concentration by inserting a metal-oxide gas sensor in sand. The robot tries to locate a chemical source by tracking the spatial chemical concentration gradient in the sand, which is estimated from the gas sensor readings obtained at a sequence of locations on the robot's path. CRABOT in Ref. [51] has a real burrowing mechanism mimicking a mole crab underground.

6.5.5 *Gas Distribution Mapping*

As described in Section 6.2.1, the goal of gas distribution mapping is to build a reliable map from fluctuating gas sensor responses measured in a turbulent airflow field at coarse grid points. Interpolation algorithms often produce artifacts in gas distribution maps because of the highly intermittent nature of the turbulent plume [1]. Therefore, a kernel-based extrapolation algorithm was proposed [76]. In this approach, a gas concentration grid map is created from a sequence of gas sensor readings collected by a mobile robot while scanning a given environment. Suppose that a gas sensor reading r_t is obtained at time t when the robot is at the location $\mathbf{x}_t$. If another measurement is conducted at $\mathbf{x}_t + \Delta\mathbf{x}$, which is in the neighbor of $\mathbf{x}_t$, it is likely that a similar gas sensor reading is obtained. To model this likelihood, a two-dimensional Gaussian weighting function is defined as

$$f(\Delta\mathbf{x}) = \frac{1}{2\pi\sigma^2}\exp\left(-\frac{|\Delta\mathbf{x}|^2}{2\sigma^2}\right) \tag{6.1}$$

where σ defines the size of the Gaussian kernel. For the grid cell at the ith row and the jth column, the weight can be calculated as

$$w_t^{(i,j)} = \begin{cases} f\left(\mathbf{x}^{(i,j)} - \mathbf{x}_t\right) & \left|\mathbf{x}^{(i,j)} - \mathbf{x}_t\right| \le R_C \\ 0 & \left|\mathbf{x}^{(i,j)} - \mathbf{x}_t\right| > R_C \end{cases} \tag{6.2}$$

where $\mathbf{x}^{(i,j)}$ represents the center of the grid cell (i, j) and R_C the cutoff radius. Then, two temporary values maintained for each grid cell are updated. One is the total sum of the weights for all sensor readings obtained so far, which is defined as

$$W_t^{(i,j)} = \sum_{t'=0}^{t} w_{t'}^{(i,j)}. \tag{6.3}$$

The other is the total sum of weighted gas sensor readings:

$$WR_t^{(i,j)} = \sum_{t'=0}^{t} r_{t'} w_{t'}^{(i,j)}. \tag{6.4}$$

If $W_t^{(i,j)}$ exceeds the threshold value $W_{\min}$, the value of the grid cell (i, j) is set to

$$c_t^{(i,j)} = \frac{WR_t^{(i,j)}}{W_t^{(i,j)}} \quad W_t^{(i,j)} \geq W_{\min}. \tag{6.5}$$

The robot used to try gas distribution mapping in various environments is shown in Figure 6.16 [28]. It is equipped with a three-dimensional ultrasonic anemometer (Model 81000, Young) and an e-nose consisting of six metal-oxide gas sensors. An example of a gas distribution grid map is shown in Figure 6.17 [28]. In the experiment, the robot followed a predefined sweeping trajectory covering the area of interest. Along its path, the robot stopped at predefined positions and carried out a sequence of measurements for 30 s. The robot was driven at a maximum speed of 0.05 m/s between the stops. The sweeping motion was performed once in each direction. The gas source used in the experiment was a small cup filled with liquid ethanol. The gas distribution map shown in Figure 6.17 was obtained in an enclosed indoor area consisting of three rooms partially separated by protruding walls. The area covered by the sweeping path of the robot was approximately 14 m × 6 m. The map was created using the readings of TGS 2600 sensor (Figaro Engineering) in the e-nose system with a kernel size

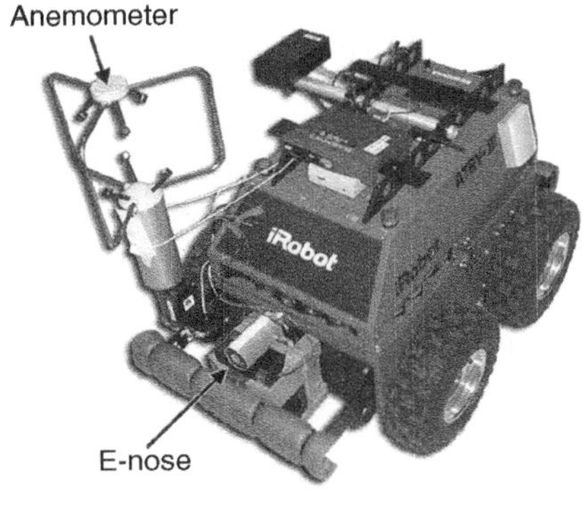

Figure 6.16 Rasmus robot equipped with an electronic nose and an anemometer (Source: From Ref. [28], figure 1, p. 2211. Reproduced with permission from IEEE)

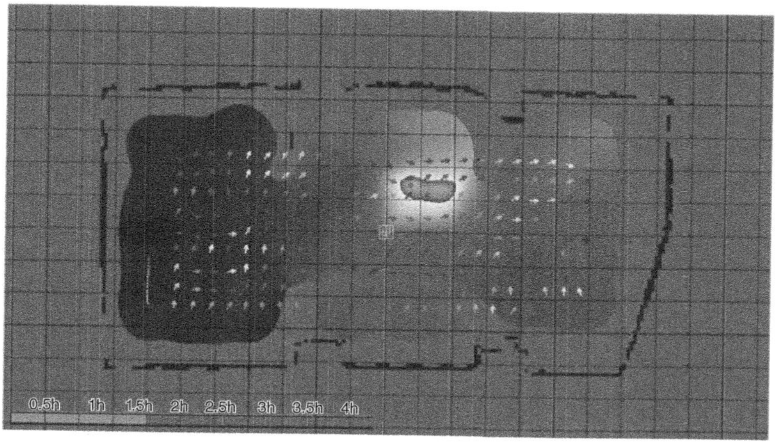

Figure 6.17 Gas distribution map obtained in an enclosed indoor space overlaid with arrows indicating the airflow direction. The occupancy grid map showing walls and obstacles in the environment is also overlaid (Source: From Ref. [28], figure 4, p. 2215. Reproduced with permission from IEEE)

of 0.6 m. The occupancy grid map showing the walls of this indoor space and arrows indicating the airflow vectors measured at the stops are overlaid on the gas distribution map. The maximum of the gas distribution map (highlighted in Figure 6.17) occurred near the gas source location. Despite the wide passage between the three rooms, the gas concentration in the leftmost room was still much lower than in other two rooms.

The kernel gas distribution mapping method was later improved so that not only the mean gas distribution but also its predictive variance could be estimated [77]. It was also shown that the gas distribution can be estimated with greater accuracy if the kernel shape is modified to take airflow measurements into account [78]. Three-dimensional gas distribution mapping was also reported in the literature. A wheeled mobile robot having multiple gas sensors at different heights [78] and a blimp robot flying randomly in a room [12] were used for collecting gas concentration measurements at various locations. Robotic gas distribution mapping can also be applied to measure a citywide air-pollutant distribution [79].

6.6 Future Prospective

Over the past two decades, various research efforts were made to realize mobile robot olfaction. As a result, it has been shown that mobile robots equipped with chemical sensors could perform, at least in simple environments, chemical trail following, chemical source localization, and chemical distribution mapping. Now the research on mobile robot olfaction seems to be getting into a next phase in which the research efforts are more directed toward real applications. It may be too ambitious to develop a versatile gas source localization robot that can be used in any environment. Generally, it is much easier to build a robot specialized for a specific target application. Based on this idea, algorithms to search for gas sources in narrow but branched cave-like environments were proposed ([80, 81]). The development of gas source localization algorithms can be expedited if the applicability of the proposed algorithms to various environments can be quickly tested through simulations. However, to provide realistic models of complicated chemical distributions in real environments to the simulator is a challenging task. So far, a filamentous plume model in Ref. [82] is most widely used. However, little is known about how the complicated airflow field and gas distributions as shown in Figures 6.8 and 6.17 are developed. Further research efforts are required to collect airflow/gas distribution data in a larger variety of environments to better characterize each environment. One of the promising applications for gas sensing mobile robots is to find methane leaks in landfill sites [8]. Methane gas produced from solid waste is not only hazardous but also known to be a greenhouse gas. Periodical monitoring of methane gas concentration at a landfill site is often required by laws and appears to be an ideal task for mobile robots. We believe that the day we see chemical sensing robots working in real applications is not so far away.

Supplemental materials of Chapter 6 (video clips) are provided online at URL.

Acknowledgment

The authors would like to thank Takashi Ushiku, Daisuke Kobayashi, Marco Trincavelli, Matteo Reggente, Yuichiro Fukazawa, and Yuta Wada for their contributions in the gas source localization and gas distribution mapping experiments described in this chapter. Hiroshi Ishida and Haruka Matsukura are thankful for the financial support under JSPS KAKENHI Grant Number 25289055.

References

[1] A. J. Lilienthal, A. Loutfi, and T. Duckett, Airborne chemical sensing with mobile robots, *Sensors*, vol. 6, no. 11, pp. 1616–1678, 2006.

[2] W. C. Agosta, *Chemical Communication: The Language of Pheromones (Scientific American Library)*. New York, USA: W. H. Freeman, 1992.

[3] D. B. Dusenbery, *Sensory Ecology: How Organisms Acquire and Respond to Information*. New York, USA: W. H. Freeman, 1992.

[4] R. Deveza, D. Thiel, A. Russell, and A. Mackay-Sim, Odor sensing for robot guidance, *Int. J. Robot. Res.*, vol. 13, no. 3, pp. 232–239, 1994.

[5] R. A. Russell, *Odour Detection by Mobile Robots*. Singapore: World Scientific, 1999.

[6] W. J. Bell and T. R. Tobin, Chemo-orientation, *Biol. Rev.*, vol. 57, no. 2, pp. 219–260, 1982.

[7] E. A. Arbas, M. A. Willis, and R. Kanzaki, Organization of goal-oriented locomotion: Pheromone-modulated flight behavior of moths, in *Biological Neural Networks in Invertebrate Neuroethology and Robotics*, R. D. Beer, R. E. Ritzmann, and T. McKenna, Eds. San Diego, CA, USA: Academic Press, 1993, pp. 159–198.

[8] V. Hernandez Bennetts, A. J. Lilienthal, P. P. Neumann, and M. Trincavelli, Mobile robots for localizing gas emission sources on landfill sites: Is bio-inspiration the way to go?, *Front. Neuroroeng.*, vol. 4, article 20, 2012.

[9] A. T. Hayes, A. Martinoli, and R. M. Goodman, Distributed odor source localization, *IEEE Sensors J.*, vol. 2, no. 3, pp. 260–271, 2002.

[10] H. Ishida, Y. Wada, and H. Matsukura, Chemical sensing in robotic applications: A review, *IEEE Sensors J.*, vol. 12, no. 11, pp. 3163–3173, 2012.

[11] M. Wandel, A. Lilienthal, T. Duckett, U. Weimar, and A. Zell, Gas distribution in unventilated indoor environments inspected by a mobile robot, in Proceedings of the IEEE International Conference on Advanced Robotics (ICAR 2003), Coimbra, Portugal, June 30–July 3, 2003, pp. 507–512.

[12] H. Ishida, Blimp robot for three-dimensional gas distribution mapping in indoor environment, *AIP Conf. Proc.*, vol. 1137, pp. 61–64, 2009.

[13] J. Murlis, J. S. Elkinton, and R. T. Cardé, Odor plumes and how insects use them, *Annu. Rev. Entomol.*, vol. 37, pp. 505–532, 1992.

[14] A. Göth, I. G. McLean, and J. Trevelyan, Part 1. How do dogs detect landmines? A summary of research results, in *Mine Detection Dogs: Training, Operations and Odour Detection*, I. G. McLean, Ed. Geneva, Switzerland: International Centre for Humanitarian Demining, 2003, pp. 195–208.
[15] B. H Sandler, L. Nikonova, W. S. Leal, and J. Clardy, Sexual attraction in the silkworm moth: Structure of the pheromone-binding-protein–bombykol complex, *Chem. Biol.*, vol. 7, no. 2, pp. 143–151, 2000.
[16] J, Yinon, Detection of explosives by electronic noses, *Anal. Chem.*, vol. 75, no. 5, pp. 98A–105A, 2003.
[17] S. Larionova, N. Almeida, L. Marques, and A. T. de Almeida, Olfactory coordinated area coverage, *Auton. Robot.*, vol. 20, no. 3, pp. 251–260, 2006.
[18] G. Sandini, G. Lucarini, and M. Varoli, Gradient driven self-organizing systems, in Proceedings of the 1993 IEEE/RSJ International Conference on Intelligent Robots and Systems (IROS 1993), Yokohama, Japan, July 26–30, 1993, vol. 1, New York: IEEE pp. 429–432.
[19] A. Lilienthal and T. Duckett, Experimental analysis of gas-sensitive Braitenberg vehicles, *Adv. Robot.*, vol. 18, no. 8, pp. 817–834, 2004.
[20] H. Ishida, K. Suetsugu, T. Nakamoto, and T. Moriizumi, Study of autonomous mobile sensing system for localization of odor source using gas sensors and anemometric sensors, *Sensors Actuat. A*, vol. 45, no. 2, pp. 153–157, 1994.
[21] H. Ishida, Y. Kagawa, T. Nakamoto, and T. Moriizumi, Odor-source localization in the clean room by an autonomous mobile sensing system, *Sensors Actuat. B*, vol. 33, nos. 1–3, pp. 115–121, 1996.
[22] H. Ishida, G. Nakayama, T. Nakamoto, and T. Moriizumi, Controlling a gas/odor plume-tracking robot based on transient responses of gas sensors, *IEEE Sensors J.*, vol. 5, no. 3, pp. 537–545, 2005.
[23] H. Ishida, T. Ushiku, S. Toyama, H. Taniguchi, and T. Moriizumi, Mobile robot path planning using vision and olfaction to search for a gas source, in Proceedings of Fourth IEEE International Conference on Sensors (IEEE Sensors 2005), Irvine, CA, USA, October 31–November 3, 2005, pp. 1112–1115.
[24] T. Lochmatter, X. Raemy, L. Matthey, S. Indra, and A. Martinoli, A comparison of casting and spiraling algorithms for odor source localization in laminar flow, in Proceedings of the 2008 IEEE International Conference on Robotics and Automation (ICRA 2008), Pasadena, CA, USA, May 19–23, 2008, pp. 1138–1143.
[25] J. G. Li, Q. H. Meng, Y. Wang, and M. Zeng, Odor source localization using a mobile robot in outdoor airflow environments with a particle filter algorithm, *Auton. Robot.*, vol. 30, no. 3, pp. 281–292, 2011.
[26] Q. H. Meng, J. G. Li, Y. Wang, and M. Zeng, Odor source searching with a mobile robot in outdoor open areas, in Proceedings of International Symposium on Olfaction and Electronic Nose (ISOEN 2013), Daegu, Korea, July 2–5, 2013, 2 pages.
[27] L. Marques, U. Nunes, and A. T. de Almeida, Olfaction-based mobile robot navigation, *Thin Solid Films*, vol. 418, no. 1, pp. 51–58, 2002.
[28] M. Trincavelli, M. Reggente, S. Coradeschi, A. Loutfi, H. Ishida, and A. J. Lilienthal, Towards environmental monitoring with mobile robots, in Proceedings of the 2008 IEEE/RSJ International Conference on Intelligent Robots and Systems (IROS 2008), Nice, France, September 22–26, 2008, pp. 2210–2215.
[29] G. Ferri, E. Caselli, V. Mattoli, A. Mondini, B. Mazzolai, and P. Dario, SPIRAL: A novel biologically-inspired algorithm for gas/odor source localization in an indoor environment with no strong airflow, *Robot. Auton. Syst.*, vol. 57, no. 4, pp. 393–402, 2009.
[30] R. A. Russell, D. Thiel, R. Deveza, and A. Mackay-Sim, A robotic system to locate hazardous chemical leaks, in Proceedings of the 1995 IEEE International Conference on Robotics and Automation (ICRA 1995), Nagoya, Japan, May 21–27, 1995, vol. 1, pp. 556–561.
[31] E. Stella, F. Musio, L. Vasanelli, and A. Distante, Goal-oriented mobile robot navigation using an odour sensor, in Proceedings of the Intelligent Vehicles ’95 Symposium, Detroit, MI, USA, September 25–26, 1995, pp. 147–151.
[32] S. Kazadi, R. Goodman, D. Tsikata, D. Green, and H. Lin, An autonomous water vapor plume tracking robot using passive resistive polymer sensors, *Auton. Robot.*, vol. 9, no. 2, pp. 175–188, 2000.
[33] R. A. Russell, A. Bab-Hadiashar, R. L. Shepherd, and G. G. Wallace, A comparison of reactive robot chemotaxis algorithms, *Robot. Auton. Syst.*, vol. 45, no. 2, pp. 83–97, 2003.
[34] Y. Kuwana, S. Nagasawa, I. Shimoyama, and R. Kanzaki, Synthesis of the pheromone-oriented behaviour of silkworm moths by a mobile robot with moth antennae as pheromone sensors, *Biosens. Bioelectron.*, vol. 14, no. 2, pp. 195–202, 1999.
[35] D. J. Harvey, T. Lu, and M. A. Keller, Comparing insect-inspired chemical plume tracking algorithms using a mobile robot, *IEEE Trans. Robot.*, vol. 24, no. 2, pp. 307–317, 2008.
[36] D. J. Harvey, T. Lu, and M. A. Keller, Effectiveness of insect-inspired chemical plume-tracking algorithms in a shifting wind field, *IEEE Trans. Robot.*, vol. 24, no. 1, pp. 196–201, 2008.

[37] A. J. Rutkowski, S. Edwards, M. A. Willis, R. D. Quinn, and G. C. Causey, A robotic platform for testing moth-inspired plume tracking strategies, in Proceedings of the 2004 IEEE International Conference on Robotics and Automation (ICRA 2004), New Orleans, LA, USA, April 26–May 1, 2004, vol. 4, pp. 3319–3324.
[38] M. Trincavelli, S. Coradeschi, and A. Loutfi, Classification of odours with mobile robots based on transient response, in Proceedings of the 2008 IEEE/RSJ International Conference on Intelligent Robots and Systems (IROS 2008), Nice, France, September 22–26, 2008, pp. 4110–4115.
[39] M. Trincavelli, S. Coradeschi, and A, Loutfi, Odour classification system for continuous monitoring applications, *Sensors Actuat. B*, vol. 139, no. 2, pp. 265–273, 2009.
[40] A. Loutfi, S. Coradeschi, A. J. Lilienthal, and J.González Jimenez, Gas distribution mapping of multiple odour sources using a mobile robot, *Robotica*, vol. 27, no. 2, pp. 311–319, 2009.
[41] V. Hernandez Bennetts, E. Schaffernicht, V. Pomareda, A. J. Lilienthal, S. Marco, and M. Trincavelli, Combining non selective gas sensors on a mobile robot for identification and mapping of multiple chemical compounds, *Sensors*, vol. 14, no. 9, pp. 17331–17352, 2014.
[42] T. R. Consi, J.Atema, C. A. Goudey, J. Cho, and C. Chryssostomidis, AUV guidance with chemical signals, in Proceedings of the IEEE Symposium on Autonomous Underwater Vehicle Technology (AUV '94), Cambridge, MA, USA, July 19–20, 1994, pp. 450–455.
[43] F. W. Grasso, T. R. Consi, D. C. Mountain, and J.Atema, Biomimetic robot lobster performs chemo-orientation in turbulence using a pair of spatially separated sensors: Progress and challenges, *Robot. Auton. Syst.*, vol. 30, nos. 1–2, pp. 115–131, 2000.
[44] F. W. Grasso and J. Atema, Integration of flow and chemical sensing for guidance of autonomous marine robots in turbulent flows, *Environ. Fluid Mech.*, vol. 2, no. 1–2, pp. 95–114, 2002.
[45] M. Ohashi, Y. Kagawa, T. Nakatsuka, and H. Ishida, Crayfish robot that generates flow field to enhance chemical reception, *J. Sensor Technol.*, vol. 2, no. 4, pp. 185–195, 2012.
[46] R. M. Arrieta, J. A. Farrell, W. Li, and S. Pang, Initial development and testing of an adaptive mission planner for a small unmanned underwater vehicle, in Proceedings of the ASME 22nd International Conference on Offshore Mechanics and Arctic Engineering (OMAE 2003), Cancun, Mexico, June 8–13, 2003, New York: The American Society of Mechanical Engineers, pp. 609–617.
[47] W. Li, J. A. Farrell, S. Pang, and R. M. Arrieta, Moth-inspired chemical plume tracing on an autonomous underwater vehicle, *IEEE Trans. Robot.*, vol. 22, no. 2, pp. 292–307, 2006.
[48] P. Pyk, S. Bermúdez i Badia, U. Bernardet, P. Knüsel, M. Carlsson, J. Gu, E. Chanie, B. S. Hansson, T. C. Pearce, and P. F. M. J. Verschure, An artificial moth: Chemical source localization using a robot based neuronal model of moth optomotor anemotactic search, *Auton. Robot.*, vol. 20, no. 3, pp. 197–213, 2006.
[49] P. P. Neumann, V. Hernandez Bennetts, A. J. Lilienthal, M. Bartholmai, and J. H. Schiller, Gas source localization with a micro-drone using bio-inspired and particle filter-based algorithms, *Adv. Robot.*, vol. 27, no. 9, pp. 725–738, 2013.
[50] R. A. Russell, A ground-penetrating robot for underground chemical source location, in Proceedings of the 2005 IEEE/RSJ International Conference on Intelligent Robots and Systems (IROS 2005), Edmonton, Canada, August 2–6, 2005, pp. 1879–1884.
[51] R. A. Russell, CRABOT: A biomimetic burrowing robot designed for underground chemical source location, *Adv. Robot.*, vol. 25, nos. 1–2, pp. 119–134, 2011.
[52] Y. Wada, M. Trincavelli, Y. Fukazawa, and H. Ishida, Collecting a database for studying gas distribution mapping and gas source localization with mobile robots, in Proceedings of the International Conference on Advanced Mechatronics (ICAM 2010), Osaka, Japan, October 4–6, 2010, pp. 183–188.
[53] D. E. Jackson, M. Holcombe, and F. L. W. Ratnieks, Trail geometry gives polarity to ant foraging networks, *Nature*, vol. 432, pp. 907–909, 2004.
[54] J. O. Hinze, *Turbulence*, 2nd Ed. New York, USA: McGraw-Hill, 1975.
[55] A. J. Lilienthal, T. Duckett, H. Ishida, and F. Werner, Indicators of gas source proximity using metal oxide sensors in a turbulent environment, in Proceedings of the First IEEE/RAS-EMBS International Conference on Biomedical Robotics and Biomechatronics (BioRob 2006), Pisa, Italy, February 20–22, 2006, pp. 733–738.
[56] A. Mafra-Neto and R. T. Cardé, Fine-scale structure of pheromone plumes modulates upwind orientation of flying moths, *Nature*, vol. 369, pp. 142–144, 1994.
[57] D. B. Dusenbery, Optimal search direction for an animal flying or swimming in a wind or current, *J. Chem. Ecol.*, vol. 15, no. 11, pp. 2511–2519, 1989.
[58] H.Ishida, Odor-source localization by mobile robot and inter-robot communication using odors, (in Japanese), in *Olfactory Display: Multimedia Tool for Presenting Scents*, T. Nakamoto, Ed. Tokyo, Japan: Fragrance Journal, 2008, pp. 191–198.

[59] A. Marjovi and L. Marques, Optimal swarm formation for odor plume finding, *IEEE Trans. Cybern.*, vol. 44, no. 12, pp. 2302–2315, 2014.
[60] D. Martinez and L. Perrinet, Cooperation between vision and olfaction in a koala robct, in Report on the 2002 Workshop on Neuromorphic Engineering, Telluride, CO, USA, June 30–July 21, 2002. pp. 51–53.
[61] H. Ishida, H. Tanaka, H. Taniguchi, and T. Moriizumi, Mobile robot navigation using vision and olfaction to search for a gas/odor source, in Proceedings of the 2004 IEEE/RSJ International Conference on Intelligent Robots and Systems (IROS 2004), Sendai, Japan, September 28–October 2, 2004, pp. 313–318.
[62] T. Lochmatter and A. Martinoli, Theoretical analysis of three bio-inspired plume tracking algorithms, in Proceedings of the 2009 IEEE International Conference on Robotics and Automation (ICRA 2009), Kobe, Japan, May 12–17, 2009, pp. 2661–2668.
[63] P. P. Neumann, V. Hernandez Bennetts, A. J. Lilienthal, and M. Bartholmai, From insects to micro air vehicles—A comparison of reactive plume tracking strategies, in Proceedings of the 13th International Conference on Intelligent Autonomous Systems (IAS-13), Padova, Italy, July 15–19, 2014, pp. 1533–1548.
[64] H. Ishida, T. Nakamoto, and T. Moriizumi, Remote sensing of gas/odor source location and concentration distribution using mobile system, *Sens. Actuators B*, vol. 49, nos. 1–2, pp. 52–57, 1998.
[65] J. Matthes, L. Groll, and H. B. Keller, Source localization by spatially distributed electronic noses for advection and diffusion, *IEEE Trans. Signal Process.*, vol. 53, no. 5, pp. 1711–1719, 2005.
[66] T. Ushiku, N. Satoh, H. Ishida, and S. Toyama, Estimation of gas-source location using gas sensors and ultrasonic anemometer, in Proceedings of the Fifth IEEE Conference on Sensors (IEEE Sensors 2005), Daegu, Korea, October 22–25, 2006, pp. 420–423.
[67] M. Vergassola, E. Villermaux, and B. I. Shraiman, 'Infotaxis' as a strategy for searching without gradients, *Nature*, vol. 445, pp. 406–409, 2007.
[68] E. M. Moraud and D. Martinez, Effectiveness and robustness of robot infotaxis for searching in dilute conditions, *Front. Neurorobot.*, vol. 4, 1, 2010.
[69] J. A. Farrell, S. Pang, and W. Li, Plume mapping via hidden Markov methods, *IEEE Trans. Syst. Man Cybern. B*, vol. 33, no. 6, pp. 850–863, 2003.
[70] S. Pang and J. A. Farrell, Chemical plume source localization, *IEEE Trans. Syst. Man Cybern. B*, vol. 36, no. 5, pp. 1068–1080, 2006.
[71] G. Kowadlo and R. A. Russell, Using naïve physics for odor localization in a cluttered indoor environment, *Auton. Robot.*, vol. 20, no. 3, pp. 215–230, 2006.
[72] W. Jatmiko, K. Sekiyama, and T. Fukuda, A pso-based mobile robot for odor source localization in dynamic advection-diffusion with obstacles environment: Theory, simulation and measurement, *IEEE Comput. Intell. Mag.*, vol. 2, no. 2, pp. 37–51, 2007.
[73] G. Ferri, M. V. Jakuba, A. Mondini, V. Mattoli, B. Mazzolai, D. R. Yoerger, and P. Dario, Mapping multiple gas/odor sources in an uncontrolled indoor environment using a Bayesian occupancy grid mapping based method, *Robot. Auton. Syst.*, vol. 59, no. 11, pp. 988–1000, 2011.
[74] T. Breithaupt, Fan organs of crayfish enhance chemical information flow, *Biol. Bull.*, vol. 200, no. 2, pp. 150–154, 2001.
[75] G. S. Settles, Sniffers: Fluid-dynamic sampling for olfactory trace detection in nature and homeland security, *J. Fluids Eng.*, vol. 127, no. 2, pp. 189–218, 2005.
[76] A. Lilienthal and T. Duckett, Building gas concentration gridmaps with a mobile robot, *Robot. Auton. Syst.*, vol. 48, no. 1, pp. 3–16, 2004.
[77] A. J. Lilienthal, M. Reggente, M. Trincavelli, J. L. Blanco, and J. Gonzalez, A statistical approach to gas distribution modelling with mobile robots—The kernel DM+V algorithm, in Proceedings of the 2009 IEEE/RSJ International Conference on Intelligent Robots and Systems (IROS 2009), St. Louis, MO, USA, October 10–15, 2009, pp. 570–576.
[78] M. Reggente and A. J. Lilienthal, The 3D-Kernel DM+V/W algorithm: Using wind information in three dimensional gas distribution modelling with a mobile robot, in Proceeding of the Ninth IEEE Conference on Sensors (IEEE Sensors 2010), Waikoloa, HI, USA, November 1–4, 2010, pp. 999–1004.
[79] M. Reggente, A. Mondini, G. Ferri, B. Mazzolai, A. Manzi, M. Gabelletti, P. Dario, and A. J. Lilienthal, The DustBot system: Using mobile robots to monitor pollution in pedestrian area, *Chem. Eng. Trans.*, vol. 23, pp. 273–278, 2010.
[80] R. A. Russell, Tracking chemical plumes in constrained environments, *Robotica*, vol. 19, no. 4, pp. 451–458, 2001.
[81] A. Marjovi and L. Marques, Multi-robot olfactory search in structured environments, *Robot. Auton. Syst.*, vol. 59, no. 11, pp. 867–881, 2011.
[82] J. A. Farrell, J. Murlis, X. Long, W. Li, and R. T. Cardé, Filament-based atmospheric dispersion model to achieve short time-scale structure of odor plumes, *Environ. Fluid Mech.*, vol. 2, no. 1–2, pp. 143–169, 2002.

7

Olfactory Display and Odor Recorder

Takamichi Nakamoto
Precision and Intelligence Laboratory, Tokyo Institute of Technology, Tokyo, Japan

7.1 Introduction

A human interface typically has both input and output. Recently devices not for vision and audition but for other senses are often studied. In case of olfaction, an olfactory display is an output from a machine, whereas an odor sensing system is its input [1]. They work in the same manners as those of a microphone and a speaker in auditory sense. Although the odor sensing system has long history, the olfactory display is relatively new. However, the first international congress of Digital Olfaction and its second congress were held in 2013 and 2014, respectively. Since a variety of olfactory display were demonstrated at that meeting, olfactory displays became known to many people. We explain how the olfactory display works in this chapter since the odor sensing systems often called electronic noses are described in other chapters.

Another topic is an odor recorder which reproduces a detected odor. The odor recorder has both input and output of human olfactory interface. Thus, we describe the odor recorder in this chapter.

Moreover, we describe a teleolfaction system. In teleolfaction system, odor information obtained from an odor sensing system is transmitted to an olfactory display via the Internet. Thus, the teleolfaction is related to both input and output.

This chapter mainly covers odor presentation technique, whereas an odor sensing system is described in Chapter 3.

7.2 Principle of Olfactory Display

A history of an olfactory display is relatively new in comparison with an odor sensing system. However, the first proposal to present smells was performed by M. Heilig in 1960s when he developed Sensorama [2]. He tried to simulate experience when he rode on a motorbike to go

Essentials of Machine Olfaction and Taste, First Edition. Edited by Takamichi Nakamoto.

Companion website: www.wiley.com/go/nakamoto/olfaction

to various places around a city. The simulated experience related to olfaction such as food scents was included. Later Toyota Motor Corporation presented scents at the showroom. Shiseido Co., Ltd, fragrance company, made the equipment called Kaori on demand to emit scents. These were aimed to send scents to remote sites through the web.

On the other hand, an olfactometer has been studied as a tool to give olfactory stimuli to a subject [3]. It is aimed to give the olfactory stimuli accurately. Since it is very expensive and bulky, a cheap apparatus easy even for a novice to handle is preferable. Thus, an olfactory display has been studied in virtual reality.

Although the number of papers related to olfactory display is not so large at the current stage, they can be classified into several categories. Their categories are device research, olfactory display related to spatial odor distribution, and temporal intensity change and interaction with other senses. Thus, these researches are described in sequence.

7.2.1 *Olfactory Display Device*

Kaye made simple perfume bottles with solenoids. Since they could be activated by electrical signals, he constructed a device to present plural smells according to the status of the stock market [4].

Kim *et al.* developed the olfactory display using the reversible phase transition of a functional polymer [5]. Its device is shown in Figure 7.1. The temperature-responsive hydrogel, functional polymer, has two phases such as sol and gel in the manner similar to ON and OFF of a switch in an electronic circuit. An aroma can be released from a card made of this polymer in the sol phase, whereas its release is stopped in the gel phase. The phase transition between sol and gel depends upon temperature controlled by a Peltier device. The authors claimed that the aroma release from this card was easy and soundless.

Two-dimensional array of aroma release devices was proposed [6]. It consists of cone-shaped polydimethylsiloxane (PDMS) micro chamber where odorant solution is stored in advance as is shown in Figure 7.2. There is a matrix composed of these devices with its rows

Figure 7.1 Aroma card. Reproduced with permission from Ref. [5]. Copyright IGI Global, 2013

and columns made up of heater lines located below micro chambers. Since the micro chamber has a small hole at the top of the cone, it can release an odor owing to the pressure increase caused by heating. Although the pressure increase due to a single heater line is not enough, its increase caused by two heater lines enables the odor release through the small hole. Thus, only the micro chamber specified by row and column addresses can release the odor. In case of Figure 7.2, the micro chamber specified by the address (X address: 2, Y address: 2) releases it. Although Figure 7.2 indicates the concept with a large array of odor-releasing devices, they only fabricated the system with two micro chambers.

Another work using MEMS (microelectromechanical system) technology was reported since MEMS technology can miniaturize device size as is shown in Figure 7.3 [7]. Tiny amount of liquid is introduced into a heater. It is released through a micro valve, which enables the elimination of scent leakage. Thin-film heater consists of Cr, Au, and Ti with its pattern formed

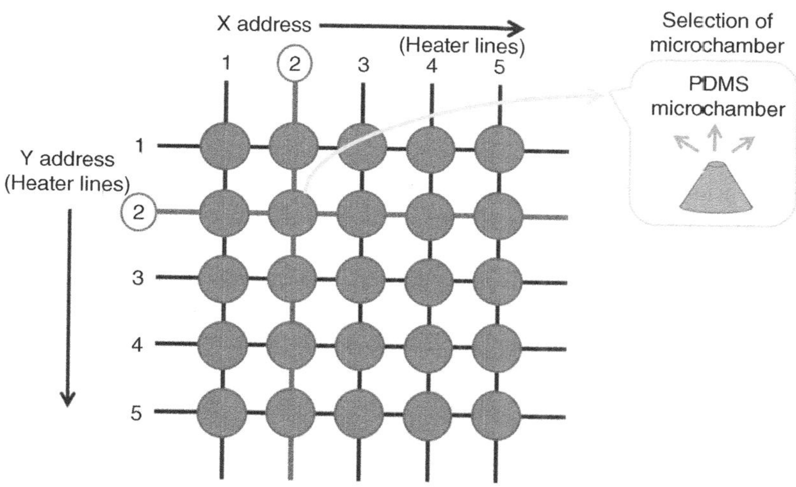

Figure 7.2 X–Y addressable odor-releasing system

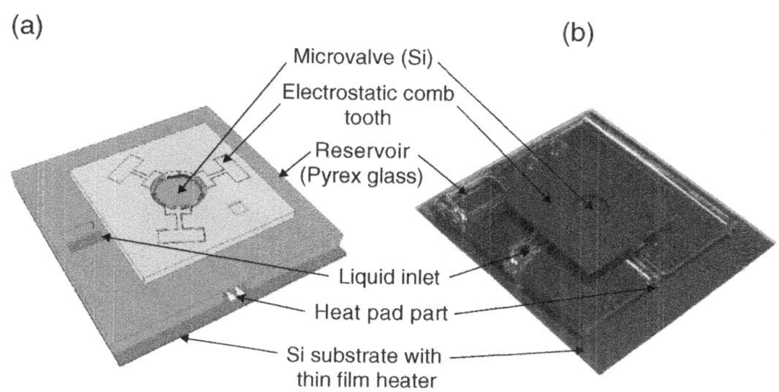

Figure 7.3 Odor generator based upon MEMS (a) schematic sketch and (b) actual structure. Reproduced with permission from Ref. [7]. Copyright IEE of Japan, 2012

by wet etching. The reservoir of odorant liquid was formed by wet etching of Pyrex glass. The fundamental device characteristics have been confirmed.

Chaku Perf made an odor-releasing gadget compatible with a mobile phone. Since it can be attached to the mobile phone, a scent can be released according to commands from a remote site [8]. The ultrasonic vibrator is used to emit the scent. Although its principle is very simple, it is commercially available. Scentcom, an Israeli company, presented ultrasonic micro plug array technology at digital olfaction congress [9]. They claimed that their technique enabled the ultrasonic nebulization with effective self-cleaning.

Sakairi *et al.* made an odor-emitting apparatus coupled with chemical capsules made of alginic acid polymer. It consists of chemical capsule cartridge, valve and temperature control unit. By changing temperatures of chemical capsules and the valve sequences, odors can be changed in strength. Its reproducibility was confirmed by a real-time mass spectrometry [10].

7.2.2 Olfactory Display Related to Spatial Distribution of Odor

Yanagida *et al.* developed an equipment to deliver an odor to a specified person using vortex rings generated by an air canon as is shown in Figure 7.4[11]. The experiment on vortex ring is often presented to children at scientific event. The air canon enables an odor to be delivered to only a single person without spreading it in space. People around him/her do not perceive odors, whereas a target person can sniff it.

However, there is a problem that a subject perceives wind, that is, airflow, together with the odor. The wind might let the subject feel the impression different from that of the only odor. Yanagida *et al.* used two air cannons to collide two vortex rings in front of a subject so that the wind from the air canon could be reduced. Moreover, they made the system for tracking the target subject when he/she moved [12]. Matsuo *et al.* performed simulation of vortex ring

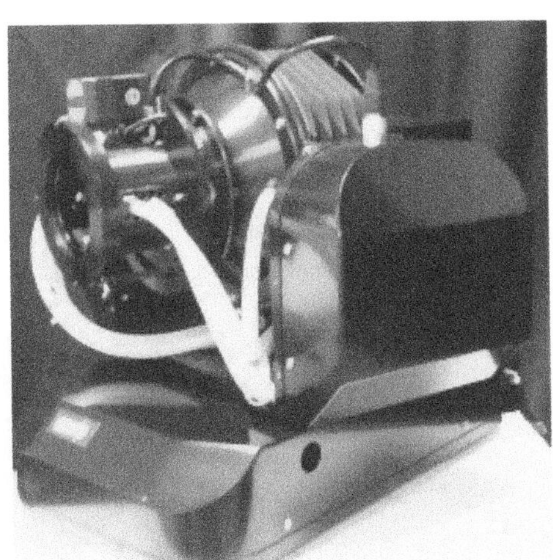

Figure 7.4 Scent projector using vortex rings generated by air canon. Reproduced with permission from Ref. [11]. Copyright IEEE, 2006

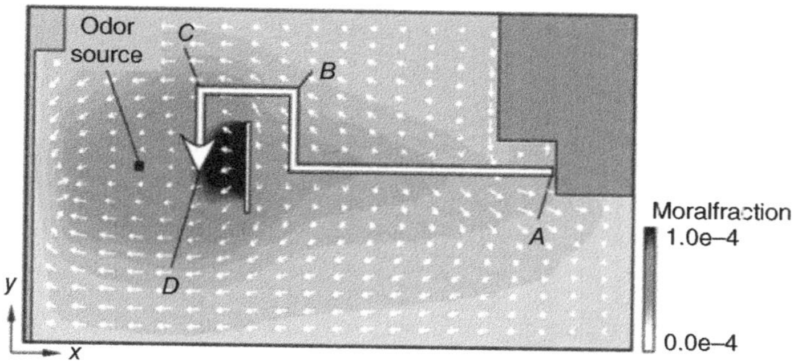

Figure 7.5 Airflow field and distribution of odor concentration calculated using CFD simulation

from the air canon. They claimed that the shape of the air canon nozzle influenced the efficiency of the odor delivery [13].

Kim *et al.* recently developed the gadget called micro aroma shooter [14]. They made micro airflow source in which porous materials such as calcium silicate soaked in odorant liquid are encapsulated. This gadget allows an odorant to be carried by directional airflow.

Hirose *et al.* proposed the wearable olfactory display. A subject with that olfactory display walked around to look for a virtual odor source [15]. When a subject approaches the virtual odor source, its perceived intensity increases. On the other hand, its intensity decreases when the subject goes away from the virtual source. The subject was requested to find the location of the virtual odor source. The subjects could find the virtual source although the time for looking for it varied from person to person. Although the actual odor distribution is complicated, they adopted a simple model with isotropic diffusion.

Ishida *et al.* realized the virtual odor distribution using computational fluid dynamics (CFD) [16]. The odor distribution in virtual space was calculated in advance using CFD. Then, a user can sniff a smell with its calculated concentration at the point in virtual space using an olfactory display. Figure 7.5 shows the airflow field and distribution of odor concentration in virtual space. The odor source is located behind a screen. CFD shows that the odor concentration just behind the screen becomes high. The user experiences that phenomenon when he/she walks in virtual space from point A to point D.

Moreover, they studied the olfactory display called smelling screen [17]. It has four fans on the four corners of the screen as is shown in Figure 7.6. The airflows that are generated by these fans collide to create an airflow that is directed toward the user from a certain position in front of the screen. A user perceives odor distribution as if the odor source were on the screen.

7.2.3 *Temporal Intensity Change of Odor*

7.2.3.1 Problem of Smell Persistence

One of the important problems of an olfactory display is smell persistence. It should release an odor according to the temporal change of the odor intensity programmed in advance. However, it is difficult to make an arbitrary profile of the odor intensity if there is the smell

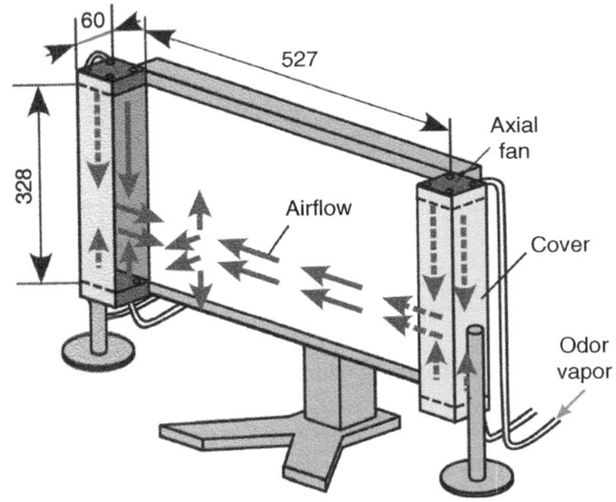

Figure 7.6 Concept of smelling screen. Reproduced with permission from Ref. [17]. Copyright IEEE, 2013

persistence. When the odor passes through a tube, it persists inside the tube due to chromatographic behavior. The smell persistence becomes dominant when the boiling point of the odorant is high. Such an odorant is called a low-volatile compound. However, low-volatile compounds cannot be ignored since their perception thresholds are very low. In other words, a human olfaction is sensitive to low-volatile compounds.

Several experiments were performed to check the smell persistence. Our group used a sample flow system and quartz crystal microbalance (QCM) gas sensor to evaluate it. The sample used in this experiment is alpha-hexyl cinnamic aldehyde with its boiling point of 305°C. The bottle with a sample liquid was connected via Teflon tube (inner diameter: 1 mm) to a sensor cell where eight QCM sensors (20 MHz, AT cut) were located. Those sensor coatings were tricresyl phosphate (TCP), Apiezon L (Ap-L), polyphenyl ether (6 rings) (PPE(6)), polyethylene glycol 1000 (PEG1000), 1,2,3-tris(2-cyanoethoxy)propane (TCEP), diethylene glycol succinate (DEGS), dinolenoyl phosphatidyl choline (DNnPC), and (R)-(+) Bis(diphenylphosphino)-1,1′-binaphthyl ((R)-BINAP). The sample flow rate was 300 ml/min, and the experiment was performed at room temperature.

The sensor responses to alpha-hexyl cinnamic aldehyde are shown in Figure 7.7 when the Teflon tube length was 22 cm, relatively short. Since alpha-hexyl cinnamic aldehyde is low volatile, it took much time for an odorant to reach the sensors, whereas their responses were very large after its arrival at the sensors. It seemed that the slow response was due to the delay at the tube.

Then, the sensor responses were again measured when the Teflon tube length was 470 cm. Its result is shown in Figure 7.8. It was found from this figure that the sensor responses were much slower than those in Figure 7.7. Since the tube length influences the delay of odorant delivery, it can be concluded that smell persists inside the tube.

Another experiment was performed using a mass spectrometer. Since the mass spectrometer is originally gas chromatograph/mass spectrometry (GC/MS) equipment, it has long stainless tube (10 m) prior to the inlet of the mass spectrometer. Although long stainless tube is typically used as a column, this stainless tube does not have coating. Since the temperature

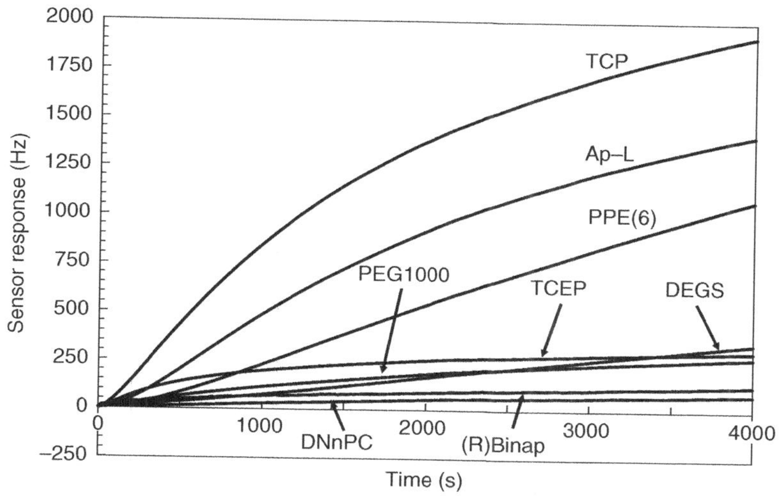

Figure 7.7 QCM sensor responses to alpha-hexyl cinnamic aldehyde (tube length: 22 cm)

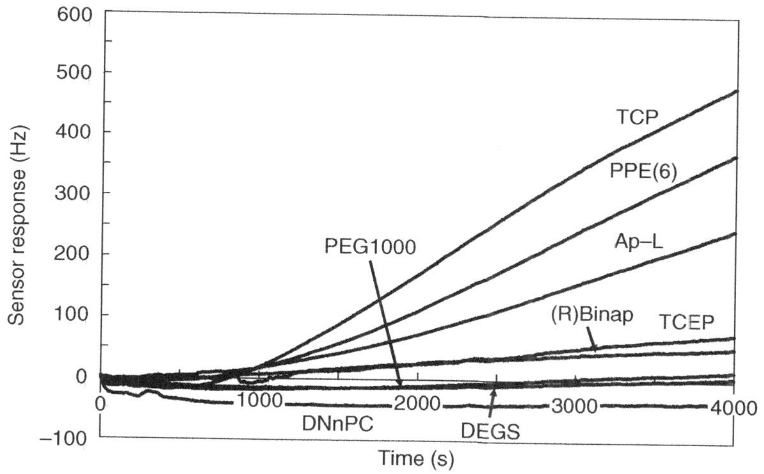

Figure 7.8 QCM sensor responses to alpha-hexyl cinnamic aldehyde (tube length: 470 cm)

of the whole stainless tube can be controlled, the influence of the tube temperature was examined. The sample used here was beta-ionone with its boiling point of 239°C. Four milliliters of beta-ionone in the gas phase was injected into the inlet of GC/MS system (Shimadzu, QP5050A) with its stainless tube (column) kept at room temperature. The temporal change of the detector output is shown in Figure 7.9. Total amount in the figure is the sum of the detector output over all m/z region. Although the mass spectrum of beta-ionone has a peak at $m/z = 177$, it could not be observed since it was widened and became flat. Then, its temperature was raised to 150°C, and its temporal change of the detector output is shown in Figure 7.10. The peak at $m/z = 177$ was observed since the peak was not widened at the long stainless tube.

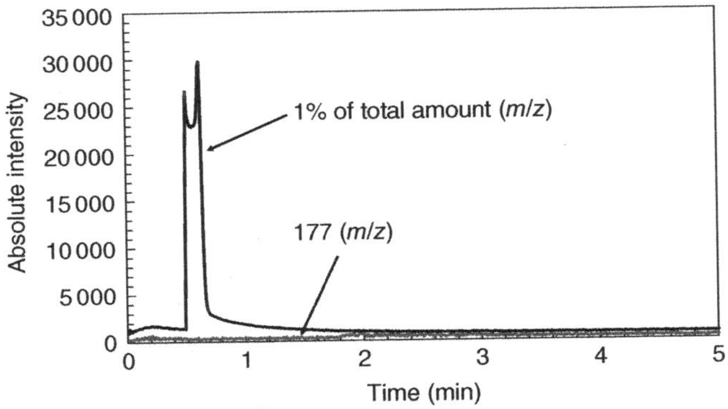

Figure 7.9 Temporal change of detector output with its tube at room temperature (sample: beta-ionone)

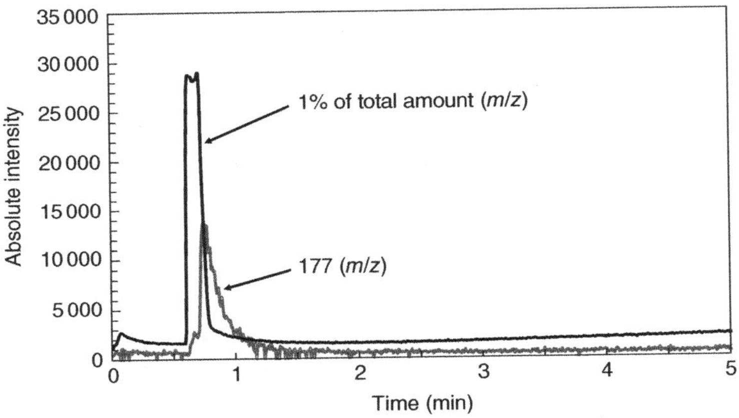

Figure 7.10 Temporal change of detector output with its tube at 150°C (sample: beta-ionone)

It can be concluded that the smell persistence does not occur at the tube kept at high temperature. However, it is very difficult to keep the whole odor delivery system at high temperature in the typical olfactory display.

7.2.3.2 Olfactory Display Using Inkjet Device

An inkjet device was used to solve the problem of smell persistence. Although an inkjet device is typically used for a printer, its applications to other fields such as printed circuit board, medical device, and microoptics are spreading. Since an inkjet device can spout a tiny liquid droplet, it can generate the smell. Although it seems appropriate to generate a low-volatile scent, only an inkjet device is not sufficient. A tiny liquid droplet still remains without evaporation after spouting it. Thus, the droplet should be forcibly evaporated by other methods such as heating. Thus, the method using the inkjet device together with a mesh heater was

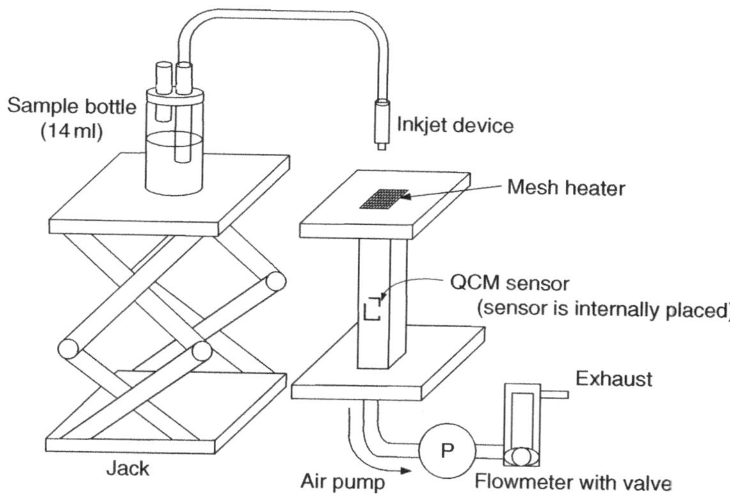

Figure 7.11 Olfactory display using inkjet device together with mesh heater

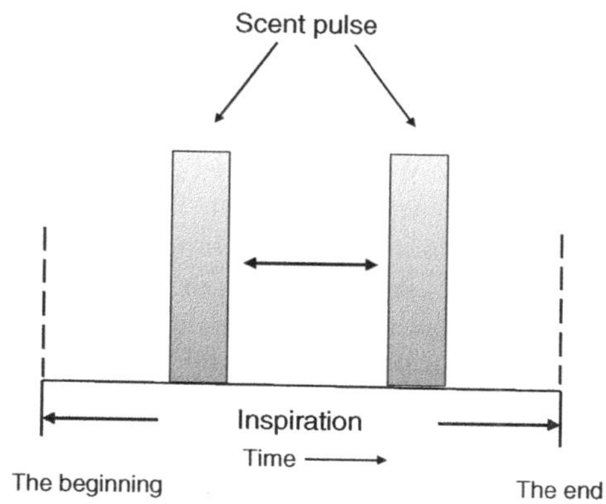

Figure 7.12 Two scent pulses during one inhalation

proposed [18]. The apparatus is shown in Figure 7.11. An inkjet device spouts tiny liquid droplets to a stainless mesh heater, and the evaporated vapor is obtained. A rapid sensor response to low-volatile compound was confirmed using the experimental setup in Figure 7.11. However, the integration of the multiple inkjet devices is not so easy since the liquid pressure at the tip of the inkjet device should be carefully adjusted to obtain the reproducibility.

Wallace *et al.* applied an inkjet device to microdispenser [19]. They also used a heater to evaporate liquid particles from an inkjet device. They did the experiment on determining human olfactory thresholds, which might be influenced by diseases.

Later Okada *et al.* studied the inkjet-based olfactory display [20]. Since their inkjet device ejects a droplet with the volume of a few picoliters, extremely narrow scent pulse can be generated. They made pulse ejection technique which enables a subject to sniff scents twice during one inhalation as is shown in Figure 7.12.

They also studied synchronization of scent pulse with inhalation using breath sensor. However, they concluded that the pulse ejection technique without a breath sensor could have the odor recognition rate similar to that using a breath sensor if the ejection period was adequately set [21].

7.2.4 *Multicomponent Olfactory Display*

The olfactory displays described previously were aimed to generate a single scent or generate several scents by switching them. However, it is essential to blend multiple odor components so that a variety of scents can be generated. Thus, multicomponent olfactory displays have been developed.

A perfume manufacturer developed a virtual aroma synthesizer for exploring flavors [22]. However, it was too bulky to be used in the virtual environment. We have several methods to blend multiple odor components as depicted in Figure 7.13a–d. Those methods are explained in this section.

7.2.4.1 Mass Flow Controller

It is possible to blend multiple odors when the flows of odor components with controlled flow rates join together. For example, the composition of the binary mixture is 8:2 when odor component 1 flow rate is 80 ml/min and odor component 2 one 20 ml/min.

Thus, several mass flow controllers (MFCs) are used to blend odor components [23], whereas MFC systems can be used to evaluate sensor characteristics [24]. An MFC is a device used to adjust its mass flow rate electronically. When you adjust gas flow rate, the most typical device is a volume flow meter with manual valve. However, it is sometimes inaccurate since the volume of gaseous fluid depends upon its pressure. On the other hand, mass flow rate is not influenced by pressure since it indicates mass of a gaseous fluid passing through a given surface per unit time. Its mass is usually converted to volume at standard state (e.g., 25°C and 1 atm.).

An MFC detects mass flow rate and then the error between the actual flow rate and specified one is fed back to the internal valve to adjust the flow rate to the specified one. Although the flow rate is typically specified by analogue voltage, recent MFC model accepts digital format. Since a user can adjust the flow rate using open-loop control as if MFC worked without feedback, its operation is easy. When several flows join together without MFCs, the flow rate of one component is typically influenced by others. MFCs can eliminate this interference.

The problem of MFC system is that they are bulky and expensive. MEMS technique might be able to reduce its size if its stability and robustness are improved [25].

7.2.4.2 Automatic Sampler

An automatic sampler is often used to sample liquid for liquid chromatography (LC). It can be applied to odor blender in the liquid phase. Odor components in the liquid phase are put into glass vials on a tray placed on a turntable as is shown in Figure 7.13b. The syringe in the automatic sampler takes the specified amount of liquid from the specified vial. Then, it moves to the specified vial and transfers the accumulated liquid sample to that vial. When this procedure is repeatedly performed, the sample blended with many components at arbitrary ratio can be obtained [26].

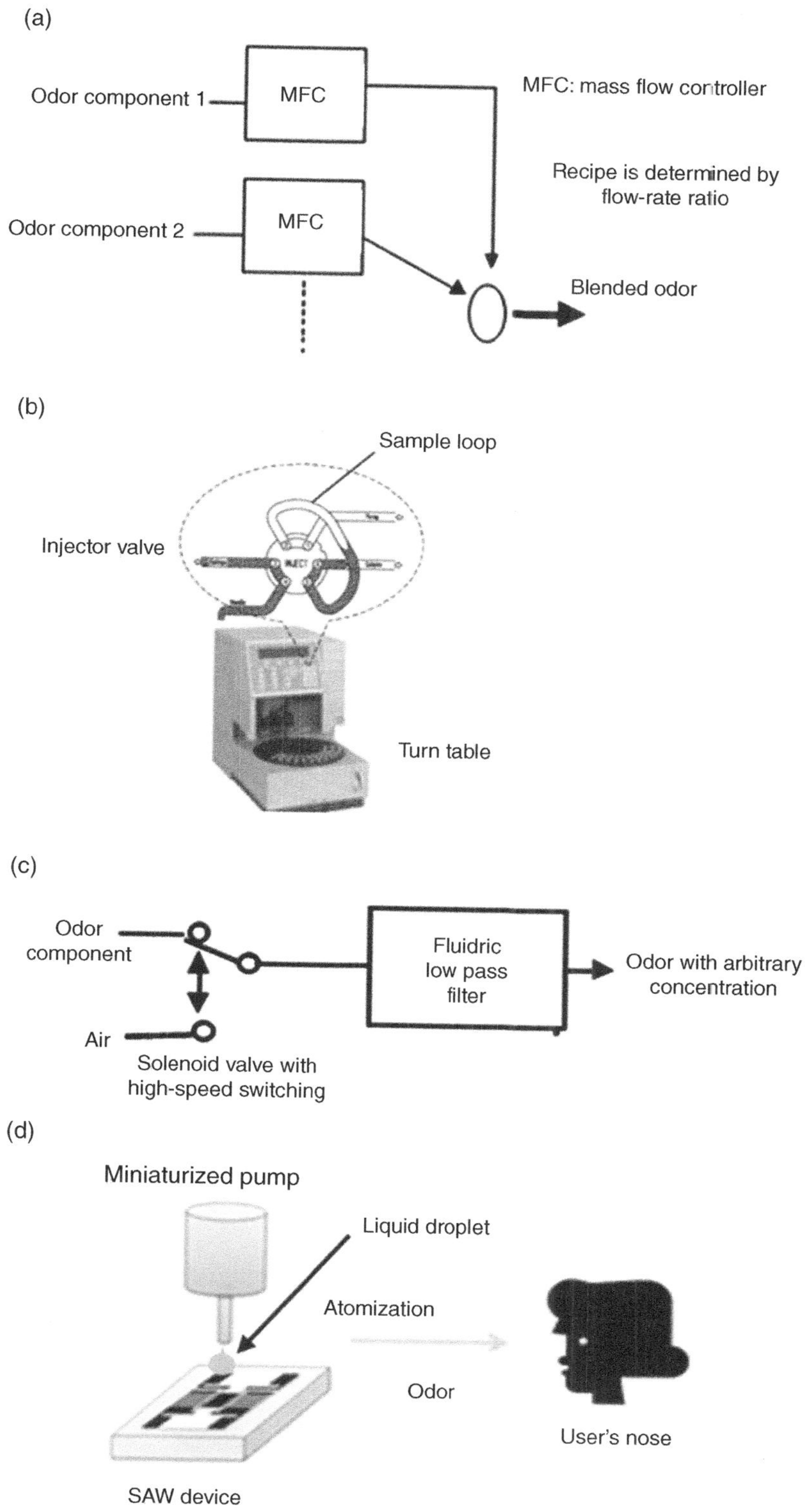

Figure 7.13 Principles of multicomponent olfactory display (a) mass flow controller, (b) autosampler, (c) solenoid valve, and (d) micropumps and SAW device

Then, a tiny amount of liquid with volume of a few microliter is precisely taken by a sample loop and is evaporated using a microheater so that a human or a sensor can be exposed to the blended odor. The advantage of this method is to blend many components. Although the number of components depends upon the tray size of the automatic sampler, it is typically around 100. Disadvantage is that it consumes long time for blending components since a single syringe transfers only one component at the same time. Since it blends one by one in sequence, it is not possible to blend components in real time.

7.2.4.3 Solenoid Valve

A solenoid valve is a fluidic switching device as is shown in Figure 7.13c. It can switch the flow path in the same manner as that of an electrical switch. When it rapidly switches the air and a sample, the intermediate concentration between zero and full scale can be realized. Thus, 1-bit analogue to digital conversion technique such as delta sigma modulation [27] or pulse width modulation technique in an electronic circuit is applicable to the fluidic system [28]. The odor concentration approaches its full scale as frequency of ON increases.

The first version of multicomponent olfactory display is shown in Figure 7.14. The switching between the air and the component using 3-way solenoid valve enables the specified relative concentration of odor. Then, multiple flows of odor components join together for blending. An odor component and empty bottle (air) in each flow path were alternately switched according to second-order delta sigma modulation. The olfactory display consists of two paths, path to a user's nose and bypath. Bypath is necessary since the flow in the sample bottle cannot be stopped to keep its concentration even when the odor is not supplied to the user [29].

Although this olfactory display works well, the problems appeared when the number of odor components was extended. First, there are too many empty bottles. Its number should be reduced. Secondly, the full-scale odor intensity is reduced as the number of odor components increases. Thus, odor blending algorithm was changed as is shown in Figure 7.15.

Only one component among all the components including the air is presented to a user at the same time. Thus, the duty cycle of each odor component is the ratio of that component [30]. In

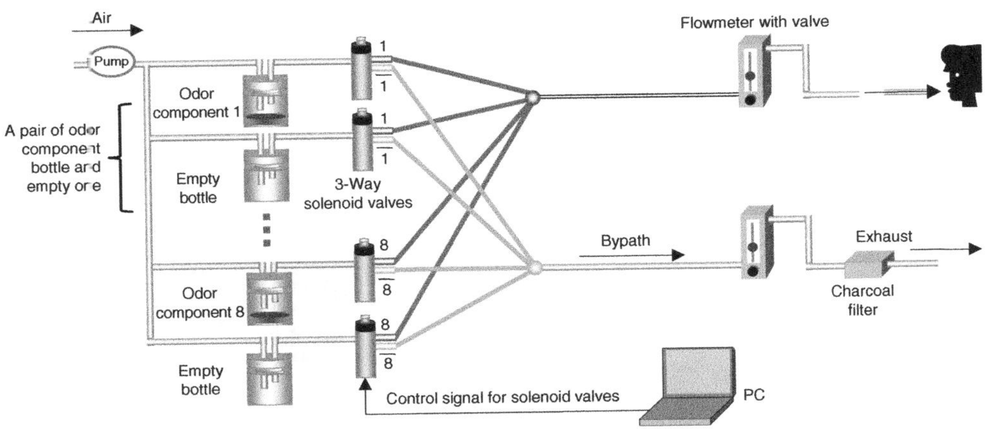

Figure 7.14 Schematic diagram of olfactory display using solenoid valves at initial stage

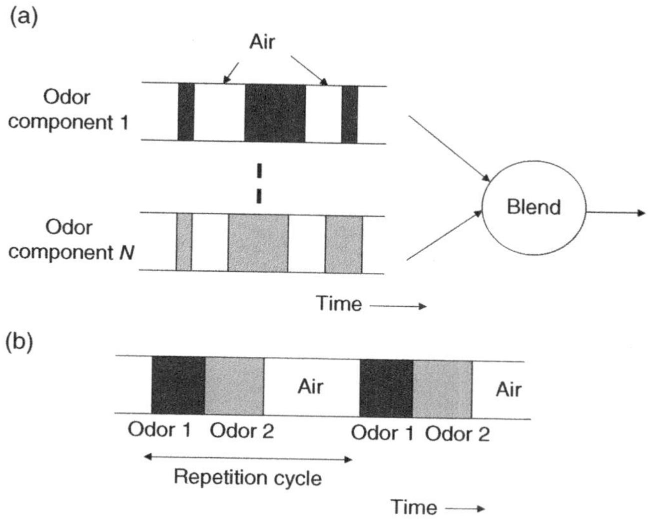

Figure 7.15 Improvement of odor blending algorithm, (a) initial stage and (b) improved one

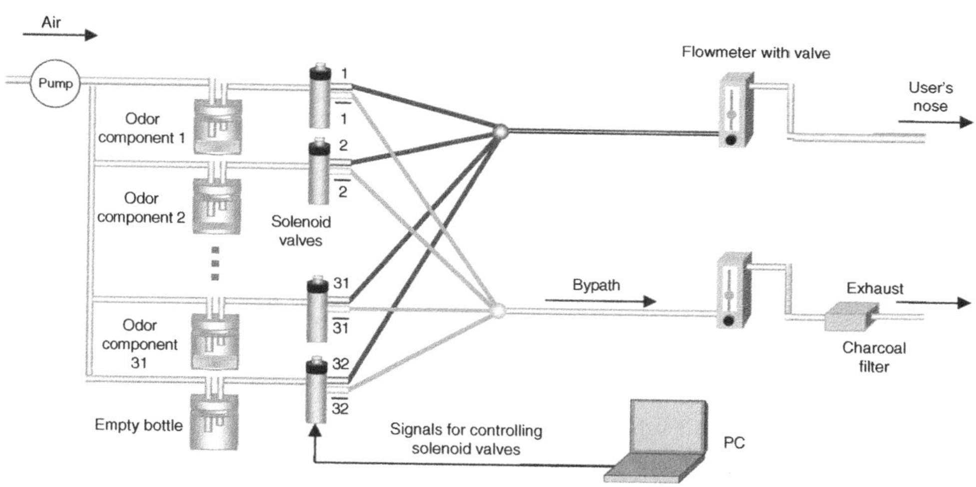

Figure 7.16 Modified olfactory display using solenoid valves

this configuration, only one empty bottle is required for the air. Moreover, the full-scale concentration of each component does not depend upon the number of odor components. Figure 7.16 shows the schematic diagram of modified olfactory display using solenoid valves. Up to 32 odor components can be blended at arbitrary recipe. The repetition cycle is 1 s and the minimum switching speed of the solenoid valve is 2 ms. Photo of the version commercially available for 13 components is shown in Figure 7.17. Its size is smaller than a laptop computer. It works together with an air pump and a laptop computer. A user can bring it in any place to do the demo using this version of the olfactory display.

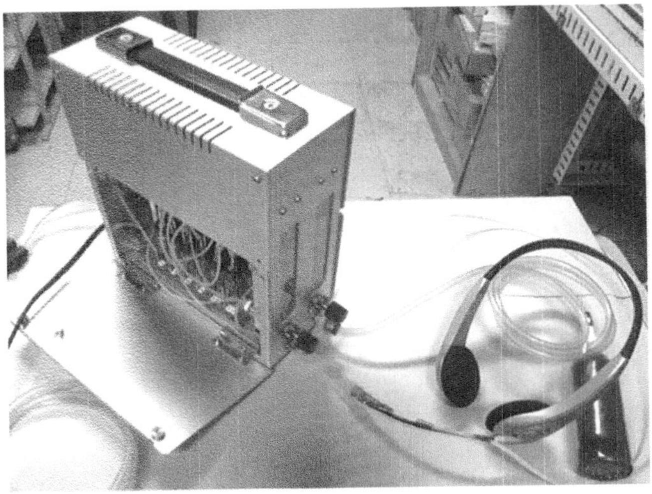

Figure 7.17 Compact olfactory display using 13 components (available from Ono Denki Inc.)

The olfactory display using solenoid valves works stably and reliably. However, its problem is smell persistence as discussed before. It deteriorates the quality of scents when low-volatile compounds are included. Many tubes for plumbing causes smell persistence. Although the increase in temperature at tubes reduces the influence of smell persistence, it is difficult to use solenoid valve under high temperature. Thus, another type of olfactory display was introduced in the next section.

7.2.4.4 Micropumps and Surface Acoustic Wave Atomizer

Principle of the olfactory display is illustrated in Figure 7.13d. Tiny liquid droplet is introduced to surface acoustic wave (SAW) device in order to atomize it [31]. When a liquid droplet is placed on a SAW device, the acoustic energy converted from SAW to longitudinal wave in the liquid droplet enables the atomization of the liquid droplet [32]. Since the atomization occurs instantaneously, a user can sniff the scents immediately even if they are low-volatile compounds. Smell persistence does not occur since no tubing is used to deliver gaseous fluid.

Micropumps used here were electroosmotic (EO) pumps [33]. Its size was 6 mm × 11.5 mm including liquid reservoir. It works soundlessly due to no mechanically moving part, and its power is strong. Since it has self-priming property, it is not necessary to fill the whole pump with liquid manually in advance although many liquid pumps do not have this property. It can drive ethanol or deionized water. Since it cannot drive odorants directly, they are diluted with ethanol. As the flow rate increases linearly with electric driving voltage, the flow rate can be easily controlled by adjusting the applied voltage. Otherwise the amount of liquid can be controlled by the duration time of applying voltage to it. Besides, the flow is not disturbed by bubble unlike the widely used inkjet device.

Using eight EO pumps and SAW device, the olfactory display was assembled as is shown in Figure 7.18. Its size is 37.4 × 56.7 × 7.5 mm smaller than a credit card. Thin stainless tube

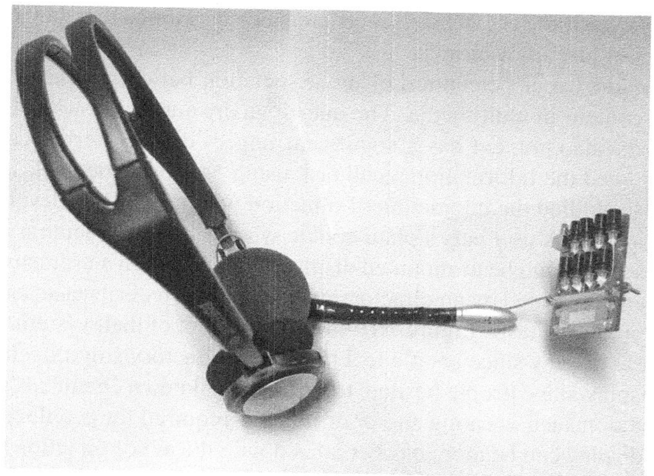

Figure 7.18 Photo of olfactory display using eight EO pumps and SAW device

with an inner diameter of 0.13 mm was used to introduce liquid droplet from the EO pump into the surface of SAW device. Using this olfactory display, nine participants sniffed low-volatile fragrance called Eau de Givenchy. When the movie for 20 s was presented to participants together with scents, all the participants sniffed them at the scene only when the fragrance was supplied. Thus, the olfactory display using micropumps and SAW device is effective to solve the problem of smell persistence.

Although the olfactory display size in Figure 7.18 does not include peripheral circuits, miniaturized olfactory display including the circuits was developed in 2014 [34]. It is expected to be applied to a variety of fields.

7.2.5 Cross Modality Interaction

Most of researches focus on interaction of olfaction with vision. Kawai *et al.* studied psychological effect of image with fragrance [35]. They carried the sensory test using 3D image together with fragrance and claimed possibility of producing psychological effect owing to scents for relaxation.

Then, they used only one kind of fruit scent made up of several ester compounds [36]. They presented various images together with this fruit scent to a user and ask him/her their impressions. Since the impression of the scent is much influenced by image, the scent impression is different from image to image even if they used a single fruit scent. This experimental result suggests that a variety of scent impression can be obtained together with images even if the number of scents is actually small.

Hirose *et al.* studied visual–olfactory display to make pseudoolfactory effect using vision [37]. They made olfactory map of 18 kinds of fruit flavors according to sensory test and isometric multidimensional scaling (MDS) [38]. They categorized four groups based upon olfactory map and picked up one as a representative in each group. They presented the representative smell together with image. If the drawing effect caused by vision occurs, people can sniff pseudo smell even if the actual smell is different. They claimed that they can make various smells virtually from just a few aromas.

Ghinea *et al.* studied user perception of the association between olfactory media content and video media content in multimedia. The questionnaire survey shows that the association between scent and video content has a significant impact on the user-perceived experience [39]. They also studied the information recall task using olfactory cue [40].

Hirose *et al.* also studied the interaction of olfaction with taste. They developed the system called meta cookie [41]. A user eats a plain cookie where chocolate, lemon, etc., are visually overlaid through see-through head-mounted display (HMD). When a user eats it together with corresponding scent generated by an olfactory display, he/she feels its taste as if he/she ate the cookie with corresponding taste. Figure 7.19 shows a concept of their system. They called this system pseudo taste display since people feel taste from olfactory stimulus. It is very difficult to make a taste display since people hesitate to drink/eat unknown chemicals synthesized by a machine. Active action such as eating and/or drinking is required for people to perceive tastes. However, a taste display can be more easily realized with the aid of olfaction since people can naturally and passively sniff smells. Moreover, people enjoy a variety of flavors through olfaction although the taste itself just covers basic tastes such as sweetness, salt, sourness, bitterness, and umami. A pseudo-gustatory display might be a realistic way to perceive taste in virtual environment [42].

Iwata *et al.* did the experiment on expressing food texture [43]. They developed the haptic interface that presents biting force. They tried to apply it to the food texture expressions of cracker and cheese. Although they claim that they succeeded in the presentation of physical property of food in terms of hardness, it does not seem easy to apply it to every food.

Tomono *et al.* studied eye catching and memory during smell presentation using gaze point analysis [44]. The gaze moves actively in order to receive a lot of information from the entire image. However, a subject had attention to even a small object related to a scent when a scent was presented together with the image. There was tendency that a gaze remains in the narrow area of a smelling object. Moreover, memory related to the object was enhanced when the associated scent was presented together with the image [45].

Zellner and Whitten reported that the perceived odor intensity was influenced by the color intensity (clear, light, medium, and dark) of the object. They claimed that the odor intensity was influenced by color intensity independent upon the color and odor types [46].

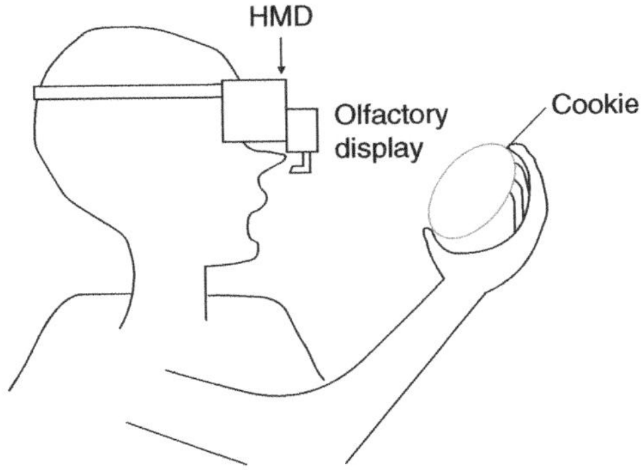

Figure 7.19 Concept of pseudo-gustatory display

Spence *et al.* focused on interaction between olfactory and auditory senses [47]. They reported that odors were preferentially matched to musical features. It was found that the odors of candied orange and iris flower were matched to significantly higher pitches than the odors of musk and roasted coffee.

Ishibashi *et al.* made Ikebana (flower arrangement) system using olfactory and haptic devices in the virtual environment [48]. A user can hold a flower, adjust the length of the held flower's stem with a pair of scissors, and impale the flower on a flower pinholder. When a user approaches the flower, he/she perceives the smell of the flower diffused by using an olfactory display. Although haptic and olfactory devices were used together, the interaction between the two is the future problem.

Cross modality between olfaction and vision has been studied as is described earlier. Cross modality between olfactory and gustatory senses can be utilized to make a gustatory display. However, interaction between olfaction and haptic has been seldom studied.

7.3 Application of Olfactory Display

A variety of olfactory displays are expected. They are classified into entertainment, art, advertisement, medical field, etc. They are described here in detail.

7.3.1 Entertainment

The most typical application of olfactory display is a movie with scents. Tomono *et al.* investigated how many scents were required for one cinema. The subjects watched 10 cinemas and were asked which scene they wanted to have smells. It was revealed from this investigation that only 3–15 types of odors are required for one cinema. Thus, the number of required odors is realistic when odors corresponding to appropriate scenes are prepared in advance [49].

Then, Nakamoto *et al.* made movie with scents using an olfactory display based upon solenoid valves [50]. The Japanese cinema *Spirited Away* was compressed into 4 min although it takes 2 h to watch the original one. The compressed version has 11 scenes as is listed in Table 7.1.

Seventeen subjects watched the movie without smell and then watched it with smells. The questionnaire survey revealed that the change of the offensive smell to pleasant smell enhanced the impression of the corresponding scene very much. This effect is similar to the case where large contrast of blue area with yellow one is expected in a picture. Thus, we can control to which scene people have attention to some degree using appropriate smells.

Chihara *et al.* made a game with scents called "Fragra" that enables players to enjoy scents together with images interactively [51]. A player is requested to answer whether olfactory cue agrees with visual one. A scent is presented to the player equipped with HMD to flow the scent from the cartridge to a human nose using the air compressor. Since the overall system is bulky, a player can walk around the limited area. A player tries to grasp a food and take it in front of the nose in virtual space. Then, a smell comes to the player. He/she should say whether the visual object matches its smell.

Nakamoto and Nagahama presented a cooking game with scents [52]. In the movie with scents, the information flow is just one way. On the other hand, an interactive olfactory display takes important role in a game since interactive behavior is important. This interactivity might enhance reality in the virtual environment.

Table 7.1 Scenes in digest of *Spirited Away* and smells attached to corresponding scenes

Number	Scene	Odor
1	Parents were selecting foods	Food1
2	Mother had her mouth full of meat	Food1 + Food2
3	Mother became a big	Smell of pig
4	Chiriro asked Haku if her parents did not become pigs	No smell
5	She was told that important and delicious foods were eaten by parents	No smell
6	She went to a pigsty through flower garden	A few scents of flowers
7	She told her parents to help them at pigsty	Smell of pig
8	She remembered her name at flower garden	Scent of flower
9	She was told to take care of stink god	No smell
10	Stink god appeared	Offensive smell
11	Stink god was soaked in medicated bath	Offensive smell and herbal one

Figure 7.20 Cooking game with scents. Reproduced with permission from Ref. [52]. Copyright IEEE, 2008

In a cooking game, a player puts ingredients such as butter, meat, onion, garlic, wine, curry roux, and spices into a pan sequentially in the virtual environment as is shown in Figure 7.20. After he/she puts each ingredient, he/she experiences a smell and movie simultaneously. The smell of any added ingredient is blended with those already existing in the pan.

Since the operation is very easy, even a small kid can enjoy the cooking game. This game became popular, and many people experienced it. People felt hungry after the game. Thus, this game might be used for the advertisement of food or cooking.

7.3.2 *Olfactory Art*

Kakehi *et al.* studied artistic expression using scents [53]. Scent in the surrounding environment was input to a computer system through metal oxide gas sensors. Since that information was converted to visual and/or auditory information, people could perceive it synchronously with visual/auditory sensation. In the interactive art called "hanahanahana," the participant soaks a leaf-shaped piece of paper in perfume and then shakes it at a place close to the sensors. A flower image then appears in each bud-like device. The color darkness of the flower changes gradually according to the strength of the ambient scent, while the color and shape of the flower also vary according to the sort of fragrance in the paper.

Iseki *et al.* made contents called virtual ice cream shop based upon vision, sound, and olfaction [54]. When people visit the cyber virtual ice cream shop, they can select favorite flavors with electronic timbre that they feel most appropriate. The scenery when a participant enjoyed the virtual ice cream shop is shown in Figure 7.21. People can feel harmony between the scent and sound in this content.

Moreover, they tried to make contents by only using scents [55]. The question here was whether people could evoke a scene by only presenting a set of scents to them. Examples of

Figure 7.21 Experience on virtual ice cream shop

scenes were seashore, forest, firework, toothbrushing, drinking liquor at a pub, etc. The correct association with scents highly depends upon personal experience. However, people can enjoy the variation of scents even without other senses since it is possible to program odor variation using the olfactory display.

Artists involved in olfactory art are not so many at the current stage. However, M. Ueda makes artistic contents [56]. For example, she made the content of *invisible white* in collaboration with an architect. As a participant walks into the dark space, he/she cannot see anything at instance, and then he/she uses the senses of auditory, tactile, and olfactory senses. As he/she gets used to the darkness, smells navigate him/her in the space. He/she can have mysterious experience through this content.

7.3.3 Advertisement

Although most of advertisements for commercial products are based upon visual and/or auditory information, we perceive too much sensory information from television, radio, signboard at a street, etc., in our daily life. Since people get used to that information, they usually do not notice it. The scent presentation is good way to let people be aware of the commercial product.

Smells related to corresponding foods are effective to increase their sales. For example, curry smell at a supermarket contributes to collecting customers. The addition of an appropriate smell to a product also contributes to promoting sales such as coffee, chocolate, herb, tea, ice cream, cookie, cake, fruit, flower, cheese, shampoo, baby powder, perfume, and so on.

One of the business-to-business experiments by NTT communication was performed in an underground shopping mall beneath Tokyo Station. The company installed an aroma-emitting digital signage unit at the entrance to a beer hall. They hoped that mall visitors looking for a place to eat and drink would be attracted by the images and the fragrance and perhaps decide to enter the beer hall [57].

France Telecom made a computer-controlled scent diffuser connected to the Internet [58]. Then, Exhalia, a spinout company from France Telecom, developed the smell diffuser which presents scents by blowing air through scent cartridges [59]. They made the contents with scents in collaboration with cooking school.

Smells are also used for their stage effects at wedding ceremony, fashion show, etc. They were used to improve the impression of hotel lobby, lobby at the airport, game center, showroom, and seminar room. Moreover, several companies adopt corporate scents for representing impressions of companies or shops.

7.3.4 Medical Field

From medical point of view, an important application of olfactory display is to measure human olfactory threshold. Hayes *et al.* used inkjet-based microdispenser to measure human olfactory threshold [60]. Decrease in olfactory capability has been reported to be an early symptom of diseases such as Alzheimer's and Parkinson's diseases [61, 62]. Since patients of neurodegenerative disease such as Alzheimer and Parkinson tend to have higher olfactory threshold than normal persons, there is possibility that this microdispenser may be used for early screening of neurodegenerative diseases through olfactory threshold determination.

Okada *et al.* also studies the measurement of human olfactory threshold using their inkjet-based olfactory display [63]. Unlike other measurement techniques, using a very small amount of droplet for scent presentation reduces lingering scent in the air and the influence of adaptation. Measurements were finished within approximately 5 min using this system even if the measurements were kept accurate.

Another application of an olfactory display to medical field is a surgery simulator for training medical students. Reality of simulator is enhanced if smell is added to the scene of the operation. Since the training and educational tools are important applications in virtual reality, the olfactory display for the medical simulator will be developed.

Moreover, patients even in a hospital can enjoy the olfactory display since foods for patients are often restricted. Since tastes of food are deeply coupled with olfaction, patient can enjoy food tastes using a pseudo-gustatory display mentioned previously.

In this subsection, an application of olfactory display was described. Since the field of olfactory display is relatively new, more and more applications might appear in the near future.

7.4 Odor Recorder

7.4.1 Background of Odor Recorder

An odor recorder is an apparatus to reproduce odor as well as reproduce it. Vision and audition among five senses can be nowadays easily recorded and reproduced under multimedia environment. However, we have never had such a kind of device for olfaction although the methods of vision and audition have been so far mature. The study of recording olfactory information is very challenging.

Although odors have been analyzed using GC/MS for long time, odor sensing systems often called electronic noses described in Chapter 3 have been studied for last three decades [64]. Its principle is based upon the pattern recognition of the outputs of the multiple sensors with partially overlapping specificities in the same manner as that of a biological olfactory system described in Chapter 2. Many researchers study the odor sensing system nowadays.

On the other hand, an olfactory display, a device for smell presentation, was recently studied in virtual reality as is explained in the preceding text. Since an olfactory display is a relatively new field, its research community is currently small. As described earlier, an odor sensing system is an input of machine (computer), whereas an olfactory display is its output. In an odor recorder, we use both techniques of an odor sensing system and olfactory display. These two techniques are tightly coupled together when an odor recorder is developed, whereas the odor sensing system and the olfactory display have been so far independently studied.

An odor recorder with the capability of reproducing smells as well as recording them in the same manner as that realized in digital video disc (DVD) recorder was proposed [65]. There are many consumer products related to smells such as food, beverage, toothpaste, cosmetics, air fresheners for breath, room and bathroom, and so on. It is possible to apply the odor recorder to a variety of fields such as game, e-commerce, virtual reality, cinema, and so on.

Moreover, the odor recorder can be used for the research on cultural anthropology of the senses. The smells about historically and ethnologically important objects can be preserved and be reproduced. Objects with scents around the world such as Asian, African, Caribbean, and European areas can be recorded. Furthermore, there is historical meaning since they can be reproduced even in many years. Especially, cooking is strongly related to scents. A variety of foods and beverages all over the world accompanied with scents can be reproduced using the odor recorder even if many years pass after recording.

Next, the researches related to the odor recorder are described. Although there have been many reports of odor sensing systems, the reports aimed for odor reproduction are just a few ones. When we blend several odor components, the range of the smell to be covered becomes much larger. Although odor components are typically prepared in advance, Davide *et al.* proposed the framework of the olfactory display, which not only blends odors but generates the new odor using a precursor [66]. Since this is just the concept and has not been so far realized, the following portion of this chapter deals with blending odor components prepared in advance

Matsushita *et al.* developed the equipment composed of conducting polymer sensor array and an odor generator using motor-driven syringes and a heater [67]. They just used linear regression method to obtain the recipe of the mixture. After the flavors were injected according to the recipe using the syringes, the flavor mixture was vaporized using a heater. Although its principle is close to an odor recorder, the feedback in active sensing, which absolutely enhances the flexibility and the accuracy of the system, is not included.

Carmel *et al.* reported the algorithm to determine the mixture composition so that the QCM sensor array output pattern of the blended odor could match that of the target odor. They obtained the mixture composition of the odor made up of five components. Although they would like to synthesize the target odor, they only did the mixture quantification [68].

Hayashi *et al.* focused on partial structure of an odorant to express an odor. They expressed a single compound by blending several odor components. Odor components were selected using charge distribution and molecular structure, and the mixture composition was calculated using quantitative structure–affinity relationship (QSAR) method. They claimed that the blended odor approached the target odorant in comparison with any single odor component [69].

The trend of the research toward odor recorder is summarized as follows. Although the matching between target and reproduced odors should be ideally achieved in sensory space, it consumes much time and labor. Thus, the sensory test should be replaced with a sensing system. However, there have been only a few reports of the odor reproduction using a sensing system, whereas many researchers focus on only odor sensing systems. The odor recorder might be a next target in the field of electronic nose. In the next section, the principle of odor recorder using the active sensing is introduced.

7.4.2 Principle of Odor Recorder

First, an ideal case in sensory space is explained. In the development of food product, food prototype is repeatedly modified together with the sensory-test evaluation so that the prototype can approach the final product. It takes much time to develop food product because of the sensory-test evaluation. The sensory test cannot be automated.

In the process of food development, the subjects evaluate foods by sniffing them. The typical method of evaluation is semantic differential (SD) method where subjects determine scores for several evaluation items. Score takes a few discrete values, for example, from 1 to 5. Evaluation items should express the feature of the product. For example, sweetness, sourness, richness, freshness, flouriness, bitterness, and total balance are considered in the process of orange juice development.

The result of SD method is analyzed using principal component analysis (PCA) and can be plotted on a scattering diagram as is shown in Figure 7.22. Positive direction of first principal axis shows the impression of sweetness while that negative direction expresses bitterness. After modifying the prototype by changing its recipe several times from A to D, the acceptable product is finally completed. Although it is possible to make a target product in this way, it consumes much time and labor during the development of the food product. Moreover, nonlinear relationship between recipe space and sensory space makes it difficult to predict an appropriate recipe within small number of repetitions. In an odor recorder, this laborious task can be automated, and the time required for exploring the target becomes as short as possible.

Although the recipe exploration is performed in sensory space in this example, an odor recorder performs it in sensor space. In the odor recorder, the odor quality is represented as the recipe of the multiple odor components [70]. The recipe can be determined so that the similarity of the blended odor to the original one can be maximized. Thus, the mixture quantification technique is indispensable although the classification is focused on in most of odor sensing system.

An odor recorder is based upon an active sensing. An active sensing is a concept to raise sensing capability, efficiency, and flexibility. A conventional sensor works passively to convert physical or chemical quantities into electrical signals. The information flows in the order of detection, recognition, judgment, and action, and the information direction is only one way. However, the exploration behavior prior to recognition is important, and the information circulation of exploratory behavior, detection, and recognition enables the remarkable enhancement of sensing system capability.

One of the examples of active sensing is an odor source localization. A robot equipped with chemical sense searches for the odor source. The exploration behavior is repeatedly performed to search for it. In case of an odor recorder, it looks for a target in sensor response space.

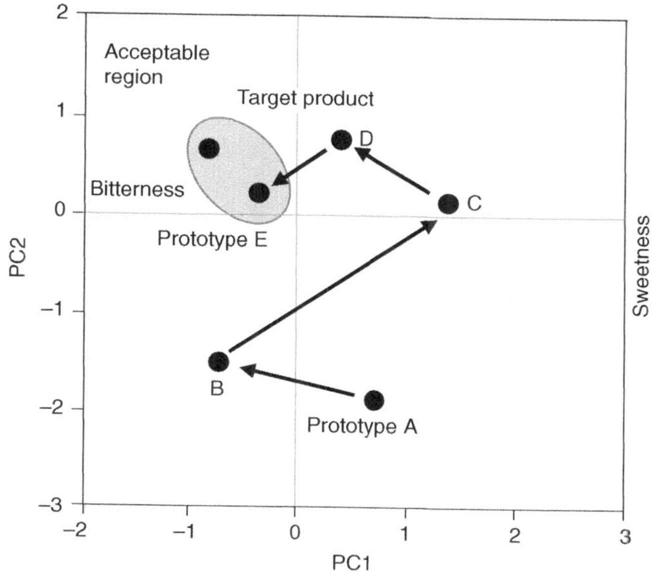

Figure 7.22 Process of food development using sensory test. Repetition is performed from A to D

The principle of odor recorder based upon active sensing is illustrated in Figure 7.23. First a target odor is introduced into a sensor array and its output pattern is memorized. Then, the array is exposed to the mixture of multiple odor components from an odor blender to obtain the sensor array pattern of the blended odor. The pattern matching, that is, the distance between the target and the blended odors in sensor feature space, is calculated and the recipe of the blended odor is modified so that the pattern matching index can be improved. The recipe modification and the measurement of the sensor responses to the blended odor and the pattern matching calculation are repeatedly performed until the sufficient pattern matching is achieved. After convergence, the recipe of the target odor is indirectly obtained from that of the blended one. This is one type of relative method to compare the target odor with the internally blended odor. Once the recipe is obtained, it can be transferred to the remote site via the Internet so that the corresponding scent can be reproduced.

In a sensor array, a variety of sensors such as metal oxide semiconductors, QCM gas sensors, SAW gas sensors, conducting polymer sensors, and even mass spectrometry are used. Another important part in the odor recorder is the odor blender, which can be used as an olfactory display [71] in the odor reproduction phase. Its principle has been already described.

The sensor feature space has smaller number of dimensions than the sensor array dimension. It should express the feature of the sensor data, for example, principal axes in PCA, variates in canonical discrimination analysis, and so on.

The odor recorder is based upon an active odor sensing system since it explores the space of odor recipe in addition to the pattern matching index. It deals with the mixture quantification problem which many researchers tried to solve using partial least squares (PLS), principal component regression (PCR), neural net, etc., described later. The odor recorder is flexible

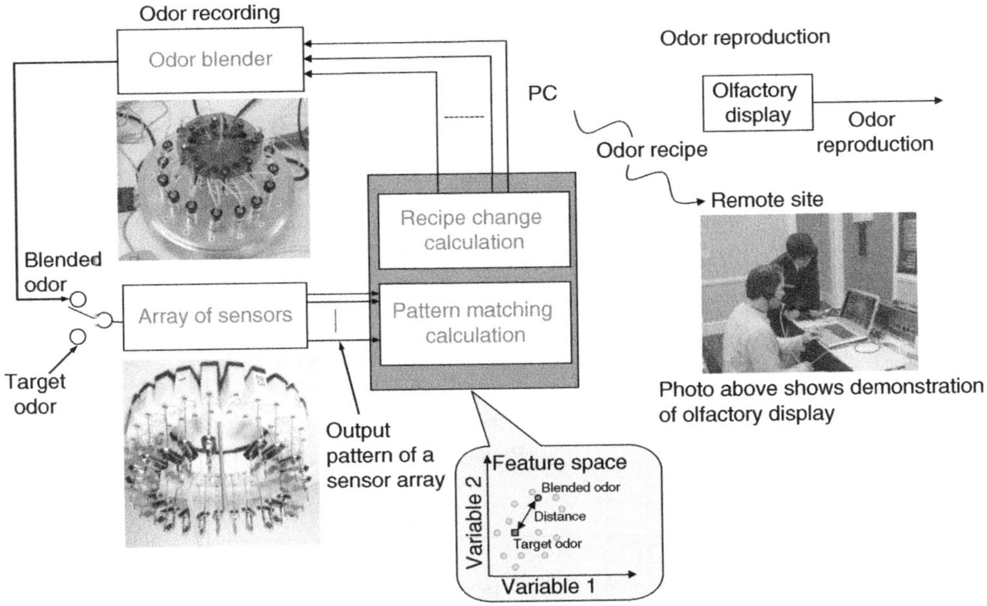

Figure 7.23 Principle of odor recorder

and is robust against nonlinearity, drift, and aging often encountered in a chemical sensor since its measurement is based upon relative comparison.

The linear superposition theorem is not valid for most of chemical sensors. Thus, nonlinear model of a sensor response explained by the concentrations of multiple odor components is required. However, its model should be built again if its characteristic changes due to drift and aging are often encountered in most of chemical sensors. Many data points are again necessary to rebuild the model. The active sensing is useful to overcome that problem since it does not require the model but does the feedback.

There are several ways to modify an odor recipe in the exploratory phase. The method such as multiinput multioutput (MIMO) feedback [72] for obtaining static recipe and the real-time reference method for dynamically changing recipe are described later. Another important issue is the selection of odor components. It will be also explained later.

7.4.3 Mixture Quantification Method

Although an odor recorder actually performs the mixture quantification, we discuss about conventional algorithms for obtaining the mixture quantification. Although many researchers study the odor classification using an odor sensing system, the mixture quantification is a little different from the classification problem. The simplest method is multiple linear regression (MLR) [73]. The linear regression model with the dependent variable Y_i and independent variables such as $X_2, X_3, \ldots, X_p$ can be written as

$$Y_i = \beta_1 + \beta_2 X_2 + \beta_3 X_3 + \cdots \beta_p X_p + \varepsilon_i, \qquad (7.1)$$

where β_1 denotes the intercept, $\beta_2 \ldots \beta_p$ are the coefficients, and ε_i is the residual of the ith measurement. For the sensor array in an odor sensing system, $X_2, X_3, \ldots, X_p$ correspond to sensor responses to a target odor and Y_i is the concentration of the ith vapor in the mixture, assuming that the linear superposition is valid for each gas sensor response. In MLR, $\beta_2 \ldots \beta_p$ are obtained using the observed data so that the sum of the residual can be minimized. However, the linear superposition is not actually valid in most of actual gas sensors.

The important problem to be solved in the mixture quantification is the collinearity problem. Let us assume the two-sensor case to obtain the composition of the binary mixture as is illustrated in Figure 7.24. Each line shows the sensor characteristic, and the intersection point of the two lines expresses the composition. However, these two lines become almost parallel if two sensor characteristics are very similar. Then, the obtained composition becomes instable even if the measurement data fluctuate just a little. In Figure 7.24, the location of the intersection point drastically changes with small noise. This problem is called collinearity problem and is the same as that of the calculation of the inverse matrix with almost zero determinant.

There are several methods to reduce the collinearity in chemometrics where statistical and mathematical techniques are used to analyze chemical data [74, 75]. PCR and PLS are often adopted [76]. PCR is the combination of PCA with MLR. PCA is often used to reduce the dimension with minimum information loss. Since the principal component with its small contribution can be regarded as noise and can be ignored, the instability due to the collinearity is reduced. Thus, a small number of principal components obtained by PCA are projected onto

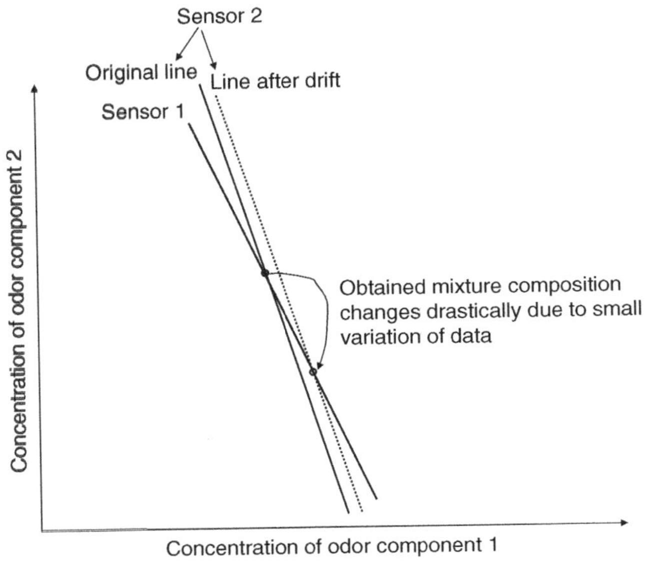

Figure 7.24 Explanation of collinearity problem. Case of two sensors for binary mixture

dependent variables using the MLR. These principal components exhibit the feature of the sensor data and are called latent variables.

PLS picks up latent variables both for independent and dependent variables, whereas PCR extracts latent variables only for independent variables of MLR. Latent variables in space of independent variables are projected onto space of dependent variables using regression equation. The number of latent variables is an important parameter for raising estimation accuracy. Thus, the cross-validation technique is used to determine the appropriate number of the latent variables.

Carey *et al.* compared MLR with PLS when the binary and ternary mixtures were quantified using the data of the quartz-resonator sensor array [77]. They showed that the estimation accuracy of PLS was better than that of MLR. PLS is currently a gold standard of the technique to quantify the mixture. Vergara *et al.* applied PLS to the data of temperature-modulated micro-hotplate gas sensor array to quantify binary gas mixture [78].

Sasaya *et al.* used PLS method for the mixture analysis of halitosis with the data from three electrochemical gas sensors combined with a preconcentrator [79]. Typical halitosis consists of three volatile sulfide compounds (VSC) such as hydrogen sulfide, methyl mercaptan, and dimethyl sulfide. The ternary mixtures with various compositions were used to build a PLS model. Then, the leave-one-out method was used to evaluate the accuracy of the estimated mixture composition as is shown in Figure 7.25. The good estimation accuracy is obtained if the plot is close to the diagonal line. Figure 7.25 shows that the concentrations of the three components could be almost estimated using PLS method. It was possible to quantify the mixture composition with the sub-ppm range concentration using this system.

Zeller showed the method of extended disjoint principal component regression (EDPCR) to extend PCR [80]. First, the principal component for calibration data of each gas is extracted. Then, the sensor responses to the gas mixture are regarded as the sum of the

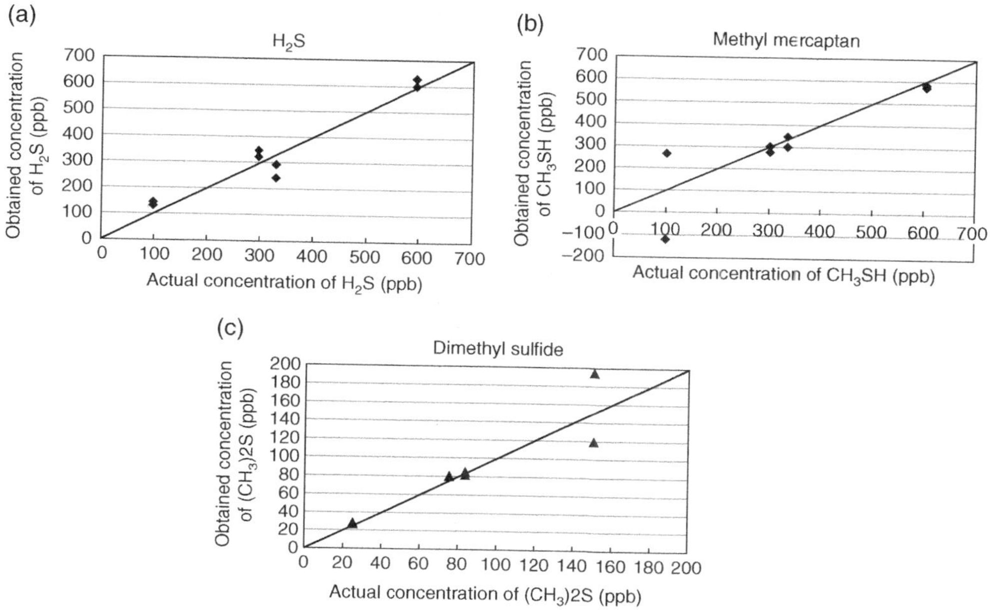

Figure 7.25 Quantification of ternary mixture composition using PLS method. (a) Hydrogen sulfide, (b) methyl mercaptan, and (c) dimethyl sulfide

principal component vector for each gas. The coefficient for each principal component vector, the ratio of the corresponding component, is determined so that the residual can be minimized. Although this method is straightforward and is easy to understand, the accuracy deteriorates if each sensor has nonlinearity.

Wang *et al.* proposed locally weighted regression (LWR) [81] where linear analysis method is partially applied in spite of the nonlinearity for the whole concentration range. The sensor response around the measurement point is approximated using linear function and calibration data. The accuracy of the quantification depends upon the number of calibration data. This method is useful when large amount of the data is available.

Those methods mentioned earlier are linear ones. However, it is expected to raise the accuracy of the mixture quantification when nonlinear method is used. Sensors such as metal oxide gas sensors have highly nonlinear properties. Although QCM gas sensors or SAW gas sensors have relatively linear characteristics, the accuracy of the mixture quantification is still influenced by the nonlinear portion. Thus, a neural network is useful to overcome the accuracy deterioration due to the nonlinear characteristics [82]. One of the typical methods of neural networks is the multilayer perceptron (MLP) composed of input layer, hidden layer, and output layer. Since it was found that it was possible to realize arbitrary mapping with arbitrary accuracy using MLP with sufficient number of hidden layer neurons, MLP is useful for nonlinear mapping. The typical algorithm for training MLP is an error backpropagation often called BP method [83]. In the quantification phase, each sensor signal is fed into each neuron in the input layer, and then the concentration of each gas component is the output at each neuron in the output layer. In the training phase, the neural network is repeatedly trained when the concentration of each gas component is presented as target input of each neuron in the output layer, followed by the synaptic weight modifications using BP method.

Sundgren *et al.* trained MLP neural network using BP method so that the gas mixture composition could be obtained from the MOSFET gas sensor array [84]. Although linear superposition is not valid for a MOSFET gas sensor, they reported that they obtained good quantification accuracy in comparison with PLS method.

Another example of nonlinear mapping is the mapping from mass spectrometry data space onto sensory data [85]. The mass spectra of mint flavors were transformed into sensory-test scores obtained from quantitative description analysis (QDA) method, using self-organizing map (SOM) [86]. It was found that the estimation accuracy using SOM was better than that using a linear method.

Z. Wang *et al.* proposed a neural network called ChemNets especially designed for metal oxide gas sensors, where the model of a metal oxide gas sensor is combined with MLP [87]. A priori knowledge of gas sensor enhances the network capability to estimate each gas concentration in the gas mixture. The rough model of the metal oxide gas sensor reduces the number of parameters in the neural network, resulting in the reduction of training epochs and the improvement of generalization capability.

Although several methods work well to quantify the mixture [88], it is indispensable to build a model for mapping sensor space onto concentration space. It is necessary to have many data points to obtain the accurate model because of nonlinear mapping. Moreover, the model should be reconstructed when we encounter the sensor drift or aging problem. The active sensing method described previously is suitable for solving those problems.

7.5 Algorithm of Odor Recipe Exploration

There are two types of odor recorder. One determines the static odor recipe kept constant during recording. The other is the odor recorder for obtaining the recipe of dynamically changing the odor. Here, these two types of the odor recorder are reviewed here.

When we quantify odors composed of many components, mass spectrometry is useful. However, the selection of odor components among the huge number of candidates is necessary to perform the quantification with high accuracy. This technique is also described.

7.5.1 Odor Approximation

Although the odor recognition and classification have been so far performed by a human panel, their results are not completely objective due to the influence of human characteristics such as mood, age, health condition, and psychological state. In the process of flavor development as is shown in Figure 7.22, the scores obtained from the sensory test are often instable. Thus, an odor sensing system can be used instead of sensory test to evaluate the developed flavor.

One of the examples is a fruit flavor approximation using small number of ingredients [89, 90] The orange flavor consisted of 14 ingredients as is given in Table 7.2. The experiment on orange flavor approximation was performed only using three ingredients. In this experiment, an array of QCMs (20 MHz, AT cut) coated with sensing films were used. Those coatings were divaleroyl phosphatidyl choline (DVPC), cardiolipin (CL), cerebrosides (CS), Apiezon L,

Table 7.2 Recipe of orange flavor used in experiment on odor approximation

Number	Component	Ratio (%v/v)
1	Perillaldehyde	0.05
2	Citronellal	0.1
3	Geranyl acetate	0.25
4	Neryl acetate	0.5
5	Citronellol	1.0
6	Nerol	1.0
7	Octyl acetate	1.2
8	Nonanal	2.0
9	Terpineol	5.0
10	Dodecanal	5.5
11	Citral	11.0
12	Octanal	12.4
13	Decanal	18.0
14	Linalool	42.0

(S)-(+)-2,2′-Bis(diphenylphosphin)-1-1′-binaphthyl ((S)-BINAP), and ethyl cellulose (EC). The steady-state sensor responses to the odor evaporated from the liquid droplets were measured in the static chamber. The recipe of the approximated odor was explored in six-dimensional space composed of six sensor responses.

Three components such as linalool (LL), citral (CT), and decanal (DC) were selected to approximate the orange flavor since these three components were likely to cover the point of the target odor in sensor space. Then, the odors with various recipes of these three components were measured, and their data were plotted on a scattering diagram of PCA as is shown in Figure 7.26. Plot (orange calc) is calculated using 14 components based upon superposition principle.

First, the binary mixture of LL and CT with various compositions (LL:CT in Figure 7.26) was measured. Their compositions (LL:CT) were 1:9, 2:8, 5:5, 8:2, and 9:1. The point that is the closest to the target one had the composition of 8:2. Then, the third component, DC, was added to the closest binary mixture. The proportions of DC were 5, 15, 25, and 35%. Finally, the point with the composition of LL:DC:CT=60:25:15 was found to be the closest to the target orange flavor.

Although the odor with the obtained recipe was very close to the target odor in sensor space, it should be confirmed whether these two are similar in sensory space. Thus, a triangle test, one of sensory tests, was performed. In the test, three samples made up of two same and one different ones are presented to a human panel. The participant in the panel is requested to pick up one different from the other two. If two samples are very close, its classification approaches 33.3%.

In this test, the number of participants was 23. The classification rate was 30.9%, very close to 33%. The sensory test revealed that participant could not distinguish the approximated orange flavor from the original one.

This experiment indicates that the odor can be approximated using the small number of odor components using the sensors if the appropriate odor components are blended at the appropriate recipe. However, the recipe exploration was still a hard task in the manner similar

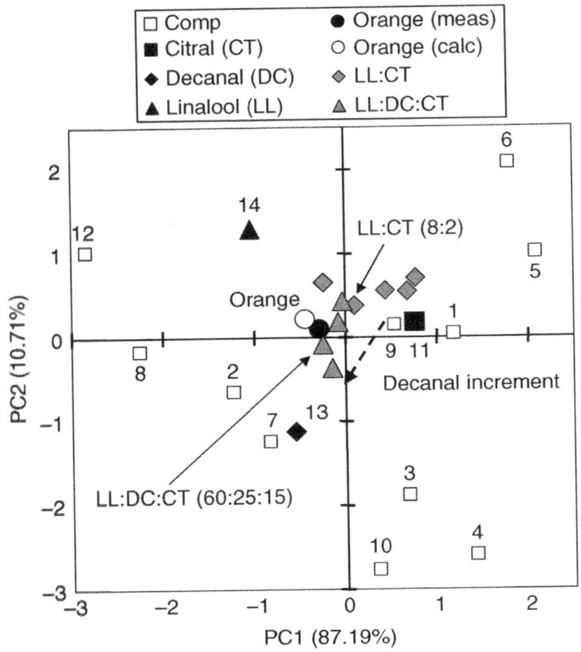

Figure 7.26 Scattering diagram of principal component analysis when odors with various recipes were measured

to the case only using sensory test as is shown in Figure 7.22. The prediction of the recipe of the approximated odor is not so easy due to the nonlinearity. Thus, the algorithm for the recipe exploration is indispensable.

7.5.2 MIMO Feedback Method

The mapping of sensor space onto recipe space is generally nonlinear. Thus, nonlinear model should be built up. Several researchers made nonlinear model as is described in Section 7.4.3. However, a chemical sensor often changes its characteristic due to drift and aging. Once a sensor characteristic changes, its model should be built again. It is a hard task since nonlinear model needs many calibration points. Thus, a feedback technique is useful since it does not need an accurate model.

When the sensor response is smaller than that to the target, the odor concentration should be increased. On the other hand, the odor concentration should be decreased when the sensor response is larger than that to the target one. Although the case of multiple sensors exposed to multiple odor components is complicated, MIMO feedback is used in an odor recorder.

In this subsection, it is described how MIMO feedback such as optimal feedback control is realized when the number of the sensors is n and the number of the components is m, respectively. The sensor response vector and the concentration-change vector of the blended odor at time kT are expressed by

$$\mathbf{s}_k = \left[s_1(k), s_2(k), \ldots, s_i(k), \ldots, s_n(k)\right]^T \tag{7.2}$$

and

$$\begin{aligned}\mathbf{u}_k &= \left[u_1(k), u_2(k), \ldots, u_j(k), \ldots, u_m(k)\right]^T \\ &= \left[c_1(k)-c_1(k-1), c_2(k)-c_2(k-1), \ldots, c_j(k)-c_j(k-1), \ldots, c_m(k)-c_m(k-1)\right]^T,\end{aligned} \tag{7.3}$$

where $s_i(k)$, $u_j(k)$, $c_j(k)$ are the ith sensor response, the jth component concentration change and the jth component concentration, respectively. The state-space equation is expressed by

$$\mathbf{s}_{k+1} = \mathbf{F}\mathbf{s}_k + \mathbf{G}\mathbf{u}_k. \tag{7.4}$$

The parameter matrices F and G were estimated in advance from the sensor responses to certain temporal profiles of odor component concentrations using the least squares method. Since the odor blender can control the temporal profile of each odor component, the sensor responses to the blended odor with programmed time-varying recipe are available.

In the optimal feedback control, the index value

$$J = \sum_{k=0}^{p-1}\left\{\left(\mathbf{s}_{k+1} - \mathbf{s}_{\text{target}}\right)^T \mathrm{Q}\left(\mathbf{s}_{k+1} - \mathbf{s}_{\text{target}}\right) + \mathbf{u}_k^{\,T}\mathrm{R}\mathbf{u}_k\right\} \tag{7.5}$$

is minimized by modifying the recipe of the blended odor where p is the number of the concentration change during the odor recipe exploration. The first term in Equation 7.5 expresses the difference of the sensor response vector between the target odor and the blended one weighted by the diagonal matrix Q, and the second term does the difference of the concentration-change vector weighted by a diagonal matrix R. The second term is necessary to suppress the oscillation behavior. The recipe of the blended odor is iteratively changed to decrease the index J using optimal control theory. Its procedure is described in the Ref. [91]. The method to determine Q and R is described in another Ref. [92].

Then, the experiment was performed according to this method. The experimental setup is shown in Figure 7.27. The odor blender in the figure is the same as multicomponent olfactory display based upon high-speed switching of solenoid valves in the previous section. The method to control the concentration of each odor component was the second-order delta sigma modulation. The concentration generated at the odor blender with the unit %RC is expressed as the concentration relative to the full scale of each component. A QCM sensor works together with oscillator (OSC) with its frequency shift measured by a frequency counter. Its resonance frequency decreases with the mass of odorants adsorbed onto the coatings due to mass loading effect [93, 94]. The QCM sensor responses were collected through the multichannel frequency counter where sampling interval was set to be 6 s. The sensor coatings used in this experiment were selected according to the condition number indicating the degree of ill conditionedness. They were thermol-1, PEG1000, TCP, EC, and Versamid 900.

The flavor made up of five compounds was used in the experiment. This is similar to the mixture composition quantification as is described in Section 7.4.3. Since the quantification

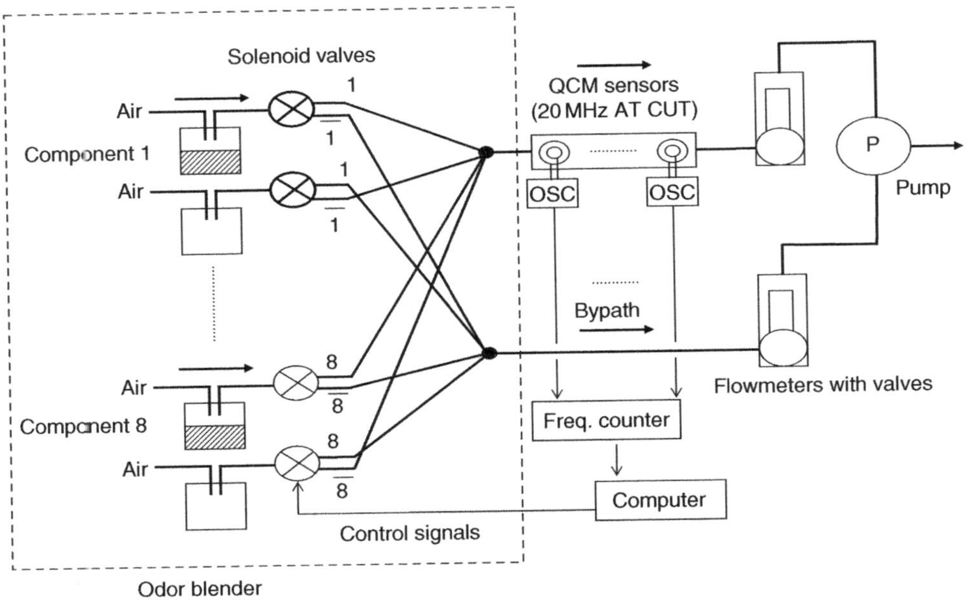

Figure 7.27 Experimental setup for recording recipe using MIMO feedback

of the five-component mixture is a hard task, most of the other researchers used up to ternary mixtures. They were *trans*-2-hexenyl acetate (component 1), *trans*-2-hexenal (component 2), isobutyric acid (component 3), ethyl valerate (component 4), and propionic acid (component 5). The target recipe was (components 1, 2, 3, 4, 5) = (50%, 10%, 50%, 50%, 50%). The result of MIMO feedback is shown in Figure 7.28. The solid lines indicate the recipe of the target odor. The plot was the relative concentration of each component during the MIMO feedback process. Figure 7.28 shows that only the relative concentration of component 1 approached 10%, whereas the other four components did 50% within 240 s. Thus, it can be said that the obtained (recorded) recipe was close to that of the target one after the convergence.

7.5.3 *Method to Increase Number of Odor Components*

Although many odor components are required to record many kinds of odors, the odor recording becomes difficult as the number of odor components increases because of the collinearity problem as is explained in Section 7.4.3. The two methods such as singular value decomposition (SVD) problem and two-level quantization methods were developed to overcome this problem.

7.5.3.1 SVD Method

SVD is one of the techniques to reduce the collinearity [95]. When the sensor characteristics are close to those of other sensors, the solution of the mixture composition is very sensitive to the small variation of the sensor response caused by noise. It is essential to solve the

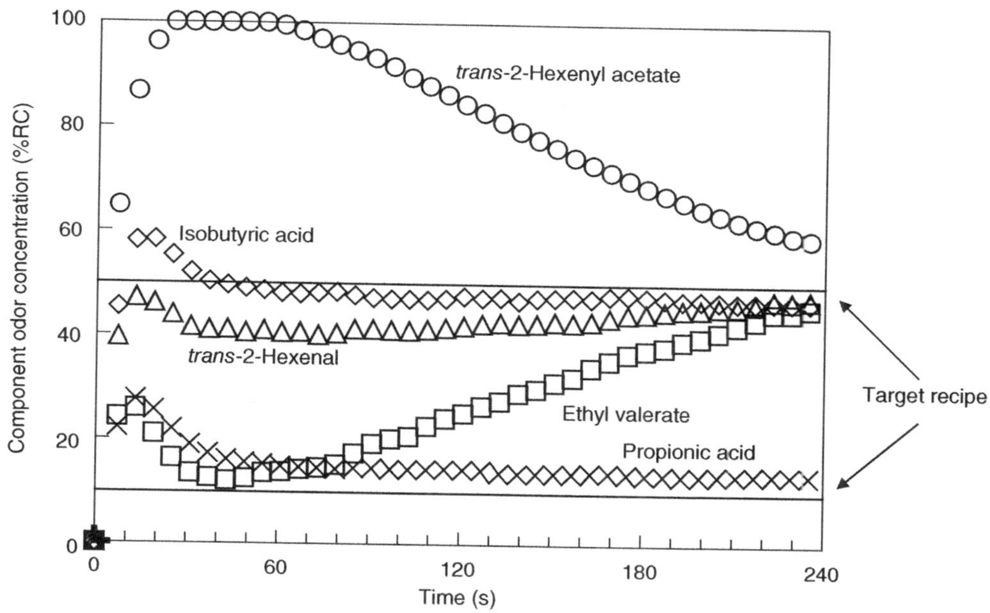

Figure 7.28 Recipe exploration of apple flavor composed of five components using MIMO feedback. Reproduced with permission from Ref. [92]. Copyright Elsevier, 2002

collinearity problem so that the stable solution can be obtained. Its decomposition process is similar to PCA. Its derivation is briefly described here.

The sensor responses *s* is

$$\mathbf{s} = A\mathbf{c} \tag{7.6}$$

where **c** expresses the m-dimensional vector with its elements of the relative concentration of odor components. In this situation, linear superposition theorem is assumed.

An $n \times m$ matrix can be decomposed into

$$\mathrm{A} = UWV^T \tag{7.7}$$

according to SVD theorem. Then, **s** is

$$\mathbf{s} = UWV^T\mathbf{c}, \tag{7.8}$$

where U is an $n \times m$ column orthogonal matrix and V an $m \times m$ square orthogonal matrix. W is an $m \times m$ diagonal matrix

$$W = \begin{bmatrix} w_1 & & 0 \\ & \ddots & \\ 0 & & w_m \end{bmatrix} \quad w_1 \geq \cdots \geq w_m, \tag{7.9}$$

where the diagonal elements are the singular values. When the m-dimensional vectors

$$\mathbf{x} = V^T \mathbf{c} \tag{7.10}$$

and

$$\mathbf{y} = U^T \mathbf{s} \tag{7.11}$$

are used instead of **s** and **c**, the relationship between **y** and **x** is

$$\mathbf{y} = W\mathbf{x} \tag{7.12}$$

using Equations 7.8, 7.10, and 7.11. Since the elements of **x** corresponding to the small singular values in the matrix W seldom contribute to **y**, those elements are set to 0s to suppress the influence of noise. The elimination of the small singular values contributes to the stabilization of the solution, that is, the recipe of a target odor, because of better ill conditionedness. The recipe of the blended odor is modified so that the following modified index value at time k can be minimized instead of Equation 7.5:

$$J_k = \left(\mathbf{y}_{k+1} - \mathbf{y}_{\text{target}}\right)^T Q\left(\mathbf{y}_{k+1} - \mathbf{y}_{\text{target}}\right) + \left(\mathbf{x}_k - \mathbf{x}_{k-1}\right)^T R\left(\mathbf{x}_k - \mathbf{x}_{k-1}\right) \tag{7.13}$$

where $\mathbf{y}_{\text{target}}$ is the transformed sensor response vector of the target odor by SVD method [82].

Then, the experiment on the exploration of the recipe made up of eight odor components was performed. The target odor was the apple flavor made up of 9 odor components such as *trans*-2-hexenyl acetate, *trans*-2-hexenal, isobutyric acid, ethyl valerate, propionic acid, 1-hexanol, 1-butanol, butyl isobutyrate, and butyl propionate. In this experiment, the first 8 compounds were used as odor components in the blended odor. The 8 sensors used here were the QCMs (20 MHz, AT cut) with their coatings of Apiezon L, thermol-1, PPE(6), TCP, Versamid 900, PEG1000, CS, and EC.

The experimental result of the apple flavor recording using 8 odor components is depicted in Figure 7.29. The unit %RC in the figure means the concentration relative to the full scale of each odor component in an odor blender as is described in Section 7.5.2. The four largest singular values were adopted in this experiment since the subspace from w_1 to w_4 occupies 95% information of the original space. It was found from the figure that the stable convergence was obtained even when the 8 odor components were used, whereas the number of the components had been at most four in the report of quantification of the gas mixtures by other researchers [96].

Then, the evaluation by sensory test was performed. The triangle test described earlier was performed to evaluate the approximated odor. The odors generated by the odor blender were filled with the sampling bag made of fluorine-contained resin. The number of the panelists was totally 36. The selection rate was 33.3%, indicating that the approximated odor was identical to the target one. On the other hand, the approximation only using five odor components (components 1–5 in Section 7.5.2) resulted in the selection rate of 47.2%. Thus, the sensory test reveals that SVD method is better than conventional MIMO feedback since more odor components can be used.

7.5.3.2 Two-Level Quantization Method

Another method to reduce the collinearity problem is two-level quantization method [97]. In general, the human impressions on the odors mostly depend on the kinds of key components rather than the accurate mixture composition. Thus, it is possible to quantize each odor

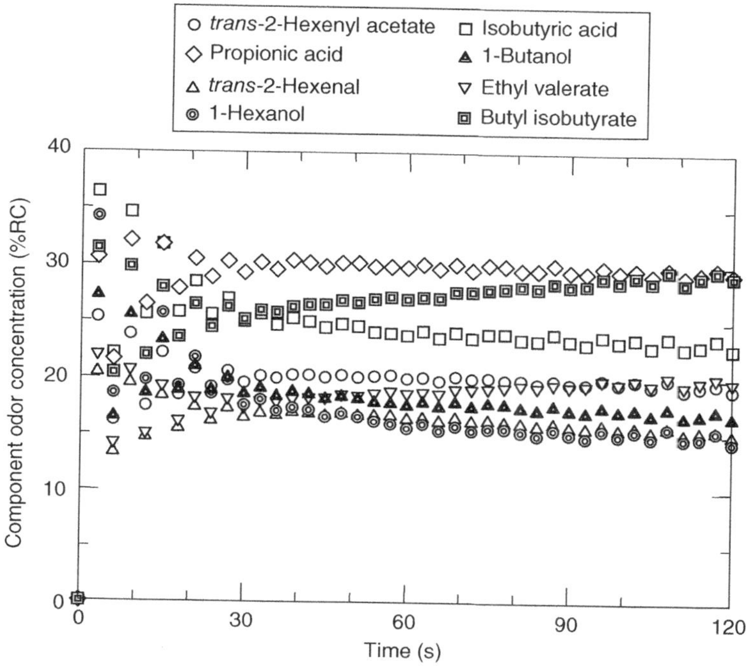

Figure 7.29 Recipe exploration process of eight odor components using SVD method. Reproduced with permission from Ref. [95]. Copyright IEEE, 2003

component concentration into several levels. Although we can take three or more levels to quantify the mixture, two-level method was adopted here for simplicity. This quantization is effective to restrict the space to be explored. When this approach is taken, the binary quantization is ultimately adopted. In other words, an odor can be approximated by the combination of the odor components as is illustrated in Figure 7.30. We take the advantage of using many odor components in two-level quantization method rather than the accurate recipe. When we use two-level quantization, a smell can be expressed as a combination of the components.

An odor recorder typically requires an odor blender with the capability of realizing any mixture composition. On the other hand, the odor blender for the two-level quantization algorithm requires just two levels such as ON and OFF. The solenoid valve can be used as fluidic ON/OFF switch.

When we use m odor components to approximate odors, there are 2^m-1 recipe combinations. Since it is difficult to test all the combinations when the number of the odor components is increased, the sophisticated algorithm to explore the recipe of the target odor is required.

Next, the procedure for two-level quantization method is described. First, each odor component is measured, and the combination of the odor components with its response pattern the closest to that of the target odor is estimated under the assumption of the linear superposition. The deviation of the target odor from the estimation by the linear superposition is calculated after measuring the point among several ones around the ideal point. Then, the point in sensor data space to be explored next time is updated taking the deviation into account. This procedure is repeatedly performed until the convergence.

One of the examples of the experiment on two-level quantization is briefly described [98]. Five fruit flavors such as apple, orange, lemon, banana, and peach were recorded using 13 odor components listed in Table 7.3. The concentration in the table means the ratio of each component volume to the total one since the sample diluted with solvent (pentane) was injected into the measurement chamber. The diluted sample in the liquid phase was evaporated by a heater to obtain sensor responses in the vapor phase. The concentration of each odor component was adjusted so that the magnitude of each sensor response vector could be the same. Each odor was automatically blended in the liquid phase using the autosampler [26].

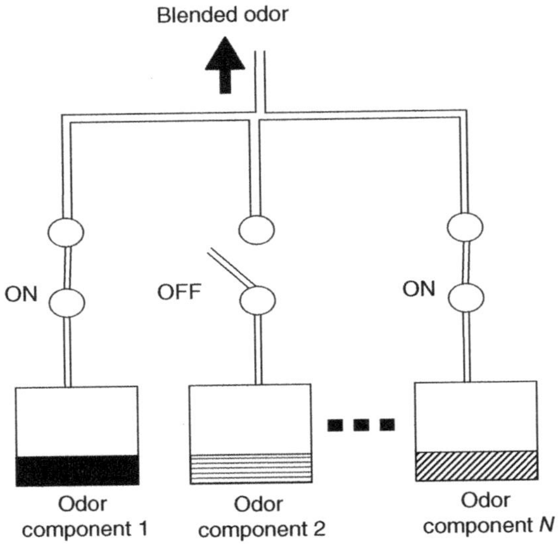

Figure 7.30 Concept of two-level quantization technique

Table 7.3 Odor components used for approximating five fruit flavors

Number	Component	Concentration (liquid, v/v) (%)
1	*trans*-2-Hexenyl acetate	16.50
2	Propionic acid	2.10
3	1-Butanol	45
4	Isoamyl acetate	55
5	Allyl caproate	15
6	Eugenol	1.64
7	Isoamyl-*n*-butyrate	14
8	Linalool	3.30
9	Citral	4
10	Octanal	15
11	Benzaldehyde	10
12	Gamma	2.20
13	Geraniol	1.75

The autosampler is useful to blend a large number of odor components, whereas it takes much time to blend odor components. 13 QCM sensors (20 MHz, AT cut) with different coatings were used here. Moreover, not only steady-state responses but also time constants both of response and recovery phases extracted by autoregressive (AR) method [99] were used as variables. An optimal set of variables made up of 13 steady-state frequency shifts and four time constants were selected according to linear discriminant analysis (LDA) technique. After 17-dimensional data were transformed into 3-dimensional space by canonical discrimination analysis technique, the recipe was explored in this 3-dimensional space.

The result of recipe exploration is shown in Figure 7.31. The circle in Figure 7.31 indicates the response pattern of the target odor and the symbol x does the response pattern of the recorded odor. The numbers in the parentheses mean the odor component number in Table 7.3. The gray plots are predicted ones using linear superposition. As is shown in the figure, the blended odor of each fruit flavor was close to the target one. Moreover, it was confirmed that the smell of the blended odor was similar to that of the target one using the sensory test. This result suggests that more odor components can be used to explore the recipe rather than the other methods described earlier even if the accurate recipe is not obtained.

7.5.4 *Dynamic Method*

Although an odor can be statically recorded in the previous way, the odor in the actual environment dynamically varies. The two methods such as real-time reference method and concurrent recording method to record the dynamic change of odor are introduced. Although

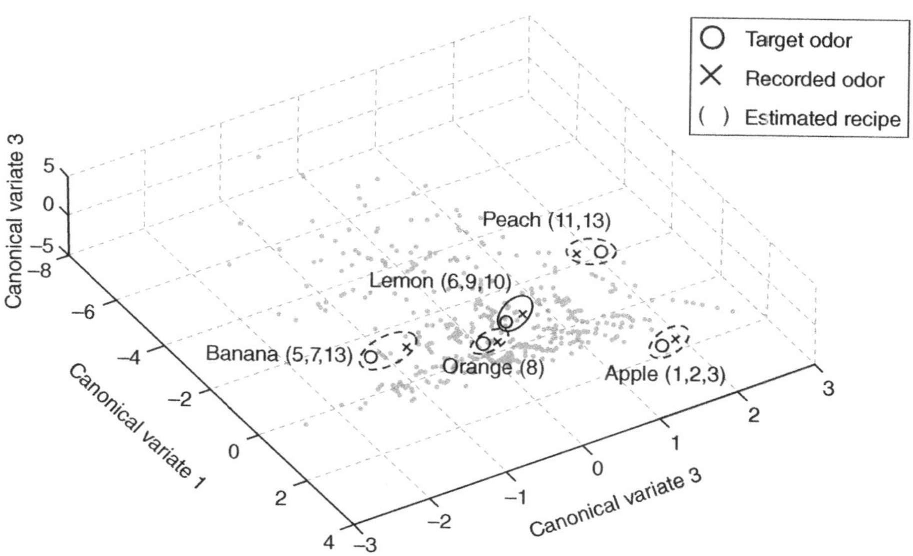

Figure 7.31 Recorded sensor response patterns in LDA space and estimated recipes of five target flavors when two-level quantization technique was used for recipe exploration. Reproduced with permission from Ref. [26]. Copyright Elsevier, 2007

the dynamical change in the composition of the binary mixture was quantified using a sensor array together with a recurrent-type neural network in 1995, the change in the composition was too slow compared with the actual environmental change [100].

7.5.4.1 Real-Time Reference Method

Real-time reference method is a relative method to frequently compare the blended odor with the target odor. It can be used to compensate for the change of the environment such as temperature and humidity as well as to record the dynamic change of odor [101, 102].

The comparison of real-time method with conventional static method is illustrated in Figure 7.32. Although only single sensor for the single odor component is shown in the figure for simplicity, the multiple sensors for the multiple odor components are actually used.

In the conventional static method in Figure 7.32a, a steady-state response to the target odor with constant concentration was measured first. Then, the recipe of the blended odor was repeatedly adjusted to match the sensor response to the blended odor with that to the target

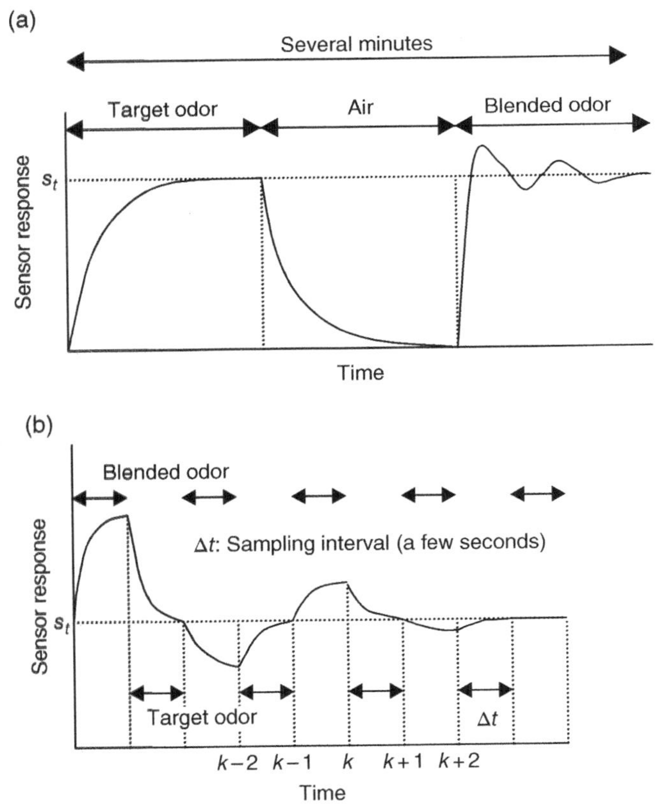

Figure 7.32 Comparison of real-time method with conventional static method in odor recorder. (a) Conventional method and (b) real-time reference method

odor. The recipe change during the recipe exploration process cannot be obtained since it takes a few minutes to determine the recipe or concentration. It is too slow to detect the dynamical change of odor.

On the other hand, a sensor is alternately exposed to the target and blended odors at the interval of a few seconds in the real-time reference method in Figure 7.32b. The sensor response to the blended odor does not match that to the target odor at first. However, it soon approaches the response to target odor. Once the sensor response to the blended odor matches that to the target odor, it tracks that to the target odor even if its recipe or concentration dynamically changes. The explanation in Figure 7.32b in the case of single odor component and single sensor can be extended to the case of the multiple sensors for the multiple odor components.

In the real-time reference method, the index value J

$$J = (\mathbf{s}_{k+2} - \mathbf{s}_{k+1})^T Q(\mathbf{s}_{k+2} - \mathbf{s}_{k+1}) + (\mathbf{c}_{k+1} - \mathbf{c}_{k-1})^T R(\mathbf{c}_{k+1} - \mathbf{c}_{k-1}) \quad (7.14)$$

is used, where k is an even number. The first term in Equation 7.14 indicates the difference between the sensor responses to the target odor and those to the blended odor. The second term indicates the odor component concentration changes in the blended odor. The smaller the first term is, the better the pattern matching is. The second term minimizes the concentration changes in the blended odor. Once the recipe of the blended odor converges to that of the target odor, it tracks that of the target odor even if that of the target odor changes dynamically. Since the target odor is frequently referred, the real-time reference method achieves the time resolution of a few seconds to record the dynamic changes of odor. Moreover, the real-time reference method is also useful to compensate for the rapid environmental changes encountered during the process of the recipe exploration in addition to recording the dynamic change of the odor.

Then, the experiment on recording a dynamically changing odor using real-time reference method is described. The sample used here is an apple flavor made up of four components such as *trans*-2-hexenyl acetate (green note, Comp1), *trans*-2-hexenal (smell of grass, Comp2), isobutyric acid (sour sweet, Comp3), and ethyl valerate (fruity, Comp4). Although the apple flavor in the previous experiment has 9 components, their quaternary mixture has the impression of the apple since those four components are major ones. The sensors used here were four QCMs (20 MHz, AT cut) coated with polyphenyl ether, PEG1000, tricresyl phosphate, and Apiezon L. Two odor blenders, one for the target odor and the other for the blended odor, were prepared for the experiment. The target and the blended odors were alternately introduced into the sensor array every 4 s.

The experimental result is shown in Figure 7.33. The concentrations of four odor components temporally changed as is shown in the figure. The plots in the figures are relative concentrations with the unit of [%RC]. The four concentrations were increased to 80 [%RC] and were kept for a while. Then, the concentration of Comp4 was first gradually reduced from 80 to 20 [%RC]. The concentrations of Comp3, Comp2, and Comp1 were in sequence gradually reduced from 80 to 20 [%RC]. Figure 7.33 shows that the concentration of each odor component in the blended odor almost tracked that in the target odor.

The next experiments using the conventional method and the real-time reference method were performed when the humidity was suddenly changed as is shown in Figure 7.34. The concentrations of the four components were kept constant during this experiment. However,

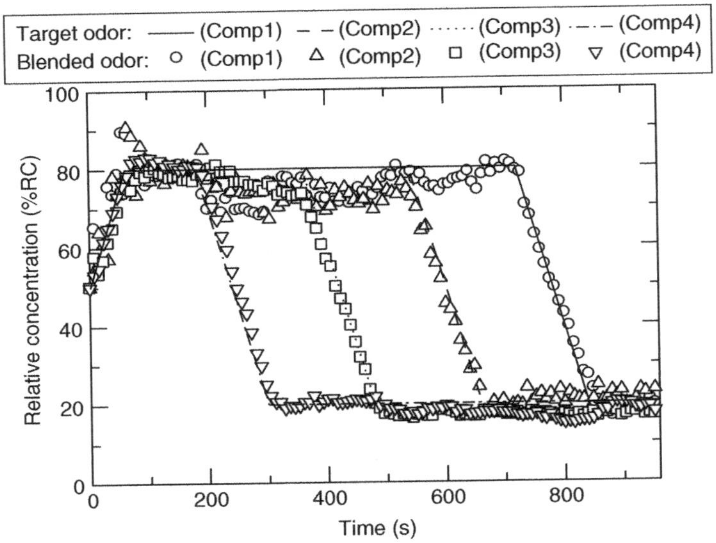

Figure 7.33 Experimental result of recording dynamically changing recipe of the apple flavor using real-time reference method. Reproduced with permission from Ref. [101]. Copyright Elsevier, 2003

the sudden humidity disturbance, that is, the approximately 10%RH increase in humidity, occurred at 100 s in the conventional method at 200 s in the real-time reference method. The conventional method means the one based upon Equation 7.5, the original version of MIMO feedback. In the conventional method shown in Figure 7.34a, the recipe of the blended odor almost converged to that of the target odor. However, it suddenly deviated from the target odor recipe after increase in humidity since humidity affects sensor responses.

On the other hand, the concentration of each odor component was kept almost constant even if the humidity was suddenly increased during the process as is shown in Figure 7.34b. Although the sensor responses were much changed after the humidity change, the concentration of each odor component was maintained. Thus, the real-time reference method enables the sensing robust against the disturbance due to the environmental change even if the sensor characteristics were much influenced by the disturbance.

Then, both temperature and humidity were intentionally changed during the experiment. The change of the air conditioner mode (DRY/COOL) caused 2.5°C of temperature change and 20%RH of the humidity change during the experiment. The concentration of each odor component was temporally changed during the experiment. The experimental result is shown in Figure 7.35a and b. The solid and dashed lines indicate the odor component concentrations in the target odor, and the plots do the recorded concentrations of odor components in the blended odor. The target odor was generated using the odor blender in the same way as that of the previous method. Since the recorded odor concentration of each odor component almost agreed with that of the target odor, it can be said that the real-time reference method achieves the record of dynamical change of the odor even if both humidity and temperature are changed.

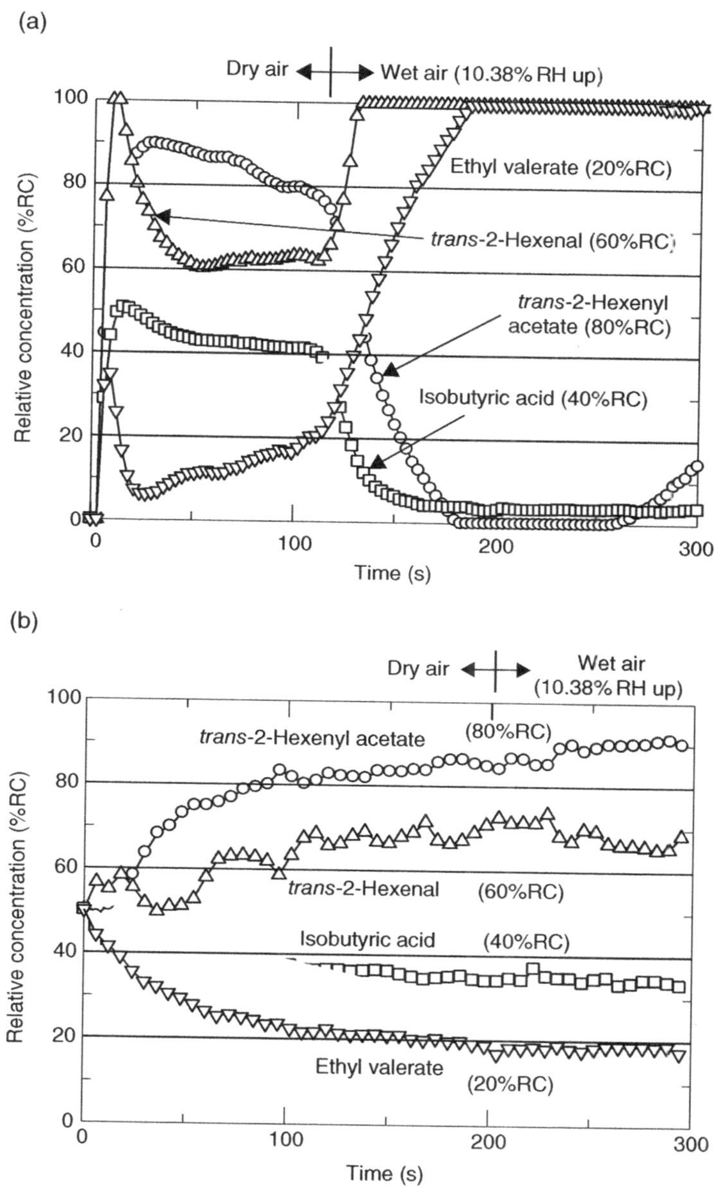

Figure 7.34 Influence of humidity change upon odor recording. (a) Conventional method and (b) real-time reference method

7.5.4.2 Concurrent Method

The real-time reference method can be speeded up when the blended odor is measured simultaneously with the target odor. It is necessary to have two identical sensor arrays to achieve the simultaneous measurements. Its concept is illustrated in Figure 7.36. In the real-time reference

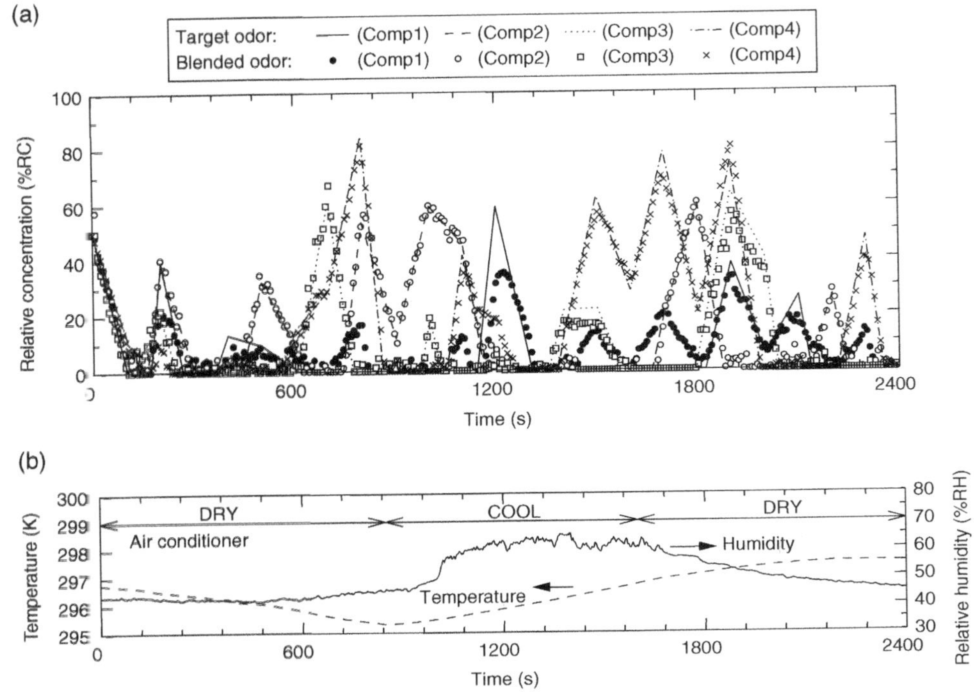

Figure 7.35 Recorded dynamic change of concentration of each odor component in the target odor. (a) Comparison of target odor with blended odor, line: concentration of each odor component in the target odor, plot: recorded concentration of each odor component, and (b) temperature and humidity change during the experiment. Reproduced with permission from Ref. [101]. Copyright Elsevier, 2003

method in Figure 7.36a, a single sensor array is alternately exposed to the target and blended odors. The target odor and blended odor cannot be measured simultaneously. However, the target and blended odors can be measured simultaneously when two identical sensor arrays are available as is shown in Figure 7.36b. The sensor array response pattern for the target odor is measured by the first sensor array, and the data are immediately used for determining the blended odor recipe at the next step. This method is called concurrent method. Since the time for exchanging odor is not required in the concurrent method, more rapid changes in odor can be measured [103].

The important point of the concurrent method is that the characteristics of two sensor arrays should be as close as possible. However, it is not so easy to fabricate multiple QCM sensor arrays with exactly the same characteristics. Although spray coating method is often used to make QCM sensors, it is difficult to coat QCMs with uniform sensing films. Although coating technique using an atomizer may improve the uniformity of the sensing film [104, 105], one can compensate for them using the state equation [103].

Then the experiment was performed. The two odor samples, *trans*-2-hexenyl acetate and isobutyric acid were used as odor components. The sensor coating used here were PEG1000 and polyphenyl ether. They were stationary phase materials for gas chromatography. The sampling interval of the QCM sensor was 4 s.

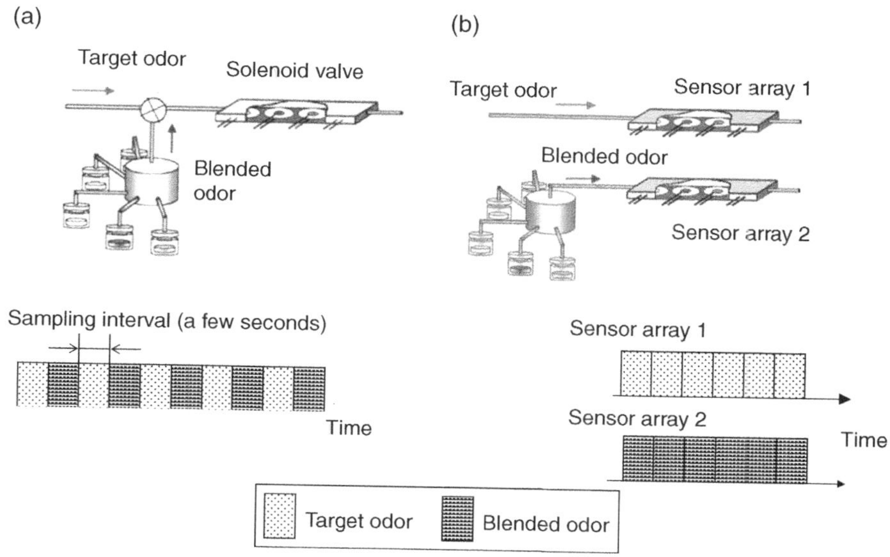

Figure 7.36 Concurrent recording of odor. (a) Real-time reference method and (b) concurrent recording

Using the concurrent system, the experiment on recording the recipe change in the atmosphere was conducted. The difficulty in recording odor increases in the presence of turbulent airflow, which causes the rapid change of the odor concentration. The odor samples of *trans*-2-hexenyl acetate, isobutyric acid, and their binary mixture were prepared in three petri dishes. The composition of the binary mixture was 1:1. The odors evaporated from the petri dishes in the ambient air were used as the target odor. The petri dish was placed inside a simple wind tunnel. The experimental result is shown in Figure 7.37a and b.

The first odor sample, *trans*-2-hexenyl acetate, was placed under the inlet of a Teflon tube connected to the sensor cell from 30 s to 90 s (time 1). The petri dish of *trans*-2-hexenyl acetate was removed from the wind tunnel at 90 s. After supplying the air from 90 to 150 s, the second sample, isobutyric acid, was placed under the inlet in the wind tunnel from 150 to 210 s (time 2). The petri dish was removed from the wind tunnel at 210 s. The third sample, the binary mixture, was placed under the inlet from 270 to 330 s (time 3). Its petri dish was removed at 330 s.

It was found from Figure 7.37a that the sensor responses were rapidly and irregularly changed due to the turbulent flow in the ambient air. Furthermore, it is clear from Figure 7.37b that the odors in the atmosphere were almost successfully recorded since the recorded odors during times 1, 2, and 3 were pure *trans*-hexenyl acetate, pure isobutyric acid, and their mixture, respectively. Thus, it can be said that the proposed system can be used in recording the rapid dynamic changes in odors even in the atmosphere.

7.5.5 Mixture Quantification Using Huge Number of Odor Candidates

In the previous section, the number of odor components used for odor recording was at most 13. However, more odor components are required to cover wide range of odors. Thus, mass spectrometry was introduced in this section.

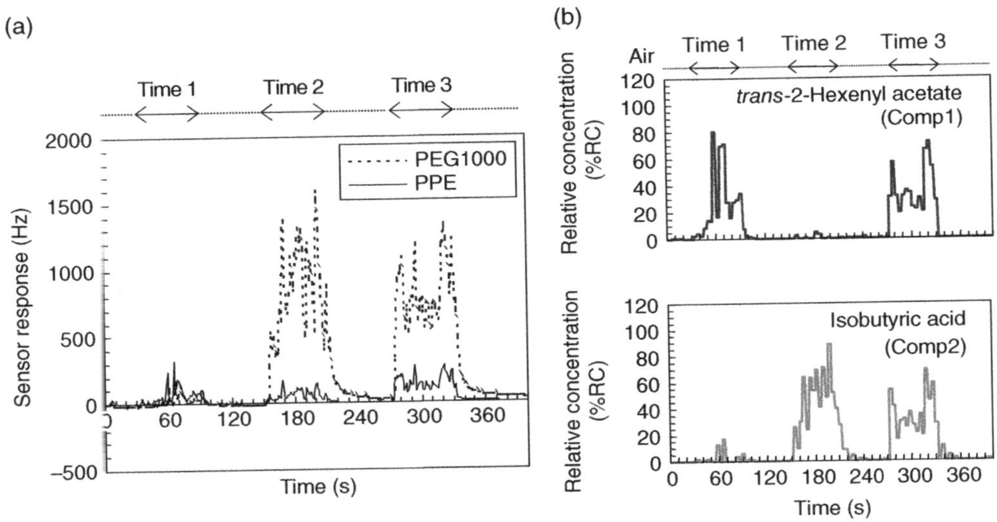

Figure 7.37 Results of recording dynamic odor changes in the atmosphere using the concurrent system. (a) Sensor responses and (b) the obtained temporal profile of concentration change. (time 1) pure *trans*-2-hexenyl acetate, (time 2) pure isobutyric acid, and (time 3) mixture of *trans*-2-hexenyl acetate and isobutyric acid. Reproduced with permission from Ref. [103]. Copyright Elsevier, 2003

It was reported that MS without GC was useful to classify smells for short time [106]. MS provides a variety of mass spectra of many compounds using more than 100 *m*/*z*s, which correspond to elements in a sensor array. The advantage of MS over other types odor sensing systems is the capability of recording smell composed of many components since high-dimensional data are available. Moreover, stable responses can be obtained compared with other chemical sensors. Thus, MS was used to quantify the recipe of an orange flavor composed of 14 components [107].

In this section, we aimed to record smells using MS and more odor components. Since the large database of MS is available, we can have the MS data composed of a huge number of odor components. Both simulation and experiment are described here.

Although linear superposition theorem is not valid in many gas sensors, that theorem is almost valid in MS [107]. Thus, the linear least squares method under the constraint that the concentration of any odor component should not be negative. However, the iterative procedure is still required to select odor components among the huge number of candidates as is described later.

The number of odor component candidates extracted from NIST database (NIST05) was 10 000. The intensity of the mass spectrum at each *m*/*z* is included in that database. Moreover, the recipe of the essential oil can be obtained from the database (ESO2006). The mass spectrum of the target odor is calculated as the sum of the inner product of the odor component mass spectrum and the recipe vector.

The typical apple flavor consists of only 9 odor components in Table 7.4. That table shows the recipe calculation result using 1040 odor candidates among 10 000 ones. Other candidates could be excluded since its intensity was not zero at *m*/*z* where the intensity of the target odor at the

Table 7.4 Result of recipe estimation using 10000 components before component selection

Odor component	Apple recipe (%)	Estimated recipe (%)
trans-2-Hexenyl acetate	12.72	0.1343
trans-2-Hexenal	1.69	0.002735
Isobutyric acid	9.05	0.435
Ethyl valerate	8.14	6.434
Propionic acid	12.9	12.22
1-Hexanol	15.74	4.698
1-Butanol	28.32	25.87
Butyl isobutyrate	1.67	0.9743
Butyl propionate	9.77	7.274
Others		61.97

corresponding *m/z* was zero. However, the recipe calculation result of the apple flavor was quite different from the actual recipe as is shown in Table 7.4. The error became larger when too many odor components were used because of numerical calculation problem. Most of them had small values in the composition although they should be zero. The accumulation of small errors of many components influenced the composition of the actual 9 components. Thus, it is indispensable to select appropriate odor components. Nakamoto *et al.* proposed the method of selecting odor components among 10000 candidates using the following procedure [108]:

Step 1. Eliminate the candidates with nonzero *m/z* which the target odor does not include.

Step 2. The constrained least squares method is applied to remaining odor components. The odor components with their ratios less than 10% in the odor component recipe are removed. This step is repeatedly performed until the number of remaining odor components is less than 40.

Step 3. At step 2, there is possibility that the odor components actually included are removed. Thus, an odor candidate removed at step 2 is temporally added to the set of remaining odor components again and the constrained least squares method is applied to that set. After this procedure is performed for all the odor components removed at step 2, the odor candidate with the minimum residual error among them is determined to join the set of the remaining odor components.

Step 3 is repeatedly performed until the minimum residual error increases.

The number of remaining odor components at step 2 was determined to be 40 because most of the samples in ESO database have less than 40 components. The recipe of the apple flavor was correctly obtained after step 3 as is demonstrated in Table 7.5.

Then, the experiment on recipe estimation was conducted. MS (QP5050A, Shimadzu) were used to estimate the odor recipes. The ionization method was electron ionization (EI) with its energy of 70eV in the same manner as that used to obtain the database data. Although the equipment used in the present study has the function of gas chromatography, no coating was used inside the column. Thus, only the mass spectrometer without gas chromatography was used. The detector output was integrated over 3min at each *m/z*, and it was used as an element of a mass spectrum vector.

Table 7.5 Result of recipe estimation using 10 000 components after the component selection together with measurement data (target odor: apple)

Odor component	Apple recipe (%)	Estimated recipe (%)	Estimated recipe (%) (experiment)
trans-2-Hexenyl acetate	12.72	12.72	13
trans-2-Hexenal	1.69	1.69	0
Isobutyric acid	9.05	9.05	6.1
Ethyl valerate	8.14	8.14	5.4
Propionic acid	12.9	12.9	13.2
1-Hexanol	15.74	15.74	12.4
1-Butanol	28.32	28.32	20.8
Butyl isobutyrate	1.67	1.67	1.9
Butyl propionate	9.77	9.77	9.2
Others		0	18

The range of m/z was from 10 to 250 except 43 and 58, the main peaks of the solvent (acetone). Thus, the dimension of the mass spectrum vector was 239. Eighty-two samples were automatically measured using an autosampler (AO-5000, Shimadzu). Every sample was diluted with acetone (dilution ratio: 25%).

Every target odor and odor component was 10 times measured, and the averaged data were used to estimate the recipe. Since the number of odor candidates was small compared with the simulation, step 1 was skipped. The estimated recipe is shown in Table 7.5. Although *trans*-2-hexenal existing in the target odor was not selected as an odor component, the recipe of the odor components was estimated within the error of 6%. Even if the recipe with the components slightly different from the actual one was obtained, the sensory test revealed that the reproduced smell was similar to the actual one.

In conclusion, the selection of the actual odor components among huge candidates is essential, whereas the application of the huge number of candidates without selection causes large estimation error.

7.6 Exploration of Odor Components

7.6.1 Introduction of Odor Components

In the previous sections, we review the odor recording technique for determining the recipe of the odor. However, the capability of recording odor depends upon the selection of odor components. One of the fundamental issues is to determine an appropriate set of odor components.

You can think about the aspect from the physiology of olfaction. Buck and Axel reported the multigene family of G-protein-coupled olfactory receptors (ORs) in 1991, and then molecular biology of olfaction rapidly progressed [109]. However, primary smells have not been so far discovered although the stereochemical theory was proposed by Amoore [110].

Amoore proposed seven primary smells mainly based upon molecular shapes. Unlike the primary colors in the vision, however, the combination of several odor components does not cover the whole range of the smell. It was reported that the number of OR types is approximately 350 [111]. However, it might be possible to express smells using a smaller number of odor components since specificities of ORs are overlapping. Thus, the technique to select odor components so that the range of the smell can be maximized is described in this section.

When we explore odor components, the best way is based upon sensory test. Although a large-scale database on the sensory test was previously made in the United States [112], and its data were analyzed by J.B. Castro *et al.* [113], the number of the data available is still limited.

Thus, a database of mass spectrometry was focused on. A mass spectrometer can be used without gas chromatography to classify odors and quantify mixture composition [114, 115]. A library of mass spectra with a huge number of data is available. Furthermore, the selectivity and stability of the mass spectrometer are good compared with other those of other electronic noses.

In addition to the database, mathematical method for extracting basis vectors corresponding to odor components is indispensable. Moreover, basis vectors must be synthesized using the vectors of available compounds. The nonnegative matrix factorization (NMF) method was used to extract basis vectors [116], and the nonnegative least squares method was used to determine the mixture compositions of the odors corresponding to the basis vectors.

The NMF method was useful since any element of the recipe and any element of the mass spectrum are nonnegative. This nonnegativity constraint leads to part-based representation relatively easy to interpret. Although PCA can be used in the similar way, it does not work because it is free of the nonnegativity constraint.

7.6.2 *Procedure for Odor Approximation*

When an odorant molecule is injected into the inlet of the mass spectrometer, it is ionized and detected in the order of m/z (ratio of mass to charge) as is shown in Figure 7.38.

When EI method is used to ionize gas molecules, multiple fragments are obtained in addition to molecular ion peak as is shown in the figure. Thus, each m/z works as a sensor element and mass spectrum can be used in the same manner as response pattern from an array of sensors used in machine olfaction. Since the dimensionality of mass spectrum is much larger than that of a typical electronic nose, diversity of the sensor response pattern among various kinds of compounds is achieved.

The procedure for odor approximation using odor components is illustrated in Figure 7.39. First, the data matrix is obtained after collecting a large amount of data using mass spectrometry. Then, NMF can be applied to the data whose elements are all nonnegative. The basis vectors are extracted from the data matrix using NMF method. The number of basis vectors, r, should be given in advance and is equal to the number of odor components.

Although basis vectors are mathematically obtained, there is no guarantee that they can be realized using available odor samples. Thus, they are approximated using the data of available samples and the nonnegative least squares method. Then, we can blend odor samples to make odor components based upon the result of nonnegative least squares. We can approximate a variety of odors after preparing for odor components. Its recipe is obtained using nonnegative least squares method again.

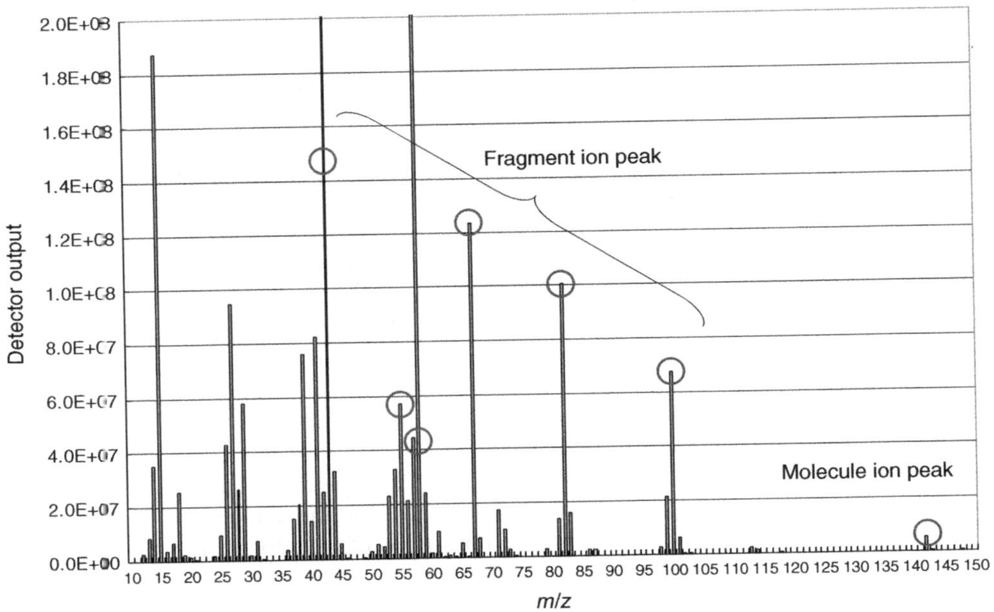

Figure 7.38 Example of a mass spectrum (sample: *trans*-2-hexenyl acetate)

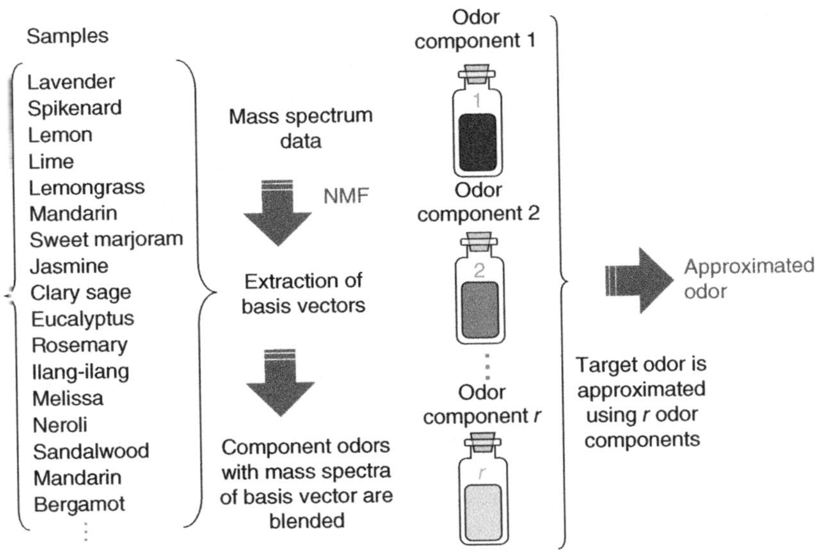

Figure 7.39 Procedure for odor approximation using odor components

We can write this procedure using mathematical expressions. Let **V** be an $n \times m$ data matrix, **WH** an approximated matrix, **W** an $n \times r$ basis matrix, and H an $r \times m$ coefficient matrix. **V** is factorized as

$$\mathbf{V} \cong \mathbf{WH} = \mathbf{W} \times \mathbf{H}, \tag{7.15}$$

where n is the number of m/z values and m is the number of the mass spectrum data, respectively. The r columns of **W** are the basis vectors corresponding to the odor components. The elements of W and H are initially set at random. A set of basis vectors are obtained after the iterative procedure shown in the Ref. [116]. The encoding is composed of the coefficients by which the mass spectrum vector of the target odor is represented with a linear combination of basis vectors corresponding to the odor components.

Since **W** is determined regardless of the actual mass spectra of existing compounds, the basis vectors should be approximated using mass spectra of existing compounds. Thus, nonnegative least squares method is used to approximate each basis vector.

The nonnegative least squares method is an optimization method used to solve linear equations with the constraint that all the elements in the solution vector are nonnegative. In this method, the m-dimensional recipe vector $\mathbf{x}_i$ for the ith n-dimensional basis vector $\mathbf{d}_i$ extracted by NMF is obtained so that the minimized index

$$I = \min_{\mathbf{x}_i} \left| \mathbf{V}\mathbf{x}_i - \mathbf{d}_i \right| \tag{7.16}$$

can be obtained. We obtained $\mathbf{x}_i$ that minimizes the residual error under the constraint that all the elements in $\mathbf{x}_i$ are nonnegative.

Then, we consider the $n \times r$ matrix

$$\mathbf{C} = \left[d_1, d_2, \ldots, d_j, \ldots, d_r \right], \tag{7.17}$$

whose jth column is equal to d_j corresponding to the jth odor component.

Once a set of odor components are fixed, we can approximate $\mathbf{d}_{\text{target}}$, mass spectrum of a target odor. The nonnegative least squares method is again used to obtain the recipe vector $\mathbf{x}_{\text{target}}$ for the minimized index

$$I = \min_{\mathbf{x}_{\text{target}}} \left| \mathbf{C}\mathbf{x}_{\text{target}} - \mathbf{d}_{\text{target}} \right| \tag{7.18}$$

under the constraint that all the elements in $\mathbf{x}_{\text{target}}$ are nonnegative. Then, a variety of odors can be made by blending odor components.

7.6.3 Simulation of Odor Approximation

First, the simulation of extracting basis vector was performed [117]. The mass spectrum database (NIST05) is commercially available. All the mass spectra were measured at an ionization energy of 70 eV. Each mass spectrum is normalized by its maximum intensity. The molecular weight of each volatile compound is at most 300. Since 9987 data were extracted from the

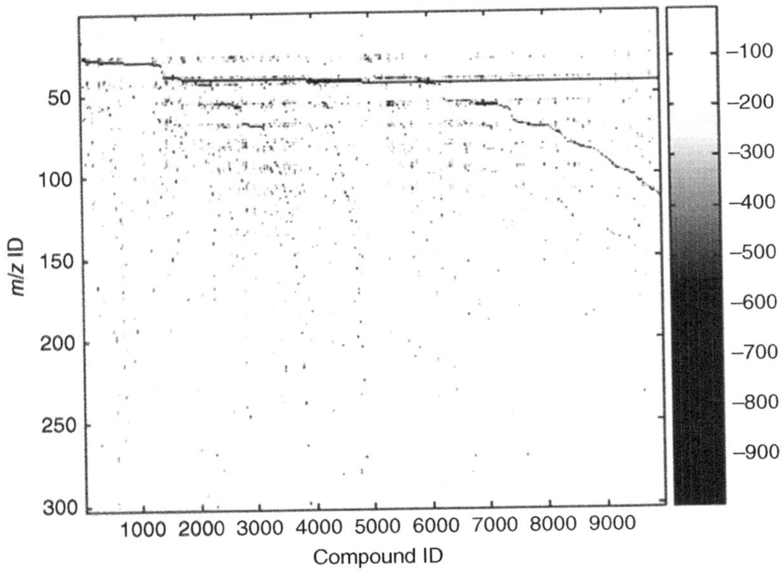

Figure 7.40 Original 10 000 mass spectra extracted from database. Reproduced with permission from Ref. [117]. Copyright IEEE, 2009

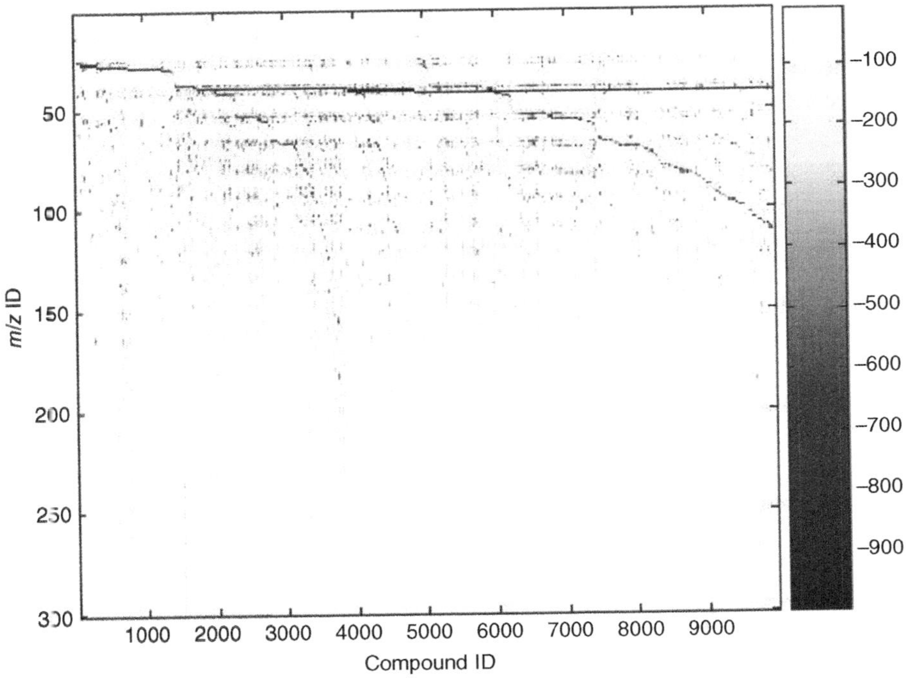

Figure 7.41 Approximated 10 000 mass spectra when number of basis vectors is 100. Reproduced with permission from Ref. [117]. Copyright IEEE, 2009

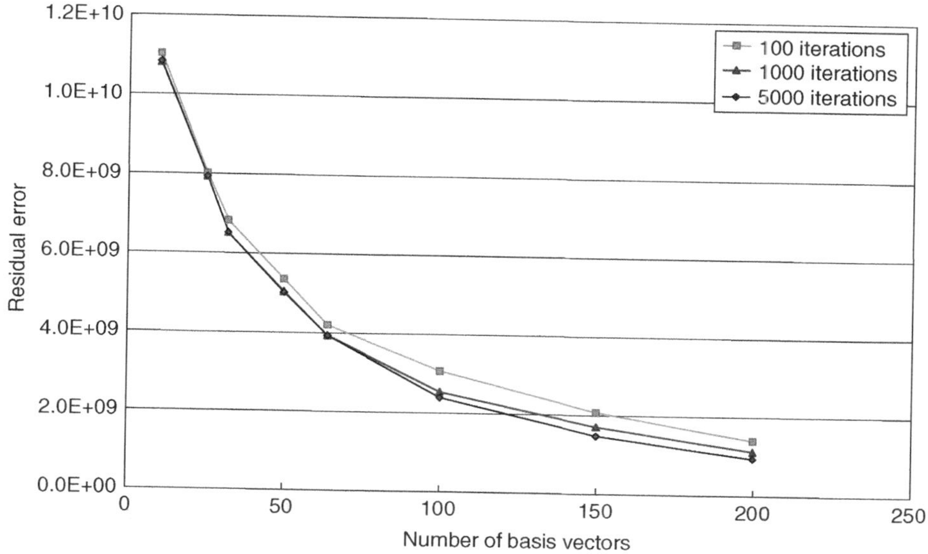

Figure 7.42 Relationship between number of basis vectors and residual error as a parameter of the number of epochs

database for simulation, the size of the data matrix V in Equation 7.15 was 300×9987. Figure 7.40 shows the original data of V from the database, whereas Figure 7.41 shows the approximate matrix WH after the application of NMF to V in case of $r = 100$. The number of iterations for updating W and H was 5000. It was found from the figures that NMF enables the mass spectrum approximation with tolerable accuracy.

Next, the residual error was evaluated as a function of r, number of basis vectors. The residual error is the sum of the squares between the elements of V and those of WH. The relationship between the number of basis vectors and the residual error as a parameter of number of epochs is shown in Figure 7.42. It was found that the approximation accuracy increased as the number of basis vectors. Moreover, just 1000 iterations were sufficient for NMF calculation. It was also found that the error became small when r exceeds 64.

7.6.4 Experiment on Essential Oil Approximation

Since the simulation result seems promising, the experiment on approximating essential oils using odor components was performed [118]. Samples used here were 158 essential oils as is tabulated in Table 7.6. Since each sample was diluted to 10% *v*/*v* with ethanol, an *m*/*z* range from 50 to 250 was used here to eliminate the influence of solvent, that is, ethanol.

Each raw mass spectrum was preprocessed to eliminate the noise. Negligible amount of the intensity at each *m*/*z* was replaced with zero. The odor approximation was performed using 12 and 30 odor components. Three samples with their scents well known to everyone were selected to evaluate the odor approximation. They were organic orange, peppermint, and black pepper. The mass spectra of approximated odors as well as the original ones are shown in Figure 7.43a–c.

Table 7.6 Essential oils used in odor approximation experiment

Number	Essential oil	Number	Essential oil	Number	Essential oil	Number	Essential oil
1	Angelica root	41	Lavandine	81	Goodnight	121	Perilla
2	Basil	42	Lavender	82	Headaid	122	Davana
3	BAY	43	Lemon	83	MOBILITY	123	Mint marigold
4	Benzoin	44	Lemongrass	84	Noel	124	Narcissus
5	Bergamot	45	Lime	85	Refresh	125	Honeysuckle
6	Black pepper	46	Mandarin	86	Relaxation	126	Japanese cypress
7	Cajeput	47	MANUKA	87	Romance	127	Hyacinth
8	CAMPHOR (WHITE)	48	May chang	88	Travel	128	Fir tree
9	CARAWAY	49	Melissa	89	Cedarwood	129	Yuzu
10	Cardamom	50	Myrtle	90	Litsea	130	Anise
11	Carrot seed	51	Neroli	91	Rosewood	131	Inula
12	Cedarwood (Atlas)	52	Niaouli	92	Wintergreen	132	Turmeric
13	Cedarwood (Himalayan)	53	NUTMEG	93	Ajowan	133	Exotic
14	Cedarwood (Virginian)	54	Organic orange	94	Atlas fern	134	Cajeput
15	CELERY SEED	55	Palmarosa	95	Mentha arvensis	135	Calophyllum inophyllum
16	Chamomile (German)	56	Patchouli	96	Oregano	136	Cumin
17	Chamomile (Maroc)	57	Peppermint	97	Calamint	137	Mediterranean cypress
18	Chamomile (Roman)	58	Petitgrain	98	Khella	138	Corn mint
19	Cinnamon (leaf)	59	Pine	99	*Cinnamosma fragrans*	139	*Copaifera officinalis*
20	Citronella	60	RABENSARA	100	Star anise	140	Santolina
21	Clary sage	61	Rose otto	101	Spikenard	141	*Cistus ladaniferus*

22	Clove bud	62	Rosemary	102	Celery	142	Siam wood
23	Coriander	63	Sage (Spanish)	103	Blue tansy	143	White birch
24	Cypress	64	Sandalwood	104	Tarragon	144	Spike lavender
25	ELEMI	65	Spanish marjoram	105	Dill	145	Savory
26	Eucalyptus (citriodora)	66	Spearmint	106	Balsam fir	146	Gentle breeze
27	Eucalyptus (globulus)	67	Spikenard	107	Himalayan cedar	147	Tangerine
28	Eucalyptus (radiata)	68	Sweet marjoram	108	Black spruce	148	*Artemisia arborescens*
29	Fennel (sweet)	69	Sweet orange	109	Helichrysum	149	Tonic
30	FIR (Siberian)	70	Tea tree	110	Bergamot mint	150	Tropical
31	FIR BALSAM	71	Thyme	111	Ponderosa pine	151	*Lippia citriodora*
32	Frankincense	72	Valerian	112	Mastic tree	152	Parsley
33	GALBANUM	73	Vetiver	113	European red pine	153	Hyssop
34	Geranium	74	Yarrow	114	Labrador tea	154	*Thujopsis dolabrata*
35	Ginger	75	Ylang Ylang	115	Lantana	155	Province
36	Grapefruit	76	Breath easy	116	Laurel	156	Penny royal mint
37	Helichrysum	77	Daybreak	117	Rock rose	157	*Chenopodium ambrosioides*
38	Ho wood	78	De-stress	118	Wild carrot	158	Refreshing
39	Jasmine	79	Equilibrium	119	Apricot		
40	Juniper berry	80	FOCUS	120	Oakmoss		

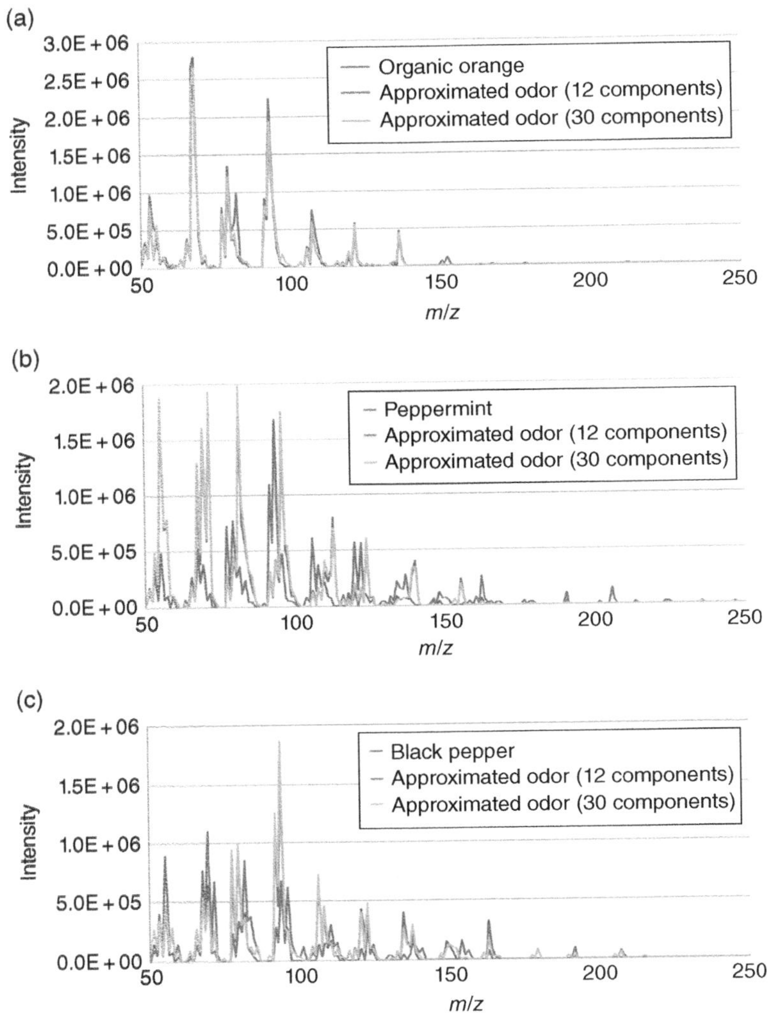

Figure 7.43 Comparison of mass spectra of original essential oils with those of approximated ones using 12 and 30 odor components. (a) Organic orange, (b) peppermint, and (c) black pepper. Reproduced with permission from Ref. [118]. Copyright IEEE, 2012

The approximated mass spectra of organic orange agreed quite well with the original one using both 12 and 30 odor components. Although the approximation accuracies of peppermint and black pepper using 12 odor components were worse than that of organic orange, they were improved using 30 odor components.

Then, a sensory test was performed to check the actual accuracy of the odor approximation. 172 subjects classified the three original essential oils. Thereafter, they classified the three oils approximated with 12 odor components and did the three ones approximated with 30 odor components as is shown in Tables 7.7a–c. Almost all subjects correctly classified

Table 7.7 Classification rates of essential oils in sensory tests: (a) original essential oils, (b) essential oils approximated with 12 odor components, and (c) essential oils approximated with 30 odor components

Predicted odor	Original odor		
	Organic orange (%)	Peppermint (%)	Black pepper (%)
(a) Original odor			
Organic orange	98.68	0.66	0.66
Peppermint	0.66	96.05	3.29
Black pepper	0.66	3.29	96.05
(b) Approximated odor ($r=12$)			
Organic orange	82.89	10.53	6.58
Peppermint	9.21	51.97	38.82
Black pepper	7.89	36.18	54.61
(c) Approximated odor ($r=30$)			
Organic orange	95.39	3.29	1.32
Peppermint	2.63	87.50	9.87
Black pepper	1.97	9.21	88.82

Source: Reproduced with permission from Ref. [118]. Copyright IEEE, 2012.

the original essential oils. A classification rate deteriorated a little in case of 30 odor components still in spite of satisfactory result. However, the confusion between peppermint and black pepper occurred when 12 odor components were used. Thus, 12 odor components were not sufficient to approximate essential oils, whereas 30 odor components brought sufficient result.

The mass spectrum at high *m/z* region is important in spite of their small peaks. The high *m/z* region is considered to greatly contribute to human olfaction since the signals in the high *m/z* region originate from chemicals with large molecular weights. Since compounds with large molecular weights tend to be sensitive to human, more sophisticated mathematical method to extract information from high *m/z* region is indispensable.

7.6.5 *Comparison of Distance Measure*

Several types of distance measures [119] were used and compared from the viewpoint of extracting information from high *m/z* region [120]. Three types of distance measures such as Euclidean distance, Kullback–Leibler (KL) divergence and Itakura–Saito (IS) divergence [108] were taken into account. Three distance measures are formulated below.

The Euclidean distance $D_{\mathrm{EU}}(V|WH)$ is

$$D_{\mathrm{EU}}(V|WH)=\sqrt{\sum_{i,j}\left(V_{ij}-WH_{ij}\right)^2}. \qquad (7.19)$$

The matrices V and WH appeared in Equation 7.15.

KL divergence $D_{\mathrm{KL}}(V|WH)$ is

$$D_{\text{KL}}(V|WH) = \sum_{i,j}\left(V_{ij}\log\frac{V_{ij}}{WH_{ij}} - V_{ij} - WH_{ij}\right). \tag{7.20}$$

IS divergence $D_{\text{IS}}(V|WH)$ is

$$D_{\text{IS}}(V|WH) = \sum_{i,j}\left(\frac{V_{ij}}{WH_{ij}} - \log\frac{V_{ij}}{WH_{ij}} - 1\right). \tag{7.21}$$

Figure 7.44 shows the comparison of three types of distance measures in one-dimensional case for simplicity. The three types of distance between x and y are shown as a function of x when y is fixed to 10. When the Euclidean distance is used, the curve is symmetric about $x=10$. The curves for KL and IS divergences are asymmetric about $x=10$. The cost of KL divergence becomes larger when x decreases from 10, whereas the increase in distance is very small for x larger than 10. This characteristic is more remarkable for IS divergence. Thus, the contribution of small peak at high m/z region can be enhanced when we use IS divergence.

When IS divergence was used, the rule for updating the matrices W and H was changed according to the Ref. [121]. The 135 mass spectra of the essential oils were applied to NMF with IS divergence and KL divergence, respectively. Figure 7.45a shows the comparison of approximated odor with target odor using KL divergence, whereas Figure 7.45b does the comparison of the approximated odor with the target odor using IS divergence. The sample was one of the blended essential oils (Refresh). The preprocessing of negligible peak removals was performed in the same manner as the previous one. The number of odor components was 30, and the number of training epochs was 10000. You can see that the approximation accuracy

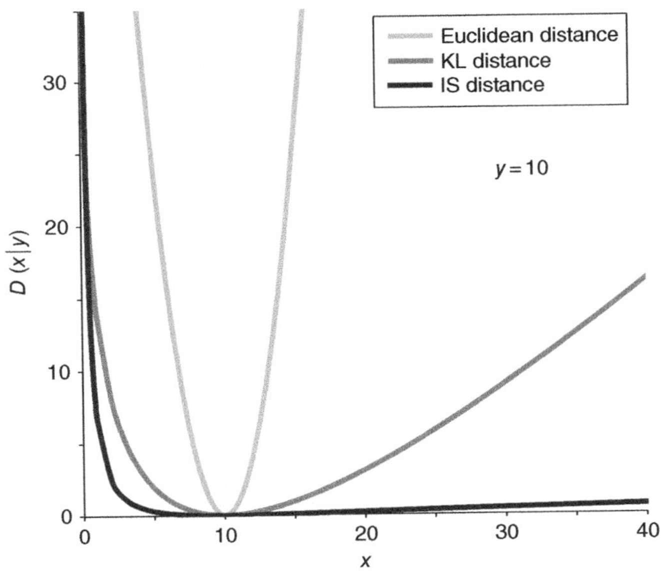

Figure 7.44 Comparison of three types of distance measures used in NMF method

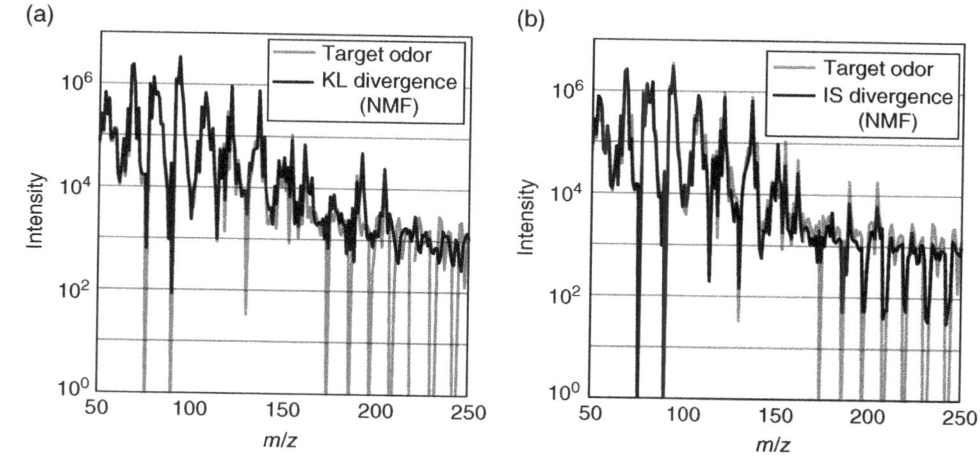

Figure 7.45 Comparison of approximated odor with target odor: Sample Refresh (blended essential oil: Refresh) (a) KL divergence, (b) IS divergence

with IS divergence was better than that with KL divergence at high *m/z* region even if the accuracies with two divergences were almost the same at low *m/z* region. Especially, the accuracy of zero point was much improved using IS divergence. This improvement was very useful since the residual error at high *m/z* region influences to the actual odor quality.

7.6.6 Improvement of Odor Approximation

Further improvement of approximation accuracy was performed [122]. Since NMF is an iterative method, its solution is not unique. Thus, the initial values of the basis vectors were set using cluster analysis here.

Figure 7.46 shows the procedure to set initial values. The data matrix is divided into four clusters using k-means method [123] since four main clusters were obtained by hierarchical clustering. NMF is performed for each cluster to extract basis vectors. The elements of W and H are initially set at random. Then, the basis vectors of four clusters are merged into a single matrix, as an initial matrix of W. Then, NMF was again applied to V to obtain the basis vectors. NMF used here was based upon KL divergence.

Hundred and forty-eight samples made up of food flavors were used in this experiment. When NMF was first applied to the data in each cluster, the relationship between the number of basis vectors and the residual was investigated so that the number of basis vectors for each cluster could be determined. The number of the basis vectors in the four clusters were fixed to 6, 4, 4, and 6, respectively, since the decrease in residual errors above those numbers became slow. These samples were approximated using 20 odor components and were compared with those obtained by the conventional one.

The result of the approximation accuracy is shown in Figure 7.47 as a histogram of correlation coefficient between each approximated and target odors. The number of samples with the correlation coefficients above 0.8 was increased except the range between 0.95 and 1 although the number of the samples with the correlation coefficients below 0.8 was

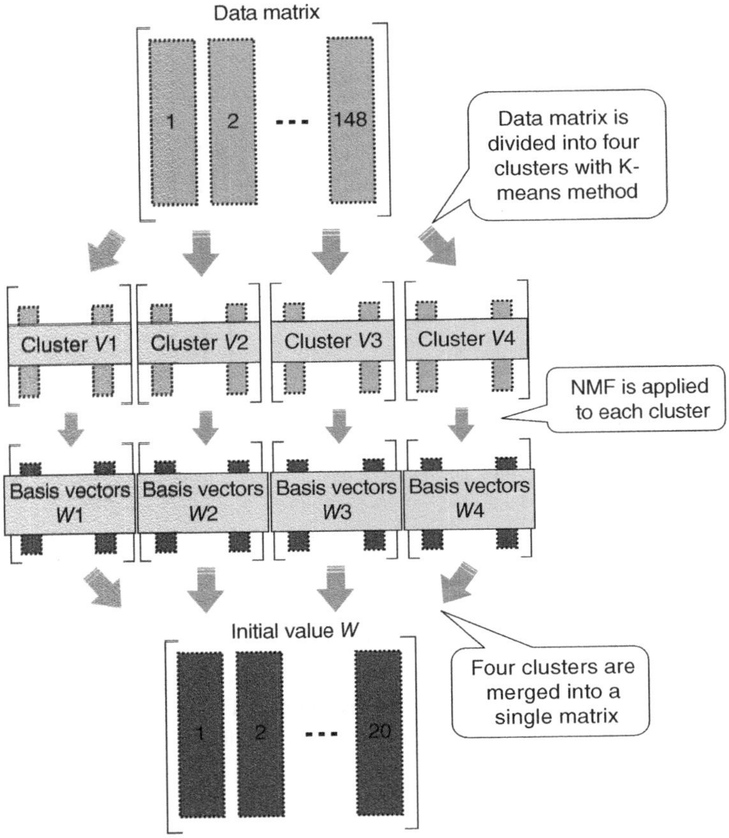

Figure 7.46 Concept of initial value setting method for NMF. Reproduced with permission from Ref. [122]. Copyright IEEE, 2013

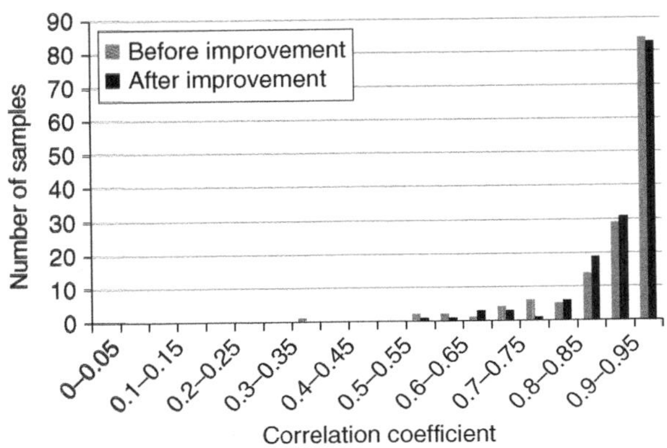

Figure 7.47 Histogram of correlation coefficient between approximated and target odors. The left bar shows the number of samples within specified correlation coefficient range without initial value setting method, whereas the right bar shows that using initial value setting method. Reproduced with permission from Ref. [122]. Copyright IEEE, 2013

remarkably decreased. The number of samples with low approximation accuracy was much smaller than that obtained without the initial value setting method.

Thus, it was revealed that the initial value setting method based upon the cluster analysis can improve the approximation accuracy, especially for the samples difficult to obtain high accuracy using the conventional method.

7.7 Teleolfaction

7.7.1 Concept of Teleolfaction

Nowadays it is very easy to send visual and auditory information to a remote site. A variety of multimedia forms based upon visual and auditory information are available. However, it is not sufficient to show our daily experience only using visual and auditory information in spite of large information originating from vision.

When we watch a television commercial for advertising soap, detergent, air refreshener, food, and so on, the existences of their smells are intended to be expressed by vision, for example, virtual moving particles. However, it is obvious that the way to express smells by vision is insufficient to give us reality. Thus, the smell reproduction with movie was studied [124, 125].

Here teleolfaction system is described. It can reproduce scents at remote place in real time. Its concept is illustrated in Figure 7.48. Teleolfaction system consists of odor sensing system and olfactory display far away from the sensing system. Users can sniff the smell in real time even if the odor source is located at another place. They can watch video together with smell transferred from the recording system via the Internet. Smells should be synchronized with video. Users can perceive much sensation as if the object were here. This effect is the teleexistence of the olfaction.

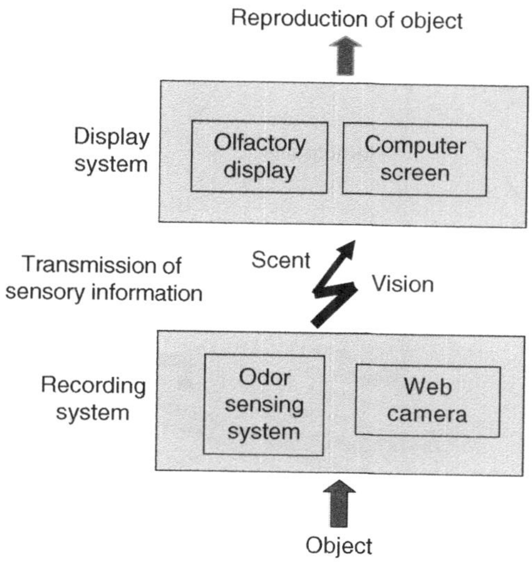

Figure 7.48 Concept of teleolfaction

It was proposed that the odor classified at one site was reproduced at remote site [126]. The output pattern of a sensor array composed of nine semiconductor gas sensors, humidity, and temperature sensors was recognized using MLP [82]. The classification result was transmitted to the remote site via the Internet, and the same smell was generated there. Although they did the experiment on odor classification, the actual experiment on the smell generation was not performed. Moreover, they did not blend odors but aimed to generate the same odor identified at the sensing site. F. Davide proposed the concept of virtual olfactory interfaces connected via network [66]. However, the actual implementation and experiment were not performed.

7.7.2 Implementation of Teleolfaction System

The teleolfaction system was first made in 2008 [127]. Although there are several versions due to its modification, only the latest version is described here. The main parts of the system are an odor sensing system, a learning vector quantization (LVQ) circuit implemented into a field-programmable gate array (FPGA) [128], an olfactory display, and the two tablet PCs. The whole system is shown in Figure 7.49.

Odor samples are soaked in a paper with its location remotely controlled by a linear actuator and a stepping motor. When an air inlet sucks the odor from its source, it reaches a QCM sensor array so that their oscillation frequency changes are obtained. They are input to LVQ circuit via the measurement circuit. LVQ circuit works as a pattern classifier after training. The information of odor is sent to a tablet PC using ZigBee module attached to the FPGA

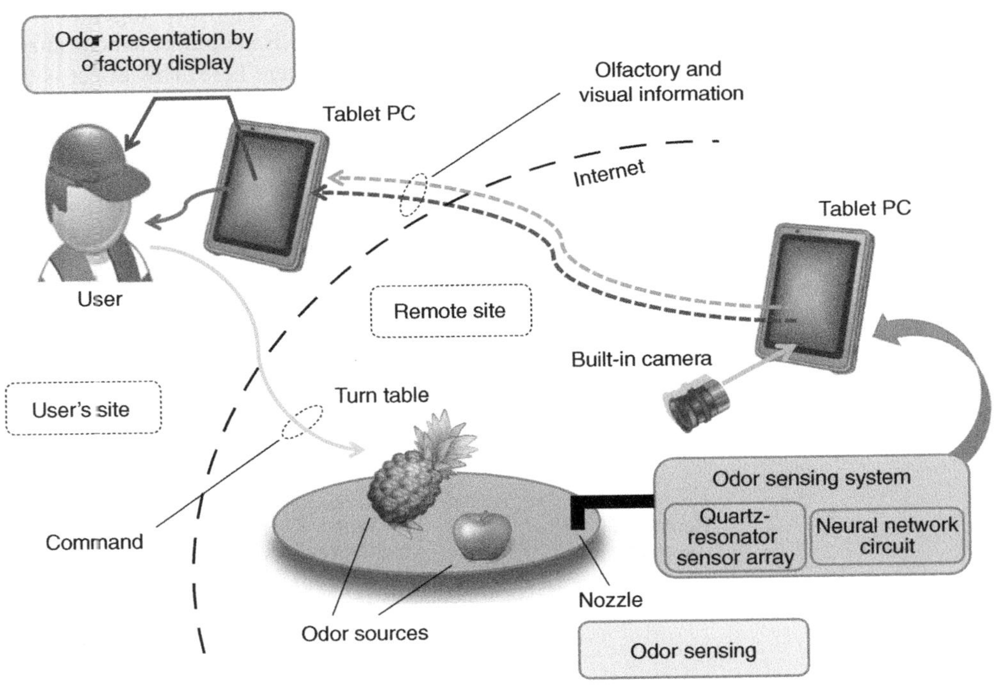

Figure 7.49 Actual block diagram of teleolfaction system

board. The odor information is combined with visual information taken by its built-in camera and is transmitted to a user's site via the Internet. At the user's site, the odor information is separated from the visual one and is presented using an olfactory display, whereas the visual information is presented at a monitor of a tablet PC. A user can rotate a turn table to sniff the smell at an arbitrary point at remote place. Moreover, a linear actuator in addition to the turn table is used to search for the odor source two-dimensionally.

The QCM measurement circuit consists of reciprocal counters to achieve high sampling measurement rate without the deterioration of the frequency resolution. Since the odor concentration in the ambient air is highly fluctuated, the feature extraction from the temporal profile of the signal is very important. short-time Fourier transform (STFT) circuit was developed [129]. Frequency components together with signal amplitudes are input to LVQ circuit. Moreover, the algorithm to determine the actual odor existence was implemented into a CPU core in the FPGA since it takes some time for the sensor response to become zero after the sensing nozzle goes away from the odor source. The odor classification stops during the period without the odor since the response pattern changes from that of the actual odor. LVQ works in the form of hardware, whereas most of the cases are the software implementations.

The hardware implementation accelerates the training speed even in an embedded environment different from the desktop PC environment. Moreover, it is expected to reduce the power consumption essential in an embedded system.

The olfactory display used here was the conventional one with high-speed switching of solenoid valves described earlier. It might be replaced with the latest one made up of micropumps and a SAW atomizer [34].

In tablet PCs, Java was adopted to do the communication and the user interface. Moreover, the interface for wireless communication was implemented to increase the flexibility and the mobility. ZigBee modules were adopted on both sensing and user sides. An image of a user operating the teleolfaction system is shown in Figure 7.50. It is quite easy for a novice to use it.

7.7.3 *Experiment on Teleolfaction*

The teleolfaction system was demonstrated at an international conference, International Conference on Artificial Reality and Telexistence (ICAT) 2013 [130]. The sensing side was located in the suburbs of Yokohama, whereas the users stayed at Odaiba, Tokyo. The distance of the sensing side to the users was about 30 km.

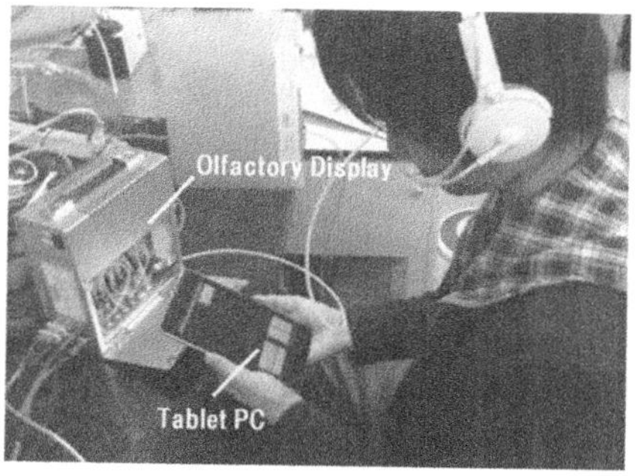

Figure 7.50 Image for a user to operate teleolfaction system

Table 7.8 Classification rate of teleolfaction system. (a) Classification rate of people using olfactory display, (b) classification rate of odor sensing system, and (c) confusion matrix of users at remote site

	Apple	Mint	Melon	Citrus
(a)				
Apple	100%	0	0	0
Mint	0	100%	0	0
Melon	0	10%	90%	0
Citrus	0	0	0	100%
(b)				
Apple	90%	5%	5%	0
Mint	0	90%	5%	5%
Melon	5%	10%	80%	5%
Citrus	5%	15%	0	80%
(c)				
Apple	6	0	1	0
Mint	0	8	0	0
Melon	0	2	5	0
Citrus	0	1	1	4

The sensors used here were four QCMs (20 MHz, AT cut) coated with TCP, OV-17 (silicone OV-17), Apiezon L (Ap-L), and PEG1000. Moreover, 16 reference vectors per category were used in LVQ circuit. The samples used here were apple, mint, melon, and citrus.

First, 10 people sniffed four samples via the olfactory display to classify them as is shown in Table 7.8a. The classification rate was more than 90%. Then, the classification of the sensing system itself was investigated as is tabulated in Table 7.8b. Each sample was classified 20 times. The overall classification rate was 85%. The classification rate in the ambient air was smaller than that in a closed system due to the disturbance even if several methods to improve the classification rate were introduced as described earlier.

However, the teleolfaction system enhances the reality even if the misclassification rate at the sensing system slightly increases. Finally, the classification result of 16 users at the remote site is shown in Table 7.8c. The classification rate was close to that at the sensing system itself. Although there is still room for the improvement, soundness of the basic idea was validated.

Although the current teleolfaction system is still primitive, it can be much extended if the odor recorder technique is applied. A variety of odors can be reproduced at a remote site using the odor recorder.

7.8 Summary

Technologies related to olfactory display and odor recorder are described in detail in this chapter. Since an olfactory display and an odor recorder are relatively new fields, systematic ways to describe them are quite important. This chapter offers the comprehensive knowledge to understand these fields.

In the near future, an odor recorder should be used in teleolfaction. From that point of view, the improvement of sensing devices is important to suppress the collinearity problem. One of the methods is real-time mass spectrometer used in the gas phase. However, its disadvantage of a real-time mass spectrometer is the remarkable decrease in its sensitivity at high *m/z* region. Thus, the sensing device improvement is very important. Two approaches such as ion mobility spectrometry (IMS), a portable analytical instrument and a biosensor originating from ORs [131, 132] might be promising. A lot of works should be done to realize a practical odor recorder and teleolfaction system.

References

[1] T. Nakamoto (Ed.) (2013). *Human olfactory displays and interfaces*. Hershey, PA: IGI-Global.

[2] M. L. Heilig, Sensorama simulator, U.S. Patent 3,050,870 (1962).

[3] U. Livneh and R. Paz, An implicit measure of olfactory performance for non-human primates reveals aversive and pleasant odor conditioning, *Journal of Neuroscience Methods*, 192 (2010) 90–95.

[4] J. J. Kaye, Making scents, *Interactions*, 11 (January/February), 2004, pp. 49–61.

[5] D. W. Kim, Y. H.Cho, K.Nishimoto, Y.Kawakami, S.Kunifuji, and H.Ando, Development of aroma-card based soundless olfactory display, Proceedings of the 16th IEEE International Conference on Electronics, Circuits, and Systems, ICECS 2009, Yasmine Hammamet, Tunisia, December 2009, pp. 703–706.

[6] H. Kim, J. Park, K. Noh, C. J. Gardner, S. D. Kong, J.Kim, and S.Jin, An X–Y addressable matrix odor-releasing system using an On–Off switchable device, *Angewandte Chemie International Edition*, 50 (2011) 6771–6775.

[7] K.Kashiwagi, K.Terao, T.Suzuki, H.Takao, F.Shimokawa, and T.Nakamoto, Heat-driven-type micro device odor generator with MEMS actuator, IEE of Japan Sensor Symposium, October 22–24, 2012, Kitakyushu, Japan, SPLN-9.

[8] K. Tsubouchi, Retransmission/restitution device for iPhone, Abstract Book of Digital Olfaction Congress April 11–12, 2013, 2013, Berlin, Germany, pp. 28.

[9] Y. Haran, Ultrasonic micro-plugs array technology, Abstract Book of Digital Olfaction Congress 2013, Berlin, Germany, 2013, pp. 31.

[10] M. Sakairi, A. Nishimura, and D. Suzuki, Olfaction presentation system using odor scanner and odor-emitting apparatus coupled with chemical capsules of alginic acid polymer, *IEICE Transactions on Fundamentals of Electronics, Communications and Computer Sciences*, E92-A (2009) 618–629.

[11] F. Nakaizumi, Y. Yanagida, H. Noma, and K. Hosaka, Spotscents: A novel method of natural scent delivery using multiple scent projectors, Proceedings of IEEE Virtual Reality 2006, Arlington, VA, USA, March 25–29, 2006, pp. 199–206.

[12] K. Murai, T. Serizawa, and Y. Yanagida, Localized scent presentation to a walking person by using scent projectors, Proceedings of the First IEEE International Symposium on Virtual Reality Innovation (ISVRI), Singapore, March, 2011, pp. 67–70.

[13] N. Mastuo and T. Komiriya, Computer simulation for efficient delivery of fragrances with air volute rings, *Aroma Research*, 8(2007) 144–147.

[14] D. W. Kim and H. Ando, Development of directional olfactory display, Proceedings of the Ninth ACM SIGGRAPH Conference on Virtual-Reality Continuum and its Applications in Industry, December 12–13, 2010, Seoul, Republic of Korea. ACM, New York, pp. 143–144.

[15] T. Yamada, S. Yokoyama, T. Tanikawa, K. Hirota, and M. Hirose, Wearable olfactory: Using odor in outdoor environment, Proceedings of the IEEE Virtual Reality Conference, March 25–29, 2006, Alexandria, VA, USA, pp. 207–212.

[16] H. Ishida, H. Matsukura, H. Yoshida, and T. Nakamoto, Application of computational fluid dynamics simulation to olfactory display, Proceedings of the 18th International Conference on Artificial Reality and Telexistence, Yokohama, Japan, December 1–3, 2008, pp. 285–288.

[17] H. Matsukura, T. Yoneda, and H. Ishida, Smelling screen: Development and evaluation of an olfactory display system for presenting a virtual odor source, *IEEE Transactions on Visualization and Computer Graphics*, 19 (2013) 606–615.

[18] T. Nakamoto, H. Takigawa, and T. Yamanaka, Fundamental study of odor blender using inkjet devices for low-volatile scents, *IEICE Transactions on Electronics*, E87-C (2004) 2081–2086.

[19] D. B.Wallace, D. Taylor, B. V.Anthohe, I. Achiriloaie, N. Comparini, R. M.Stewart, and M. K. Sanghera, Determination of the olfactory threshold using a piezoelectric microdispenser for neurodegenerative disease diagnostics, *Measurement Science and Technology*, 17 (2006) 3102–3109.

[20] J. Sato, K. Ohtsu, Y. Bannai, and K. Okada, Effective presentation technique of scent using small ejection quantities of odor, Proceedings of the IEEE Virtual Reality Conference, Lafayette, LA, USA, March 14–18, 2009, pp. 151–158.

[21] K. Ohtsu, J. Sato, Y. Bannai, and K. Okada, Scent presentation technique of pulse ejection synchronized with breathing, Proceedings of the Ninth Annual International Symposium on Applications and the Internet, SAINT '09, Bellevue, WA, USA, July, 2009, pp. 125–128.

[22] P. Berry, C. Crawford, T. Glancy, D. Luzuriago, and M. J. W. Murphy, A taste of things to come, *Perfumer&Flavorist*, 31 (2006) 36–42.

[23] T.Nakamoto, S. Utsumi, N. Yamashita, T. Moriizumi, and Y. Sonoda, Active gas sensing system using automatically controlled gas blender and numerical optimization technique, *Sensors and Actuators B: Chemical*, 20 (1994) 131–137.

[24] J. W.Grate, D. S. Ballantine, and H. Wohltjen, An automated vapor-generation and data collection instrument for the evaluation of chemical microsensors, *Sensors and Actuators*, 11 (1987) 173–188.

[25] S Shoji and M Esashi, Microflow devices and systems, *Journal of Micromechanics and Microengineering* 4 (1994) 157.

[26] P. Somboon, B. Wyszynski, and T. Nakamoto, Novel odor recorder for extending range of recordable odor, *Sensors and Actuators: B. Chemical*, 121 (2007) 583–589.

[27] S. R.Norsworthy, R.Schreier, and G. C.Temes (1992). *Delta-sigma data converter*. New York: IEEE Press.

[28] T. Yamanaka, R. Matsumoto, and T. Nakamoto, Study of odor blender using solenoid valves controlled by delta-sigma modulation method, *Sensors and Actuators: B. Chemical*, 87 (2002) 457–463.

[29] T. Nakamoto, H. Ishida, and H. Matsukura (2012). Olfactory display using solenoid valves and fluid dynamic simulation, in G. Ghinea, F. Andres, and S. Gulliver (Eds.), *Multiple sensorial media advances and applications* (pp. 140–163), IGI Global, Hershey, PA.

[30] T. Nakamoto, Pham Hai Dinh Minh, Improvement of olfactory display using solenoid valves, Proceedings of the IEEE Virtual Reality Conference, Charlotte, NC, USA, March 10–14, 2007, pp. 171–178.

[31] Y. Ariyakul and T. Nakamoto, Improvement of odor blender using electroosmic pumps and SAW atomizer for low-volatile scents, *IEEE Sensors Journal*, 13 (2013) 4918–4923.

[32] M. Kurosawa, T. Watanabe, and T. Higuchi, Surface acoustic wave atomizer with pumping effect, Proceedings of the IEEE International Conference on Micro Electro Mechanical Systems, 1995, MEMS '95, Amsterdam, the Netherlands, January 29–February 2, 1995, pp. 25–30.

[33] Y. Takamura, H. Onoda, H. Inokuchi, S. Adachi, A. Oki, and Y. Horiike, Low-voltage electroosmosis pump for stand-alone microfluidics devices, *Electrophoresis*, 24, 1–2 (2003) 185–192.

[34] T. Nakamoto, K. Hashimoto, T. Aizawa, and Y. Ariyakul, Multi-component olfactory display with a SAW atomizer and micropumps controlled by a tablet PC, Proceedings of the IEEE International Frequency Control Symposium (FCS), Taipei, May 19–22, 2014, pp. 1–4.

[35] T. Kawai and K. Noro, Psychological effect of stereoscopic 3-D images with fragrances, *Ergonomics*, 39 (1996) 1364–1369.

[36] T. Kawai and T. Takatsu, Development and evaluation of fragrance presentation system using visual information, *Aroma Research*, 3/4 (2002) 26–30.

[37] A. Nambu, T. Narumi, K. Nishimura, T. Tanikawa, and M. Hirose, Visual-olfactory display using olfactory sensory map, Proceedings of the IEEE Virtual Reality Conference, Waltham, MA, USA, March 20–24, 2010, pp. 39–42.

[38] W. R. Dillion and M. Goldstein (1984). *Multivariate analysis* (pp. 107–156). New York: John Wiley & Sons, Inc.

[39] G.Ghinea and O. Ademoye, User perception of media content association in olfaction-enhanced multimedia, *ACM Transactions on Multimedia Computing, Communications, and Applications*, 8 (2012) 1–19.

[40] O. Ademoye and G. GHINEA, Information recall task impact in olfaction-enhanced multimedia, *ACM Transactions on Multimedia Computing, Communications, and Applications*, 9 (2013) 1–16.

[41] T. Narumi, T. Kajinami, S. Nishizaka, T. Tanikawa, and M. Hirose, Pseudo-gustatory display system based on cross-modal integration of vision, olfaction and gustation, Proceedings of the IEEE Virtual Reality Conference, Singapore, March 19–23, 2011, pp. 127–130.

[42] Y. Kobayashi and H. Ikezaki (2013). Advanced taste sensors based on artificial lipid membrane, in K. Toko (Ed.), *Biochemical sensors* (pp. 5–44). Singapore: Pan Stanford Publishing.

[43] H. Iwata, H. Yano, T. Uemura, and T. Moriya, Food texture display, Proceedings 12th International Symposium on Haptic Interfaces for Virtual Environment and Teleoperator Systems, 2004. HAPTICS '04, March 27–28, 2004, Chicago, IL, USA, pp. 310–315.

[44] A. Tomono (2013). Display technology of images with scents and its psychological evaluation, in T. Nakamoto (Ed.), *Human olfactory displays and interfaces* (pp. 429–445). IGI Global, Hershey, PA.
[45] A. Tomono, K. Kanda, and S. Otake, Effect that smell presentation has on an individual in regards to eye catching and memory, *IEEJ Transactions on Sensors and Micromachines*, 128-E (2008) 478–486 (in Japanese).
[46] D. A. Zellner and L. A. Whitten, The effect of color intensity and appropriateness on color-induced odor enhancement, *The American Journal of Psychology*, 112 (1999) 585–604.
[47] A. S. Crisinel, C. Jacquier, O. Deroy, and C. Spence, Composing with cross-modal correspondences: Music and odors in concert, *Chemosensory Perception*, 6 (2013) 45–52.
[48] P. Huang, Y. Ishibashi, N. Fukushima, and S. Sugawara, Remote Ikebana with olfactory and haptic media in virtual environments, Proceedings of the 21st International Conference on Artificial Reality and Telexistence, Osaka, Japan, November 28–30, 2011, pp. 133–133.
[49] A. Tomono, Y. Yanagida, and N. Tetsutani, A sense-of-smell stimulus presentation for making visual media with scent, The Meeting Abstract of System Integration Division, SICE, 2003, pp. 373–374.
[50] T. Nakamoto and K. Yoshikawa, Movie with scents generated by olfactory display using solenoid valves, *IEICE Transactions on Fundamentals of Electronics, Communications and Computer Sciences*, 89, 11, 3327–3332.
[51] A. Mochizuki, T. Amada, S. Sawa, T. Takeda, S. Motoyashiki, K. Kohyama, M. Imura, and K. Chihara, Fragra: A visual-olfactory VR game, Proceedings of the SIGGRAPH '04 ACM SIGGRAPH 2004 Sketches, Los Angeles, CA, USA, August 8–12, 2004, pp. 123–123.
[52] T. Nakamoto, S. Otaguro, M. Kinoshita, M. Nagahama, K. Ohnishi, and T. Ishida, Cooking up an interactive olfactory game display, *IEEE Computer Graphics and Application*, 28 (2008) 75–78.
[53] Y. Kakehi, M. Chikamori, and K. Kunoh, hanahanahana, Proceedings of the ACM SIGGRAPH 2010 Art Gallery, July 25–29, 2010, Los Angeles, CA, USA, pp. 396–397.
[54] M. Iseki, Virtual ice cream shop. Abstract of Digital Olfaction Congress, April 11–12, 2013, Berlin, Germany, pp. 29–29.
[55] M. Iseki and T. Nakamoto, Cyber incense smelling with olfactory display, Proceedings of the International Conference on Artificial Reality and Telexistence, Tokyo, Japan, 2013, Research demo, D13.
[56] Ueda, M. Atelier Maki Ueda. Available at: http://www.ueda.nl/index.php?lang=en (accessed on September 15, 2015).
[57] IEEE Spectrum, NTT becomes a smell-o-phone company. Available at: http://spectrum.ieee.org/consumer-electronics/portable-devices/ntt-becomes-a-smellophone-company (accessed on September 7, 2015).
[58] Messager, J. The diffusion of fragrances in a multimedia environment, Paper Presented at the Third Aroma Science Forum, Tokyo, Japan (in Japanese).
[59] Exhalia, scent, the new dimension of multimedia 4D. Available at: http://www.exhalia.com/ (accessed on September 15, 2015).
[60] D. J. Hayes, D. B. Wallace, D. Taylor, B. V. Antohe, I. Achiriloaie, and N. Comparini, A digital olfactometer for smell threshold measurements in neurodegenerative disease diagnostics, Proceedings of the 2007 IEEE Dallas Engineering in Medicine and Biology Workshop, Dallas, TX, USA, November 11–12, 2007, pp. 47–50.
[61] T. Baba, A. Kikuchi, K. Hirayama, Y. Nishio, Y. Hosokai, S. Kanno, T. Hasegawa, N. Sugeno, M. Konno, K. Suzuki, S. Takahashi, H. Fukuda, M. Aoki, Y. Itoyama, E. Mori, and A. Takeda, Severe olfactory dysfunction is a prodromal symptom of dementia associated with Parkinson's disease: A 3-year longitudinal study, *Brain*, 135, 1, 2012 161–169.
[62] P. J. Moberg, R. L. Doty, R. N. Mahr, R. I. Mesholam, S. E. Arnold, B. I. Turetsk, and R. E. Gur, Olfactory identification in elderly schizophrenia and Alzheimer's disease, *Neurobiology of Aging*, 18, 2, 1997 163–167.
[63] A. Fukasawa, R. Suzuki, and K. Okada, Olfactory measurement system to quantify ability to smell using pulse ejection, Proceedings of the 2013 IEEE International Conference on Healthcare Informatics, Philadelphia, PA, USA, September 9–11, 2013, pp. 99–106.
[64] T. C. Pearce, S. S. Schiffman, H. T. Nagle, and J. W. Gardner (Eds.) (2003). *Handbook of machine olfaction*. Weinheim: Wiley-VCH.
[65] T. Nakamoto, Y. Nakahira, H. Hiramatsu, and T. Moriizumi, Odor recorder using active odor sensing system, *Sensors and Actuators B: Chemical*, 76 (2001) 465.
[66] F. Davide, M. Holmberg, and I. Lundstrom (2001). Virtual olfactory interfaces: Electronic noses and olfactory displays, in G. Riva and F. Davide (Eds.), *Communications through virtual technologies* (pp. 193–220). Amsterdam: IOS Press.
[67] Y. Matsushita, A virtual space with aroma and wind, *Aroma Research*, 3 (2002) 42–49 (in Japanese).
[68] L. Carmel and D. Harel, Mix-to-mimic odor synthesis for electronic noses, *Sensors and Actuators: B. Chemical*, 125 (2007) 635–643.

[69] Y. Iwasa, R. Izumi, K. Hayashi, and K. Toko, Synthesis of odor based on substructures and physicochemical properties of odorants, Technical report of IEICE, MBE2004-80, 2004, pp. 29.
[70] T. Nakamoto, Odor recorder, *Sensor Letters*, 3 (2005) 136–150.
[71] T. Nakamoto, H. Ishida, and H. Matsukura (2011). Olfactory display using solenoid valves and fluid dynamics simulation, in G. Ghinea, F. Andres, and S. Gulliver (Eds.), *Multiple sensorial media advances and applications: New developments in MulSeMedia* (pp. 140–163). IGI Global, Hershey, PA.
[72] S. Skogestad and I. Postlethwaite (1996). *Multivariable feedback control*. Chichester: John Wiley & Sons, Ltd.
[73] W. R. Dillon and M. Goldstein (1983). *Multivariate analysis* (pp. 23–250). New York: John Wiley & Sons, Inc.
[74] K. R. Beebe, R. J. Pell, and M. B. Seasholz (1998). *Chemometrics: Practical guide*. New York: John Wiley & Sons, Inc.
[75] M. A. Sharaf, D. L. Illman, and B. R. Kowalski (1986). *Chemometrics*. New York: John Wiley & Sons, Inc.
[76] P. Geladi and B. R. Kowalski, Partial least-squares regression: A tutorial, *Analytica Chimica Acta*, 185 (1986) 1–17.
[77] W. P. Carey, K. R. Beebe, and B. R. Kowalski, Multicomponent analysis using an array of piezoelectric crystal sensors, *Analytical Chemistry*, 59 (1987) 1529–1534.
[78] A. Vergara, E. Llobet, J. Brezmes, P. Ivanov, C. Cané, I. Gràcia, X. Vilanova, and X. Correig, Quantitative gas mixture analysis using temperature-modulated micro-hotplate gas sensors: Selection and validation of the optimal modulating frequencies, *Sensors and Actuators B: Chemical*, 123 (2007) 1002–1016.
[79] Y. Sasaya and T. Nakamoto, Study of halitosis-substance sensing at low concentration using an electrochemical sensor array combined with a preconcentrator, *IEEJ Transactions on Sensors and Micromachines*, 126 (2006) 292–296.
[80] E. T. Zellers, S. A. Batteman, M. Han, and S. J. Patrish, Optimal coating selection for the analysis of organic vapor mixtures with polymer-coated surface acoustic wave sensor arrays, *Analytical Chemistry*, 67 (1995) 1092.
[81] Z. Wang, T. Isaksson, and B. R. Kowalski, New approach for distance measurement in locally weighted regression, *Analytical Chemistry*, 66 (1994) 249–260.
[82] R. O. Duda, P. E. Hart, and D. G. Stork (2001). *Pattern classification* (pp. 282–349). New York: Wiley-Interscience.
[83] D. E. Rumelhart, J. L. McClelland, and PDP Research Group. (1986). *Parallel distributed processing* (pp. 318). Cambridge, MA: MIT Press.
[84] H. Sundgren, F. Winquist, and I. Lundstrom (1991). Artificial neural networks and statistical pattern recognition improve MOSFET gas sensor array calibration, International Conference on Solid-State Sensors and Actuators, 1991. Digest of Technical Papers, TRANSDUCERS '91, San Francisco, CA, June 24–27, 1991, pp. 574–577.
[85] T. Nakamoto, Y. Hirota, and J. Ide, Record of mint flavor using mass spectrometry, Proceedings of the IEEE Sensors Conference, Irvine, CA, USA, ctober 30–November 3, 2005, pp. 393–396.
[86] T. Kohonen (1988). *Self-organizing and associative memory*. New York: Springer-Verlag.
[87] Z. Wang, J. Hwang, and B. R. Kowalski, ChemNets: Theory and application, *Analytical Chemistry* 67 (1995) 1497–1504.
[88] E. L. Hines, E. Llobet, and J. W. Gardner, Electronic noses: A review of signal processing techniques, *IEE Proceedings. Circuits, Devices, and Systems*, 146 (1999) 297–310.
[89] S. Munoz, T. Nakamoto, and T. Moriizumi, Study of quartz crystal microbalance odor sensing system for apple and banana flavors, *IEICE Transactions on Electronics*, E85-C (2002) 1291–1297.
[90] S. Munoz, A. Yoshino, T. Nakamoto, and T. Moriizumi, Odor approximation of fruit flavors using a QCM odor sensing system, *Sensors and Actuators B: Chemical*, 123 (2007) 1101–1106.
[91] T. Nakamoto, N. Okazaki, and T. Moriizumi, High speed active gas/odor sensing system using adaptive control theory, *Sensors and Actuators B: Chemical*, 41 (1997) 183–188.
[92] T. Yamanaka, R. Matsumoto, and T. Nakamoto, Study of recording apple flavor using odor recorder with five components, *Sensors and Actuators B: Chemical*, 89 (2003) 112–119.
[93] G. Sauerbrey, Verwendung von schwingquarzen zur mikrowagung, *Zeitschrift für Physik*, 155 (1959) 289.
[94] T. Nakamoto and T. Moriizumi, A theory of a quartz crystal microbalance based upon a mason equivalent circuit, *Japanese Journal of Applied Physics*, 29, 5 (1990) 963–969.
[95] T. Yamanaka, R. Matsumoto, and T. Nakamoto, Fundamental study of odor recorder for multi-component odor using recipe exploration based on singular value decomposition, *IEEE Sensor Journal*, 3 (2003) 468–474.
[96] M. E. H. Amrani, R. M. Dowdeswell, P. A. Payne, and K. C. Persaud, An intelligent gas sensing system, *Sensors and Actuators B: Chemical*, 44 (1997) 512–516.

[97] T. Yamanaka, R. Matsumoto, and T. Nakamoto, Odor recorder for multi-component odor using two-level quantization method, *Sensors and Actuators B: Chemical*, 89 (2003) 120–125.
[98] P. Somboon, B. Wyszynki, and T. Nakamoto, Realization of recording wide range of odor by utilizing both of transient and steady-state sensor responses in recording process, *Sensors and Actuators B: Chemical*, 124 (2007) 557–563.
[99] M. Nakamura, I. Sugimoto, H. Kuwano, and R. Lemos, Chemical sensing by analyzing dynamics of plasma polymer film-coated sensors, *Sensors and Actuators B: Chemical* 20 (1994) 231–237.
[100] M. Schweizer-Berberich, J. Goeppert, A. Hierlemann, J. Mitrovics, U. Weimar, W. Rosenstiel, and W. Goepel, Application of neural-network systems to the dynamic response of polymer-based sensor arrays, *Sensors and Actuators B: Chemical*, 26–27 (1995) 232–236.
[101] T. Yamanaka and T. Nakamoto, Real-time reference method in odor recorder under environmental change, *Sensors and Actuators B: Chemical*, 93 (2003) 51.
[102] T. Nakamoto, Record of dynamic changes of odors using an odor recorder, *Sensors and Materials*, 17 (2005) 365–383.
[103] T. Yamanaka, K. Yoshikawa, and T. Nakamoto, Improvement of odor-recorder capability for dynamical change of odor, *Sensors and Actuators B: Chemical*, 99 (2004) 367.
[104] S. Munoz, T. Nakamoto, and T. Moriizumi, Study of deposition of gas sensing films on quartz crystal microbalance using an ultrasonic atomizer, *Sensors and Actuators B: Chemical*, 105 (2005) 144–149.
[105] B.Wyszynski, A. G. Galvez, and T. Nakamoto, Improvement of ultrasonic atomizer method for deposition of gas-sensing film on QCM, *Sensors and Actuators B: Chemical*, 127 (2007) 253–259.
[106] B. Dittman, S. Nitz, and G. Horner, A new chemical sensor on a mass spectrometric basis, *Advance in Food Sciences (CMTL)*30 (1998) 115.
[107] T. Miura, T. Nakamoto, and T. Moriizumi, Study of odor recorder using mass spectrometry, *IEEJ Transactions on Sensors and Micromachines* 456-E (2003) 513–518 (in Japanese).
[108] T. Nakamoto, Odor recorder using mass spectrometry, *Sensor Letters*, 6 (2008) 1–5.
[109] L. Buck and R. Axel, A novel multigene family may encode odorant receptors: Molecular basis for odor recognition, *Cell*, 65 (1991) 175–187.
[110] J. E. Amoore (1970). *Molecular basis of odor*. Spring field, IL: Charles C Thomas Publisher.
[111] S. S. Sciffman and T. C. Pearce (2003). Introduction to olfaction: Perception, anatomy, physiology and molecular biology, in T. C. Pearce (Ed.), *Handbook of machine olfaction*. New York: Wiley-VCH.
[112] D. Andrew (1985). *Atlas of odor character profiles*. Philadelphia, PA: ASTM.
[113] J. B.Castro, A.Ramanathan, and C. S.Chennubholta, Categorical dimensions of human odor descriptor space revealed by non-negative matrix factorization, *PloS One*, 8 (2013) 1–16.
[114] D.Cozzolino, H. E.Smyth, W.Cynkar, R. G.Dambergs, and M.Gishen, Usefulness of chemometrics and mass spectrometry-based electronic nose to classify Australian white wines by their varietal origin, *Talanta*, 68 (2005) 382–387.
[115] A. Mamun and T. Nakamoto, Recipe estimation using mass spectrometer and large-scale data. *IEEJ Transactions on Sensors and Micromachines*, 128 (2008) 467–471.
[116] D. D. Lee and H. S. Seung, Learning the parts of objects by non-negative matrix factorization, *Nature*, 401 (1999) 788–791.
[117] T. Nakamoto and K. Murakami, Selection method of odor components for olfactory display using mass spectrum database, Proceedings of the IEEE Virtual Reality Conference, Lafayette, LA, USA, March 14–18, 2009, pp. 159–162.
[118] T. Nakamoto, M. Ohno, and Y. Nihei, Odor approximation using mass spectrometry, *IEEE Sensors Journal*, 12 (2012) 3225–3231.
[119] A. Cichocki, R. Zdunek, A. H. Phan, and S. Amari (2009). *Nonnegative matrix and tensor factorizations* (pp. 81–130). Chichester: John Wiley & Sons, Ltd.
[120] Y. Harada, K. Honda and T. Nakamoto, Study of similarity measures for odor approximation by using nonnegative matrix factorization method, *The Japanese Journal of Taste and Smell Research*, 20 (2013) 411–414.
[121] C. Févotte, N. Bertin, and J.-L. Durrieu, Nonnegative matrix factorization with the Itakura-Saito divergence: With application to music analysis, *Neural Computation*, 21 (2009) 793–830.
[122] T. Nakamoto and Y. Nihei, Improvement method of odor approximation using mass spectrometry, *IEEE Sensors Journal*, 13 (2013) 4305–4311.
[123] W. R. Dillon and M. Goldstein (1984). *Multivariate analysis* (pp. 186–190). New York: John Wiley & Sons, Inc.

[124] T. Nakamoto and T. Yamanaka (2011). Odor reproduction with movie and its application to teleolfaction, in E. L. Hines and M. S. Leeson (Eds.), *Intelligent systems for machine olfaction* (pp. 126–152). Hershey, PA: IGI-Global.

[125] T. Yamanaka, N. Nitikarn, and T. Nakamoto, Concurrent recording and regeneration of visual and olfactory information using odor sensor, *Presence*, 16 (2007) 307–317.

[126] P. E. Keller, R. T. Kouzes, L. J. Kangas, and S. Hashem (1995). Transmission of olfactory information for telemedicine, in K. Morgan, R. M. Satava, H. B. Sieburg, R. Matteus, and J. P. Christensen (Eds.), *Interactive technology and the new paradigm for healthcare* (pp. 168–172). Amsterdam: IOS Press and Ohmsha.

[127] T. Nakamoto, N. Nimsuk, B. Wyszynski, H. Takushima, M. Kinoshita, N. Cho, Experiment on teleolfaction using odor sensing system and olfactory display synchronous with visual information, Proceedings of the International Conference on Artificial Reality and Telexistence (ICAT), Yokohama, Japan, December 1–3, 2008, pp. 85–92.

[128] F. Yoshino and T. Nakamoto, Study of odor recognition system using embedded LVQ circuit, *Sensors and Materials*, 26 (2014) 137–148.

[129] N. Nimsuk and T. Nakamoto, Study on the odor classification in dynamical concentration robust both against humidity and temperature changes, *Sensors and Actuators B: Chemical*, 234 (2008) 252–257.

[130] Y. Nozaki, F. Yoshino, B. Wyszynski, and T. Nakamoto, Demonstration of improved teleolfaction system, ICAT2013, D11, December 11–13, 2013, Tokyo, Japan.

[131] H. Mitsuno, T. Sakurai, S. Namiki, H. Mitsuhashi, and R. Kanzaki, Novel cell-based odorant sensor elements based on insect odorant receptors, *Biosensors and Bioelectronics*, 65 (2014) 287–294.

[132] T. Nakamoto, M. Kakizaki, Y. Suzuki, H. Mitsuno, and R. Kanzaki, Response analysis of odor sensor based upon insect olfactory receptors using image processing method, Proceedings of the Technical Digest of IEEE Sensors, Valencia, November 2–5, 2014, pp. 1014–1017.

8

Summary and Future Perspectives

Takamichi Nakamoto
Precision and Intelligence Laboratory, Tokyo Institute of Technology, Tokyo, Japan

Technologies of machine olfaction and taste are described from basics to advanced matters. We tried to describe it systematically so that novices as well as experts can understand them. People know the fundamentals and recent progress of machine olfaction and taste after they read this book.

These technologies are not still complete at the current stage and are now changing. The latest odor sensing technology has two aspects such as biological sensor and large-scale sensor array. These aspects might be a breakthrough toward the new generation of electronic nose. The pattern recognition technology should give an electronic nose more robustness against disturbance. The drift and aging problems can be overcome by development of both sensor device and the algorithm.

The technology related to olfactory display is relatively new. It is important to realize cyber-space with chemical sense. If a single tiny device with the capability of emitting a variety of odorants is available, it will open a variety of applications. Moreover, an odor recorder for general purpose should be developed. The progress of odor sensing technology enables the odor recorder to a wide range of its applications.

The taste sensor has been on the market. It should spread into wide area so that anyone can use it. Moreover, the taste display in combination with other senses is expected to be developed.

In the twentieth and early twenty-first centuries, the progress of hardware such as LSI, PC, and mobile phone was remarkable. The software implemented into them had also the progress. Although those technologies are important, these are kinds of infrastructure to enrich human life. From now on, human interface technology directly facing with humans will become more important. More technologies along with humans should be developed instead of the performance competition of LSI, PC, and mobile phone.

Machine olfaction and taste are key technologies to realize the advanced human interface technology. More researchers and industry people are expected to enter the field of machine olfaction and taste.

Essentials of Machine Olfaction and Taste, First Edition. Edited by Takamichi Nakamoto.

Companion website: www.wiley.com/go/nakamoto/olfaction

Index

acid-sensing ion channel, 19, 20
active pharmaceutical ingredients, 90, 110, 141, 144, 147, 150, 154–156, 158, 159
active sensing, 269
adenylyl cyclase, 12, 13, 15
adsorbed on the membrane, 116, 117, 152, 153
advertisement, 266
aftertaste, 98, 99, 110, 116, 125–127, 130, 132, 139, 140, 147–150, 152, 159, 160
air canon, 250
air quality, 179, 206, 212, 214–216
AL *see* antennal lobe
Alzheimer, 266
amino acid, 88, 89, 92, 132, 138
ant, 221
antennal lobe, 4
antibody, 122, 123
APIs *see* active pharmaceutical ingredients
application
  beer, 90–92, 105, 111, 124, 126, 140
  coffee, 88, 92, 105, 107, 110, 111, 120, 127–132, 134, 135, 138–140
  fried chicken, 133, 138, 139
  meat, 105, 107, 111, 132, 133
  medicine, 87, 118, 141, 143
artificial neural network, 92
artificial nose, 49
artificial olfactory devices, 58
artificial sniffers, 51
ASIC *see* acid-sensing ion channel
atomization, 260
atomizer, 288
automatic sampler, 256
autoregressive (AR) method, 283

backpropagation neural network, 91
basal cell, 11, 12
basic characteristics, 103
basic taste, 17, 18, 21, 262
basis matrix, 295
basis vector, 293
BBPA *see* bis(1-butylpentyl)adipate
BEHS *see* bis(2-ethylhexyl)sebacate
binary quantization, 281
bipolarons, 61
bis(1-butylpentyl)adipate, 94, 101, 102, 144, 154
bis(2-ethylhexyl)sebacate, 94, 102
1-bit analogue to digital conversion, 258
bitterness score, 141–142
blending ratio, 129, 132
blimp, 238
BmPBP1 *see Bombyx mori* pheromone-binding protein 1
Bombyx mori pheromone-binding protein 1, 6
BP method, 273
breath sensor, 256
business-to-business experiment, 266
bypath, 258

Essentials of Machine Olfaction and Taste, First Edition. Edited by Takamichi Nakamoto.

Companion website: www.wiley.com/go/nakamoto/olfaction

calcium silicate, 251
calcium-gated chloride channel, 12
calibration transfer, 176, 208–210, 216
calmodulin, 14, 15
cAMP *see* cyclic adenosine monophosphate
canonical discrimination analysis, 270
carrageenan, 146
casting, 233
cetirizine hydrochloride, 147–151
CFD *see* computational fluid dynamics
change of membrane potential caused by adsorption of chemical substances, 97, 98, 101, 102, 104–108, 110, 111, 116, 117, 121, 141–146, 152, 154–160
characteristic diffusion length, 222
chemical analyte delivery system, 68
chemical capsule, 250
chemical distribution mapping, 222
chemical finding, 221
chemical gas sensor, 52
chemical source declaration, 221
chemical source localization, 221
chemical tracking, 221
chemical trail following, 221
chemical trail-following robot, 231
chemiresistor, 60
ChemNet, 274
chemometrics, 271
cinema, 263
circumvallate papillae, 17
classification, 175, 178, 184, 188, 190–196, 200, 204, 205, 207, 210–217
classification trees, 196
clustering, 189, 190, 211, 214
coefficient matrix, 295
collinearity, 271
color intensity, 262
computational fluid dynamics, 251
computational models, 50
concept drift, 178, 187, 206, 207
concurrent method, 288
condition number, 277
conducting organic polymer sensor, 60
conducting polymer sensor, 55, 268
contact pheromone, 8
cooking game, 264
CPA *see* change of membrane potential caused by adsorption of chemical substances
CRABOT, 239
crayfish, 238
cross-sensitivity, 176
CTZ *see* cetirizine hydrochloride
cultural anthropology, 267
cyber-physical system, 2
cyberspace, 315
cyclic adenosine monophosphate, 12, 14, 15
cyclic nucleotide-gated channel, 12, 13, 19
cyclotrimethylenetrinitramine, 27

DAG *see* diacylglycerol
definition of information, 98
DEG/ENaC *see* degenerin/epitherial Na^+ channel
degenerin/epitherial Na+ channel, 11
delta sigma modulation, 258
dependent variable, 271
diacylglycerol, 15
diffusion constant, 222
digital signage, 266
dioctyl phenylphosphonate, 101, 102, 105, 154
dipyridamole, 156, 157, 159, 160
discriminant analysis, 92
disodium guanylate, 88, 89, 107
disodium inosinate, 88, 89
dog, 224
DOPP *see* dioctyl phenylphosphonate
drift and aging, 271
dynamic change of odor, 283

EAG *see* electroantennogram
EDPCR *see* extended disjoint principal component regression
EI *see* electron ionization
EIS *see* electrochemical impedance spectroscopy
electroantennogram, 23, 26
electrochemical impedance spectroscopy, 36, 38
electron ionization, 291
electronic nose, 225, 267
electronic tongue, 90, 92, 160
electroosmotic pump, 260
ENaC *see* epithelial Na^+ channel
e-nose *see* electronic nose
environmental change, 286
EO pump *see* electroosmotic pump
epithelial Na^+ channel, 10, 18, 19, 21
error backpropagation, 273
essential oil, 297, 300, 301
Euclidean distance, 301
extended disjoint principal component regression, 272

feature extraction, 180, 181, 184, 186, 188, 198, 211, 212, 214, 216
feature selection, 184, 200, 202, 214
feedback, 271
FET *see* field effect transistor
field effect transistor, 36
field programmable gate array, 64, 306
five senses, 89, 161–163
fluidic switching device, 258
foliate papillae, 17

food texture, 262
FPGA *see* field programmable gate array
functional polymer, 248
fungiform papillae, 17

G protein-coupled receptor, 4, 8, 15, 18, 20, 21, 38
GA *see* genetic algorithm
gallic acid, 94, 105, 107, 108
game with scents, 263
gas chromatograph/mass spectrometry, 252
gas concentration gradient, 232
gaze point analysis, 262
GCaMP3, 30
GC/MS *see* gas chromatograph/mass spectrometry
gel phase, 248
genetic algorithm, 90, 129, 134–137, 141
global selectivity, 88, 98, 106–108, 116, 117, 160, 161
glomerulus, 8, 11, 12
GMP *see* disodium guanylate
Golf, 12
Gouy-Chapman theory, 93
GPCR *see* G protein-coupled receptor
GR *see* gustatory receptor
GRNs *see* gustatory receptor neurons
gustatory nerve, 89, 90
gustatory receptor, 9
gustatory receptor neurons, 9
gustducin, 19

halitosis, 272
haptic interface, 262
HCN *see* hyperpolarization-activated cyclic nucleotide-gated channel
HEK293, 27, 34
1-hexadecanol, 94
hierarchical clustering, 303
hopping mechanism, 61
hOR *see* human olfactory receptor
human interface, 315
human olfactory receptor, 27, 36
humidity disturbance, 286
hyperpolarization-activated cyclic nucleotide-gated channel, 19, 20

ICP sensor *see* intrinsically conducting polymer sensor
ill conditionedness, 277
3D image, 261
IMP *see* disodium inosinate
independent variable, 271
indirect competitive assay, 123
inhalation, 255
inkjet device, 254
insect antenna, 225
interaction, 88, 93, 97–99, 103, 105, 110, 116, 122, 150, 152–154, 156
intrinsically conducting polymer sensor, 55
ion detector, 225
ionotropic receptor family, 10
IR family *see* ionotropic receptor family
Itakura–Saito divergence, 301
IS divergence *see* Itakura–Saito divergence

kernel gas distribution mapping, 241
KL divergence *see* Kullback–Leibler divergence
K_2P (two-pore domain potassium), 19
Kullback–Leibler divergence, 301

labella, 8
LAPS *see* light-addressable potentiometric sensor
LDA *see* linear discriminant analysis
large sensor array, 50
large-scale sensor array, 59
latent variable, 272
learning vector quantization, 306
leave-one-out method, 272
light-addressable potentiometric sensor, 23, 27
Limonene, 27
linear discriminant analysis, 283
linear least squares, 290
linear regression model, 271
lipids concentration, 99–101, 103
lobster, 238
locally weighted regression, 273
low-volatile compound, 252, 260
LWR *see* locally weighted regression
LVQ *see* learning vector quantization

machine learning, 175, 190
machine olfaction, 1
mass flow controller, 256
mass spectrometer, 252
mass spectrometry, 289
MCF-7 cell, 40
MDS *see* multidimensional scaling
MEA *see* microelectrode array
mechanical vane, 225
membrane potential, 96–100, 111, 116
MEMS *see* microelectromechanical system
mesh heater, 254
metal oxide gas sensor, 53, 224, 265
MFC *see* mass flow controller
micro chamber, 248
microdispenser, 255
microdrone, 238
microelectrode array, 23, 27
microelectromechanical system, 249
microheater, 258

MIMO feedback *see* multiinput multioutput feedback
miniaturized taste sensor, 117
poly-(hydroxyethyl methacrylate) hydrogel, 118
portable taste sensor, 118–120
sensor chip, 118–121
taste map, 120
mitral cell, 12
mixture quantification, 269, 271
MLP *see* multilayer perceptron
MLR *see* multiple linear regression
mobile robot, 219
mole crab, 239
MOLE I robot, 239
molecular diffusion, 222
monosodium glutamate, 88–89, 96–97, 99–100, 102–104, 106–107, 110, 162
moth, 224
movie with scents, 263
MSG *see* monosodium glutamate
multicomponent olfactory display, 256
multidimensional scaling, 261
multiinput multioutput feedback, 271, 276
multilayer perceptron, 273
multiple linear regression, 91, 271
m/z, 293

natural convection, 229
neural network, 180, 193, 198, 200, 203, 205, 209, 212, 215, 216, 273
2-nitrophenyl octyl ether, 94, 99
NMF method *see* nonnegative matrix factorization method
nonlinear regression, 91
nonnegative least squares method, 293
nonnegative matrix factorization method, 293
nonselective chemical sensors, 49, 50
NPOE *see* 2-nitrophenyl octyl ether

OBP *see* odorant binding protein
obstacle avoidance, 237
occupancy grid, 237
odor adaptation, 13–15
odor approximation, 274
odor blender, 256, 270
odor components, 292
odor delivery system, 50
odor-gated anemotaxis, 233
odor-gated rheotaxis, 233
odor recognition, 51
odor recorder, 1, 247, 267
odor-releasing gadget, 250
odor segmentation process, 75
odor sensing system, 247
odorant binding protein, 6, 40, 42
odorant-gated ion channel, 6
odorant molecule, 49
odorant receptor-derived peptide, 36
odorant receptor gene, 11, 15
odorant receptors, 7, 11–13, 15–17, 30, 36, 38
ODRs *see* odorant receptors
ODTs *see* orally disintegrating tablets
olfactory art, 266
olfactory bulb, 11, 12
olfactory cilia, 11, 12
olfactory cue, 262
olfactory display, 1, 247
olfactory epithelium, 11–13, 15, 50
olfactory mucosa, 64
olfactory sensillum, 4
olfactory sensory neuron, 26, 27
olfactory system, 49
olfactory threshold, 266
one receptor-one neuron rule, 15
optimal feedback control, 276
ORs *see* odorant receptors
orally disintegrating tablets, 146, 150
Orco, 6
ORI7, 27, 28, 36, 38, 40, 41
ORN, 3, 11–16, 26, 27, 34, 42
ORP *see* odorant receptor-derived peptide
OSN *see* olfactory sensory neuron

PADE *see* phosphoric acid di-*n*-decyl ester
pannexin 1, 20
PARC engine *see* pattern recognition engine
Parkinson, 266
partial least squares, 270
partial least squares regression analysis, 68
particle filter, 237
pattern matching, 270
pattern recognition engine, 50
PCA *see* principal component analysis
PCR *see* principal component regression
PDMS *see* polydimethylsiloxane
permeation tubes, 68
pheromone trail, 221
phospholipase C, 15, 18, 19
phosphoric acid di-*n*-decyl ester, 94, 101–102, 144
phosphoric acid tris(2-ethylhexyl)ester, 94, 102
physiology of olfaction, 292
piezoelectric sensors, 54
PKD1L3, 19–21
PKD2L1, 19–21
PLC *see* phospholipase C
PLS *see* partial least squares
plume, 223
Poisson-Boltzmann equation, 93
polaron, 61
polydimethylsiloxane, 248

polymer-based conductometric sensor, 225
polyvinyl chloride, 93, 118
prediction of bitterness, 144
primary smells, 292
principal component analysis, 53, 68, 87, 90–92, 98, 134, 144, 146, 269
principal component regression, 270
proboscics, 8
procedure, 93, 97, 116, 121, 123, 129, 133, 136, 145, 157–158
protein kinase C, 7
pseudo-gustatory display, 267
pseudo olfactory effect, 261
pseudo smell, 262
pseudo taste display, 262
PTEH *see* phosphoric acid tris(2-ethylhexyl)ester
pulse width modulation, 258
pungency, 88–89, 122, 124, 160
pungent sensor, 122
PVC *see* polyvinyl chloride
P2X2, 20
P2X3, 20

QCM *see* quartz crystal microbalance
QDA *see* quantitative description analysis
QSAR *see* quantitative structure–affinity relationship
quantitative description analysis, 274
quantitative structure–affinity relationship, 268
quartz crystal microbalance, 3, 27, 36, 38, 40, 54, 225, 252

radial basis function, 193–195, 198, 199, 207
Rasmus robot, 240
RAT *see* Reactive Autonomous Testbed
RBF *see* radial basis function
RDX *see* cyclotrimethylenetrinitramine
Reactive Autonomous Testbed, 232
real-time reference method, 271, 284
receptor, 89, 122
  epithelial sodium channel, 89
  polycystic kidney disease 2-like 1 protein, 89
  sweetness receptors, 89
  taste-2 receptors, 89
  transient receptor potential, 122
  umami receptors, 89
recipe, 270
redundancy, 58
regression, 190, 191, 196–200, 204, 205, 209, 210, 213, 214
remote environmental monitoring units, 238
REMUS *see* remote environmental monitoring units
resistive arrays, 63
response mechanism, 93, 96–97, 101, 105, 154
RoboLobster, 238
Robo-moth, 238

SAM *see* self-assembled monolayer
sample loop, 258
SAW *see* surface acoustic wave
SAW atomizer, 307
scent diffuser, 266
SD method *see* semantic differential method
see-through head-mounted display, 262
selectivity, 175, 176, 202
self-assembled monolayer, 35, 36, 40, 41, 123
self-organizing map, 274
self-priming property, 260
semantic differential method, 268
semisupervised learning, 206, 207
sensor design, 98–99, 103, 143
sensor drift, 176, 202, 204–208
sensory score, 141–144, 154–157
sensory test, 268
sexual pheromone, 224
Sf21 cell, 27, 30
short-time Fourier transform, 307
signal transduction, 11–13, 15, 18, 19
silkworm moth, 226
singular value, 279
singular value decomposition, 278
smell, 49
smell persistence, 251, 260
smell reproduction, 305
sniffer dog, 219
sol phase, 248
solenoid valve, 258
SOM *see* self-organizing map
specificity, 176
SPR *see* surface plasmon resonance
SPR immunosensor, 122–124, 160
SSL *see* semisupervised learning
steady-state response, 283
STFT *see* short-time Fourier transform
suboesophageal ganglion, 9
support vector machines, 191
suppression effect, 110–111, 141, 143
supporting cell, 11, 12
surface acoustic wave, 3, 36, 40, 54, 260
surface plasmon resonance, 27, 36, 40, 41
surgery simulator, 267
SVD *see* singular value decomposition
SVM *see* support vector machines
synergistic behavior, 58
syringe, 256

Tarsi, 9
taste information, 89, 98, 109–110, 113, 115–116, 129, 132, 134–140, 155, 162
taste map, 147
taste qualities
  astringency, 88–89, 98, 103–110, 116, 119, 121, 122

taste qualities (*cont'd*)
tannic acid, 102, 108, 110, 116–117, 121
tannin 88–89
bitterness, 88–89, 97–101, 103, 105–111, 116, 119–120, 124–127, 130–132, 134–136, 139–157, 159–161
caffeine, 88–89
humulone, 88
quinine, 88–89, 93, 96–97, 99, 101–102, 104–111, 116–117, 141–142, 144–145, 149, 155, 156, 160, 162
theobromine, 88–89
pungency, 88–89, 122
allyl isothiocyanate, 88–89, 122, 124
capsaicin, 122–124, 160
piperine, 88–89, 122, 124
saltiness, 88–89, 93, 98–99, 103–104, 106, 109–110, 119, 126, 130, 132, 134–136, 139, 149, 151
potassium chloride, 93, 96–98, 102, 109–111, 160
sodium chloride, 93, 151, 160
sourness, 88–89, 93, 98–99, 101, 103–104, 106, 110, 119–120, 125–126, 130–132, 134–136, 139, 147–148, 150–151
acetic acid, 88, 160
citric acids, 88
hydrochloric acids, 88
sweetness, 88–89, 93, 103, 105, 107, 110, 134, 135–136, 143
artificial sweetener, 88
glucose, 8
sucrose, 88, 104–107, 110–111, 160, 162
umami, 88–89, 96–99, 101, 103–104, 106–107, 110, 119, 126, 130, 132, 139, 140, 162
disodium guanylate, 88, 89, 107
disodium inosinate, 88, 89
GMP see disodium guanylate
IMP see disodium inosinate
monosodium glutamate, 88–89, 96–97, 99–100, 102–104, 106–107, 110, 162
MSG see monosodium glutamate
taste-1 receptor, 20, 21
taste-2 receptor, 20, 21
taste sensillum, 9
taste sensing system, 88, 92–93, 95
taste sensor electrode
astringency sensor AE1, 103–110
bitterness sensor AC0, 103–105, 110, 142, 143, 152–153
bitterness sensor AN0, 103–105, 110, 144, 152–153
bitterness sensor BT0, 154–157
bitterness sensor C00, 103–105, 110
sourness sensor CA0, 101, 103–104, 110, 119, 124–125, 148
sweetness sensor GL1, 103–107, 110
saltiness sensor CT0, 103–104, 109–110, 119
umami sensor AAE, 103–104, 107, 110, 119–110
TBAC *see* tributyl *o*-acetylcitrate
TDAB *see* tetradodecyl ammonium bromide
teleexistence, 305
teleolfaction, 1, 305
ternary mixture, 278
tetradodecyl ammonium bromide, 94, 99, 101
thermistor-type airflow sensor, 225
threshold, 98, 99, 106, 108–109, 126, 144–146
time constant, 283
TMSPM *see* 3-(trimethoxysilyl)propyl methacrylate
TNT *see* trinitrotoluene
TOTM *see* trioctyl trimellitate
T1R *see* taste-1 receptor
T2R *see* taste-2 receptor
transient receptor potential channel, 15, 19, 20
triangle test, 275
tributyl *o*-acetylcitrate, 94, 102, 144
3-(trimethoxysilyl)propyl methacrylate, 94, 102
trinitrotoluene, 27
trioctyl trimellitate, 94, 102, 144
trioctylmethylammonium chloride, 94
TRPC *see* transient receptor potential channel
TRPC2, 15
TRPM5, 19
TUAT mobile robot, 234
tufted cell, 12
turbulent airflow, 289
turbulent flow, 223
two-level quantization method, 280
two-pore domain potassium leak channel, 19

ultrasonic anemometer, 225
ultrasonic micro plug array, 250
upwind surge, 233

VCs *see* volatile compounds
vial, 256
virtual odor source, 251
virtual olfactory interface, 306
VNO *see* vomeronasal organ
VOC *see* volatile organic compound
volatile compounds, 23, 24
volatile organic compound, 23, 40
volatile sulfide compound, 272
vomeronasal organ, 11, 12, 15
vomeronasal sensory neuron, 11, 12, 15
vortex ring, 250
V1R, 15
V2R, 15
VSC *see* volatile sulfide compound
VSN *see* vomeronasal sensory neuron

3-way solenoid valve, 258
Weber–Fechner law, 109
wind tunnel, 226

Xenopus laevis, 27, 28

yellow cameleon-2, 27

ZigBee module, 306

Printed and bound by CPI Group (UK) Ltd, Croydon, CR0 4YY
17/09/2026
14964267-0003